Molecular Biology in Medicinal Chemistry

Edited by Th. Dingermann, D. Steinhilber and G. Folkers

Molecular Biology in Medicinal Chemistry. Edited by Th. Dingermann, D. Steinhilber, G. Folkers

ISBN: 3-527-30431-2

Methods and Principles in Medicinal Chemistry

Edited by R. Mannhold, H. Kubinyi, G. Folkers

Recently Published Volumes:

G. Molema, D. K. F. Meijer (eds.)

Drug Targeting
Vol. 12

2001, ISBN 3-527-29989-0

D. Smith, D. Walker, H. van de Waterbeemd

Pharmacokinetics and Metabolism in Drug Design
Vol. 13

2001, ISBN 3-527-30197-6

T. Lengauer (ed.)

Bioinformatics – From Genomes to Drugs
Vol. 14

2001, ISBN 3-527-29988-2

J. K. Seydel, M. Wiese

Drug-Membrane Interactions
Vol. 15

2002, ISBN 3-527-30427-4

O. Zerbe (ed.)

BioNMR in Drug Research
Vol. 16

2002, ISBN 3-527-30465-7

P. Carloni, F. Alber (eds.)

Quantum Medicinal Chemistry
Vol. 17

2003, ISBN 3-527-30456-8

H. van de Waterbeemd, H Lennernäs, P. Artursson (eds.)

Drug Bioavailability
Vol. 18

2003, ISBN 3-527-30438-X

H.-J. Böhm, G. Schneider (eds.)

Protein-Ligand Interactions
Vol. 19

2003, ISBN 3-527-30521-1

R. E. Babine, S. S. Abdel-Meguid (eds.)

Protein Crystallography in Drug Discovery
Vol. 20

2003, ISBN 3-527-30678-1

Molecular Biology in Medicinal Chemistry

Edited by Th. Dingermann, D. Steinhilber and G. Folkers

WILEY-VCH Verlag GmbH & Co. KGaA

Series Editors:

Prof. Dr. Raimund Mannhold
Biomedical Research Center
Molecular Drug Research Group
Heinrich-Heine-Universität
Universitätsstraße 1
40225 Düsseldorf
Germany
raimund.mannhold@uni-duesseldorf.de

Prof. Dr. Hugo Kubinyi
BASF AG Ludwigshafen
c/o Donnersbergstraße 9
67256 Weisenheim am Sand
Germany
kubinyi@t-online.de

Prof. Dr. Gerd Folkers
Department of Applied Biosciences
ETH Zürich
Winterthurerstr. 190
8057 Zürich
Switzerland
folkers@pharma.anbi.ethz.ch

Volume Editors:

Prof. Dr. Theodor Dingermann
Institute of Pharmaceutical Biology
Johann Wolfgang Goethe-University
Marie-Curie-Str. 9
60439 Frankfurt
Germany
dingermann@em.uni-frankfurt.de

Prof. Dr. Dieter Steinhilber
Institute of Pharmaceutical Chemistry
Johann Wolfgang Goethe-University
Marie-Curie-Str. 9
60439 Frankfurt
Germany
steinhilber@em.uni-frankfurt.de

Prof. Dr. Gerd Folkers
Department of Applied Biosciences
ETH Zürich
Winterthurerstr. 190
8057 Zürich
Switzerland
folkers@pharma.anbi.ethz.ch

Library of Congress Card No. applied for
British Library Cataloguing-in-Publication Data: A catalogue record for this book is available from the British Library
Die Deutsche Bibliothek – CIP Cataloguing-in-Publication-Data: A catalogue record for this publication is available from Die Deutsche Bibliothek

Printed in the Federal Republic of Germany.
Printed on acid-free paper

Typesetting: Asco Typesetters, Hong Kong
Printing: betz-druck gmbh, Darmstadt
Bookbinding: Litges & Dopf Buchbinderei GmbH, Heppenheim

ISBN 3-527-30431-2

Contents

Preface

Why address molecular biology and related technologies in a series named "Methods and Priciples in Medicinal Chemistry"? It was the advent of the silicon chip and the detection of DNA processing enzymes that jointly started an evolutionary track in the 1970s which boosted the whole variety of methodologies in what is today known as the life sciences. Also, the classical field of medicinal chemistry has been augmented and today comprises a huge range of techniques and methodologies from QSAR and structure-based design to the recently developed "high-throughput" synthesis and screening. A paradigmatic change in the 1990s gave rise to a focus on the molecular level of drug action and hence demanded the development of appropriate biological assay technology. This is the point where the present book starts.

In the first part, molecular targets are dealt with, going deep into cellular assay technologies. Cell-based assays imply not only the "simple" detection of one cellular product, but the tracking of a variety of metabolic processes, finally resulting in a multidimensional phenotypic characterization of cellular behavior. In a hierarchical step, the second chapter introduces the "gene knock-out" models, a technique that allows to "design" a disease model within a complex organism to generate a more relevant analytical tool for medicinal chemistry. The subsequent chapter deals with a fascinating readout technology for molecular assays, the so-called reporter genes.

The recent elucidation of the human genome has provided another boost to the whole field. Suddenly, a huge amount of targets was available for study. The question, however, which still remained was: What does the target do within the cellular biochemistry and how is it controlled? Those are the questions tackled in chapter 4, which deals with orphan receptors of the GPCR type and shows the challenges and the opportunities for finding new ligands with hitherto unknown biological activity.

The second part of the book is devoted to synthesis. Two important fields can benefit tremendously from molecular biology and its techniques: Stereoselective synthesis of natural compounds and of their mimics, and synthesis of DNA-derived drugs or protein drugs. The first chapter within this section gives a comprehensive overview about the use of enzymes in stereoselective synthesis, emphasizing recombinant technologies, which greatly enhance the selection of working tools. The fascinating field of nucleic acid drugs, the design and synthesis, their mimics and their mechanisms of action are the topics of chapter 6.

The third part deals with questions of analysis. Invaluable contributions have come from use of proteins for enantioseparation and affinity chromatography. The use of NMR and associated techniques for structure elucidation is another analytical topic to read about in this section.

Kinetics, metabolism, toxicology, and the very rapidly growing fields of pharmacogenomics and toxicogenomics form the contents of the final part of this book which is unique in its presentation of today's entanglement of the biosciences with medicinal chemistry.

The editors are indebted to the volume editors and the authors, whose work and motivation adds an highly important and fascinating facet to the series and gratefully acknowledge the ongoing support from Frank Weinreich, Wiley-VCH, during the whole project.

September 2003

Raimund Mannhold, Düsseldorf
Hugo Kubinyi, Weisenheim am Sand
Gerd Folkers, Zürich

Foreword

Where to start and where to end was the initial question when we selected the subjects which should be included into a volume dealing with the application of molecular biology in medicinal chemistry. The enormous progress in molecular biology during the last decades has led to the development of many methods with impact on drug discovery and drug development as well.

Modern target identification and validation on the one hand, and drug development and characterization on the other hand require an overlapping spectrum of methods and assays which goes far beyond the classical methodological repertoire of medicinal chemistry. The detailed molecular characterization of interactions of drugs with their targets is as important and demanding as detailed knowledge on drug interactions with the entire physiological system. Concepts and assays are available that provide us with information on drug transport, metabolism and stability. But we also can determine and predict how a system will react upon the application of a drug, even if this drug is considered to act very specifically at a certain target. Methods and concepts of modern pharmacogenomics and toxicogenomics have clearly demonstrated this.

Considering all this we finally decided to organize this volume in four parts: (I) Molecular Targets, (II) Synthesis, (III) Analysis and finally (IV) Kinetics, Metabolism and Toxicology.

The first chapter in part I sets the biological stage by providing an up-to-date description of available "Cellular Assays" and their use and impact on "Drug Discovery". Assays for membrane proteins and fast cellular responses, assays for gene and protein expression profiling in high-throughput formats, spatio-temporal assays and subpopulation analysis as well as phenotypic assays are introduced.

More complex systems are addressed in the second chapter of part I, which deals with gene knockout mice and techniques available for the generation of such animals.

G-protein coupled receptors are addressed in the third and fourth chapter. In the first of these two chapters, the focus lies on the characterization of G-protein coupled receptors and the application of reporter genes as read out systems, while the subsequent chapter, as a reference to the postgenomic era, discusses strategies for the identification of ligands for orphan G-protein coupled receptors.

Part II of the volume covers several aspects of drug synthesis. A classical overlap between organic synthesis and biotechnology is the stereoselective synthesis of drugs with the help of recombinant enzymes. The first chapter in this part gives an overview on this topic and provides many examples, underscoring the impact of such strategies.

Nucleic acid drugs eventually will come of age. Their attractiveness as potentially very specific ligands was always in conflict with numerous pharmacokinetic problems. However, various concepts for stabilizing these molecules, the fascinating potential of RNAi and the first approved drugs were strong reminders of these molecules, e.g. as manipulators of cellular signaling. The chapter by Engels and Parsch touches many aspects, including synthesis and application of this type of compounds, including the RNAi technology.

Part III of the book focuses on analytical aspects. Enantioseparation of chiral drugs and affinity chromatography are extremely important tools in drug development and comprehensive reviews on these topics are presented in chapters 7 and 8.

Analytical methods related to structural biology are included in a series of three articles. Two of these papers deal with NMR technologies that have strongly developed during the last decades. This led to the establishment of NMR as an relevant tool for the determination also of macromolecular structures and for the detection and characterization of ligand-target interactions. Chapters 9 and 10 summarize the application of NMR in drug discovery and describe techniques for ^{13}C- and ^{15}N-isotopic labeling of proteins.

Rational drug design depends on exact structure knowledge, which is still best provided by X-ray crystallography. Despite of extreme methodological improvements in this field, structures of membrane receptors, which represent the most important drug targets, are not available. A strong move towards solving this problem might be the use of antibody fragments as crystallization enhancers. Details of this exciting new technique are described in the final chapter of part III.

Pharmacogenomics and toxicogenomics are new fields with considerable impact on future drug development. The last section of the book covers these hot topics, which will eventually initiate the change from the "one size fits all" concept to the "right drug, right size, right person" concept.

The editors would like to gratefully acknowledge the contributions of all authors. They also thank Dr. Frank Weinreich and Wiley-VCH for a steady support during the ongoing project.

September 2003

Theo Dingermann, Frankfurt
Dieter Steinhilber, Frankfurt
Gerd Folkers, Zürich

Contributors

Dr. Hugo Albrecht
Discovery Partners International AG
Gewerbestrasse 16
4123 Allschwil
Switzerland

Dr. Remko A. Bakker
Leiden/Amsterdam Center for Drug Research
Department of Medicinal Chemistry
Vrije Universiteit Amsterdam
De Boelelaan 1083
1081 HV Amsterdam
The Netherlands

Prof. Dr. Annette G. Beck-Sickinger
Institute of Biochemistry
University of Leipzig
Talstrasse 33
04103 Leipzig
Germany

Dr. Frank Bernhard
Institute of Biophysical Chemistry
Johann Wolfgang Goethe-University
Marie-Curie-Strasse 9
60439 Frankfurt
Germany

Dr. Daniela Brodbeck-Hummel
Discovery Partners International AG
Gewerbestrasse 16
4123 Allschwil
Switzerland

Prof. Dr. Ingolf Cascorbi
Institute of Pharmacology
Ernst-Moritz-Arndt-University Greifswald
Friedrich-Loeffler-Strasse 23D
17487 Greifswald
Germany

Prof. Dr. Bezhan Chankvetadze
Institute of Pharmaceutical and Medicinal Chemistry
University of Münster
Hittorfstrasse 58–62
48149 Münster
Germany

Dr. Michaela C. Dinger
Institute of Biochemistry
University of Leipzig
Talstrasse 33
04103 Leipzig
Germany

Prof. Dr. Theo Dingermann
Institute of Pharmaceutical Biology
Johann Wolfgang Goethe-University
Marie-Curie-Strasse 9
60439 Frankfurt
Germany

Prof. Dr. Joachim Engels
Institut of Organic Chemistry and Chemical Biology
Johann Wolfgang Goethe-University
Marie-Curie-Strasse 11
60439 Frankfurt
Germany

Dr. Christina Fischer
Institute of Biophysical Chemistry
Johann Wolfgang Goethe-University
Marie-Curie-Str. 9
60439 Frankfurt
Germany

Prof. Dr. Gerd Folkers
Institute of Pharmaceutical Sciences
ETH Zürich
Wintherthurerstrasse 190

8057 Zürich
Switzerland

Dr. Ulrich L. Günther
Institute of Biophysical Chemistry
Johann Wolfgang Goethe-University
Marie-Curie-Str. 9
60439 Frankfurt
Germany

Dr. Wilbert H. M. Heijne
TNO Food and Nutrition Research
Department of Explanatory Toxicology
Utrechtseweg 48
P.O. Box 360
3600 AJ Zeist
The Netherlands

Dr. Michael Hoever
Discovery Partners Interational AG
Gewerbestrasse 16
4123 Allschwil
Switzerland

Dr. Carola Hunte
Max-Planck-Institute of Biophysics
Department of Molecular Membrane Biology
Marie-Curie-Strasse 13–15
60439 Frankfurt
Germany

Dr. Kurt Kesseler
Industriepark Höchst
Building H 823
65926 Frankfurt am Main
Germany

Dr. Christian Klammt
Institute of Biophysical Chemistry
Johann Wolfgang Goethe-University
Marie-Curie-Strasse 9
60439 Frankfurt
Germany

Prof. Dr. Heyo K. Kroemer
Institute of Pharmacology
Ernst-Moritz-Arndt-University Greifswald
Friedrich-Loeffler-Strasse 23D
17487 Greifswald
Germany

Prof. Dr. Rob Leurs
Leiden/Amsterdam Center for Drug Research
Department of Medicinal Chemistry
Vrije Universiteit Amsterdam
De Boelelaan 1083
1081 HV Amsterdam
The Netherlands

Dr. Andreas Liese
Institute of Biotechnology 2
Research Center Jülich GmbH
P.O. Box 1913
52425 Jülich
Germany

Dr. Cornelia Münke
Max-Planck-Institute of Biophysics
Department of Molecular Membrane Biology
Marie-Curie-Strasse 13–15
60439 Frankfurt
Germany

Dr. Beatrice Nickel
Discovery Partners International AG
Gewerbestrasse 16
4123 Allschwil
Switzerland

Dr. Ben van Ommen
TNO Food and Nutrition Research
Department of Explanatory Toxicology
Utrechtseweg 48
P.O. Box 360
3600 AJ Zeist
The Netherlands

Dr. Jörg Parsch
Beilstein Chemiedaten und Software GmbH
Trakehner Strasse 7–9
60487 Frankfurt
Germany

Dr. Nagaraj Rao
Rane Rao Reshamia Laboratories Pvt. Ltd.
Plot 80, Sector 23
Turbhe
Navi Mumbai 400705
India

Dr. Urs Regenass
Discovery Partners Interational AG
Gewerbestrasse 16
4123 Allschwil
Switzerland

Prof. Dr. Heinz Rüterjans
Institute of Biophysical Chemistry
Johann Wolfgang Goethe-University
Marie-Curie-Strasse 9
60439 Frankfurt
Germany

Prof. Dr. Peter Ruth
Institute of Pharmacy
University of Tübingen
Auf der Morgenstelle 8

72076 Tübingen
Germany

Dr. Matthias Sausbier
Institute of Pharmacy
University of Tübingen
Auf der Morgenstelle 8
72076 Tübingen
Germany

Prof. Dr. Gerhard K. E. Scriba
Institute of Pharmacy
Friedrich-Schiller-University Jena
Philosophenweg 14
07743 Jena
Germany

Prof. Dr. Dieter Steinhilber
Institute of Pharmaceutical Chemistry
Johann Wolfgang Goethe-University
Marie-Curie-Strasse 9
60439 Frankfurt
Germany

Dr. Rob H. Stierum
TNO Food and Nutrition Research
Department of Explanatory Toxicology
Utrechtseweg 48
P.O. Box 360
3600 AJ Zeist
The Netherlands

Part I
Molecular Targets

Molecular Biology in Medicinal Chemistry. Edited by Th. Dingermann, D. Steinhilber, G. Folkers

ISBN: 3-527-30431-2

1
Cellular Assays in Drug Discovery

Hugo Albrecht, Daniela Brodbeck-Hummel, Michael Hoever, Beatrice Nickel and Urs Regenass

1.1
Introduction

1.1.1
Positioning Cellular Assays

Cell-based assays allow to study the function of pharmaceutical or disease targets within complex environments and in the overall biological context. The application of techniques in molecular biology together with the introduction of new tools and analytical devices has moved readouts of cellular assays from phenotypic endpoints such as cell proliferation, cell death, respiration and functional differentiation to the cellular analysis of functional states of specific signaling molecules and metabolic components.

This transition of cellular assays from "black box" systems to specific target associated, mechanistic measurements allows the identification of chemical compounds with target-specific or closely associated modulatory function. It can therefore be assumed that cellular assays will play an increasingly important role in early drug discovery, in particular in discovering compounds for target validation ("chemical genetics"; [1] for review) and in hit-to-lead selection as well as in lead optimization. In contrast to biochemical testing of chemicals for modulatory activity on molecular targets, cellular systems can deliver additional information with respect to drug transport, metabolism and stability. Most importantly, pharmacological targets remain in their natural environment when probed for modulation and data should therefore have a higher predictive value (Fig. 1.1).

However, it remains to be said that, in order to use cellular assays in high throughput, cells need to be taken out of their natural habitat and put into culture. This will inevitably lead to alterations in the repertoire of gene and protein expression, and therefore in their phenotype. Cells used in assays should therefore be characterized with respect to the functionality of their intracellular networks, responses to external stimuli, at least for the signaling pathway of interest, and compared with responses and signaling events at the tissue or organ level. In many cases this is unfortunately not yet fully possible.

Molecular Biology in Medicinal Chemistry. Edited by Th. Dingermann, D. Steinhilber, G. Folkers

ISBN: 3-527-30431-2

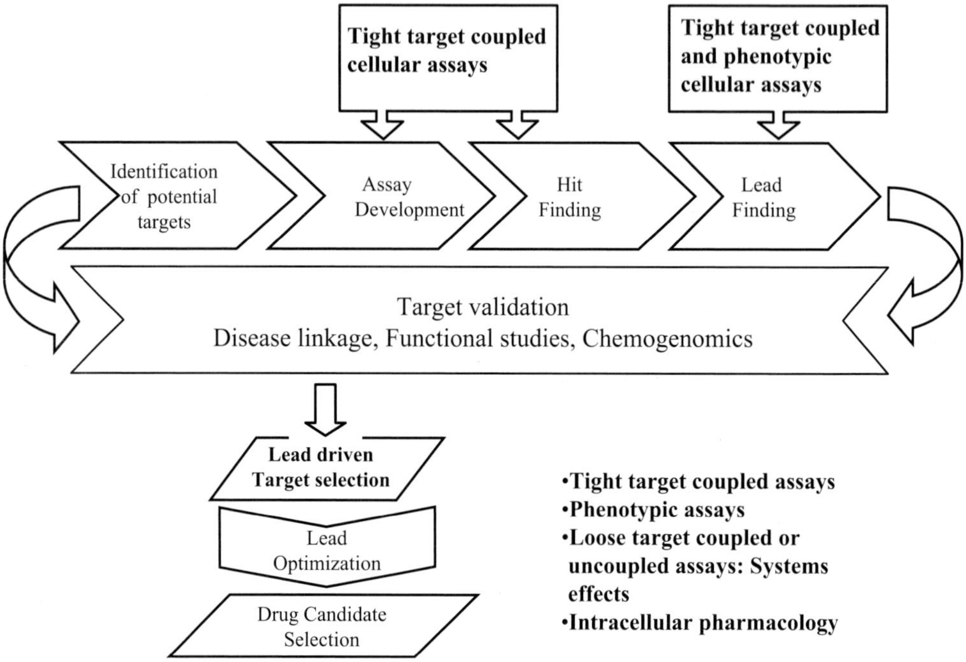

Fig. 1.1 Chemistry-driven target selection. The increased number of potential pharmaceutical targets emerging from the functional annotation of the human genome requires an efficient process to select those that are valid for modulating disease-relevant pathways and phenotypes. Cellular assays which are tightly coupled in the readout with the target of interest allow us to qualify the hits identified in cellular or biochemical high-throughput assays with respect to functionality in a physiologically relevant environment. Cellular assays which are not tightly coupled to the target will indicate system effects of compounds and can be interpreted as a cellular "safety" classification. Cellular assays can be used for target and compound prioritization.

1.1.2 Impact on Drug Discovery

Many compounds are lost during the discovery and development process due to undesirable effects or toxicity and lack of required efficacy. A large part of these effects can be related to interactions of the potential drug with targets other than the one initially selected to modulate disease. Cellular test systems have the potential to elucidate the multiplicity of interactions of drugs, and therefore contribute to a better understanding of drug action and drug selection.

The cell is organized in metabolic and signal transduction cascades. Most of these cascades are complex and nonlinear [2]. In addition, different cascades communicate with each other, and form intracellular circuits and domains. Most diseases cannot be explained by one altered step in a pathway, but are due to several alterations that affect networks in a complex way [3, 4]. Similarly, drugs, whether they interact specifically with one single target or with several molecules, will disturb signaling and metabolic cascades in a complex fashion. In addition, since many cell types make use of the same types of signaling molecules, in a different context, many cell types might be affected to different extents by the same drug. The recent introduction of high-content screening, high-throughput gene and protein expression, as well as metabolite analysis [5, 6], offers the possibility to study system interactions in a multiparallel fashion, and allows a better understanding of the degree and location of interference. Cell-based assays therefore provide a solution to study the modulation of a pharmaceutical target within its complex natural network as well as to assess potential effects on other networks.

Human diseases are characterized by one or several alterations in metabolic and signaling pathways important to maintain cellular homeostasis and the differentiated functions. Cellular assays can offer a representation of such alterations and therefore become relevant functional models to assess drug action.

Nowadays, cellular assays offer the opportunity to identify the intervention points that lead to phenotypic or endpoint alterations. Often, multiple intervention sites are necessary to alter the outcome of biological networks. Similar rules might apply for the number of drug targets that need to be modulated; in particular, for complex diseases. The right targets and drug combinations can be selected only in the case where each single component of a particular pathway can be individually studied, and this is only rendered possible by the use of cellular systems.

The combination of compound that lead to the desired endpoint can be identified by analyzing multiple parameters. The analysis can, moreover, facilitate the assessment of the effect of individual compounds on multiple pathways. This can be taken as a measure for "unwanted effects" (safety profile), but can also lead to the identification of applications in new indications. The analysis of targeted and focused chemical libraries in cellular systems may allow a rapid selection of compounds with various and desired profiles.

Hierarchical, spatio-temporal readouts will generate system-level insight into mechanisms of drug action (intracellular pharmacology) and allow us to predict systemic effects (Fig. 1.2). The development and application of computational tools as well as models of intracellular pathways are a prerequisite to secure data evaluation and interpretation [7, 8]. System-level understanding is only possible by seeking experimental results at the cellular level or beyond.

The recent developments of cellular high-throughput systems, which allow specific measurements of molecular events, render applications in primary screening, and in hit-to-lead selection and lead optimization, in particular, possible. This will enable the development of structure–activity relationships at the system level as a new paradigm in compound selection and decision-making.

1.1.3 Classification of Cellular Assays

Cellular assays have the advantage that the pharmacological targets do not need to be purified. However, in many instances, cellular assays require cell manipulation to allow for a specific signal readout. Overexpression of cell-surface receptors or intracellular proteins can have a substantial effect on the cell physiology and

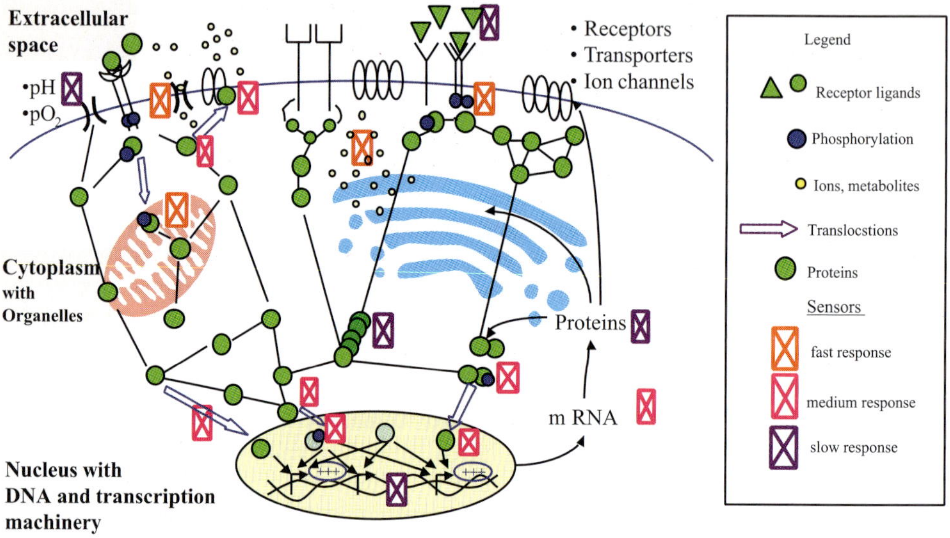

Fig. 1.2 Schematic cellular circuit and cell systems analysis concept. Cells connect to their environment with a wealth of receptors, channels and transporters. Intracellular signaling and metabolic pathways and networks react to intra- and extracellular stimuli in a spatio-temporal manner. Different cell types can react differently to the same stimuli. It can be envisaged that different cell types, cells from normal or pathological tissue, or genetically manipulated cells, can be exposed to different stimuli with and without drugs and the cellular reaction can be assessed by sensor systems in a spatio-temporal fashion, from fast responses to phenotypic changes. Multiparametric analysis will identify compounds with target-specific, target-unrelated, reversible or irreversible effects on cellular homeostasis.

have to be taken into account when using the systems for the discovery of novel drugs.

Early reactions to hormones, growth factors and neurotransmitters are often mediated via second-messenger systems or rapid influx and efflux of ions. Post-translational protein modifications and translocation of proteins are the next steps that alter the physiological state of cells. Such changes are able to lead to flow changes in metabolic pathways, and to alterations in gene and protein expression, ultimately leading to new cellular phenotypes, including cell proliferation and cell death.

Cellular assays can be classified by the temporal events occurring when a cell is exposed to alterations in the extra- and intracellular environment (Tab. 1.1). Fast responses indicated by changes in ion or second-messenger concentrations ultimately lead to functional and phenotypic changes.

Some of the cellular events can be recapitulated in so-called model systems, which might be easier to handle and to manipulate genetically than mammalian cells. Yeast has turned out to be an organism of choice, where metabolic signals and protein–protein interactions relevant for signal transduction can be translated

Tab. 1.1 Classification of functional cellular assay types (see text for references).

Second messengers and channels	Ca^{2+} levels
	ion cannels
	transporters
	cyclic nucleotides
Protein dynamics	posttranslational modifications (e.g. phosphorylation)
	protein translocations
	protein–protein interaction
Gene expression	chip technologies
	branched DNA technology
	oligonucleotide sensors
Protein expression	reporter gene technology
	ELISA assays
	protein chip arrays
Metabolites profiling	quantitative analytical profiling
Metabolic and phenotypic responses	cell proliferation, cytotoxicity, cell death/apoptosis
	respiration, acidification
	cell differentiation

into easily measurable growth response (for review, see [9, 10]). Several modifications of the yeast two-hybrid system have shown utility in finding compounds that functionally interfere with specific intracellular mechanisms. Most recently, a sensor system for drug discovery based on conformational activation of a proliferation-relevant enzyme upon drug binding has been described [11]. However, compared to mammalian cells, the yeast cell has some trade-offs, which include the cell wall, lack of the full complement of mammalian genes and alterations in its metabolic systems. The use of yeast two-hybrid and yeast "tribrid" systems [12] are not discussed further in this article.

1.1.4
Progress in Tools and Technologies for Cellular Compound Profiling

The refinement of analytical devices [13] and reader systems today enables investigators to obtain broad profiles of alterations in gene and protein expression [14] as well as metabolites [5, 15]. Unless a complete system-wide analysis is required, the present technologies allow a reasonable throughput for drug profiling. For gene expression studies, reporter gene assays can substitute for chip-based arrays, at least in selected cases [16, 17].

Several cellular imaging or high-throughput microscopy and laser-scanning systems are now becoming available together with the development of new fluorescent tags and specific antibodies. This combination allows us to study protein movements in cells. The study of protein dynamics and interactions has been largely facilitated by the identification of fluorescent proteins [6, 18, 19] and other fluorescent dyes [20] in combination with the development of fluorescence resonance energy transfer (FRET) applications.

Other technologies such as multipole coupling spectroscopy (MCS) [21] and biochips for multiparametric cell monitoring [22] are just becoming available.

The FLIPR™ (FLuorometric Imaging Plate Reader; Molecular Devices, Sunnyvale, CA) system has been available for several years to study rapid cellular second-messenger responses, such as Ca^{2+} release, and has been recently upgraded for the analysis of ion channels. The FLIPR applications are discussed in detail below.

In the following sections, we will describe a selected set of applications in more detail. The discussion is grouped according to Tab. 1.1 into fast responses or second-messenger systems, protein dynamics, alterations in gene and protein expression, and metabolic as well as phenotypic readouts.

1.2 Membrane Proteins and Fast Cellular Responses

A great variety of membrane proteins have been expressed by using recombinant DNA technology in different eukaryotic cells (e.g. mammalian cell lines, insect cells and yeast) for functional studies of cell-surface receptors, ion channels or transport proteins.

1.2.1 Receptors

Cell-surface receptors typically mediate cell signaling from the cell surface to the cytoplasm and/or nucleus involving two major principles, i.e. signaling via enzyme-linked receptors or trimeric GTP-binding proteins (G-proteins). The enzyme-linked receptors include the receptor guanylyl cyclases, receptor tyrosine kinases, tyrosine-kinase-associated receptors, receptor tyrosine phosphatases and receptor serine/threonine kinases [23]. In all cases, a cascade of events is initiated upon activation, which alters receptor status and the concentration of one or more small intracellular signaling molecules. Here, we will discuss assays monitoring signals mediated by G-proteins, which are activated by G-protein-coupled receptors (GPCRs) upon binding of a specific ligand. Subsequent to stimulation, changes of intracellular cyclic AMP (cAMP) or Ca^{2+} concentrations are induced. These two signaling molecules are known to be involved in regulatory processes of the cardiovascular and nervous systems, immune mechanisms, cell growth and differentiation, and general metabolism. Both messengers act as allosteric effectors on specific target proteins in the cell including, but not limited to, kinases (e.g. protein kinase A and C), lipases (e.g. phospholipase Cβ), Ca^{2+}-binding proteins (e.g. calmodulin) and ion channels (e.g. ionotropic glutamate receptors). The cytoplasmic concentrations of the second messengers are regulated via plasma membrane enzymes – in the cAMP pathway the enzyme adenylate cyclase directly produces cAMP and in the Ca^{2+} pathway lipases are stimulated to produce the soluble messenger inositol triphosphate (IP_3), which in turn induces the release of Ca^{2+} from the endoplasmic reticulum. In some cases, ion channels are activated by direct interaction with G-protein subunits.

Different technologies have been developed for intracellular monitoring of second messengers in living cells. Detection of Ca^{2+} is generally based on fluorescent indicators, which have the property of being able to easily penetrate cell membranes and alter their fluorescent emission when bound to Ca^{2+}. For measurement of a typical fast and transient response, the binding reaction to Ca^{2+} must be reversible, therefore providing important kinetic data which allow us to distinguish between a typically transient signal due to specific receptor activation and non-specific signals affecting Ca^{2+} homeostasis or membrane permeability.

There are currently many different systems available to detect cAMP in cell-based assays. In most cases cells are lysed after exposure to a specific ligand and the amount of cAMP is determined with a competitive immunoassay (PerkinElmer Life Sciences, Boston, MA; Molecular Devices; Biomol Research Laboratories, Plymouth Meeting, CA) or a competitive enzyme complementation assay (DiscoveRx, Fremont, CA).

Typically, GPCRs are overexpressed in eukaryotic cell lines to monitor the concentration of second messengers upon stimulation with a specific ligand [24]. In many cases, the co-expression of a suitable G-protein subunit or chimeric G-proteins is necessary in order to elicit a signal sufficient for detection [25, 26].

1.2.1.1 FLIPR Technology for Detection of Intracellular Calcium Release

The FLIPR system provides readouts for high-throughput, cell-based assays that represent information-rich kinetic data in 96- or 384-well format. In brief, cells are loaded with a fluorescent indicator (e.g. Fluo-4 AM for Ca^{2+}), which shows an absorption spectrum compatible with excitation at 488 nm with an argon-ion laser source, leading to maximum emission at 516 nm in the presence of Ca^{2+}. Fluorescence emission is induced simultaneously in all wells of a 96- or 384-well plate and kinetics of intracellular Ca^{2+} release, triggered by exposure to a specific ligand, can be monitored at a maximum of 60 sequential measurements per minute. The FLIPR system rapidly discriminates between full agonists, partial agonists and antagonists to accelerate primary and secondary screening. Figure 1.3 illustrates a typical experiment conducted with a mammalian cell line stably transfected with an expression vector encoding a GPCR gene. Positive compounds depicted in Fig. 1.3(b) were subjected to verification in duplicate measurement and confirmed hits were further tested for dose–response relationships in EC_{50}/IC_{50} experiments.

1.2.1.2 Competitive Immunoassay for Detection of Intracellular cAMP

Most assays designed for detection of intracellular cAMP are based on monoclonal antibodies specifically recognizing cAMP. Typically, cell lysates are mixed with exogenously added biotinylated cAMP to form a heterotrimer with a specific antibody and streptavidin. This complex formation is competed by endogenous non-biotinylated cAMP. Streptavidin and antibodies can be linked covalently to a donor and an acceptor fluorophore (e.g. europium cryptate and allophycocyanin), forming an interacting FRET pair upon complex formation. A decreased signal is observed with an increase in intracellular cAMP produced, due to competition for binding to the specific antibody.

(a)

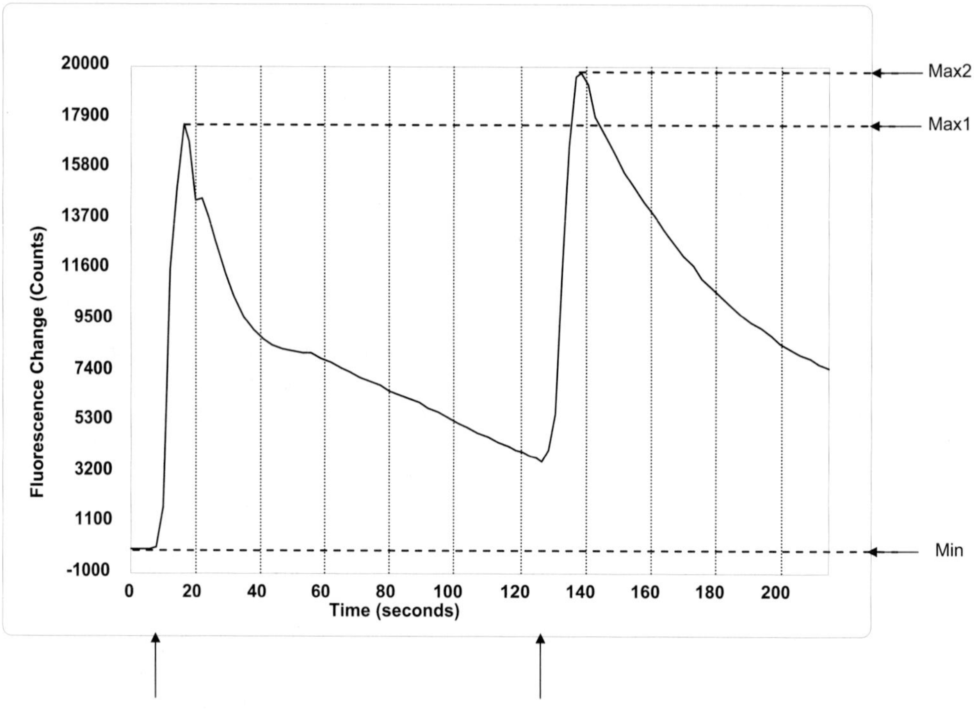

Fig. 1.3 Determination of intracellular Ca^{2+} with the FLIPR system. (a) A fluorescent trace of a typical Ca^{2+} kinetic is shown and the measured parameters used for calculation of maximal fluorescence signals are depicted. In this example the maxima were measured before and after ligand addition (Max1 and Max2 respectively) and standardized by dividing by the minima measured before compound addition (Min = baseline). The average of the positive controls was set as 100% control and the average of the negative controls was set as 0% control. For all wells, relative activities with respect to the controls were calculated and the cut-off criterion for both agonists and antagonists was 50%.

The AlphaScreen™ technology (PerkinElmer Life Sciences) is an alternative technology that allows quantitative detection of a broad range of cellular components, including cAMP. AlphaScreen relies on the use of "donor" and "acceptor" beads that are coated with a layer of hydrogel providing functional groups for bioconjugation. When the beads are brought into proximity, a cascade of chemical reactions is initiated to produce a greatly amplified signal. Upon laser excitation, a photosensitizer in the "donor" bead converts ambient oxygen to a more excited singlet state. The singlet state oxygen molecules diffuse across to react with a chemiluminescent in the "acceptor" bead that further activates fluorophores contained within the same bead. The fluorophores subsequently emit light at 520–620 nm. In the absence of a specific biological interaction, the singlet state oxygen

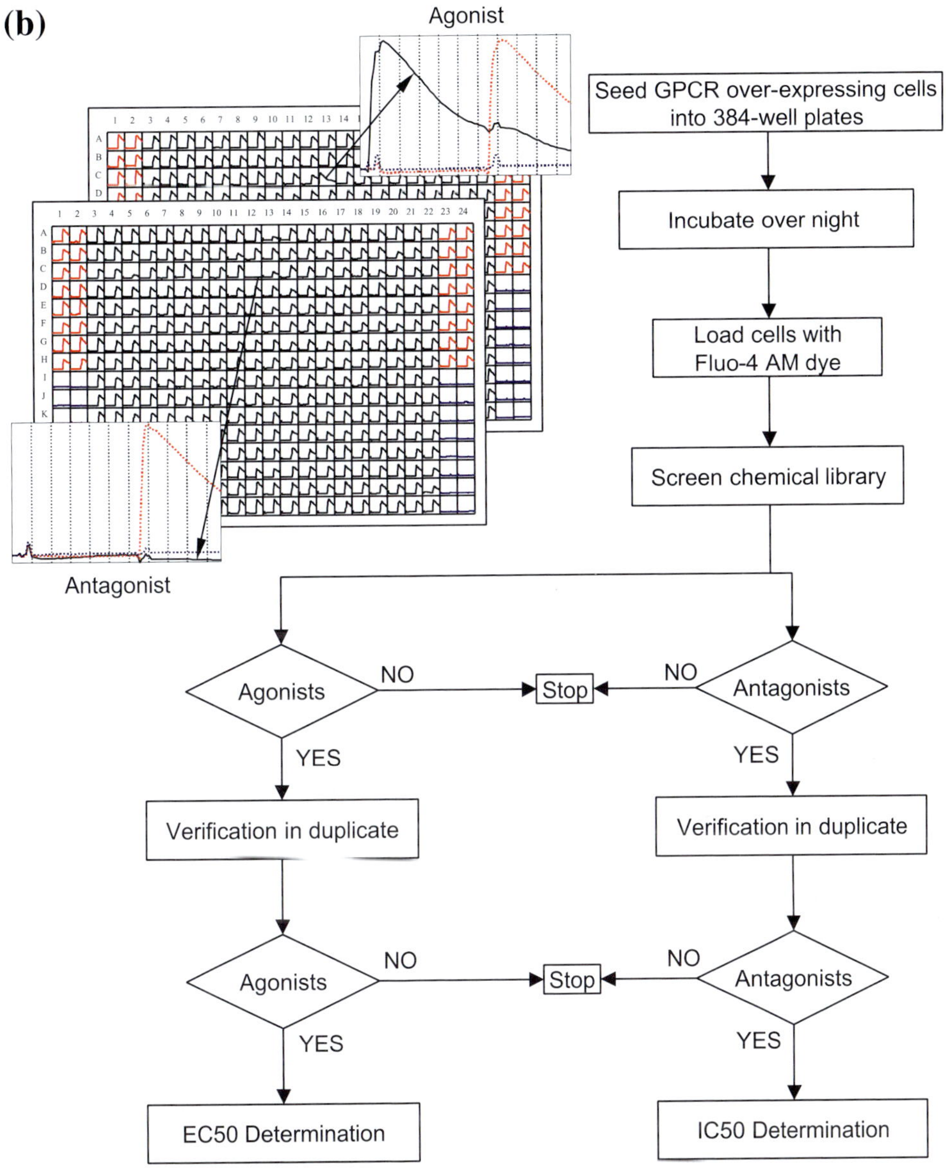

Fig. 1.3 (b) A flow diagram for a typical experiment is shown. Genetically modified cells overexpressing the target GPCR were seeded into 384-well plates overnight. The cells were washed with suitable buffer prior to adding buffer containing the Fluo-4 AM dye, followed by incubation for 45 min. After an additional wash step, the cell plates were transferred into the FLIPR machine and the chemical compounds were added followed by ligand addition. Fluorescence emission was monitored starting before compound addition and continued until a substantial decay of the signal was observed. Two 384-well plates from a typical screen are included. Examples of an agonist and an antagonist are shown as enlarged graphs. The 100% and 0% controls are overlaid as red- and blue-hatched lines, respectively.

molecules produced by the "donor" bead go undetected without the close proximity of the "acceptor" bead. AlphaScreen has been used in GPCR functional assays for detection of endogenous cAMP production.

In all the above-mentioned systems, the cells have to be subjected to lysis prior to measurement, adding not only an extra step to the assay procedure, but more importantly making kinetic measurements impossible because data can only be generated from endpoints. In addition, whole-cell FRET methods have been developed to determine the temporal and spatial dynamics of cyclic nucleotides (see Section 1.4.4).

1.2.1.3 Enzyme Fragment Complementation (EFC) Technology

EFC is based on genetically modified enzymes consisting of two fragments, i.e. the enzyme acceptor (EA) and the enzyme donor (ED). The separated fragments are inactive and only spontaneous formation of heterodimers will lead to an enzymatically active form [27]. The HitHunter™ technology (DiscoveRx) is based on a modified version of the β-galactosidase enzyme forming an EA/ED pair. The technology has been adapted for use in many different assays including, but not limited to, kinase, progesterone receptor, estrogen receptor and cAMP assays. Here, we only discuss specific application for intracellular detection of cAMP. In brief, an ED–cAMP peptide conjugate is utilized, in which cAMP is recognized by a specific antibody. This ED fragment is able to complement EA to form an active complex, but only in the absence of the specific antibody. In the assay, anti-cAMP antibody is titrated for complete inhibition of enzyme formation. Exposure of a defined antibody aliquot to cell lysate will lead to antibody–cAMP complex formation, therefore reducing the free amount of antibody. Subsequently, addition of ED conjugate in the following assay step will lead to reduced antibody–ED complex formation, therefore promoting complementation of active enzyme. Free ED–conjugate in the assay is proportional to the concentration of intracellular cAMP and the same applies for the measured β-galactosidase activity (see www.discoverx.com for additional information).

1.2.2 Membrane Transport Proteins

A great number of substances are transported in and out of complex organisms, in some cases using passive transport, driven solely by diffusion down a concentration gradient either through the paracellular or transcellular route. Absorption and distribution of many food constituents such as ions, sugars, amino acids and nucleotides in the body are mediated/facilitated by specific membrane transport proteins, which form channels across cellular membranes. Similar mechanisms are needed for active removal of xenobiotics, waste products and cell metabolites from body tissue. These specific proteins are divided into two classes, i.e. uniporters and coupled transporters, of which the former transport a single solute from one side of the membrane to the other and, for the latter, the transfer of one solute depends on the simultaneous or sequential transport of a second solute.

Coupled transporters are further divided into symporters moving compounds in the same direction and antiporters moving the compounds in opposite directions. Some specific transporters act in a passive manner moving molecules down a concentration gradient, as is the case with many nutrients such as glucose, which are found at high concentrations in the extracellular space of tissues. A similar situation is found with ligand- and voltage-gated ion channels of the nervous system, which specifically pass ions down a concentration gradient upon stimulation by neurotransmitters or membrane depolarization, respectively. In contrast, active transporters are linked to hydrolysis of ATP and move compounds in a highly specific manner against a gradient. The largest known and most diverse group of active transport proteins is the ABC transporter superfamily [28]. Over 50 ABC transporters have been described and the variety of substrates transported is large, including amino acids, sugars, inorganic ions, polysaccharides, peptides and even large proteins. One of the most cited ABC transporters, the multidrug-resistance (MDR1) protein [29], has been in the limelight of many drug discovery efforts in recent years, since overexpression in human cancer cells can render these cells simultaneously resistant to a variety of chemically unrelated drugs that are widely used in cancer chemotherapy. Other important phenomena linked to ABC transporters include resistance to chloroquine in malaria parasites, presentation of antigenic peptides on immune cells and the genetic disease known as cystic fibrosis [30–32].

1.2.2.1 **Ion Channels**

Most cells maintain an electropotential across the membrane which has been most extensively studied in neurons and muscle cells where the potential is particularly high, and reaches a value of about −70 mV. The membrane potential is maintained by the ubiquitous Na^+-K^+-ATPase, which operates as an antiporter actively pumping Na^+ out of the cell, while pumping K^+ into the cell. For every molecule of ATP hydrolyzed, three Na^+ are pumped out and two K^+ pumped in, therefore creating a negative net charge inside the cell.

Ion channel proteins form pores in cell membranes, allowing certain ions to pass through driven by a concentration gradient. A characteristic feature of ion channels is a gating mechanism that controls the movement of ions. Here we focus on ion channels, which are expressed in neurons or muscle cells, and therefore play important roles in neuronal signaling and muscle contraction. Changes of the membrane potential can be induced in a spatially limited fashion by ligand-gated Na^+ channels sensitive to neurotransmitters. Subsequent opening of voltage-gated Na^+ channels induces self-propagation of the depolarization along the cell membrane (action potential), finally inducing the opening of voltage-gated Ca^{2+} channels typically at the synapse (end plate). Influx of Ca^{2+} induces the release of neurotransmitters into the synaptic cleft, further propagating the signal on to the post-synaptic cell. Malfunctions of ion channels are the basis of many ailments, including muscle and cardiac disorders, epilepsy, migraine, depression, and ataxia.

Opening of ion channels is in most cases monitored by measuring changes in the membrane potential. The gold standard for this purpose is the patch clamp technology, offering very high sensitivity, good temporal resolution and measurement of single channel conductance at a precise voltage. To date this technology provides the best data for functional assessment of ion channels; however, the massive disadvantages include requirement of skilled operators, laborious work flow and low throughput. Currently, different companies are involved in designing novel devices to achieve automation of the patch clamping procedure. One of the most advanced tools is the IonWorks^{TM} station (Molecular Devices, Sunnyvale, CA), which uses integrated fluidics for automatic positioning of single cells on microapertures on specially fabricated supports [33]. Single ion-channel recordings were demonstrated at a throughput of up to 2000 assay points per day. Higher throughputs can be achieved by using solvatochromatic, cell-permeable dyes like DiBAC (Molecular Probes, Eugene, OR) or the more recently developed FMP dye (Molecular Devices). Combination with a FLIPR system allows monitoring of ion channel function in true high-throughput mode surpassing analysis of 30,000 compounds per day (for more details, see the following paragraph). More details including other well-established ion-channel assay technologies were described in a recent review article [34].

FLIPR Technology and Ion Channel Assays To identify compounds that modulate ion channel activity, it is imperative that rapid and economical evaluation of biological activities in a high-throughput manner is available. One of the fluorescence-based assays of membrane potential changes uses a potentiometric fluorescent probe, bis-(1,3-dibutylbarbituric acid) trimethine oxonol [$DiBAC_4(3)$], that partitions between the intracellular and extracellular compartments depending on the electropotential across the membrane. Upon depolarization of the cells, the dye moves into the cell, and subsequent binding to intracellular hydrophobic sites results in enhanced quantum yield and in an increase in fluorescence intensity. Conversely, membrane hyperpolarization triggers dye extrusion from the cells and consequently decreases fluorescence, because the quantum yield of $DiBAC_4(3)$ is low in an aqueous environment. The partitioning of $DiBAC_4(3)$ is a slow process and therefore the technology is not amenable to measure channels with fast-gating kinetics or with rapid desensitization. Significant improvement was achieved with the recent development of the FMP dye [35, 36], which has faster partitioning kinetics. Figure 1.4 shows a typical experiment conducted with differentiated myotubes (C2C12 cell line).

The LOPAC library (Sigma-Aldrich, St Louis, MO) of pharmacologically active compounds was screened for identification of compounds exerting agonistic or antagonistic effects on the nicotinic acetylcholine receptor (nAChR). In brief, the cells were washed with buffer and loaded with the FMP dye for 30 min at room temperature. Six hundred and forty compounds from the LOPAC library were added in the FLIPR followed by stimulation with carbachol. In contrast to a typical Ca^{2+} signal, the membrane depolarization does not show a transient kinetic. However data evaluation and hit identification can be carried out as described in Fig. 1.3. Identification of agonists and antagonists is possible in one experiment. Positives were subjected to EC_{50}/IC_{50} determination and, in the case of agonists, the specific AChR antagonist pancuronium was used to show reversal of the car-

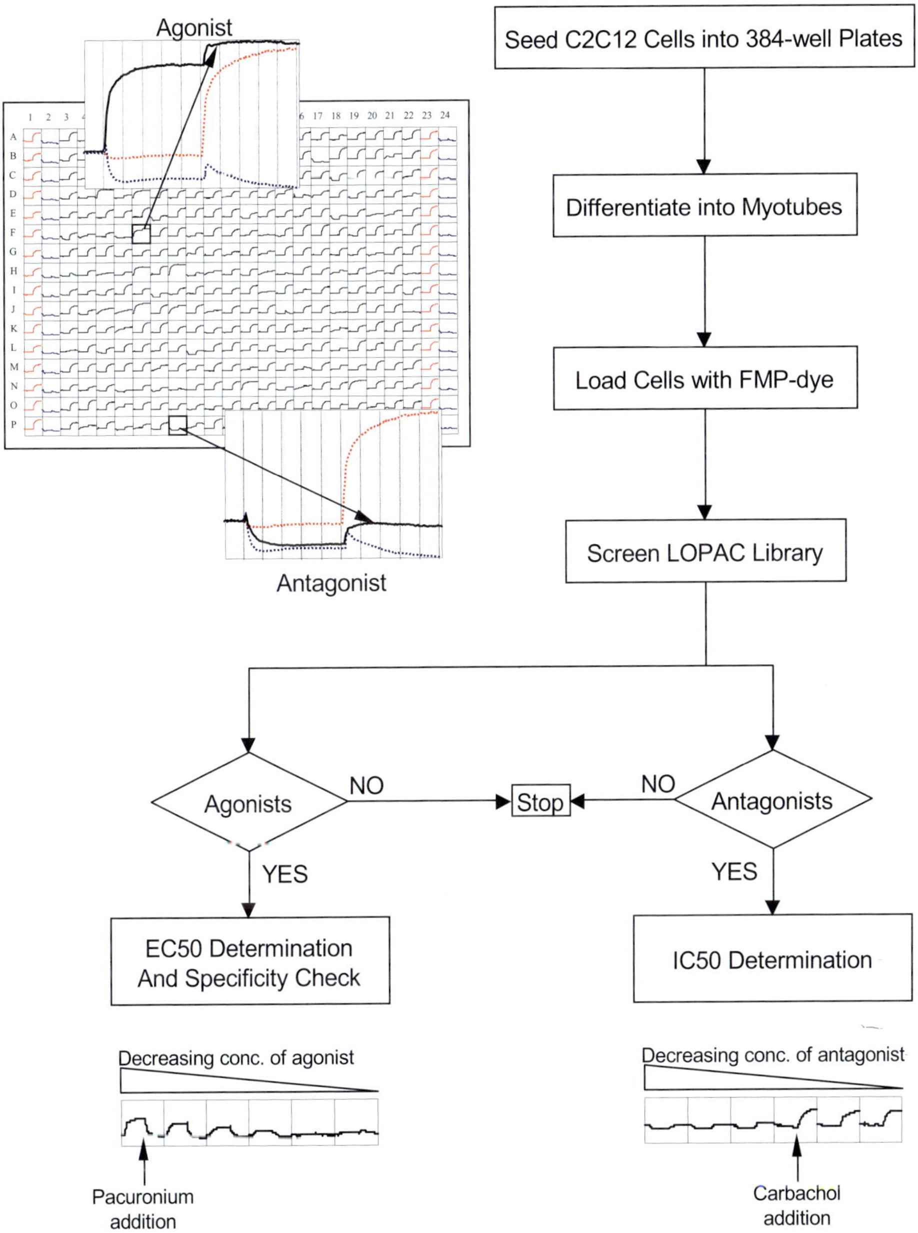

Fig. 1.4 Ion-channel activity in differentiated myotubes. The murine C2C12 muscle cell line was used to validate the FMP dye. C2C12 cells were seeded into 384-well plates and differentiated into myotubes by exposure to horse serum. Differentiated C2C12 cells strongly express nAChR and depolarization of the cells can be induced with specific receptor agonists, such as carbachol, which was applied in the outlined experiment. A 384-well plate from a typical screen for agonists and antagonists is shown. Example of an agonist and an antagonist are depicted as enlarged graphs. The 100% and 0% controls are overlaid as red- and blue-hatched lines, respectively.

bachol signal (see fluorescence traces shown for the EC_{50} determination, Fig. 1.4). Initially 21 compounds were identified as agonists and 12 compounds showed specificity for the AChR.

Voltage Ion Probe Reader (VIPR)™ Technology and Ion Channel Assays VIPR technology (Aurora Bioscience, San Diego, CA) is based on FRET and uses two fluorescent molecules. The first molecule, oxonol, is a highly fluorescent, negatively charged, hydrophobic ion that can rapidly redistribute between two binding sites on opposite sides of the plasma membrane in response to changes in membrane potential. The second fluorescent molecule, coumarin lipid, binds specifically to one face of the plasma membrane and functions as a FRET donor to the voltage-sensing oxonol acceptor molecule. When the oxonol moves to the intracellular plasma membrane binding site upon depolarization, FRET is decreased, and this results in an increase in the donor fluorescence and a decrease in the oxonol emission. This technology enables research and assay development in the field of ion channel biology. VIPR technology can deliver subsecond, real-time readouts from assays in a microplate format (see http://www.aurorabio.com/ for additional information).

1.2.2.2 MDR Proteins

MDR is mainly linked to overexpression of proteins of the ABC transporter superfamily. The most extensively studied member is the MDR1 gene product (P-glycoprotein), which is expressed in the intestine, kidneys, liver and blood–brain barrier, as well as in many tumor cells. In the latter case, high overexpression is one of the hallmarks of cell resistance to cancer chemotherapy. This observation is a result of anticancer drugs being actively pumped out of cells, resulting in decreased intracellular drug levels. Small molecules capable of reversing efflux can restore drug sensitivity in resistant tumor cells. The most commonly used assays to identify MDR pump inhibitors are based on fluorescent P-glycoprotein substrates such as Hoechst 33342, rhodamine 123 and rhodamine 6G. The cells are loaded with the substrate and extracellular dye is removed by washing prior to lysis with detergents [37]. Subsequent measurement of fluorescence emission provides a quantitative readout, which is proportional to the intracellular accumulation of the substrate. Exposure of the cells to MDR pump inhibitors will reduce efflux of the substrate and therefore leads to an increased fluorescence readout. However, these assays are not homogeneous and involve washing steps after loading of the cells with the corresponding dye. A novel MDR pump substrate was recently developed to run fully homogeneous assays on a FLIPR system.

The FLIPR Multidrug Resistance Assay (Molecular Devices) is a fully homogeneous application needing only 15 min of incubation at room temperature with a proprietary dye, which is transported by the MDR1 gene product and changes its fluorescent properties only after modification by intracellular enzymes. This allows performance of live-cell kinetic assays for screening MDR pump inhibitors in high-throughput mode. The elimination of wash steps and temperature control means that the assay is highly amenable to automation [37].

1.3
Gene and Protein Expression Profiling in High-throughput Formats

Gene and protein expression profiling can be applied in drug discovery for at least two different purposes.

The first application will be for the discovery of new potential drug leads. In this case, the target of interest is either a component of the expression machinery itself or monitoring of expression serves as an indicator for an upstream target and reflects the coupled cellular response [16]. To make high-throughput screening (HTS) possible, the promoter sequences of interest, that have been shown to be coupled to upstream cellular events, are linked to so-called reporter genes, which serve as indicators for transcriptional and translational activation.

The second application is the use of expression profiling to obtain information on the specificity of a drug candidate at the cellular level or for the identification of surrogate markers for drug action. In this case, gene or protein expression in response to a potential drug candidate of as many genes and proteins as possible should be investigated. Expression patterns can be used as an indicator for the number of interactions within the cell. Such data could be of relevance for getting an understanding of the broader impact of modulating a potential drug target on the whole cellular system. Gene expression chips have been introduced recently for the measurement of hundreds of simultaneously expressed genes at the mRNA level [38]. The utility of the approach has been recently validated in yeast, where expression profiles identified pathways of previously uncharacterized mutations or identified new potential drug targets [39]. Similarly, protein chip arrays have been introduced to capture a particular fraction of proteins and allow analysis with a reasonable sample throughput [40]. In addition, the technology to analyze the cellular proteome, or parts of it, has been recently introduced [14].

The analysis of thousands of samples becomes a prerequisite if expression profiling is to be used in hit identification. Hit-to-lead selection and lead optimization will require a throughput which is at the level of hundreds to thousands of compounds. Both can be achieved by focusing on selected indicator genes or their products, indicative of the discovery target of interest, rather than on broad expression profiling. Several targets can be addressed at once by running established single-readout assays in parallel or by using multiplexing as a tool to combine several readouts in one assay step [41].

In the following section, we will focus on examples that can be applied in early drug discovery and which allow a high sample throughput. Gene expression chips [38, 42] and proteomic technologies [43] are the subject of many recent reviews. However, due to their still limited sample throughput, they will only be mentioned briefly here.

1.3.1
Reporter Gene Assays in Lead Finding

Transcription factors can be classified according to their structural nature, but also according to their behavior to respond to modulation of signal transduction pathways. Such a classification, which is the basis for reporter gene technology and expression profiling, has recently been published by Brivanlou and Darnell [44]. According to this classification, transcription factors can be grouped into two large classes – those who are constitutively activated, and those whose activity is regulated in a spatio-temporal, quantitative and qualitative manner. Among the regulated transcription factors is the large class of signal-dependent factors, which are activated upon an intra- or extracellular signal. Genes regulated via the latter are attractive as reporter genes since they couple the signaling pathways with a transcriptional response.

Extracellular signals, after interacting with appropriate receptor molecules at the cell surface, initiate a cascade of intracellular events which often leads to alterations in the expression of a set of genes. One or several of these genes can be used as an indicator for the modulation of targets within specific signaling pathways. The activation of a certain gene can be easily identified by "tagging" the particular gene with a so-called reporter gene. For this, plasmids need to be constructed which contain the promoter or regulatory region of the gene of interest linked to the reporter gene. These plasmids are then introduced into a cell line which expresses the pathway of interest containing the pharmacological targets and couples the reporter readout with the particular pathway of interest. A number of genes have been described that can serve as reporter genes [16, 45]. Expression can be monitored by fluorescence, e.g. by using green fluorescent protein (GFP) as a reporter, or through the detection with fluorescently labeled antibodies. Alternatively, the enzymatic activity of the reporter is used, as it is the case for luciferase, which converts the substrate to a luminescent product, or the reporter is detected with well-established and characterized secondary readouts producing fluorescence, luminescence or colored reactions.

Reporter gene technology has been widely used to establish high-throughput screens for a whole variety of targets of interest in drug discovery (for a review, see [45]). For many targets, such as transmembrane receptors for which other screening methods were difficult to establish, reporter gene technology became the method of choice for finding leads. Selected examples include the use of reporter gene assays to find agonists and antagonists for a whole variety of GPCRs [17] stably transfected into Chinese hamster ovary (CHO) cells with a 6 × CRE–luciferase reporter construct. This cell line served as an indicator for GPCRs, such as the melanocortin-1 receptor, which upon stimulation modulates cAMP levels in the cell. The assay can be performed in 384-well plates, therefore offering significant savings in cost, time and cell culture efforts. Along the same lines, a HTS system was established for the metabotropic glutamate receptor mGluR7 [46]. The mGluR7 receptor is negatively coupled to adenylyl cyclase and its stimulation decreases cAMP levels. CHO cells were stably transfected with rat mGluR7 cDNA and an AMP-responsive luciferase reporter gene. Suitability for HTS was demonstrated in a line where cAMP levels induced by forskolin (an activator of adenylyl cyclase) were significantly reduced by receptor agonists. A direct method to assay receptors negatively coupled to adenylyl cyclase has been described by Xing et al. [47]. The stably transfected CHO reporter cell line contained three separate expression plasmids: (i) a chimeric $G_{q/i}$-protein, which redirects a negative G_i signal into a positive G_q signal; (ii) a G_i-coupled GPCR, in this case the μ-opioid receptor or the 5-hydroxytryptamine receptor, and (iii) the 3 × NFAT–β-lactamase reporter. Thus activation of the G_i-coupled GPCR translated into NFAT activation via the G_q-mediated signaling pathway, allowing the monitoring of β-lactamase expression in a HTS format.

A wide variety of different cell lines have been used to establish reporter gene assays, a selection of which is indicated here as examples. Human embryonic kidney (HEK) 293-EBNA cells grown in suspension culture were used to assay expression of prostanoid receptors coupled to a CRE-SEAP (secreted human placental alkaline phosphatase) reporter [48]. ECV304 cells, a spontaneously transformed immortal line derived from human umbilical vein endothelial cells, were transfected with an ICAM-1 reporter to screen for compounds which inhibit ICAM-1 expression upon stimulation with interleukin (IL)-1β [49]. The human breast cancer cell line MDA-MB-453 was used as a transfection host to establish stable reporters under the control of a mouse mammary tumor virus (MMTV) promoter to study androgen and glucocorticoid receptor agonists. MDA-MB cells express both the androgen and the glucocorticoid receptors, and therefore serve as an ideal host cell for the reporter constructs [50].

As a practical example, data obtained with a reporter gene assay under HTS conditions is presented in Fig. 1.5 and Tab. 1.2. Cells stably transfected with a luciferase reporter gene construct were seeded in 384-well plates and dosed with serial dilutions of a compound known to activate the promoter of the reporter gene construct. After an incubation time of 24 h, the cells were washed, lysed and the luciferase substrate (LucLite®; PerkinElmer Life Sciences) was added. The amount of luciferase protein expressed correlated directly with the amount of substrate metabolized to a luminescent compound, which was detected in a luminometric plate reader (TopCount® NXT; PerkinElmer Life Sciences). The data presented here provided information as to the minimal concentration of positive reference compound required as standard on each plate for the HTS of 100,000 compounds. To assess the quality of each individual assay plate during HTS, a criterion called the Z' factor is calculated from the dynamic range of the signal (difference between the means of the background and the positive control wells) and the standard deviation (SD) of the control wells on each plate [51]. Since the Z' factor takes into account both the signal intensity and the inherent variance, it gives a measure of the stability of an assay. Screening is generally considered possible if $0 < Z' < 1$ [51]; however, in practice, the minimum value to be met by the Z' factor is often set higher. As shown in Fig. 1.5 and in the values calculated in Tab. 1.2, a concentration of at least 3 μM of the agonist in the positive control would be required to obtain a $Z' > 0.5$. In effect, the positive control was used at 10 μM during HTS.

Reporter gene assays can be highly informative if the gene product of interest and the pathway of induction are well characterized. It needs to be borne in mind that the reporter gene will be modulated by any type of stimulus influencing any step of the coupled signal transduction cascade. Therefore, the effects on reporter gene expression observed in response to test compounds should always be studied

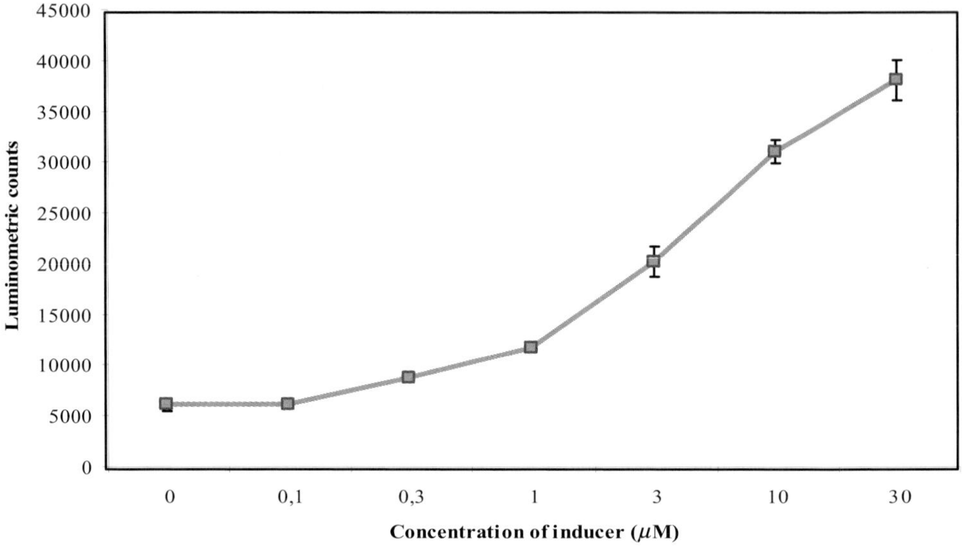

Fig. 1.5 Dose-dependent induction of luciferase reporter gene activity by an inducing substance subsequently used as an agonist control in HTS. Each value represents the average of eight individual determinations, the error bars indicate 1 SD. The mean of luminometric counts measured with TopCount NXT (PerkinElmer Life Science) is plotted against the concentration of the inducing substance (in μM). Counts obtained in the absence of compound refer to the basal expression of the luciferase reporter gene and constitute the assay background.

further in assay panels addressing specific steps in the pathway, to rule out mechanisms not directly associated with the target of interest. This has been shown recently by inhibiting the CXCR1-agonist-stimulated mitogen-activated protein (MAP) kinase reporter gene by pertussis toxin (acting at the G-protein level) and inhibitors of one MAP kinase subfamily, but not by inhibitors of a second subfamily [52].

Even though expression of the reporter genes generally does not influence homeostasis of the cell, a certain artificial component is introduced into the system.

Tab. 1.2 Luciferase reporter gene activation and assay performance.

Concentration (μM)	*Mean of 8 wells*	*SD*	*Z′ factor*
0	6364	641	
0.1	6347	314	n.d.
0.3	9028	458	−0.238
1	12006	509	0.389
3	20436	1476	0.549
10	31292	1181	0.781
30	38343	2027	0.750

First and foremost, it needs to be established whether the signal transduction pathway leading to activation of a particular promoter is present and functionally active in the cells in question. The introduction of the reporter gene itself into cells can disturb normal transcription patterns, due to scavenging of particular transcription factors or through positional effects due to integration of the plasmid during the generation of stable cell clones. Thus, great care has to be taken in choosing the promoter and the cell line in which the reporter gene assay is to be established. Limitations can occur regarding the cell lines that can be used, the transfection methods that should be employed and the efficiency of expression of the reporter gene. Nevertheless, even though the generation of well-characterized reporter gene cell lines is time-consuming and labor intensive, they constitute a highly efficient system for HTS programmes.

1.3.2
Reporter Gene Assays in Lead Optimization

The use of gene expression arrays to monitor simultaneously the expression of hundreds or thousands of genes has been mentioned before. Expression arrays are useful tools not only to characterize the expression phenotypes of cells and tissues for their classification [53], but they also allow us to monitor the impact of mutations [54] or drugs on the cellular phenotype [55] at a systems level. Arrays are widely used in toxico- and pharmacogenomic studies [56, 57]. Furthermore, the large data sets obtained with gene expression arrays have been used to construct predictive models for diagnostic purposes [58, 59].

The use of expression arrays requires the production of probes suitable for hybridization, which can be fastidious to produce. Furthermore, analysis of the large data sets obtained requires sophisticated software, and a broad special knowledge as to the significance and relevance of the multiplicity of the responses observed. In some instances, only a particular fraction of all genes measured is informative for the target in question. Thomas et al. [60] have recently analyzed the expression of 1200 liver transcripts using cDNA arrays, after dosing mice with 24 different treatments falling into five toxicological categories. A set of 12 transcripts was identified and considered diagnostic for classification of the treatments into the toxicological classes. Interestingly, addition of further transcripts lead to a decline in the predictive accuracy, and the "individuality" of the specific treatment profile began to take over. In a similar study, Bartosiewicz et al. [61] have used a set of 260 genes involved in phase I and II drug metabolism to test the effect of 10 toxicants on expression profiles in different tissues and for different exposure times in mice. Distinct groups of chemical classes could be correlated to the specific expression profile of a relatively low number of transcripts. Similar expression profiling studies performed on HepG2 cells [62] and on livers of treated rats [63] show that even with a highly focused set of genes, separate classes of chemicals can be identified through shared modulation of expression of the selected gene set.

Today, the method of choice to assess gene expression profiles *in vivo* consists in the use of oligonucleotide or cDNA microarrays or gene chips. However, due to the high cost involved and the labor-intensive sample preparation, which generally preclude a high-throughput method, these studies are done mostly in a late phase of the drug discovery process. Thus, one or a few compounds are analyzed for the induced expression patterns of many genes. To reduce late-phase attrition of potential drug candidates due to unforeseen side-effects, expression-profiling systems which are amenable to HTS are needed. An approach making use of reporter gene assays has been described [64, 65]. Albano et al. [65] selected a set of oxidative stress-responsive promoters from *Escherichia coli* genes to generate GFP–reporter gene constructs, as a tool for mechanistic screening of antitumor drugs in 96-well format. Zhu and Fahl [66] generated stable HepG2 cell clones expressing GFP under the control of the electrophile response element (EpRE), which allow HTS for induction of phase II drug-metabolizing enzymes. In addition to high-throughput, a further advantage of GFP–reporter gene constructs is that the expression can be detected in live cells, allowing easy monitoring of time courses and reversibility of induction. The use of reporter gene cell lines for expression profiling has further been documented by Farr and Todd [64], who established a panel of cell lines containing the bacterial chloramphenicol *N*-acetyltransferase (CAT) reporter gene under the control of promoters indicative of specific cellular stress responses. Induction of reporter constructs by chemical compounds can give an indication concerning the action of compounds at the subcellular level. Thus, promoter–reporter gene constructs expressed in well-characterized cell lines can provide useful model systems to profile compounds with respect to activation of canonical signal transduction pathways associated with defined cell fates or phenotypes. The assay system will detect compounds which modulate reporter gene expression directly and those which interfere with the expression elicited by external stimuli such as growth factors or cytokines. When a panel of reporter gene assays is used for compound profiling, similar expression patterns can point to comparable mechanisms of action of test compounds in intact cells. Changes of expression patterns in response to specific signaling pathways and in different cell lines can provide information on the specificity of test compounds. These systems can easily be adapted to HTS, and the analysis of time course and dose dependency of induction. Therefore, the profiling of hit collections or compound sets resulting from lead optimizations might allow the clustering of compounds into groups with pharmacologically similar mechanistic responses. For example, activation of reporter genes coupled to the p53-responsive element (p53RE) [67] would indicate that the test compound leads to DNA damage, and activation of reporters under the control of the xenobiotic response element (XRE) [68, 69] would point to the transcriptional induction of drug-metabolizing enzymes, such as cytochrome P450s and others. Thus, *in vitro* reporter gene profiling with a well-defined and validated set of genes can contribute not only to compound classification, but also to the identification of possible side-effect profiles of the compounds tested.

More recently, with the advance of novel and more sensitive techniques of monitoring gene expression, an entirely new panel of assays has been brought forward. While reporter gene assays were devised with the idea of visualizing a particular gene expression event through the coupling to a protein which can be easily monitored, the novel techniques directly monitor gene expression by quantitation of the mRNA levels for several to up to thousands of genes simultaneously. Furthermore, these techniques do not require the prior introduction of a detection system, allowing unlimited choice with respect to cell type or tissue derived from organs.

Table 1.3 gives an overview of a number of techniques developed for HTS applications that are currently available. This list is by no means complete, but should serve as a general reference to the different approaches taken.

Tab. 1.3 Differential gene expression technologies (adapted from [155]).

Technology	*Methodology*	*Target genes*	*Samples*	*References*
Open systems: no prior sequence information required				
SAGE	transcript capture by "fingerprint" SAGE tags, identification and quantitation by sequencing of serial concatemers	HT-compatible	few	70
GeneCalling (transcript profiling)	restriction endonuclease "fingerprinting" of cDNA, confirmation by competitive PCR	96 reactions	few	71
TOGA	PCR-based cDNA display, identification of mRNAs with 3′poly(A) tail by sequencing	256 reactions	few	72
Closed systems: analysis of genes with known sequence information				
DNA microarray/ oligoarrays	hybridization of total RNA to gene probes arrayed on silicone chips or glass slides	HT-compatible	few	156, 157
Quantitative RT-PCR (Taqman™)	PCR-mediated release of gene-specific fluorogenic probes; threshold of detection is inversely proportional to concentration	few probes	96-well format	73
Branched DNA (Quantigene™)	direct mRNA detection by hybridization with gene-specific linker and enzyme-linked amplification probes	1 probe	96-well format	74, 75
HPSA™	direct mRNA detection by hybridization with gene-specific capture and enzyme-linked amplification probes	2 probes	96-well format	158

With respect to the HTS suitability of these novel technologies, two major categories can be differentiated. There are methodologies like serial analysis of gene expression (SAGE) [70], GeneCalling (transcript profiling by restriction endonuclease fingerprinting) [71] and total gene expression analysis (TOGA) [72], where initial sample preparation is very fastidious, but where high-speed parallel detection of many different genes can be achieved. These techniques are collectively called "open" systems – no prior sequence information is required and virtually all genes induced by a particular compound can be detected. Open systems challenge the DNA microarray technology, providing an alternative to the high cost and high demand of starting material associated with the latter. In practice, they will prove useful if only a few samples need to be analyzed, but where the high throughput (HT) refers to the detection of the different gene products.

Quantitative reverse transcription-polymerase chain reaction (RT-PCR) [73], the branched DNA technology promoted by Bayer (Emeryville, CA) [74, 75] and the high-performance signal amplification (HPSA™ [76]) system promoted by Chromagen (San Diego, CA) can detect one or a few mRNAs at a time. They are called "closed" system technologies, as are the DNA and oligonucleotide microarrays and the reporter gene technology, since they can be applied only to genes with known sequence. However, these methodologies can be used for HTS, since the assays can be run in a microtiter plate format, allowing the simultaneous processing of many samples. Moreover, they provide an advantage to the traditional reporter gene format, because they are more flexible and can be performed on virtually any source material, and the detection of a novel gene requires only the design and synthesis of gene-specific probes. These novel technologies promise increased sensitivity, with a readout not compromised by secondary effects due to the reporter gene, but they are not yet as widely established as the reporter gene systems.

1.4
Spatio-temporal Assays and Subpopulation Analysis

Cellular assays addressing protein expression levels, spatial and temporal distribution of proteins within the cell, and intracellular signaling dynamics are currently exploited to better understand pharmaceutical target functions in cells.

To date, there is an increasing number of labeling reagents and fluorescent probes for imaging applications, including target-specific antibodies, complementing enzymes [77, 78], dyes (e.g. Alexa Fluor dye series; Molecular Probes), affinity reagents [79] and genetically encoded fluorescent proteins [18, 80, 81].

Systems that allow high-throughput analysis of cells include flow cytometry-based devices, such as the recently described high-throughput pharmacological system [82]. This system allows multifactorial characterization of cells, but does not allow us to quantify spatial rearrangements inside cells.

Recently developed microscopic imaging technologies have broad applications. They allow analysis of a wide range of cellular functions including cell motility, cell morphology, protein movement inside cells, protein localization, intracellular second messengers and endpoint measurements of signaling cascades. The development of automated imaging systems has made compound screening in high-content format possible. The great advantage of automated imaging and scanning techniques is the tremendous increase in sample throughput compared to manual microscopy. In the case of fluorescence microscopy, multiple fluorescent probes can be detected and quantified simultaneously.

The ArrayScan® II system was introduced in 1999 by Cellomics [83] (www.cellomics.com) and was one of the first automated imaging instruments for high-content screening on the market. Meanwhile, other instruments, software and hardware were developed by several companies including Acumen-Bioscience (www.acusite.co.uk), Amersham Bioscience (www.amersham.com), Applied Biosystems (www.appliedbiosystems.com), Q3DM (www.q3dm.com), CompuCyte (www.compucyte.com), Kinetics Imaging (www.kinetikimaging.com), Universal Imaging (www.imagel.com), Axon Instruments (www.axon.com) and Evotec OAI (www.evotecoai.com).

The rapid progress in the development of analytical devices for subcellular analysis in combination with labeling reagents for specific identification of the molecules of interest will allow us to analyze specific signaling and metabolic steps in a high-throughput mode.

1.4.1 Phosphorylation Stage-specific Antibodies

Of particular interest for automated imaging are activity-dependent reagents, such as phosphospecific antibodies that allow selective analysis of signaling pathways within cells. Phosphorylation and dephosphorylation of proteins by kinases and phosphatases, respectively, are part of a complex signaling network of tightly regulated dynamic processes. Protein phosphorylation frequently results in subcellular redistribution of key signaling molecules, which is critical for their biological activity [84–86]. The spatial distribution of two key signaling proteins upon phosphorylation, extracellular signal-regulated kinase (Erk) and p38 MAP kinase, was analyzed with phosphotyrosine-specific antibodies recognizing only the phosphorylated form of the protein [87]. Exposure of cells to a potent Cdc25 phosphatase inhibitor resulted in a concentration-dependent nuclear accumulation of phospho-Erk and phospho-p38, but not nuclear factor-κB (NF-κB). These experiments revealed that Cdc25 inhibition increased Erk phosphorylation and nuclear accumulation, and indicated that the phosphorylation status of Erk is regulated by a Cdc25 phosphatase.

A phospho-histone H3 antibody that targets the phosphorylation site on Ser10 of histone H3 and a phospho-histone H1 antibody were used to detect mitogen- and oncogene-stimulated phosphorylation in small nucleosomal regions that are sensitive to hyperacetylation [88, 89].

Active kinase states have been analyzed with a panel of phosphospecific monoclonal anti-kinase antibodies. The specificity of the antibodies for the specific kinase has been verified by Western blot analysis and with specific inhibitors for upstream kinases. Using multiparameter flow cytometry, Perez and Nolan [90] could identify lymphocyte subsets with as many as four different intracellular active kinases. This technique allows us for the first time to investigate simultaneously different active kinases in one individual cell. While this technology can be best applied to cells in suspension, high-throughput microscopy is the technology of choice for adherent cells. In addition microscopic applications allow analysis of intracellular spatial distributions which is not possible with flow cytometry.

Mitotic cells are characterized by an increase in phosphorylated nucleolin, a nuclear protein phosphorylated in cells entering mitosis. Using a monoclonal antibody that specifically recognizes phosphorylated nucleolin, Mayer et al. [91] have developed a high-throughput whole-cell immunoblot assay. Compounds arresting cells in mitosis by novel modes of action could be identified after sorting compounds with secondary assays.

Other protein modifications like acetylation can be detected with specific antibodies as well. The organization of transcriptionally active chromatin and activity of nuclear histone acetyltransferase and deacetylase was investigated by using an acetyl-histone H3-specific antibody [92, 93]. Compounds affecting ligand-stimulated changes in chromatin structure can be screened with such antibodies by automated imaging techniques.

1.4.2 Target-protein-specific Antibodies

Many proteins play important roles in both the cytoplasm and nucleus. Therefore, they need to move from one compartment to the other in order to perform their functions. Typical examples are latent cytoplasmic transcription factors [44] or nuclear transport factors [94]. Assays for nuclear transport factors using heterokaryon formation in combination with fluorescent proteins have been described [95].

The movement of latent cytoplasmic transcription factors upon activation to the nucleus can be monitored and quantified by automated fluorescence imaging. IL-1-induced and tumor necrosis factor-α-induced translocation of NF-κB to the nucleus has been quantified with the application of the Cellomics ArrayScan II technology [96].

Unstimulated cells contain inactive NF-κB distributed within the cytoplasm that can be detected by a NF-κB-specific antibody combined with a fluorescently labeled secondary antibody. In IL-1α-stimulated cells, most NF-κB is activated and translocates to the nucleus in a time- and dose-dependent manner (Figs 1.6 and 1.7). Upon IL-1α stimulation, HeLa cells were analyzed with the ArrayScan II system using the specific software that allows quantitative analysis of fluorescence in the nucleus and in the adjacent cytoplasm. Cells are identified as objects by the software through the nuclear staining with Hoechst 33324 or DAPI, and are overlaid with masks for nuclear and cytoplasm fluorescence measurements (Fig. 1.8, see p. 30).

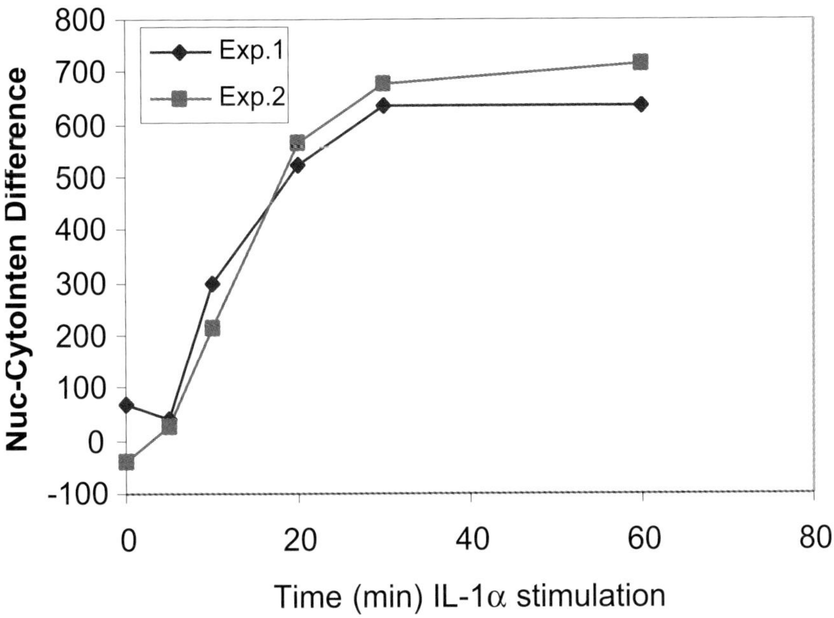

Fig. 1.6 Time course of NF-κB nuclear translocation. HeLa cells were stimulated with 1 ng/ml IL-1α for the indicated times at 37°C. Mean values of 100 counted cells per well of two different experiments (Exp.1 and Exp.2) are shown. Nuc-CytoInten = nuclear minus cytoplasmic fluorescence intensity.

1.4.3
Protein–GFP Fusions

Tagging technologies have become popular to monitor the movement or transport of molecules into cells and within cells. Extracellular ligands can be conjugated to biotin and later visualized with enzyme-conjugated streptavidin. A HTS assay has been described on this basis for epidermal growth factor (EGF) internalization [97]. Fluorescent proteins significantly contribute to the understanding of protein dynamics within the cell. They represent powerful tools and ideal tags to assess real-time analysis of molecular events in living cells in the form of reporters or fusion proteins. The most prominent fluorescent protein is GFP, originally isolated and cloned from the jelly fish *Aequorea victoria* [98]. A number of GFP variants have been developed with different excitation and emission spectra, including enhanced blue fluorescent protein (eBFP), enhanced yellow fluorescent protein (eYFP) and enhanced cyan fluorescent protein (eCFP). The nature of their structures renders these proteins extremely stable and resistant to degradation by most proteases [99]. This high level of stability does not necessarily represent an advantage for cellular applications. If changes of gene expression with transcription reporter assays are to be monitored, a shorter half-life of the reporter is desired. Therefore, short half-life

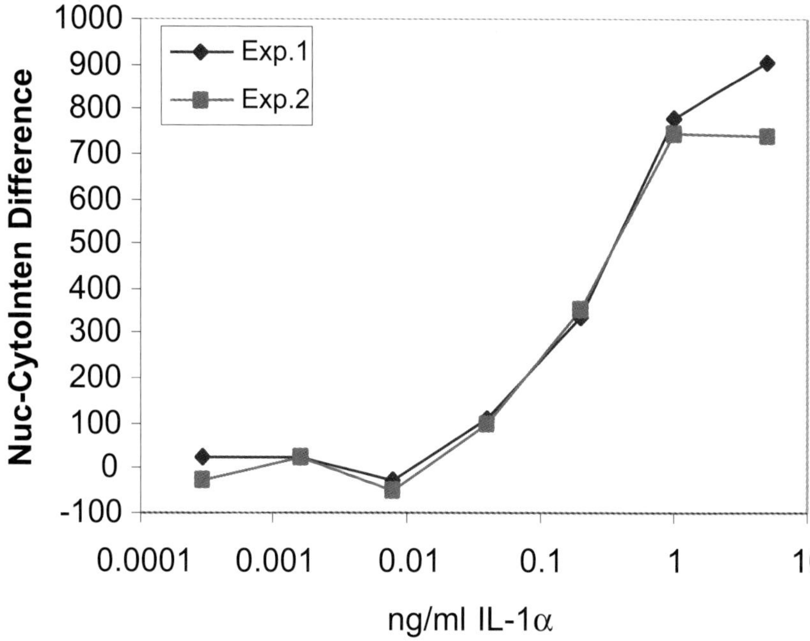

Fig. 1.7 Dose–response relationship of IL-1α-induced NF-κB nuclear translocation. HeLa cells were stimulated with indicated concentrations of IL-1α for 30 min at 37°C. Mean values of 100 counted cells per well of one block two different experiments (Exp.1 and Exp.2) are shown. Nuc-CytoInten = nuclear minus cytoplasmic fluorescence intensity.

forms of eGFP were developed in order to be used as transcription reporter proteins and these are termed destabilized eGFP [80]. Meanwhile, other fluorescent proteins from marine invertebrates have been cloned, like the fluorescent proteins from reef corals [81]. The combination of distinct spectral variants of fluorescent proteins enables a wide range of applications, such as the simultaneous analysis of multiple gene expression cascades and localization or translocation of different proteins.

Measurement of co-internalization of β-arrestin bound to GPCRs can be used to monitor receptor activation and recycling. Barak et al. [100] demonstrated, using confocal microscopy, the translocation of β-arrestin-2–GFP fusion proteins from the cytoplasm to membrane-anchored GPCRs. This was demonstrated for more than 15 different ligand-activated GPCRs in HEK-293 cells. In the absence of receptor activation, β-arrestins are distributed throughout the cytosol and are absent in the nucleus. Upon addition of ligand to the cells, β-arrestin-1–GFP fusion proteins translocate to the plasma membrane and further to endocytic vesicles [101]. The accumulation of the GFP-tagged proteins can then be quantified using a fluorescence imaging technology. With a β-arrestin-1–GFP fusion protein, co-internalization of arrestin with thyrotropin-releasing hormone receptor 1 in response to agonist was monitored and imaged [102, 103]. NORAK Biosciences

(Research Triangle Park, NC; www.norakbio.com) exploits arrestin–GFP movement under the term Transfluor™ technology as a broadly applicable, stable HTS platform to identify compounds modulation GPCR signaling. Receptor internalization and trafficking was also exploited by the use of fluorescently labeled ligands such as ligand of transferrin receptor [104]. High-content screening was conducted on the ArrayScan II system, and internalization and recycling of the transferrin receptor with bound ligand could be monitored and quantified.

GPCR activation and internalization has also been monitored using GPCR–GFP fluorescent protein fusions [104, 105]. Conway et al. [105] have used the ArrayScan high-content screening system to monitor parathyroid hormone receptor and β_2-adrenergic receptor internalization upon stimulation by specific ligands.

Other examples where fusions with fluorescent proteins were used to quantitate protein movement include nuclear translocation of calmodulin [106] or the membrane-bound transcriptional regulator tubby after phosphoinositide hydrolysis [107].

In other applications, translocation domains such as the Ca^{2+}-sensitive C2 domain of protein kinase Cδ have been fused to YFP and expressed in rat leukemia cells. Evanescent wave single-cell array technology has been applied as an imaging technology to monitor protein movement upon addition of platelet-activating factor [108].

Simultaneous analysis of mixed cell populations can be performed by transfecting each cell line with a single fluorescent fusion protein. Vectors of BFP and YFP were used to stably transfect two different colon cancer lines, the parental line with a mutant K-*ras* gene and a variant where the mutant gene has been deleted. To screen for compounds that selectively kill only the tumorigenic variant with the mutant *ras* gene, both populations were mixed and simultaneously assayed for cell survival using fluorescence intensity measurements [109]. Compounds that preferentially affect the cells with the mutant *ras* allele could be identified.

1.4.4
Fluorescence Resonance Energy Transfer (FRET)

FRET has become a popular method to study protein dynamics in cells [110]. FRET is based on the transfer of energy absorbed by one fluorescent molecule to another, causing excitation of this second molecule. To achieve FRET, the two fluorescent molecules with different excitation/emission spectra need to come into close proximity of each other. The most general strategy for creating such indicators is to sandwich the conformationally sensitive domain of interest between two mutants of GFP. In case of ligand binding or modification of the domain, the two fluorescent moieties get into close proximity, which enables FRET.

In a pioneering study, Mahajan et al. [111] used plasmid constructs that expressed the fusion proteins GFP–Bax and BFP–Bcl-2. Bax and Bcl-2 are two key modulators of apoptosis, but their direct interaction had then never been shown at the cellular level. Using FRET, the direct interaction of the two proteins in mitochondria could be demonstrated. FRET has also been applied to determine

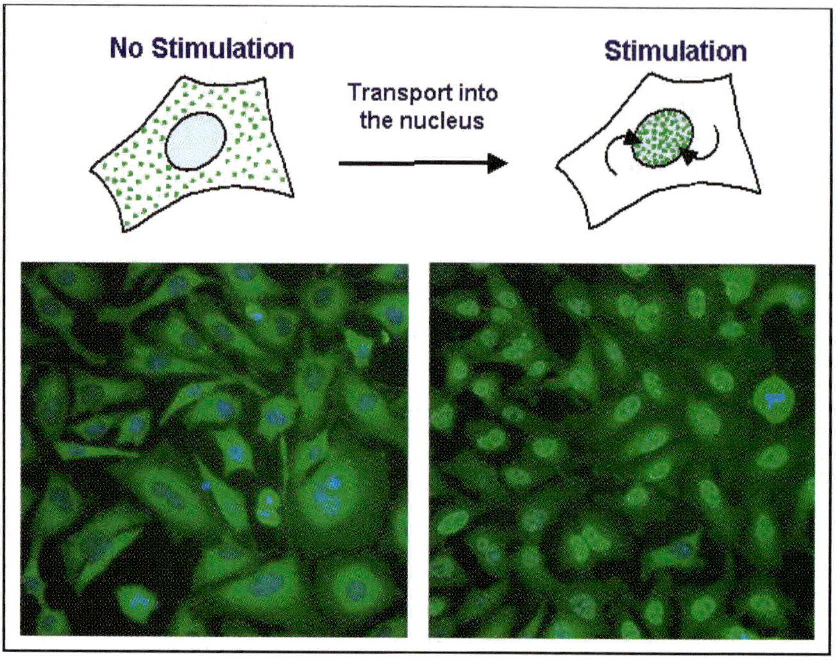

Fig. 1.8 IL-1α-induced NF-κB translocation in HeLa cells. Cells were stained with rabbit anti-p65 NF-κB antibody and goat anti-rabbit Alexa Fluor 488 antibody. Nuclear DNA was stained with Hoechst 33342. NF-κB is quantified by measurement of green fluorescence intensity; nuclei appear blue. Unstimulated cells (left) show most green fluorescence in the cytoplasm, whereas cells stimulated with 2 ng/ml of IL-1α for 30 min at 37°C (right) show translocation of NF-κB to the nucleus.

kinase activities in living cells. For this purpose, reporters were constructed, consisting of fusions of CFP, a phospho-amino acid-binding domain, a consensus substrate for the relevant kinase of interest and YFP. In cells transfected with the reporter constructs and stimulated for the particular kinase activity, the fusion protein after phosphorylation underwent a conformational change, and allowed FRET between the YFP and CFP. The change in the ratio of yellow to cyan emission could be quantified, and temporally as well as spatially determined. This technology has been successfully applied for protein kinase A [112], and the three tyrosine kinases Abl, Scr and EGF-receptor (EGF-R) [113].

In a similar approach, Honda et al. [114] demonstrated the use of a genetically encoded fluorescent indicator to investigate the dynamics of guanosine 3′,5′-cyclic monophosphate (cGMP) in single cells. The cGMP indicator was constructed by bracketing deletion mutants of the cGMP-dependent protein kinase between cyan and yellow mutants of GFP. Binding of cGMP to the fusion protein indicator altered FRET and the ratio of cyan to yellow emissions selectively over cAMP. By using this method, the temporal and spatial dynamics of intracellular cGMP levels in rat fetal lung fibroblasts and in primary rat Purkinje neurons was monitored upon addition of different stimuli. Similar sensors for cAMP have also been published [115].

FRET-based sensors were also used to demonstrate the spatio-temporal behavior of GTB-binding proteins after cellular stimulation with growth factors [116]. A fusion protein was engineered which contained H-Ras, the Ras-binding domain of Raf, and was flanked by a pair of yellow and cyan mutants of GFP. H-Ras was replaced in a second construct by the analogue protein Rap1. Activation of the Ras/Rap pathway by growth factors induced FRET. Spatio-temporal localization of the fusion protein indicated that EGF stimulation of COS-1 cells activated Ras at the peripheral plasma membrane, but Rap-1 in the perinuclear region. In addition, this study showed that growth factors activate only a subpopulation of Ras and Rap, and only at very restricted areas in the cells. This technology in combination with cellular imaging will greatly increase the sensitivity and facilitate cellular HTS approaches for specific intracellular signaling events.

1.4.5
GPCR Activation using Bioluminescence Resonance Energy Transfer (BRET)

BRET is a naturally occurring process. The jelly fish *A. victoria* or the sea pansy *Renilla reniformes* create green fluorescence by BRET. In the former blue-light-emitting case, aequorin associates with GFP, whereas in the latter case, the energy generated by degradation of coelanterazine by luciferase is transferred to GFP. This transfer, however, only occurs if luciferase and GFP form a heterodimer. BRET has been used to demonstrate the interaction of circadian clock proteins from cyanobacteria expressed in *E. coli* [117]. More recently, BRET has been applied to study GPCR activation by several groups [118–120]. To demonstrate β_2-adrenergic receptor (β_2AR) dimerization, β_2AR–*Renilla* luciferase and β_2AR–red-shifted GFP (YFP) constructs were coexpressed in HEK-293 cells. Occurrence of BRET was an indication of receptor dimerization. In the same study, β-arrestin, which is known to associate with the intracellular domain of activated G protein-coupled receptors, was expressed as an arrestin–YFP construct together with β_2AR–*Renilla* luciferase. BRET between the two proteins occurred, but was entirely dependent upon receptor activation and indicated the association of arrestin with the intracellular domain of the β_2AR in the active state of the receptor. This assay has been developed as a HTS assay technology by PerkinElmer Life Sciences.

1.4.6
Protein Fragment Complementation Assays (PCA)

The methods described above to study protein–protein interactions are associated with certain shortcomings. The yeast two-hybrid system is based on transcription, which requires proteins to be available in the nucleus. In FRET, fluorescent proteins need to be expressed in cells at relatively high levels and the interaction must be such that the proteins are in close proximity to allow energy transfer.

Enzyme complementation and its use in studying protein–protein interactions have been developed for *E. coli* β-galactosidase [121], dihydrofolate reductase (DHFR) [122] and most recently for β-lactamase [123]. In the case of β-galactosidase, two weakly complementing deletion mutants are each fused to the two proteins to be probed for interaction. Only if the two proteins interact are the two parts of the galactosidase forced to complement each other and exert enzymatic activity. The system has been developed by demonstrating FRAP and FKBP12 interaction in the presence of rapamycin. More recently, β-galactosidase complementation was used to establish a HTS system to screen for antagonists of EGF-mediated EGF-R dimerization [77, 124]. For this purpose, the complementing deletion mutants of β-galactosidase were fused to the intracellular domains of the EGF-R and cotransfected into C2C12 mouse myoblasts. Enzyme activity was reconstituted after adding EGF to the cells in culture. With this assay adapted to a 384-well format, primary hits identified in the EGF-R assay were screened for specificity with two complementing deletion mutants of β-galactosidase fused to human β_2-adrenoceptor and β-arrestin-2. By the use of sensitive chemiluminescent and fluorescent galactosidase reagents, this technology enables the measurement of protein–protein interactions in living cells for HTS applications.

The reassembly of active DHFR from rationally designed fragments was established by Pelletier et al. in a prokaryotic system [122]. *E. coli* DHFR is selectively inhibited by the antifolate trimethoprim. Murine DHFR can rescue *E. coli* cells. Complementing fragments of murine DHFR fused to homo- or heterodimerizing proteins were used to transform *E. coli*, which were then grown in the presence of trimethoprim. Survival was taken as a measure for protein interactions and enzyme complementation. Besides GCN4 leucine zipper fusions, *raf–ras* interaction, and rapamycin-induced interaction of FKBP and TOR2 were used in survival studies to demonstrate interaction and complementation. In another study, protein partners fused to complementing fragments of murine DHFR were introduced into DHFR-negative mammalian cells grown in the absence of nucleotides. Cells where the proteins associate and DHFR can complement are able to survive under these conditions [125]. As an alternative, protein interaction and DHFR complementation can be monitored by feeding cells fluorescein-conjugated methotrexate. Methotrexate only binds to DHFR when the complementary fragments are coexpressed and reassembled. Fluorescence signals can then be monitored by fluorescence microscopy, FACS or by spectroscopical devices used in HTS. The FKBP–rapamycin–FRAP complex and the activation of the erythropoietin receptor were used as model systems [125]. Most recently, DHFR-PCA assays were used to analyze whole biochemical networks in living cells [78]. Cells expressing associating protein pairs were selected and fluorescein-conjugated methotrexate, which binds to reconstituted DHFR, was applied to localize interacting protein partners and to study inhibitors preventing the interaction. The same method has been successfully applied in plant protoplasts expressing complementary fragments of DHFR fused to interacting proteins. The binding of fluorescein-conjugated methotrexate as measure of protein–protein interaction was monitored by fluorescence microscopy [126].

1.5 Phenotypic Assays

1.5.1 Proliferation/Respiration/Toxicity

Many cytotoxicity and proliferation assays for mammalian cells rely on radioactivity and measure the incorporation of radiolabeled nucleotides such as [^{3}H]thymidine into cellular DNA. Although very sensitive, the [^{3}H]thymidine assay [127] is relatively expensive and labor intensive, and requires handling and disposal of the radioactive waste. Other cytotoxicity assays, such as MTT [3-(4,5-dimethylthiazol-2-yl)-2,5-diphenyltetrazolium bromide] [128], Alamar Blue [129], lactate dehydrogenase [130] and ATP content [131], require the addition of specific reagents to generate a signal. Among those, the most commonly used is MTT [132]. This tetrazolium salt is reduced within the mitochondria of metabolically active cells to form a colored formazan dye precipitate. The broad utility of MTT is limited by the fact that it may be susceptible to interference from drugs, either by reacting with reducing groups on the drugs or by scatter of absorbance resulting from the precipitation of drugs absorbing light in the visible spectrum. Cell viability measurement by the MTT assay relies on its reaction with mitochondrial succinate dehydrogenase, which may itself perturb the cells. Furthermore, the test itself is not reversible and, because it is an endpoint reading, each time point reading requires a separate assay. It has to be pointed out that toxicity and proliferation assays are used separately and independently of each other as well as in combination. The combination of both assays allows a thorough investigation of a compounds effect on a particular cell line.

There is a considerable need for assays that are homogeneous and reversible. The Oxygen Biosensor System (OBS™; BD Biosciences, Bedford, MA) can be seen as a prototype representative. Unlike dye conversion methods, there is no need to add any reagent in order to generate a signal.

The OBS is a microplate-based assay platform that enables kinetic monitoring of dissolved oxygen. The oxygen-sensitive dye is embedded in a silicone matrix, which is permanently attached to the bottom of each well of a microplate. The gas-permeable matrix allows oxygen in the vicinity of the dye to be in equilibrium with the oxygen in the liquid medium. Oxygen quenches the ability of the dye to fluoresce in a predictable, concentration-dependent manner. Hence, as oxygen is depleted from a well (e.g. as cells grow and consume oxygen in the surrounding medium), the concentration of oxygen in the matrix decreases and the fluorescence increases. The amount of fluorescence correlates directly to oxygen consumption in the well. This, in turn, relates to the number or relative viability of cells in this particular well. Any reaction that can be linked to oxygen consumption or generation can potentially be measured by using this system (Fig. 1.9).

The signal is dynamic and is read in real-time. Samples may be read as frequently and as many times as desired. The signal is completely reversible, and will rise as the cells proliferate and then fall as they die, giving a more complete picture of proliferation and toxicological kinetics.

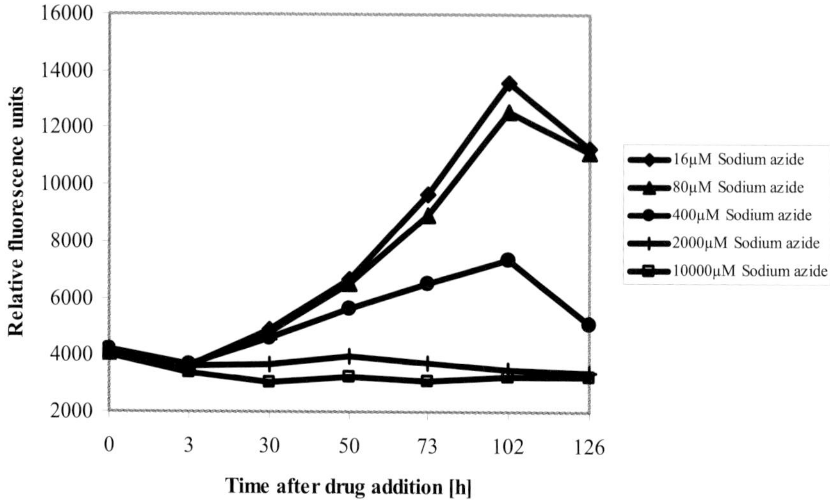

Fig. 1.9 Oxygen consumption of HeLa cells after exposure to sodium azide. Cells (1×10^5), growing on beads, were seeded in individual wells of an oxygen biosensor plate and treated with sodium azide at the concentrations indicated. Cells were cultivated and fluorescence was measured at the times indicated. Low micromolar concentrations of sodium azide did not affect the growth of HeLa cells, as judged from the increased oxygen consumption during cells growth. On the other hand, millimolar concentrations of sodium azide are toxic and the relative fluorescence is maintained at a constant (background) level. The decrease of oxygen consumption at 126 h after drug addition indicates that the cells became confluent and stopped proliferating.

1.5.2
Apoptosis

Apoptosis, or programmed cell death, is one of the most complex cellular processes. Early and late events occurring in cells undergoing apoptosis comprise activation of different caspases, changes in the cytoskeleton, changes of the mitochondrial membrane potential and mitochondrial mass [133], nuclear condensation or fragmentation [134], flipping of phosphatidylserine to the outer membrane layer and blebbing of the plasma membrane. Changes in nuclear morphology, actin reorganization, and mitochondrial potential and mass can be detected and quantified by a multiparameter apoptosis assay [135]. Increase or shrinkage of the nuclear area and chromatin distribution within the nucleus is analyzed by staining cells with a fluorescent DNA dye, such as DAPI or Hoechst 33342. Bundling of intracellular F-actin filaments and increase of the actin content in apoptotic cells [136, 137] is detected by the use of fluorescent phalloidin. This phallotoxin, isolated from the mushroom *Amanita phalloides*, binds highly specifically to F-actin and can be labeled with any fluorescent dye. However, increase in F-actin content is not observed in all cases of apoptotic cells, depending on the cell type or inducer [138, 139]. The mitochondrion-specific cationic dye Mitotracker® Red, which accumulates in active mitochondria is used to measure mitochondrial membrane potential and mass. Apoptotic cells can exhibit loss of mitochondrial membrane potential or increase of total mitochondrial mass [140]. In both cases, the quantity of mitochondrial stain is a measure for the degree of apoptosis. In the case of the multiparameter apoptosis application the ArrayScan II algorithm (see Section 1.4.2) not only quantifies the label, but also takes into account the distribution of fluorescent dyes within cellular compartments, and calculates nuclear fragmentation and mitochondrial mass increase. This algorithm has been successfully applied in detection of compound-induced apoptosis [135] and can be used for high-throughput compound screening.

1.5.3 Differentiation

Human stem cells of various types have the ability to differentiate into a number of mature cell types [141], and such stem cells are currently used in bone marrow transplants and in drug research. New drugs such as erythropoietin and granulocyte colony-stimulating factor have been developed that act directly on such cells, and are essential to the success of bone marrow transplantation [142]. The expansion of bone marrow progenitors is extremely sensitive to a wide range of chemotherapeutics and other toxic agents. In some cases, the toxic side-effects of drugs are specific to particular hematopoietic lineages [143].

In the past, use of hematopoietic progenitors for high-throughput drug discovery and toxicology screening was limited by tissue availability. Given the current availability of high-quality primary human hematopoietic progenitors, the current bottleneck in the high-throughput use of these cells is the assay of their differentiation phenotypes.

The most commonly used *in vitro* assay in order to determine the differentiation response of hematopoietic progenitor cells is the colony assay [144]. The colony assay uses selected progenitor populations from human bone marrow, umbilical cord blood or mobilized peripheral blood in clonogenic assays in a semi-solid medium. Colonies of differentiating blood cells appear after a period of 14 days and are counted manually. Different types of colonies are identified by their morphology (e.g. myeloid and erythroid) or cytochemical staining techniques (megakaryocytes). Although the parameters of these assays have been optimized to obtain quantitative data, the extremely low throughput and subjective nature of this assay cause some problems when it is used for chemical compound library screening, early drug discovery and high-throughput toxicity screening.

Differentiative responses can also be assayed by the transfection of a target cell with a receptor gene tied to the expression of a differentiation-specific marker [145]. It requires, however, the transfection of a cell containing the many intact signal transduction cascades characteristic of hematopoietic progenitors.

Another approach was described by Warren et al. [146]. The authors described a cell-based immunoassay in which the cells were transferred to a fixative-treated assay plate, and fixed in place prior to incubations with primary and secondary antibodies. Although this assay has been used successfully, transfer of cells from the culture plate to the assay plate is time-consuming and not amenable to HTS. Recently [147], this assay design was revised and improved in order to reduce the amount of manipulations required. On termination of the culture period, cell culture supernatant is removed via vacuum filtration and a labeled specific antibody is added. In order to obtain the sensitivity required, the monoclonal antibodies were conjugated to a DELFIA europium chelate [148]. In addition to its sensitivity, the relatively long emission time of the DELFIA europium chelate allowed the use of time-resolved fluorescence (TRF) to avoid the endogenous background fluorescence that is often a significant problem in biological assays. Removal of the cell culture supernatant, unbound antibodies and subsequent wash steps are performed by using a 96-well filter plate. Because there is little carryover from one wash cycle to the next, removal of unbound antibody is very efficient. Repeated washes without cell loss are possible. For signal detection, an enhancement solution was added and europium fluorescence (excitation at 340 nm and emission at 615 nm) was measured over a 400-μs time period after an initial delay of 400 μs using a suitable spectrofluorometer.

1.5.4
Monitoring Cell Metabolism

Direct determination of metabolites at the cellular or *in vivo* level provides a sensitive system for assessing drug action because metabolites represent end products of cellular regulatory processes. In addition, intracellular metabolites are generally far less abundant than proteins. In a cellular environment, signal transduction, initiation of gene expression, protein synthesis and metabolic responses to a stress factor must be essentially sequential. The fields of metabolomics – which investigates metabolic regulation and fluxes in individual cells or cell types – and metabonomics – the determination of systemic biochemical profiles and regulation of function in whole organisms by analyzing biofluids and tissues – have therefore been developed recently [15, 149].

Studying the effects of drugs on cells or whole organisms by metabonomics relies on multiparametric measurements of alterations in metabolism over time in response to a stress factor. Samples are mapped according to their biochemical composition [150].

Metabolites proved particularly useful in the characterization of mutant phenotypes as shown recently in plants [151] and in the characterization of so-called "silent mutations" in yeast [6]. Many genes are "silent", which means that if mutated they do not lead to an easily visible phenotypic change, such as alterations in growth rate. The growth rate may not be changed because the concentrations of different intracellular metabolites are adapted in a way that compensates for the effect of the mutation. Accordingly, metabolome analysis can reveal a phenotype at the level of metabolites in many mutants previously scored as silent. In addition, quantifying the relative changes of metabolite concentrations caused by a mutation

Tab. 1.4 Readouts for multiparametric cell monitoring systems.

Readout	***Corresponding cellular activity***
Changes in oxygen level	mitochondrial activity (energy metabolism; cell ageing; apoptosis)
Acidification of the medium	global indicator of metabolic activity; fast changes: cellular activation events; slow changes: cell growth or death
Changes in impedance	cell adhesion: matrix attachment, tumor metastasis; cell outgrowth, e.g. neurite differentiation
Ion flow	influx and efflux of Ca^{2+}, Na^{+}, K^{+}
Metabolites	e.g. glucose uptake; lactate excretion
Light emission	e.g. induction of reporter gene expression coupled to the generation of bioluminescence

makes it possible to identify the site of action of the altered gene product as demonstrated by Raamsdonk et al. [6]. By using nuclear magnetic resonance or mass spectrometry (MS)/liquid chromatography/MS applications, a reasonable sample throughput is possible. This will allow the use of metabolome and metabonome analysis in drug discovery as a sensitive measure of drug action [15].

An emerging approach, which allows the analysis of the metabolic state in intact cells, is the functional online analysis of living cells in physiologically controlled environments for extended periods of time ("biochips"). Efforts are being directed towards the parallel development, adaptation and integration of different microelectronic sensors into miniaturized biochips for a multiparametric cellular monitoring ("Cell Monitoring System" [22]). Some possible readouts and the corresponding cellular activities are summarized in Tab. 1.4.

Monitoring of cellular oxygen exchange [152] can be employed for the assessment of mitochondrial activities. Other sinks or sources of oxygen do not contribute significantly in most cell types. Mitochondria, apart from supplying the cell with energy, are suggested to be involved in cell ageing accelerated by reactive oxygen species, in apoptosis and in the toxic mechanisms of numerous drugs. Monitoring mitochondrial activity by lipophilic dyes such as rhodamine 123 suffers from the risk of potential artifacts, and therefore, nontoxic and noninvasive techniques are particularly valuable.

The OBS has been described above as one of the first noninvasive techniques to monitor oxygen consumption. One limitation of the OBS, however, is the requirement for a large number of cells. The system needs at least 50,000 cells per well in order to produce viable results (less cells do not consume enough oxygen in order to (over)compensate the diffusion of oxygen back from the outside environment to the oxygen biosensor). In comparison, microsensor-based technologies are so small in scale that, under ideal circumstances, signals can be obtained from a single cell.

Microsensor-based pH recording [153] is predominantly used for determinations of extracellular acidification rates. Several metabolic pathways contribute to extracellular acidification and, thus, can be regarded as a global indicator of metabolic activity. In this sense, fast changes (within minutes) tend to reflect cellular activation events (e.g. receptor-mediated signaling), while slow changes (several hours) are usually attributable to cell growth or cell death [154], respectively.

Monitoring of cell adhesion (e.g. tumor metastasis) or outgrowth (e.g. for neurites) can be accomplished by growing the cells directly on pairs of interdigitated electrodes. The cellular impedance signal results from insulation by the cell membranes. If cells are placed on the electrodes, they block the current flow in a passive way and the impedance increases. Therefore, the impedance of the system gives insight into the adhesive behavior of the cells.

This methodology is based on on-line data acquisition using multiparametric microsensor devices. A fluid system for the supply of culture media/drug solutions and for the realization of metabolic measurements is connected to the chip. The precise maintenance of *in vitro* microenvironmental conditions is also necessary for the cellular specimen. The cellular parameters currently amenable by sensor chip readings are metabolic activity (rates of extracellular acidification and cellular respiration), morphologic properties and some ion fluxes. Small-scale sensor chip areas require very small amounts of cellular specimen. This may allow the utilization use of primary cells.

Methods have been developed which allow testing of dozens of compounds, but 96-well formats might become available soon [22].

1.5.5
Other Phenotypic Assays

There is an increased need for real-time, label-free observations of proteins and cells under their natural (unperturbed) conditions due to some limitations associated with the currently existing technologies. A major drawback of today's technologies is the need to label and overexpress the protein of interest. Due to its overexpression and modification, the protein's biochemical activity and localization may not reflect the natural (unperturbed) conditions, and therefore the results have to be interpreted with caution. One technology that addresses those issues uses radiofrequency waves [21] to probe the physiological state of a protein in the context of the whole cell. MCS uses microwaves, which correspond to the natural frequency of protein dynamics, allowing the measurement of electrodynamic properties of proteins and cells. MCS does not use any tags or markers and may be performed in any environment, from simple aqueous solutions to very complex mixtures (i.e. cells), thereby allowing a direct approach to determine changes in cellular phenotypes (i.e. differentiation).

Isolated proteins in physiological buffers exposed to a range of microwave frequencies, demonstrate two properties: the spectral response (the proteins "signature") and the biophysical profile. The first of these determines the raw response of each protein over a range of frequencies. The biophysical profile describes the way in which a protein interacts with the environment – molecular volume, complexation of water, interaction with other compounds/proteins – as well as how it compares to other proteins belonging to similar classes. By scanning a range of frequencies and evaluating different environments (e.g. buffer conditions, presence of cofactors), this analysis defines a multidimensional signature to each protein even in complex (cellular) environments. This would allow the development of cell-based assays where the biophysical profile of target proteins can be analyzed in an environment that closely mimics the situation in primary disease tissues.

Acknowledgments

We thank Dr. Guy Marais for carefully reading and reviewing the manuscript.

References

1 Tan DS. 2002. Sweet surrender to chemical genetics. *Nat Biotechnol* **20**, 561–3.

2 Stephanopoulos G, J Kelleher. 2001. Biochemistry. How to make a superior cell. *Science* **292**, 2024–5.

3 Hanahan D, RA Weinberg. 2000. The hallmarks of cancer. *Cell* **100**, 57–70.

4 Somogyi R, LD Greller. 2001. The dynamics of molecular networks: applications to therapeutic discovery. *Drug Discov Today* **6**, 1267–77.

5 Glassbrook N, C Beecher, J Ryals. 2000. Metabolic profiling on the right path. *Nat Biotechnol* **18**, 1142–3.

6 Raamsdonk LM, B Teusink, D Broadhurst, N Zhang, A Hayes, MC Walsh, JA Berden, KM Brindle, DB Kell, JJ Rowland, HV Westerhoff, K van Dam, SG Oliver. 2000. A functional genomics strategy that uses metabolome data to reveal the phenotype of silent mutations. *Nat Biotechnol* **19**, 45–50.

7 Gifford DK. 2002. Blazing pathways through genetic mountains. *Science* **293**, 2049–51.

8 Karp PD. 2001. Pathway databases: a case study in computational symbolic theories. *Science* **293**, 2040–4.

9 White MA. 1996. The yeast two-hybrid system: forward and reverse. *Proc Natl Acad Sci USA* **93**, 10001–3.

10 Ito T, T Chiba, M Yoshida. 2001. Exploring the protein interactome using comprehensive two-hybrid projects. *Trends Biotechnol* **19**, S23–7.

11 Tucker CL, S Fields. 2001. A yeast sensor of ligand binding. *Nat Biotechnol* **19**, 1042–6.

12 Osborne MA, S Dalton, JP Kochan. 1995. The yeast tribrid system – genetic detection of *trans*-phosphorylated ITAM–SH2-interactions. *Biotechnology* **13**, 1474–8.

13 Straub B, E Meyer, P Fromherz. 2001. Recombinant maxi-K channels on transistor, a prototype of iono-electronic interfacing. *Nat Biotechnol* **19**, 121–124.

14 Celis JE, M Kruhoffer, I Gromova, C Frederiksen, M Ostergraad, T Thykjaer, P Gromov, J Vu, H Pàlsdottir, N Magnusson, TF Ortntoft. 2000. Gene expression profiling: monitoring transcription and translation products using DNA microarrays and proteomics. *FEBS Lett* **480**, 2–16.

15 Nicholson JK, J Connelly, C Lindon, E Holmes. 2002. Metabonomics: a platform for studying drug toxicity and gene function. *Nat Rev Drug Discov* **1**, 153–61.

16 Broach JR, J Thorner. 1996. High-throughput screening for drug discovery. *Nature* **384**, 14–6.

17 Goetz AS, JL Andrews, TR Littleton, DM Ignar. 2000. Development of a facile method for high throughput screening with reporter gene assays. *J Biomol Screen* **5**, 377–84.

18 Remington SJ. 2002. Negotiating the speed bumps to fluorescence. *Nat Biotechnol* **20**, 28–9.

19 Schaufele F. 2001. Shedding light on molecular timing. *Nat Biotechnol* **19**, 21–2.
20 Mitchell P. 2001. Turning the spotlight on cellular imaging. *Nat Biotechnol* **19**, 1013–7.
21 Hefti JP, A Kumar. 1999. Sensitive detection method of dielectric dispersions in aqueous based, surface bound macromolecular structures using microwave spectroscopy. *Appl Phys Lett* **75**, 1802–4.
22 Baumann W, M Lehmann, M Bitzenhofer, A Schwinde, M Birschwein, R Ehret, B Wolf. 1999. Microelectronic sensor system for microphysiological application on living cells. *Sensors and Acuators B* **55**, 77–89.
23 Schlessinger J. 1993. Cellular signaling by receptor tyrosine kinases. *Harvey Lect* **89**, 105–23.
24 Sullivan E, EM Tucker, IL Dale. 1999. Measurement of $[Ca^{2+}]$ using the Fluorometric Imaging Plate Reader (FLIPR). *Methods Mol Biol* **114**, 125–33.
25 Milligan G, S Rees. 1999. Chimaeric G alpha proteins: their potential use in drug discovery. *Trends Pharmacol Sci* **20**, 118–24.
26 Neer EJ. 1995. Heterotrimeric G proteins: organizers of transmembrane signals. *Cell* **80**, 249–57.
27 Zabin I. 1982. A model of protein interaction. *Mol Cell Biochem* **49**, 87–96.
28 Dean M, Y Hamon, G Chimini. 2001. The human ATP-binding cassette (ABC) transporter superfamily. *J Lipid Res* **42**, 1007–17.
29 Brinkmann U, M Eichelbaum. 2001. Polymorphisms in the ABC drug transporter gene MDR1. *Pharmacogenomics* **1**, 59–64.
30 Hwang LY, PT Lieu, PA Peterson, Y Yang. 2001. Functional regulation of immunoproteasomes and transporter associated with antigen processing. *Immunol Res* **24**, 245–72.
31 Ko YH, PL Pedersen. 2001. Cystic fibrosis: a brief look at some highlights of a decade of research focused on elucidating and correcting the molecular basis of the disease. *J Bioenerg Biomembr* **33**, 513–21.
32 Ouellette M, D Legare, B Papadopoulou. 2001. Multidrug resistance and ABC transporters in parasitic protozoa. *J Mol Microbiol Biotechnol* **3**, 201–6.
33 Schmidt C, M Mayer, H Vogel. 2000. A chip-based biosensor for the functional analysis of single ion channels. *Angew Chem Int* **39**, 3137–40.
34 Xu JZ, X Wang, B Ensign, M Li, L Wu, A Guia, J Xu. 2001. Ion-channel assay technologies: *quo vadis? Drug Discov Today* **6**, 1278–87.
35 Baxter DF, M Kirk, AF Garcia, A Raimondi, MH Holmqvist, KK Flint, D Bojanic, PS Distefano, R Curtis, Y Xie. 2002. A novel membrane potential-sensitive fluorescent dye improves cell-based assays for ion channels. *J Biomol Screen* **7**, 79–85.
36 Whiteaker KL, SM Gopalakrishnan, D Groebe, CC Shieh, U Warrior, DJ Burns, MJ Coghlan, VE Scott, M Gopalakrishnan. 2001. Validation of FLIPR membrane potential dye for high throughput screening of potassium channel modulators. *J Biomol Screen* **6**, 305–12.
37 Sarver JG, WA Klis, JP Byers, PW Erhardt. 2002. Microplate screening of the differential effects of test agents on Hoechst 33342, rhodamine 123, and rhodamine 6G accumulation in breast cancer cells that overexpress P-glycoprotein. *J Biomol Screen* **7**, 29–34.
38 Epstein CB, RA Butow. 2000. Microarray technology – enhanced versatility, persistent challenge. *Curr Opin Biotechnol* **11**, 36–41.
39 Hughes TR, MJ Marton, AR Jones, CJ Roberts, R Stoughton, CD Armour, HA Bennett, E Coffey, H Dai, YD He, MJ Kidd, AM King, MR Meyer, D Slade, PY Lum, SB Stepaniants, DD Shoemaker, D Gachotte, K Chakraburtty, J Simon, M Bard, SH Friend. 2000. Functional discovery via a compendium of expression profiles. *Cell* **102**, 109–26.

40 Weinberger SR, EA Dalmasso, ET Fung. 2002. Current achievements using ProteinChip Array technology. *Curr Opin Chem Biol* **6**, 86–91.

41 Goetz AS, J Liacos, J Yingling, DM Ignar. 1999. A combination assay for simultaneous assessment of multiple signaling pathways. *J Pharmacol Toxicol Methods* **42**, 225–35.

42 Freeman T. 2000. High throughput gene expression screening: its emerging role in drug discovery. *Med Res Rev* **20**, 197–202.

43 Pandey A, M Mann. 2000. Proteomics to study genes and genomes. *Nature* **405**, 837–46.

44 Brivanlou AH, JE Darnell, Jr. 2002. Signal transduction and the control of gene expression. *Science* **295**, 813–8.

45 Naylor LH. 1999. Reporter gene technology: the future looks bright. *Biochem Pharmacol* **58**, 749–57.

46 Terstappen GC, A Giacometti, E Ballini, L Aldegheri. 2000. Development of a functional reporter gene HTS assay for the identification of mGluR7 modulators. *J Biomol Screen* **5**, 255–62.

47 Xing H, HC Tran, TE Knapp, PA Negulescu, BA Pollok. 2000. A fluorescent reporter assay for the detection of ligands acting through G_i protein-coupled receptors. *J Recept Signal Transduct Res* **20**, 189–210.

48 Durocher Y, Perret S, Thibaudeau E, Gaumond MH, Kamen A, Stocco R, Abramovitz M. 2000. A reporter gene assay for high-throughput screening of G-protein-coupled receptors stably or transiently expressed in HEK293 EBNA cells grown in suspension culture. *Anal Biochem* **284**, 316–26.

49 Davis GF, TR Downs, JA Farmer, CR Pierson, JT Roesgen, EJ Cabrera, SL Nelson. 2002. Comparison of high throughput screening technologies for luminescence cell-based reporter screens. *J Biomol Screen* **7**, 67–77.

50 Wilson VS, K Bobseine, CR Lambright, LE Gray, Jr. 2002. A novel cell line, mda-kb2, that stably expresses an androgen- and glucocorticoid-responsive reporter for the detection of hormone receptor agonists and antagonists. *Toxicol Sci* **66**, 69–81.

51 Zhang JH, TD Chung, KR Oldenburg. 1999. A simple statistical parameter for use in evaluation and validation of high throughput screening assays. *J Biomol Screen* **4**, 67–73.

52 Rees S, DP Martin, SV Scott, SH Brown, N Fraser, C O'Shaughnessy, IJ Beresford. 2001. Development of a homogeneous MAP kinase reporter gene screen for the identification of agonists and antagonists at the CXCR1 chemokine receptor. *J Biomol Screen* **6**, 19–27.

53 Pomeroy SL, P Tamayo, M Gaasenbeek, LM Sturla, M Angelo, ME McLaughlin, JY Kim, LC Goumnerova, PM Black, C Lau, JC Allen, D Zagzag, JM Olson, T Curran, C Wetmore, JA Biegel, T Poggio, S Mukherjee, R Rifkin, A Califano, G Stolovitzky, DN Louis, JP Mesirov, ES Lander, TR Golub. 2002. Prediction of central nervous system embryonal tumour outcome based on gene expression. *Nature* **415**, 436–42.

54 Ideker T, V Thorsson, JA Ranish, R Christmas, J Buhler, JK Eng, R Bumgarner, DR Goodlett, R Aebersold, L Hood. 2001. Integrated genomic and proteomic analyses of a systematically perturbed metabolic network. *Science* **292**, 929–34.

55 Ulrich RG, SH Friend. 2002. Toxicogenomics and drug discovery: will new technologies help us produce better drugs? *Nat Rev Drug Discov* **1**, 84–5.

56 Nuwaysir EF, M Bittner, J Trent, JC Barrett, CA Afshari. 1999. Microarrays and toxicology: the advent of toxicogenomics. *Mol Carcinog* **24**, 153–9.

57 Rothberg BE. 2001. The use of animal models in expression pharmacogenomic analyses. *Pharmacogenomics* **1**, 48–58.

58 Shipp MA, KN Ross, P Tamayo, AP Weng, JL Kutok, RC Aguiar, M Gaasenbeek, M Angelo, M Reich,

GS Pinkus, TS Ray, MA Koval, KW Last, A Norton, TA Lister, J Mesirov, DS Neuberg, ES Lander, JC Aster, TR Golub. 2002. Diffuse large B-cell lymphoma outcome prediction by gene-expression profiling and supervised machine learning. *Nat Med* **8**, 68–74.

59 van't Veer LJ, H Dai, MJ van de Vijver, YD He, AA Hart, M Mao, HL Peterse, KK van der, MJ Marton, AT Witteveen, GJ Schreiber, RM Kerkhoven, C Roberts, PS Linsley, R Bernards, SH Friend. 2002. Gene expression profiling predicts clinical outcome of breast cancer. *Nature* **415**, 530–6.

60 Thomas RS, DR Rank, SG Penn, GM Zastrow, KR Hayes, K Pande, E Glover, T Silander, MW Craven, JK Reddy, SB Jovanovich, CA Bradfield. 2001. Identification of toxicologically predictive gene sets using cDNA microarrays. *Mol Pharmacol* **60**, 1189–94.

61 Bartosiewicz MJ, D Jenkins, S Penn, J Emery, A Buckpitt. 2001. Unique gene expression patterns in liver and kidney associated with exposure to chemical toxicants. *J Pharmacol Exp Ther* **297**, 895–905.

62 Burczynski ME, M McMillian, J Ciervo, L Li, JB Parker, RT Dunn, S Hicken, S Farr, MD Johnson. 2000. Toxicogenomics-based discrimination of toxic mechanism in HepG2 human hepatoma cells. *Toxicol Sci* **58**, 399–415.

63 Waring JF, R Ciurlionis, RA Jolly, M Heindel, RG Ulrich. 2001. Microarray analysis of hepatotoxins *in vitro* reveals a correlation between gene expression profiles and mechanisms of toxicity. *Toxicol Lett* **120**, 359–68.

64 Farr SB, MD Todd. 1998. Methods and kits for eukaryotic gene profiling. President and Fellows of Harvard College (Cambridge, MA, Xenometrix, Inc.). *US Patent 5,811,231.*

65 Albano CL, WE Bentley, G Rao. 2001. High throughput studies of gene expression using green fluorescent protein–oxidative stress promoter probe constructs. *J Biomol Screen* **6**, 421–8.

66 Zhu M, WE Fahl. 2000. Development of a green fluorescent protein microplate assay for the screening of chemopreventive agents. *Anal Biochem* **287**, 210–7.

67 Zhan Q, F Carrier, AJ Fornace, Jr. 1993. Induction of cellular p53 activity by DNA-damaging agents and growth arrest. *Mol Cell Biol* **13**, 4242–50.

68 Delescluse C, N Ledirac, R Li, MP Piechocki, RN Hines, X Gidrol, R Rahmani. 2001. Induction of cytochrome P450 1A1 gene expression, oxidative stress, and genotoxicity by carbaryl and thiabendazole in transfected human HepG2 and lymphoblastoid cells. *Biochem Pharmacol* **61**, 399–407.

69 Watkins RE, GB Wisely, LB Moore, JL Collins, MH Lambert, SP Williams, TM Willson, SA Kliewer, MR Redinbo. 2001. The human nuclear xenobiotic receptor PXR: structural determinants of directed promiscuity. *Science* **292**, 2329–33.

70 Bertelsen AH, VE Velculenscu. 1998. High-throughput gene expression analysis using SAGE. *Drug Discov Today* **3**, 152–9.

71 Shimkets RA, DG Lowe, JT Tai, P Sehl, H Jin, R Yang, PF Predki, BE Rothberg, MT Murtha, ME Roth, SG Shenoy, A Windemuth, JW Simpson, JF Simons, MP Daley, SA Gold, MP McKenna, K Hillan, GT Went, JM Rothberg. 1999. Gene expression analysis by transcript profiling coupled to a gene database query. *Nat Biotechnol* **17**, 798–803.

72 Sutcliffe JG, PE Foye, MG Erlander, BS Hilbush, LJ Bodzin, JT Durham, KW Hasel. 2000. TOGA: an automated parsing technology for analyzing expression of nearly all genes. *Proc Natl Acad Sci USA* **97**, 1976–81.

73 Heid CA, J Stevens, KJ Livak, PM Williams. 1996. Real time quantitative PCR. *Genome Res* **6**, 986–94.

74 Todd MD, X Lin, LF Stankowski, Jr., M Desai, GH Wolfgang. 1999.

Toxicity screening of a combinatorial library: correlation of cytotoxicity and gene induction to compound structure. *J Biomol Screen* **4**, 259–68.

75 WARRIOR U, Y FAN, CA DAVID, JA WILKINS, EM MCKEEGAN, JL KOFRON, DJ BURNS. 2000. Application of QuantiGene nucleic acid quantification technology for high throughput screening. *J Biomol Screen* **5**, 343–52.

76 CONRAD MJ. 1999. Measurement of specific gene transcripts in 96-well plate assays. *Genet Engng News* **19**.

77 GRAHAM DL, N BEVAN, PN LOWE, M PALMER, S REES. 2001. Application of beta-galactosidase enzyme complementation technology as a high throughput screening format for antagonists of the epidermal growth factor receptor. *J Biomol Screen* **6**, 401–11.

78 REMY I, SW MICHNICK. 2001. Visualization of biochemical networks in living cells. *Proc Natl Acad Sci USA* **98**, 7678–83.

79 GRIFFIN BA, SR ADAMS, RY TSIEN. 1998. Specific covalent labeling of recombinant protein molecules inside live cells. *Science* **281**, 269–72.

80 KAIN SR. 1999. Green fluorescent protein (GFP): applications in cell-based assays for drug discovery. *Drug Discov Today* **4**, 304–12.

81 MATZ MV, AF FRADKOV, YA LABAS, AP SAVITSKY, AG ZARAISKY, ML MARKELOV, SA LUKYANOV. 1999. Fluorescent proteins from nonbioluminescent *Anthozoa* species. *Nat Biotechnol* **17**, 969–73.

82 EDWARDS BS, FW KUCKUCK, ER PROSSNITZ, JT RANSOM, LA SKLAR. 2001. HTPS flow cytometry: a novel platform for automated high throughput drug discovery and characterization. *J Biomol Screen* **6**, 83–90.

83 TAYLOR DL, ES WOO, KA GIULIANO. 2001. Real-time molecular and cellular analysis: the new frontier of drug discovery. *Curr Opin Biotechnol* **12**, 75–81.

84 BRUNET A, D ROUX, P LENORMAND, S DOWD, S KEYSE, J POUYSSEGUR. 1999. Nuclear translocation of p42/p44 mitogen-activated protein kinase is required for growth factor-induced gene expression and cell cycle entry. *EMBO J* **18**, 664–74.

85 DALAL SN, CM SCHWEITZER, J GAN, J-A DECAPRIO. 1999. Cytoplasmic localization of human cdc25C during interphase requires an intact 14-3-3 binding site. *Mol Cell Biol* **19**, 4465–79.

86 LIU F, C ROTHBLUM-OVIATT, CE RYAN, H PIWNICA-WORMS. 1999. Overproduction of human Myt1 kinase induces a G_2 cell cycle delay by interfering with the intracellular trafficking of Cdc2–cyclin B1 complexes. *Mol Cell Biol* **19**, 5113–23.

87 VOGT A, T ADACHI, AP DUCRUET, J CHESEBROUGH, K NEMOTO, BI CARR, JS LAZO. 2001. Spatial analysis of key signaling proteins by high-content solid-phase cytometry in Hep3B cells treated with an inhibitor of Cdc25 dual-specificity phosphatases. *J Biol Chem* **276**, 20544–50.

88 CHADEE DN, MJ HENDZEL, CP TYLIPSKI, CD ALLIS, DP BAZETT-JONES, JA WRIGHT, JR DAVIE. 1999. Increased Ser-10 phosphorylation of histone H3 in mitogen-stimulated and oncogene-transformed mouse fibroblasts. *J Biol Chem* **274**, 24914–20.

89 BOGGS BA, CD ALLIS, AC CHINAULT. 2000. Immunofluorescent studies of human chromosomes with antibodies against phosphorylated H1 histone. *Chromosoma* **108**, 485–90.

90 PEREZ OD, GP NOLAN. 2002. Simultaneous measurement of multiple active kinase states using polychromatic flow cytometry. *Nat Biotechnol* **20**, 155–62.

91 MAYER TU, TM KAPOOR, SJ HAGGARTY, RW KING, SL SCHREIBER, TJ MITCHISON. 1999. Small molecule inhibitor of mitotic spindle bipolarity identified in a phenotype-based screen. *Science* **286**, 971–4.

92 BOGGS BA, B CONNORS, RE SOBEL, AC CHINAULT, CD ALLIS. 1996. Reduced levels of histone H3 acetylation on the inactive X chromosome in human females. *Chromosoma* **105**, 303–9.

93 HENDZEL MJ, MJ KRUHLAK, DP

Bazett-Jones. 1998. Organization of highly acetylated chromatin around sites of heterogeneous nuclear RNA accumulation. *Mol Biol Cell* **9**, 2491–507.

94 Yoneda Y. 2000. Nucleocytoplasmic protein traffic and its significance to cell function. *Genes Cells* **5**, 777–87.

95 Howell JL, R Truant. 2002. Live-cell nucleocytoplasmic protein shuttle assay utilizing laser confocal microscopy and FRAP. *Biotechniques* **32**, 80–7.

96 Ding GJ, PA Fischer, RC Boltz, JA Schmidt, JJ Colaianne, A Gough, RA Rubin, DK Miller. 1998. Characterization and quantitation of NF-kappaB nuclear translocation induced by interleukin-1 and tumor necrosis factor-alpha. Development and use of a high capacity fluorescence cytometric system. *J Biol Chem* **273**, 28897–905.

97 De Wit R, CM Hendrix, J Boonstra, AJ Verkleu, JA Post. 2000. Large-scale screening assay to measure epidermal growth factor internalization. *J Biomol Screen* **5**, 133–40.

98 Prasher DC, VK Eckenrode, WW Ward, FG Prendergast, MJ Cormier. 1992. Primary structure of the *Aequorea victoria* green-fluorescent protein. *Gene* **111**, 229–33.

99 Cody CW, DC Prasher, WM Westler, FG Prendergast, WW Ward. 1993. Chemical structure of the hexapeptide chromophore of the *Aequorea* green-fluorescent protein. *Biochemistry* **32**, 1212–8.

100 Barak LS, SS Ferguson, J Zhang, MG Caron. 1997. A beta-arrestin/green fluorescent protein biosensor for detecting G protein-coupled receptor activation. *J Biol Chem* **272**, 27497–500.

101 Oakley RH, SA Laporte, JA Holt, MG Caron, LS Barak. 2000. Differential affinities of visual arrestin, beta arrestin1, and beta arrestin2 for G protein-coupled receptors delineate two major classes of receptors. *J Biol Chem* **275**, 17201–10.

102 Groarke DA, S Wilson, C Krasel, G Milligan. 1999. Visualization of agonist-induced association and trafficking of green fluorescent protein-tagged forms of both beta-arrestin-1 and the thyrotropin-releasing hormone receptor-1. *J Biol Chem* **274**, 23263–9.

103 Groarke DA, T Drmota, DS Bahia, NA Evans, S Wilson, G Milligan. 2001. Analysis of the C-terminal tail of the rat thyrotropin-releasing hormone receptor-1 in interactions and cointernalization with beta-arrestin 1-green fluorescent protein. *Mol Pharmacol* **59**, 375–85.

104 Ghosh RN, YT Chen, R DeBiasio, RL DeBiasio, BR Conway, LK Minor, KT Demarest. 2000. Cell-based, high-content screen for receptor internalization, recycling and intracellular trafficking. *Biotechniques* **29**, 170–5.

105 Conway BR, LK Minor, JZ Xu, JW Gunnet, R DeBiasio, MR D'Andrea, R Rubin, R DeBiasio, K Giuliano, L DeBiasio, KT Demarest. 1999. Quantification of G-protein coupled receptor internalization using G-protein coupled receptor-green fluorescent protein conjugates with the ArrayScan™ high-content screening system. *J Biomol Screen* **4**, 75–86.

106 Mermelstein PG, K Deisseroth, N Dasgupta, AL Isaksen, RW Tsien. 2001. Calmodulin priming: nuclear translocation of a calmodulin complex and the memory of prior neuronal activity. *Proc Natl Acad Sci USA* **98**, 15342–7.

107 Santagata S, TJ Boggon, CL Baird, CA Gomez, J Zhao, WS Shan, DG Myszka, L Shapiro. 2001. G-protein signaling through tubby proteins. *Science* **292**, 2041–50.

108 Teruel MN, T Meyer. 2002. Parallel single-cell monitoring of receptor-triggered membrane translocation of a calcium-sensing protein module. *Science* **295**, 1910–2.

109 Torrance CJ, V Agrawal, B Vogelstein, KW Kinzler. 2001. Use of isogenic human cancer cells for high-throughput screening and drug discovery. *Nat Biotechnol* **19**, 940–5.

110 Lippincott-Schwartz J, E Snapp, A Kenworthy. 2001. Studying protein dynamics in living cells. *Nat Rev Mol Cell Biol* **2**, 444–56.

111 Mahajan NP, K Linder, G Berry, GW Gordon, R Heim, B Herman. 1998. Bcl-2 and Bax interactions in mitochondria probed with green fluorescent protein and fluorescence resonance energy transfer. *Nat Biotechnol* **16**, 547–52.

112 Zhang J, Y Ma, SS Taylor, RY Tsien. 2001. Genetically encoded reporters of protein kinase A activity reveal impact of substrate tethering. *Proc Natl Acad Sci USA* **98**, 14997–5002.

113 Ting AY, KH Kain, RL Klemke, RY Tsien. 2001. Genetically encoded fluorescent reporters of protein tyrosine kinase activities in living cells. *Proc Natl Acad Sci USA* **98**, 15003–8.

114 Honda A, SR Adams, CL Sawyer, V Lev-Ram, RY Tsien, WR Dostmann. 2001. Spatiotemporal dynamics of guanosine 3′,5′-cyclic monophosphate revealed by a genetically encoded, fluorescent indicator. *Proc Natl Acad Sci USA* **98**, 2437–42.

115 Zaccolo M, F De Giorgi, CY Cho, L Feng, T Knapp, PA Negulescu, SS Taylor, RY Tsien, T Pozzan. 2000. A genetically encoded, fluorescent indicator for cyclic AMP in living cells. *Nat Cell Biol* **2**, 25–9.

116 Mochizuki N, S Yamashita, K Kurokawa, Y Ohba, T Nagai, A Miyawaki, M Matsuda. 2001. Spatio-temporal images of growth-factor-induced activation of Ras and Rap1. *Nature* **411**, 1065–8.

117 Xu Y, DW Piston, CH Johnson. 1999. A bioluminescence resonance energy transfer (BRET) system: application to interacting circadian clock proteins. *Proc Natl Acad Sci USA* **96**, 151–6.

118 Angers S, A Salahpour, E Joly, S Hilairet, D Chelsky, M Dennis, M Bouvier. 2000. Detection of beta 2-adrenergic receptor dimerization in living cells using bioluminescence resonance energy transfer (BRET). *Proc Natl Acad Sci USA* **97**, 3684–9.

119 Ayoub MA, C Couturier, E Lucas-Meunier, S Angers, P Fossier, M Bouvier, R Jockers. 2002. Monitoring of ligand-independent dimerization and ligand-induced conformational changes of melatonin receptors in living cells by bioluminescence resonance energy transfer. *J Biol Chem* **277**, 21522–8.

120 Ramsay D, E Kellett, M McVey, S Rees, G Milligan. 2002. Homo- and hetero-oligomeric interactions between G protein-coupled receptors in living cells monitored by two variants of bioluminescence resonance energy transfer. *Biochem J* **365**, 429–40.

121 Rossi F, CA Charlton, HM Blau. 1997. Monitoring protein–protein interactions in intact eukaryotic cells by beta-galactosidase complementation. *Proc Natl Acad Sci USA* **94**, 8405–410.

122 Pelletier JN, FX Campbell-Valois, SW Michnick. 1998. Oligomerization domain-directed reassembly of active dihydrofolate reductase from rationally designed fragments. *Proc Natl Acad Sci USA* **95**, 12141–6.

123 Galarneau A, M Primeau, LE Trudeau, SW Michnick. 2002. Beta-lactamase protein fragment complementation assays as *in vivo* and *in vitro* sensors of protein–protein interactions. *Nat Biotechnol* **20**, 619–22.

124 Blakely BT, FM Rossi, B Tillotson, M Palmer, A Estelles, HM Blau. 2000. Epidermal growth factor receptor dimerization monitored in live cells. *Nat Biotechnol* **18**, 218–22.

125 Remy I, SW Michnick. 1999. Clonal selection and *in vivo* quantitation of protein interactions with protein-fragment complementation assays. *Proc Natl Acad Sci USA* **96**, 5394–9.

126 Subramaniam R, D Desveaux, C Spickler, SW Michnick, N Brisson. 2001. Direct visualization of protein interactions in plant cells. *Nat Biotechnol* **19**, 769–72.

127 Marshall ES, KM Holdaway, JH Shaw, GJ Finlay, JH Matthews, BC Baguley. 1993. Anticancer drug sensitivity profiles of new and estab-

lished melanoma cell lines. *Oncol Res* 5, 301–9.

128 Petty RD, LA Sutherland, EM Hunter, IA Cree. 1995. Comparison of MTT and ATP-based assays for the measurement of viable cell number. *J Biolumin Chemilumin* **10**, 29–34.

129 Ahmed SA, RM Gogal, Jr., JE Walsh. 1994. A new rapid and simple non-radioactive assay to monitor and determine the proliferation of lymphocytes. An alternative to [^{3}H]thymidine incorporation assay. *J Immunol Methods* **170**, 211–24.

130 Skaanild MT, J Clausen. 1989. Estimation of LC_{50} values by assay of lactate dehydrogenase and DNA redistribution in human lymphocyte cultures. *ATLA*, 293–6.

131 Crouch SP, R Kozlowski, KJ Slater, J Fletcher. 1993. The use of ATP bioluminescence as a measure of cell proliferation and cytotoxicity. *J Immunol Methods* **160**, 81–8.

132 Mosmann T. 1983. Rapid colorimetric assay for cellular growth and survival: application to proliferation and cytotoxicity assays. *J Immunol Methods* **65**, 55–63.

133 Green DR, JC Reed. 1998. Mitochondria and apoptosis. *Science* **281**, 1309–12.

134 Majno G, I Joris. 1995. Apoptosis, oncosis, and necrosis. An overview of cell death. *Am J Pathol* **146**, 3–15.

135 Woo ES, G Larocca, O Lapets, GR Bright. 1999. Multiparametric high content screening for apoptotic signals. High Throughput Screening Supplement to *Biomed Products.*

136 Levee MG, MI Dabrowska, JL Lelli, Jr., DB Hinshaw. 1996. Actin polymerization and depolymerization during apoptosis in HL-60 cells. *Am J Physiol* **271**, C1981–92.

137 Maekawa TL, TA Takahashi, M Fujihara, J Akasaka, S Fujikawa, M Minami, S Sekiguchi. 1996. Effects of ultraviolet B irradiation on cell–cell interaction; implication of morphological changes and actin filaments in irradiated cells. *Clin Exp Immunol* **105**, 389–96.

138 Endresen PC, PS Prytz, J Aarbakke. 1995. A new flow cytometric method for discrimination of apoptotic cells and detection of their cell cycle specificity through staining of F-actin and DNA. *Cytometry* **20**, 162–71.

139 van Engeland M, HJ Kuijpers, FC Ramaekers, CP Reutelingsperger, B Schutte. 1997. Plasma membrane alterations and cytoskeletal changes in apoptosis. *Exp Cell Res* **235**, 421–30.

140 Camilleri-Broet S, H Vanderwerff, E Caldwell, D Hockenbery. 1998. Distinct alterations in mitochondrial mass and function characterize different models of apoptosis. *Exp Cell Res* **239**, 277–92.

141 Anderson DJ, FH Gage, IL Weissman. 2001. Can stem cells cross lineage boundaries? *Nat Med* **7**, 393–5.

142 Henry D. 1997. Haematological toxicities associated with dose-intensive chemotherapy, the role for and use of recombinant growth factors. *Ann Oncol* **8** (Suppl. 3), S7–S10.

143 Scheding S, JE Media, A Nakeff. 1994. Acute toxic effects of 3′-azido-3′-deoxythymidine (AZT) on normal and regenerating murine hematopoiesis. *Exp Hematol* **22**, 60–5.

144 Eaves CJ. 1995. Assay of hematopoietic progenitor cells. In Beutler E, MA Lichtman, BS Collier, TJ Kipps (eds), *Williams Hematology*, 5th edn. New York: McGraw-Hill, 22–6.

145 Brasier AR, D Ron. 1992. Luciferase reporter gene assay in mammalian cells. *Methods Enzymol* **216**, 386–97.

146 Warren MK, WL Rose, LD Beall, J Cone. 1995. $CD34^+$ cell expansion and expression of lineage markers during liquid culture of human progenitor cells. *Stem Cells* **13**, 167–74.

147 Greenwalt DE, J Szabo, I Manchel. 2001. High throughput cell-based assay of hematopoietic progenitor differentiation. *J Biomol Screen* **6**, 383–92.

148 Saarinen K, K Kivisto, K Blomberg, K Punnonen, L Leino. 2000. Time-resolved fluorometric assay for leukocyte adhesion using a fluorescence enhancing ligand. *J Immunol Methods* **236**, 19–26.

149 Fiehn O. 2002. Metabolomics – the link between genotypes and phenotypes. *Plant Mol Biol* **48**, 155–71.

150 Nicholson JK, JC Lindon, E Holmes. 1999. "Metabonomics": understanding the metabolic responses of living systems to pathophysiological stimuli via multivariate statistical analysis of biological NMR spectroscopic data. *Xenobiotica* **29**, 1181–9.

151 Fiehn O. 2002. Metabolomics – the link between genotypes and phenotypes. *Plant Mol Biol* **48**, 155–71.

152 Lehmann M, W Baumann, M Brischwein, H Gahle, I Freund, R Ehret, S Drechsler, H Palzer, M Kleintges, U Sieben, B Wolf. 2001. Simultaneous measurement of cellular respiration and acidification with a single CMOS ISFET. *Biosens Bioelectron* **16**, 195–203.

153 Lehmann M, W Baumann, M Brischwein, R Ehret, M Kraus, A Schwinde, M Bitzenhofer, I Freund, B Wolf. 2000. Non-invasive measurement of cell membrane associated proton gradients by ion-sensitive field effect transistor arrays for microphysiological and bioelectronical applications. *Biosens Bioelectron* **15**, 117–24.

154 Henning T, M Brischwein, W Baumann, R Ehret, I Freund, R Kammerer, M Lehmann, A Schwinde, B Wolf. 2001. Approach to a multiparametric sensor-chip-based tumor chemosensitivity assay. *Anticancer Drugs* **12**, 21–32.

155 Rininger JA, VA DiPippo, BE Gould-Rothberg. 2000. Differential gene expression technologies for identifying surrogate markers of drug efficacy and toxicity. *Drug Discov Today* **5**, 560–8.

156 Schena M, Shalon D, Davis RW, Brown PO. 1995. Quantitative monitoring of gene expression patterns with a complementary DNA microarrays. *Science* **270**, 368–71.

157 Lockhart DJ, Dong H, Byrne MC, Follettie MT, Gallo MV, Chee MS, Mittmann M, Wang C, Kobayashi M, Horton H, Brown EL. 1996. Expression monitoring by hybridization to high-density oligonucleotide arrays. *Nat Biotechnol* **14**, 1649.

158 Gonzales F, Morita K, Seki T, Miyata N, Ueki T, Lois A, Milan L, Munday C, Garcia E, Nakao T. 2001. Simultaneous HTS measurement of heme oxygenase and GAPDH mRNAs by *multi*Star™ High Performance Signal Amplification (HPSA™). Presented at *SBS 7th Annual Conference & Exhibition*, Baltimore, MD.

2
Gene Knockout Models

Peter Ruth and Matthias Sausbier

2.1
Introduction

Targeted gene disruption in mice represents a new technology that is revolutionizing biomedical research. It is a powerful tool for producing murine models for human development and hereditary disease. While the human genome project has helped to establish the number of about 35 000 candidate genes, only a minor part of these genes have been characterized for their precise *in vivo* functions. Gene targeting has an enormous impact on our ability to delineate the functional roles of these genes. The major benefit of this technology is that it enables the analysis of the function of a protein produced from a specific gene *in vivo*. Thus, the function can be determined in all normal cell types and the complex interactions between molecules and cells are taken into account. It is possible to determine the function of gene products in the resting state of the body, in different physiological states and also in various pathological conditions. Many gene knockout mouse models mimic the phenotypes of human diseases. Genetically altered mouse models represent unique opportunities for developing and testing different therapeutic strategies before they are introduced into the clinic. In the future, genetically engineered animal models will be indispensable for gaining important insights into the molecular mechanisms underlying cellular signalling, development, as well as disease pathogenesis, diagnosis, prevention, and treatment.

2.2
Gene Knockout Mice

Animal models of human diseases are important for biomedical research, because they help us understand disease pathogenesis and allow for the development of therapeutic strategies to study the efficacy of novel treatments prior to the conduct of costly and time-consuming human clinical trials [1–5]. In some cases, the lack of an animal model has hindered progress in the development of effective therapies for debiliating and fatal genetic diseases. The laboratory mouse is rapidly

Molecular Biology in Medicinal Chemistry. Edited by Th. Dingermann, D. Steinhilber, G. Folkers

ISBN: 3-527-30431-2

becoming the most commonly used mammalian resource in biomedicine because of its small size, short reproductive cycle, known genetic information, and, most importantly, because of the possibility to introduce precise genetic alterations into mouse embryonic stem cells and transfer them to stable mouse lines. Therefore, genetically manipulated mice are now widely used as animal models to help us understand how diseases develop and are prevented and treated. Numerous other animal species, such as *Drosophila* fly [6–9], *Caenorhabditis elegans* [10–12] worm, and zebrafish [13–15] have been used for mutagenesis experiments to study developmental and signaltransduction processes. However, mammals such as mice prove to be better animal models, since they are more likely to mimic the phenotype of human diseases and disorders. The rationale for generating a gene knockout mouse bearing an inactivated or mutated gene is most often based on presumed or unknown functions of the candidate gene from its encoded protein, its expression pattern, pharmacological studies, or from the study of a chromosomal locus implicated in the cohorts of patients under study. Following the publication of the humane genome, many researchers have turned to murine knockout strategies to delineate the *in vivo* functions of genes of interest and their contributions to disease pathophysiology. The process of developing a knockout mouse with a specific gene to be deleted begins with the DNA of the gene. If the DNA sequence of the gene or a large part of the gene is not known, a useful targeting vector can not be designed. The DNA of most murine genes can meanwhile be obtained from data bases containing the genomic DNA bioinformation of the mouse genome project that has been successfully completed in 2002 [16]. The DNA information of a gene allows the construction of a targeting vector that provides the ability to manipulate the genome of embryonic stem (ES) cells to make a designer alterations in the gene of interest.

2.2.1
ES Cells

In the sixteen years since the first gene deletion experiments were performed in murine embryonic stem cells [17], a remarkable number of mouse knockout strains has been generated. A prerequisit for this achievement has been the availability of murine ES cells which have characteristics that make them easy to culture. ES cells are lines of cultured cells, originally harvested from the inner cell mass of blastocysts of a 3.5 day postcoitus embryo at the blastula stage of the development (Fig. 2.1). Because these cells are from a very early stage of development, they are pluripotent. All of their genes can be activated and all of their gene products can be synthesized. ES cells can be readily passaged with a reasonable cloning efficiency, allowing large numbers of cells to be propagated for uses such as transfection and subsequent homologous recombination. Additionally, murine ES cells can be maintained in an undifferentiated state in the presence of feeder layers and when culture medium is supplemented with leukemia inhibitory factor. In spite of this treatments, ES cells retain their ability to contribute to all types of tissue and all cell lineages when reintroduced into a host blastocyst [18].

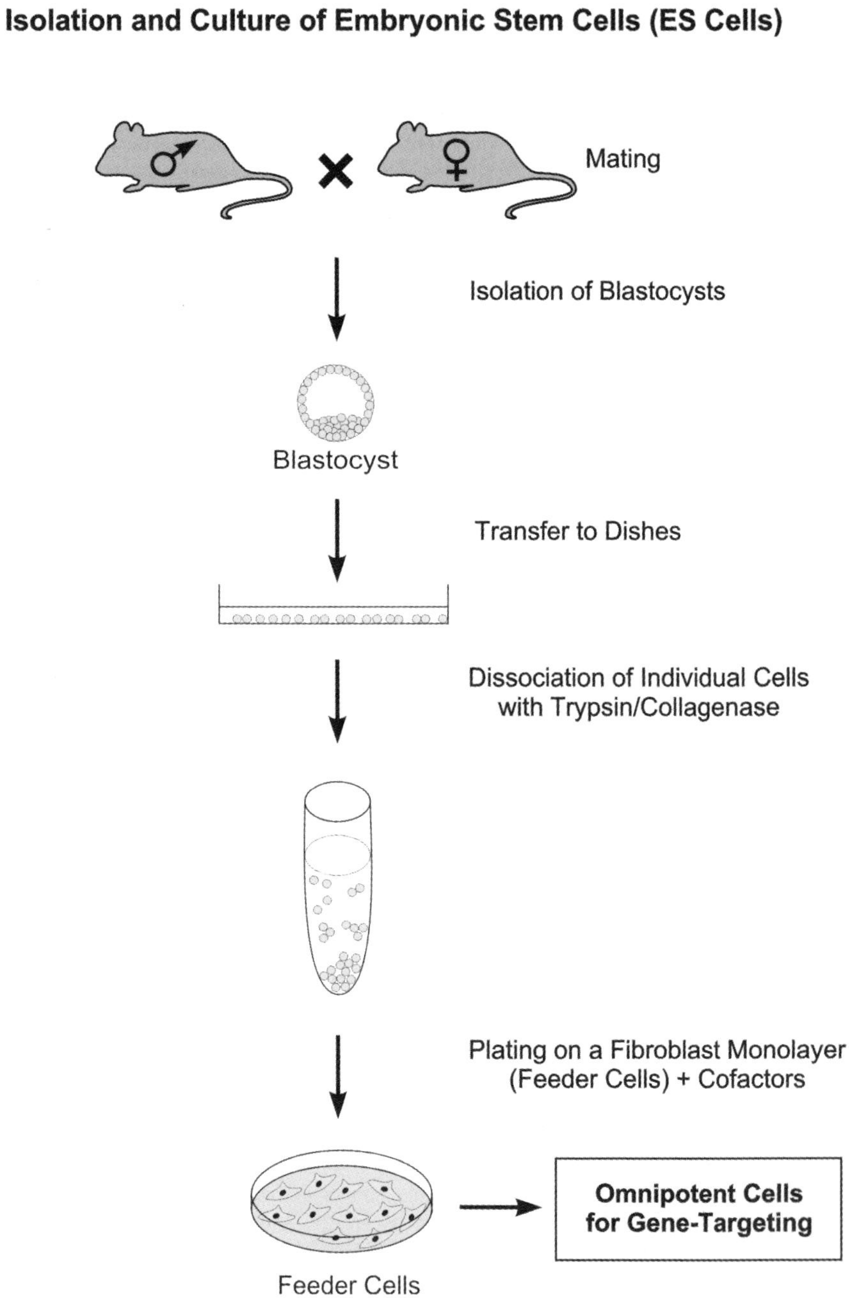

Fig. 2.1 Scheme of embryonic stem cell isolation and propagation.

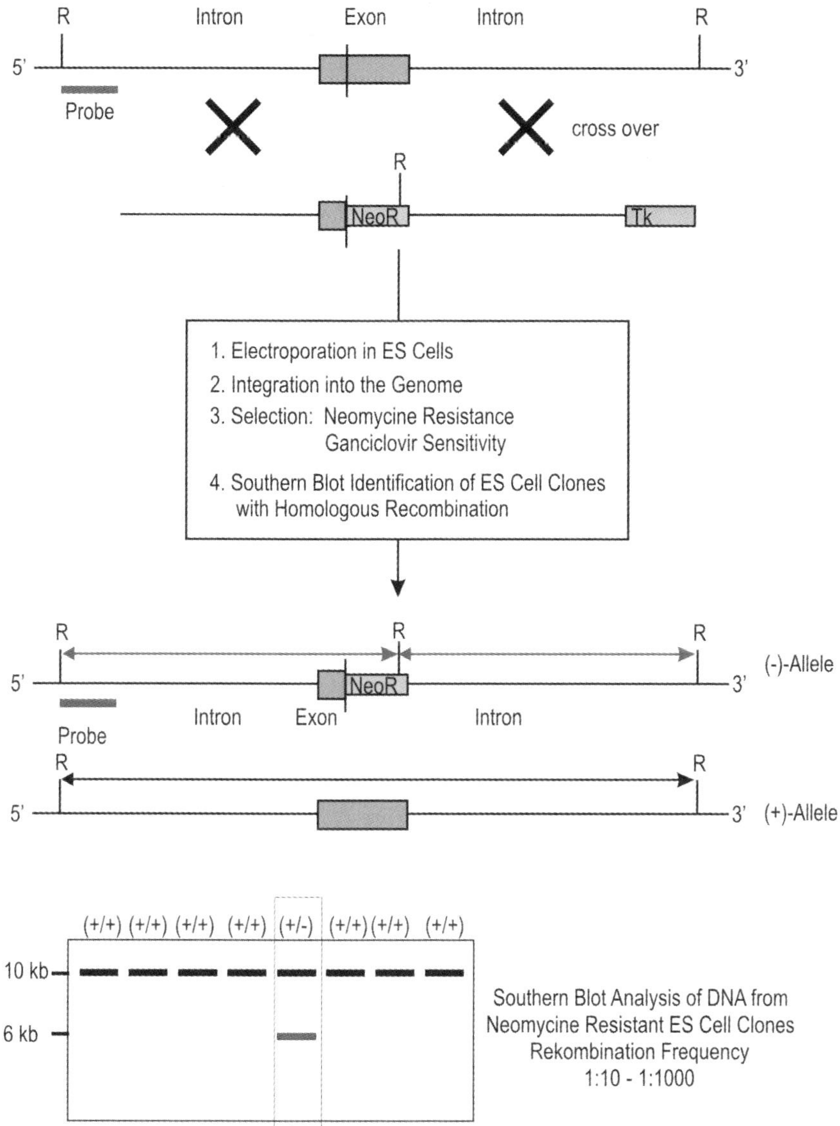

Fig. 2.2 Sequence replacement vector for positive-negative selection in ES cells. The vector includes an expression cassette of the *Neo* resistance gene and the HSV *Tk* gene. Following homologous recombination the positive drug selection cassette is incorporated into the target locus and the negative selection cassette is lost. (–) Allele indicates the mutant allele after homologous recombination. R indicates restriction enzyme sites used for cutting the DNA prior to detection of homologous recombination in Southern blotting with the help of a specific probe derived from outside of the targeted region.

2.2.2
Targeting Vector

To establish a knockout mouse model, a DNA construct termed targeting vector, containing the mutated form of the responsible gene is developed (Fig. 2.2). In its most basic form, the early targeting vector is used to disrupt target gene function by deleting specific sequences and replacing them with heterologous DNA, usually a drug selection gene or a marker gene to analyze target gene expression [19, 20]. As shown in Fig. 2.2 the targeting vector is composed of three essential DNA elements arranged in a specific order: i) a region of 5′ target gene homology, ii) a drug resistance marker gene for the positive selection of cells that integrate the targeting vector, and iii) a region of 3′ target gene homology. The DNA sequences resident in the 5′ and 3′ regions on the targeting vector are the substrates for a double reciprocal crossover or gene conversion mechanism that transfers the selection marker cassette element into the endogenous locus. The homologous sequences are derived from the cloned genomic portions of the target gene and flank the target gene sequences that are being replaced by the selection marker cassette to generate the mutant locus. Genomic DNA isogenic to that of the ES cells to be targeted should be used for the targeting construct. The specificity and length of DNA sequences flanking the targeted region should be maximized for efficient homologous recombination, i.e. repetitive sequences should be avoided. When construction has been completed, the vector is linearized at either end of the 5′ or 3′ homology regions so that the DNA ends can serve as a substrate to initiate homologous recombination with the target locus. Following the transfection of the linearized targeting vector into ES cells, a drug selection is performed to enrich for those ES cells that have integrated the targeting vector and are expressing the drug-sensitive gene. The marker gene expression cassette includes all of the regulatory sequence elements needed for transcription. The most commonly used selectable marker has been the neomycin phosphotransferase gene (*Neo*) which confers resistance to the neomycin analog G418. A widely used method to enrich for homologous recombinant ES cell clones is a gene targeting strategy that employs a vector designed for use in a double drug selection protocol known as positive-negative selection (Fig. 2.2). Apart from the positive selection marker *Neo* positioned between the 5′ and 3′ target gene homologous regions, an additional marker gene cassette for negative selection located on either end of the homology regions is included. The negative selection gene that has been most commonly used is the herpes simplex virus thymidine kinase (*Tk*) gene expression cassette, which confers sensitivity to the guanosine analog gancyclovir. The *Tk* gene product phosphorylates gancyclovir, allowing it to be incorporated into replicating DNA to act as an inhibitor of DNA synthesis (Fig. 2.3).

In a typical gene-targeting experiment, the positive/negative selection vector will randomly integrate by its linear ends into the genome of many cells, retaining both the *Neo* and *Tk* selection cassette and confering resistance to neo and sensitivity to gancyclovir. However, in recipient ES cells where homologous recombination between the targeting vector and the endogenous gene has occured, the *Neo* cas-

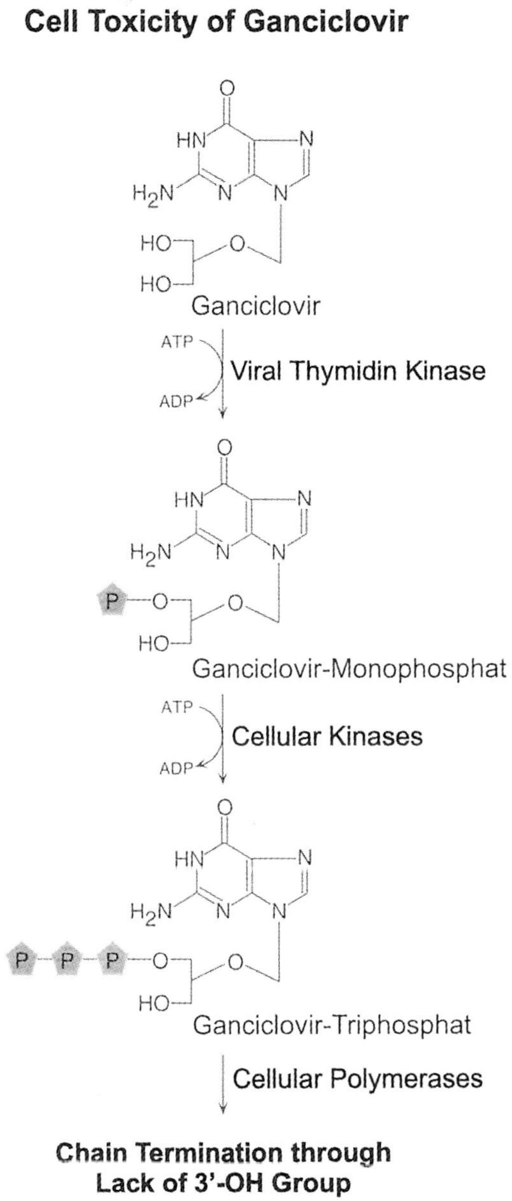

Fig. 2.3 Mechanism of cell toxicity of ganciclovir used for negative selection in the screening for homologous recombinant ES cells.

sette will be integrated into the target locus while the *Tk* cassette is lost and these cells will survive both positive and gancyclovir selection. By employing a positive/negative selection scheme, the absolute number of ES cell colonies that survive double drug selection is about 10-fold lower than the number of colonies that survive positive selection only, greatly reducing the number of colonies to be screened for the gene targeting event. Empirically, not all of the ES clones that survive positive/negative selection are recognized as being homologous recombinants by a Southern blot screen of the target locus in ES cell clones with specific probes (Fig. 2.2). The reason for this might be that the negative selection marker gene acquires inactivating mutations prior to random integration of the targeting vector into the ES cell genome.

2.2.3
Selection of Recombinant ES Cells

The linearized targeting vector is inserted into ES cells by mixing it with ES cells in an electroporation cuvette prior to building up an electric field. The electrical field opens pores in the membranes of the ES cells that permit the entry of the targeting vector. When the current is turned off, the membrane closes and returns to its normal state. The DNA construct diffuses through the cytoplasm and enters the nucleus. All of the ES cells that underwent electroporation are grown in tissue culture containing a solution of the antibiotic. After transfection, homologous recombination between the targeting vector and the endogenous locus in ES cells is a rare event, but the presence of marker genes for drug selection allows to enrich the recovery of recombinant ES cells. Visible colonies of ES cells growing in the presence of selection compounds are harvested as candidates likely to contain the mutated gene in their genome. ES cells that have undergone homologous recombination will be genetically heterozygous at the targeted chromosomal locus. Thus, in planning any gene targeting experiment it is important to have a genotyping strategy that will easily discriminate between the wild-type and mutant loci in the cells that harbor a homologous recombination event. When selecting the target gene homology regions for cloning into the targeting vector, it is important to anticipate the use of Southern blot analysis (or PCR analysis) by considering available restriction enzyme sites to generate diagnostic fragments. The success of this analysis is dependent on the selection of DNA probes taken from outside of the recombination region and the choice of restriction sites that will identify restriction fragments that are unique to the mutant and wild-type allele, respectively (Fig. 2.2).

2.2.4
Injection of Recombinant ES Cells into Blastocysts and Blastocyst Transfer to Pseudopregnant Recipients

The ES cell colonies exhibiting the expected homologous recombination are individualized by trypsination and collected by a fine-gauge needle which is con-

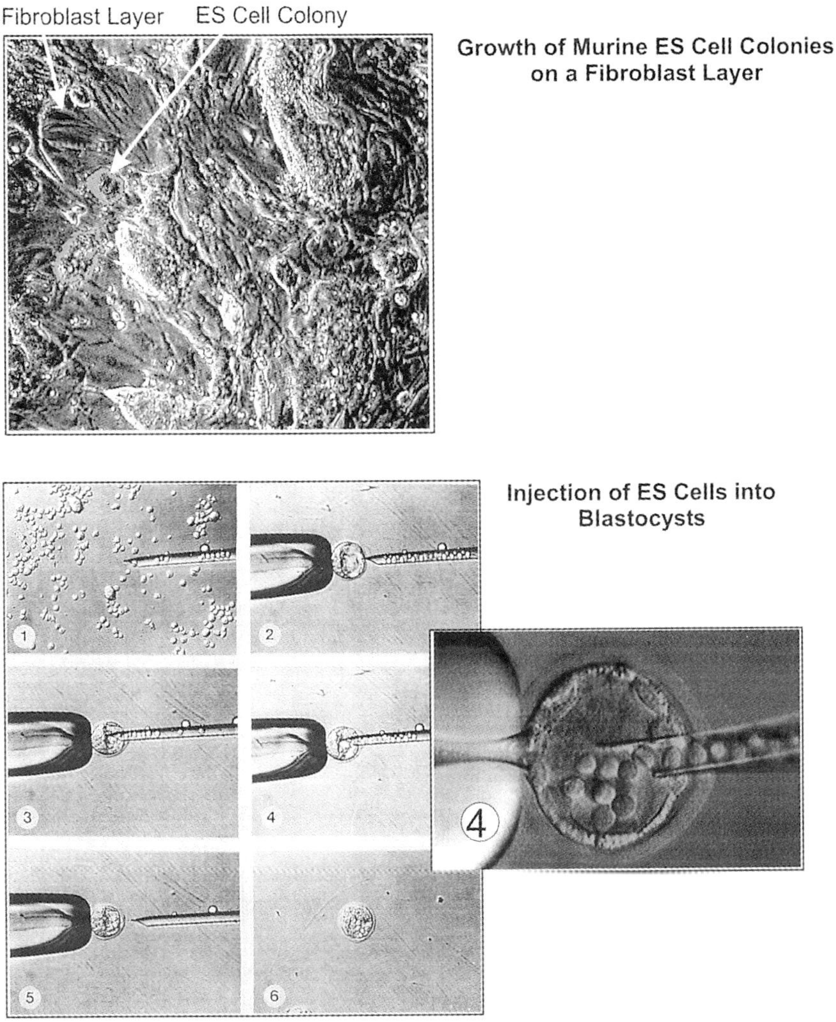

Fig. 2.4 *Upper panel.* Multiple ES cell clones growing on a fibroblast layer. The fibroblasts are from transgenic mouse embryos carrying a *Neo* selection cassette confering resistance to geneticin for the positive selection. After preparing them from embryos, the fibroblasts are irradiated to prevent that they proliferate further.

Lower panel. Injection of ES cells into a blastocyst. Individualized ES cells are loaded into the injection pipette (1), the sharp tip of the pipette penetrates into the blastocoel cavity (2), one smooth push bring the tip of the injection pipette into the embryo without collapsing or puncturing the opposite wall (3), the ES cells are expelled by positive pressure (4), the needle is withdrawn from the embryo (5), and injected blastocysts are collected for transfer to recipient females (6).

nected to micromanipulator (Fig. 2.4). Blastocysts for microinjection of ES cells are taken from donor females. The sharply grinded tip of the needle penetrates the blastocyst and releases 15–25 ES cells by positive pressure. The injected embryos must be implanted into recipient females. Successful embryo transfer depends on the quality of the host maternal enviroment. Mating of recipients with male mice is required for the hormonal changes necessary to establish pregnancy. Because recipients should not carry embryos of their own, females must be mated with vasectomized males. Mice are spontaneous ovulators and can be made pseudopregnant by mating with vasectomized males during oestrus. They will then display the hormonal profile of a normal pregnant female.

2.2.5
Chimeras and F1 and F2 Offspring

Successful microinjection of ES cells into the blastocyst results in an embryo comprised of cells from the original blastocyst plus the ES cells implanted into the blastocyst. Cells of both origins grow in concert to form one complete embryo. The resulting mouse pub has tissues and organs composed of mosaic mixture of cells derived from the original cells usually stemming from C57/BL6 mice with black coat and the ES cells usually stemming from SV129 mice with yellow/brown coat. The pub is called a chimera because it contains cells from two independent sources. The coat color of chimeric pubs is a mosaic of yellow/brown and black. If no ES cell is incorporated at the blastocyst stage, the pup will show a black coat. The appearance of the coat is a useful early marker of a successful mutation. Chimeras are identified as soon as the coat color is detectable. When incorporation of mutant cells is only in the somatic cells that develop into nonreproductive tissues, the original chimeras express the mutation, but their offspring do not. When incorporation is in cells that develop into germ-line gametes, the targeted mutation is transmitted to the next generation of the offspring of the chimera (Fig. 2.5).

To detect germ-line transmission crossing between the chimera and wild-type mice is conducted. The chimera is bred to a mouse of a normal inbred strain, such as C57/BL6. The F1 offspring of this crossing are analysed for the expression of the mutation (Fig. 2.5). An F1 offspring that receives the mutated gene from the chimeric progenitor parent is heterozygous for the mutated gene. Southern blot analysis or allele-specific PCR is performed on genomic DNA from small tissue sample from the tail of the offspring, in order to identify the positive heterozygous. Each positive heterozygote is a potential founder for a line of mutant mice. Identified F1 heterozygote offspring are mated with each other to produce an F2 generation. Theoretically, the F2 population will follow the principles of Mendelian segregation, resulting in 25% homozygous mutants, 25% homozygous control wild-type and 50% heterozygous mutants. If the gene deletion is lethal, the homozygous mutants will not survive. Finally, the absence of the gene product in homozygous mutants is tested by a specific antibody in a Western blot using tissue where the protein is usually expressed.

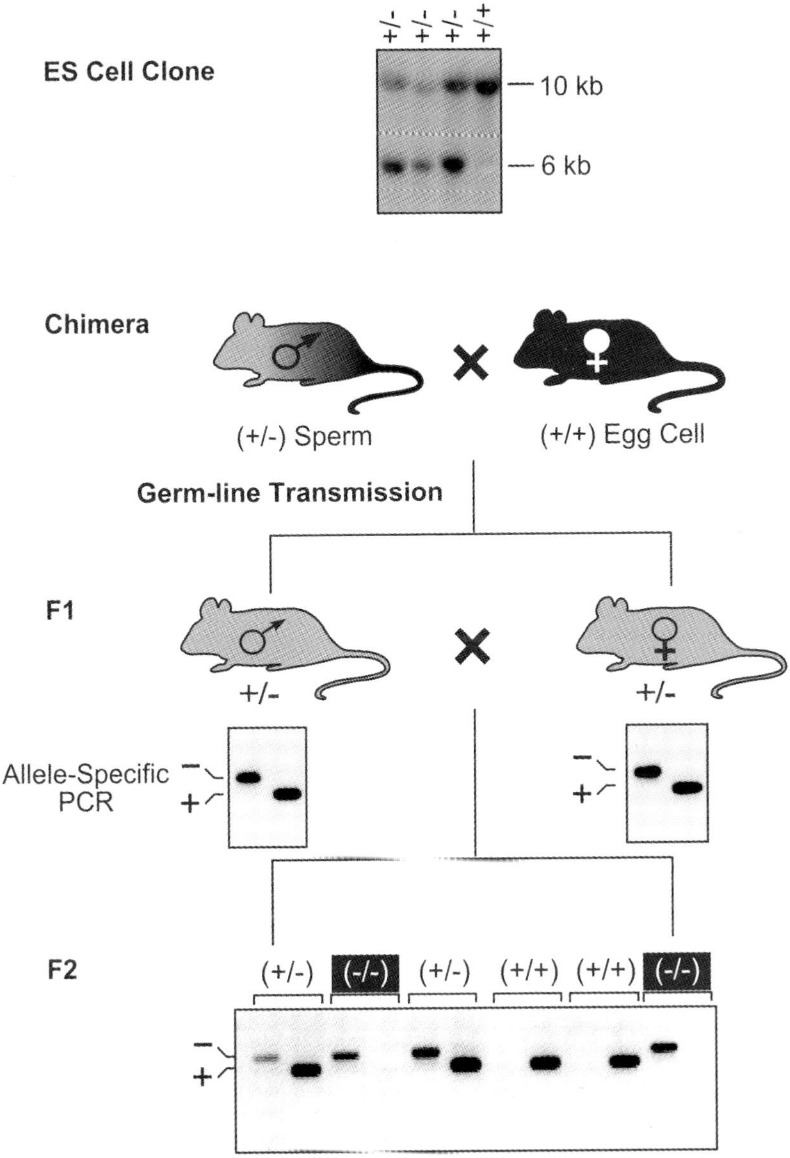

Fig. 2.5 Mating chimeras with wild-type mice for obtaining heterozygous mutants. ES cell clones carrying one wild-type and one mutant allele (+/−) are injected into blastocysts. A characterizing Southern blot of ES cell clones is shown. The injected blastocysts are transferred into female recipients which deliver chimeric pubs. Regularly ES cells carrying the dominant agouti mutation providing brown-/yellowness of the fur (strain SV129) are injected into the blastocysts from a black mouse strain (strain C57Bl6). Therefore, the degree of chimerism can easily be recognized by the contribution of brown/yellow colour to the mice's coat. Chimeras are then crossed to black C57Bl6 mice. In the case of germ line transmission of the mutation, heterozygous F1 mice (+/−) with brown/yellow fur are yielded. Mating heterozygous mice among each other will result in one-quarter of F2 homozygous mice if viability is not compromised by the mutation.

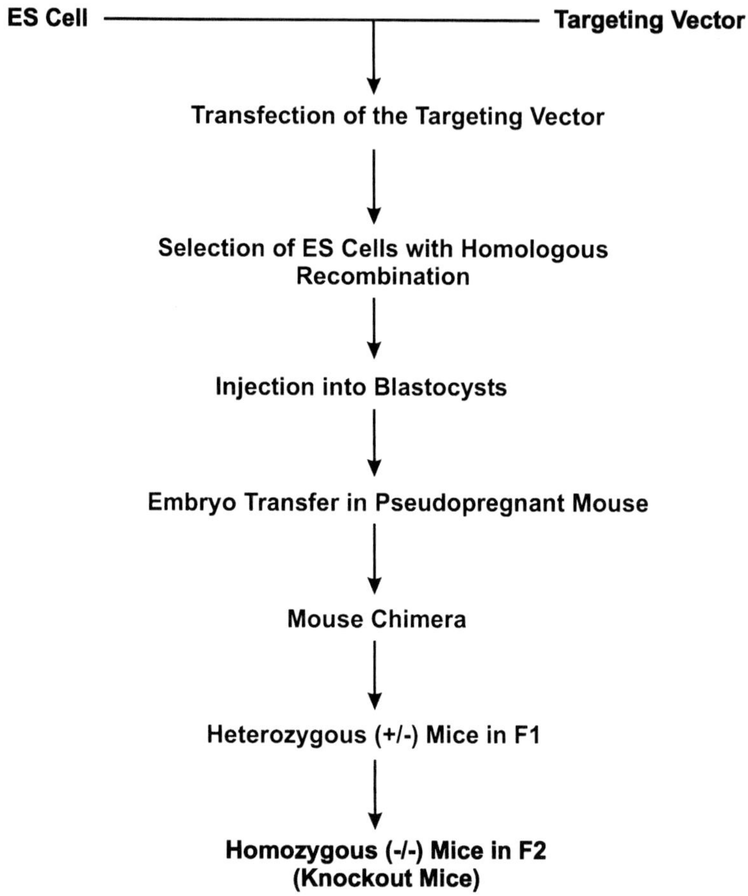

Fig. 2.6 Flow diagram of the gene knockout procedure.

A line of mice with the mutated gene is generated. The null mutant homozygous mouse is deficient in both alleles of a gene; the heterozygote is deficient in one of its two alleles for the gene. The genotype is $-/-$ for the null mutant, $+/-$ for the heterozygote, and $+/+$ for the wildtype normal control. The phenotype is the set of observed characteristics resulting from the mutation. Phenotype include biochemical, anatomical, physiological, and behavioral characteristics. Characteristics of the mutant mice are identified in comparison to normal controls. Salient characteristics relevant to human diseases are quantitated. These disease-like traits are then used as test variables for evaluating the effectiveness of treatments. Putative treatments are administered to the mutant mice. A treatment that prevents or reverses the disease traits in the mutant mice is taken for further testing as a potential therapeutic treatment for the human genetic disease.

2.3 Tissue-Specific Gene Expression

Conventional gene knockout technology is limited by embryonal or early postnatal lethality since the role of such genes can not be investigated. Further, the role of a particular gene in adult tissues may be masked in any knockout mice generated by developmental abnormalities or compensations in form of expressional up- and down-regulation of other genes. The implementation of site specific recombinases, such as the bacteriophage P1 LoxP/CRE system, enables the development of conditional knockouts that lack a particular gene only in a specific tissue or after a specific stage of development [21, 22]. The CRE enzyme is isolated from bacteriophage P1 and acts as a site specific recombinase. The sequence recombined by CRE, termed LoxP, consists of two short inverted DNA repeats (CRE recognition site) with an intervening sequence that conveys a directional nature to the LoxP site. Two LoxP sequences in the same orientation will result in excision of the intervening DNA by CRE leaving one intact LoxP site (Fig. 2.7). In contrast, where two LoxP sites are in the opposite orientation, CRE will invert the intervening segment of DNA.

Targeting constructs for the insertion of LoxP sites into genomic DNA are very similar to those used for the generation of conventional gene-knockout ES cells (Fig. 2.2). In addition to these aspects, the design of conditional gene knockout constructs requires a number of important additional considerations: Conventional knockouts are usually designed so that the *Neo* cassette disrupts the transcription of the endogenous gene. In conditional knockouts, the endogenous gene must function normally until recombined by CRE recombinase and thus, any modifications must be outside coding regions, i.e. intronic, and must not interfere with regulatory regions (Fig. 2.7). The LoxP site must be inserted in a manner where splicing will result in excision of sufficient DNA to render the gene inactive. This can be particularly difficult for genes with large introns where only small amounts of coding sequence can be removed. Thus, the LoxP sites should be positioned to remove exons that encode an important functional domain and would result in a frameshift. Resistance genes included for the selection of correctly targeted clones usually include several kilobases of foreign DNA, which has the potential to influence the expression of the targeted gene prior to recombination. To minimize this potential influence on gene expression, it is possible to splice out the resistance gene from the correctly targeted ES cell clone *in vitro*, prior to blastocyst microinjection. To remove the resistance gene, an additional LoxP site must be included to flank the resistance gene (Fig. 2.7). The modified genomic DNA would thus include three LoxP sites, two positioned flanking the resistance gene, and the third located to cause the desired deletion when recombination occurs. Transfection of the correctly targeted ES cell clone with a vector containing the cDNA of CRE recombinase produces ES cells with three alternative genome configurations in addition to non-recombined clones (Fig. 2.7). To limit the number of recombination events and to facilitate the selection of desired clones, a negative selection can

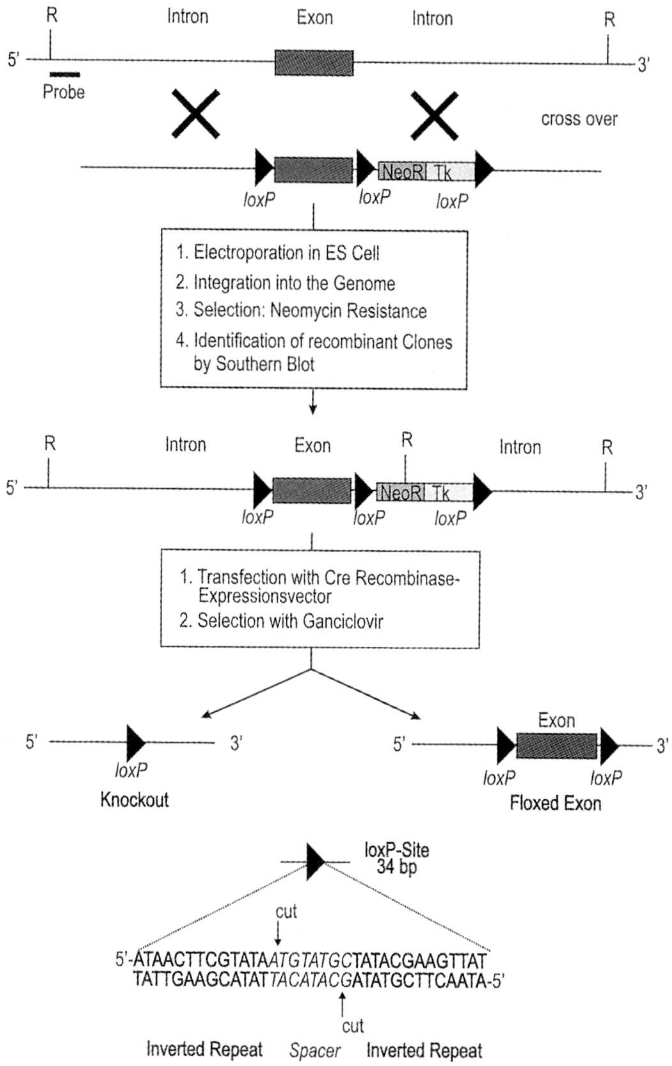

Fig. 2.7 Conditional gene knockouts using LoxP/CRE. Wild-type locus and conditional targeting construct showing three LoxP sites flanking a *Neo/Tk* tandem selection marker cassette and the exon to be deleted. In the first targeting, this construct recombinates with the desired locus. A positively recombinated clone is chosen and subjected to a second targeting, comprising the transfection with a CRE recombinase cDNA vector. Three recombination events after exposure to CRE are possible. Cell population A represents the complete excision and can be used for injection into blastocysts to generate the total knockout mice. Cell population B is the form where the selection cassette has been removed, but the gene is not yet deleted. These ES cells can be taken to establish the conditional knockout, where the floxed exon is deleted later, i.e. after crossing the resulting animals with transgenic mice expressing CRE recombinase under a tissue specific promoter. The partially palindromic nucleotide sequence of LoxP sites is presented in the lower part.

be applied by the addition of gancylovir to the ES cells. This strategy requires the use of a selection marker cassette containing both the *Tk* gene and the *Neo* resistance gene for selection. Addition of gancyclovir to the ES cell culture will then allow the growth of clones with i) the exon of interest flanked by two LoxP sites, i.e. the floxed exon for the conditional knockout, and ii) ES cell clones with a single LoxP site, ready to use for establishing the general knockout. Thus the consecutive first and second targeting using three LoxP sites and the *Neo*/*Tk* selection marker cassette inserted into the ES cell genome yields both clones for the classical, i.e. general knockout, and clones for the conditional knockout. Selection of clones where the resistance gene has been deleted and the genomic DNA, with its appropriate modifications, has been left intact can be made by Southern blot screening.

When blastocysts injection and germline transmission of the conditional knockout ES cells is achieved and mice with the floxed exon have been obtained, CRE recombinase is expressed in the tissue or cell type of interest (Fig. 2.8). The recombinase specifically removes the essential exon from the modified allele and creates a CRE-excision-dependent knockout. To accomplish this, a CRE transgenic line expressing the recombinase under a tissue- or cell type-specific promoter is crossed over the conditional knockout mouse with the floxed exon. Only in cells where the CRE recombinase is expressed, an excision of the exon takes place and results in a tissue specific gene deletion. Meanwhile, a panel of CRE transgenic mouse lines is available, each expressing the recombinase with different specificities. Thus the LoxP/CRE recombination system is a powerful tool for the conditional and cell-specific deletion of genes. The introduction of this system into transgenic mice facilitates studies on the loss of function of genes in a particular cell type. Through insertion of Lox sites via homologous recombination into the gene of interest and targeting CRE recombinase expression to a specific cell type using a tissue-specific promoter, it will be possible to introduce predetermined deletions into the mammalian genome.

2.3.1 Ligand-Activated CRE Recombinases

Within the past several years new technologies have been developed that allow for the inducible gene expression in transgenic mice that can be combined with more conventional gene-targeting approaches in order to control the timing and tissue-specificity of the desired mutation. For example, a conditional gene targeting method based on the inducible activity of an engineered CRE recombinase could allow the tissue-specific inactivation of floxed target genes at any time desired during development and in the adult mouse (Fig. 2.8, lower panel). Fusion of the ligand-binding domain (LBD) of the estrogen receptor to the CRE recombinase generates a chimeric recombinase whose activity is dependent on the presence of estrogen or the estrogen receptor modulator tamoxifen [23–24]. To achieve conditional gene targeting in mice, where endogenous estradiol is present, the CRE re-

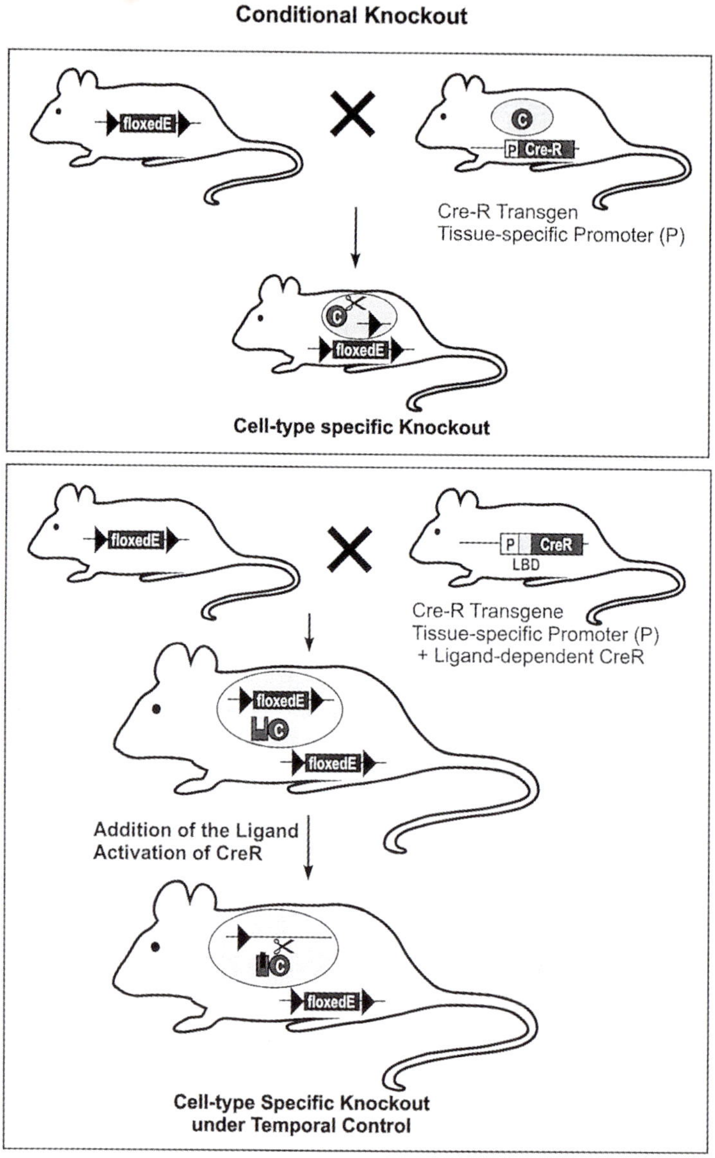

Fig. 2.8 *Upper panel.* Breeding strategy to obtain mice with a cell-type specific knockout. Mice carrying a null allele (not shown) and in addition, an allele where the gene or exon to be deleted is flanked by two LoxP sites (floxed exon) are phenotypically as heterozygous mice. When crossed with transgenic mice expressing CRE recombinase (C) under a tissue-specific promoter (P), these mice become homozygous null mutants with regard to this tissue.

Lower panel. Breeding strategy to obtain mice with a cell-type specific knockout under temporal control. The CRE recombinase transgene is fused to the cDNA of the ligand-binding domain of the estrogen receptor. The activation of the CRE recombinase is dependent of the presence of tamoxifen that can be supplied to the drinking water of mice. The addition of tamoxifen enables the knockout of the gene or exon of interest to any time point desired.

combinase has been fused to a mutated LBD of the human estrogen receptor. This mutated estrogen receptor LBD does not bind estradiol, whereas it binds the synthetic ligand tamoxifen. Transgenic mice lines have been generated expressing the CRE recombinase fused to the mutant estrogen receptor LBD under the control of different tissue-specific promoters (P) (Fig. 2.8, lower panel). Such a transgenic mouse can be mated with a mouse exhibiting a floxed exon within the gene that should be deleted. Their offsprings exhibiting both modifications, the floxed exon and the ligand-dependent CRE recombinase, are fed with tamoxifen. The floxed exon is subsequently deleted from the genome in those tissues where the CRE recombinase fusion protein is expressed. The cell-specific expression of CRE recombinase fused to a mutated estrogen receptor LBD in transgenic mice can thus be used for efficient tamoxifen-dependent, CRE-mediated recombination at loci containing LoxP sites to generate site-specific somatic mutations in a spatio-temporally controlled manner.

2.3.2
The Tetracycline/Doxycycline-Inducible Expression System

The ability to control gene expression, and ultimately gene targeting, in transgenic mice by simply administering an exogenous substance affords an additional level of control over the CRE/LoxP system for inducing gene mutations in mice. The tetracycline-responsive system is composed of the tetracycline repressor protein (TetR) fused to a herpes simplex virus transcriptional activation domain, and the tetracycline operator sequence (tetO) linked to a minimal promoter element to control the transcription of a downstream gene following TetR binding to tetO [25, 26]. The vectors for expression of TetR and tetO-minimal promoter directed gene transcription in mammalian cells are commercially available.

The tetracycline-responsive system can be used in concert with CRE/LoxP technology to allow for the temporal control of gene targeting in mice by placing the expression of the CRE gene under the control of the tetO-minimal promoter. For more specificity, a tissue-specific promoter can be used to drive the expression of the TetR gene. Moreover, there are Tet-Off and Tet-On variations for controlling tet-responsive gene expression, which depend on the binding of tetracycline to TetR. In the Tet-Off version of this system, the TetR transactivating protein, tTA, will bind to the tetO-minimal promoter in the presence of tetracycline to repress gene expression. When tetracycline administration is stopped, tTA no longer binds to the tetO-minimal promoter and transcription is induced by as much as several thousandfold in transgenic mouse tissues. As illustrated (Fig. 2.9), it is necessary to generate two mouse lines that are homozygous for the tTA gene and heterozygous for the floxed mutant locus, and homozygous for the CRE expressing transgene and heterozygous for the floxed mutant locus, respectively. By breeding the two mouse lines, 25% of the offsprings will be homozygous for the conditionally mutant locus and heterozygous for both the tTA transgene and the CRE-

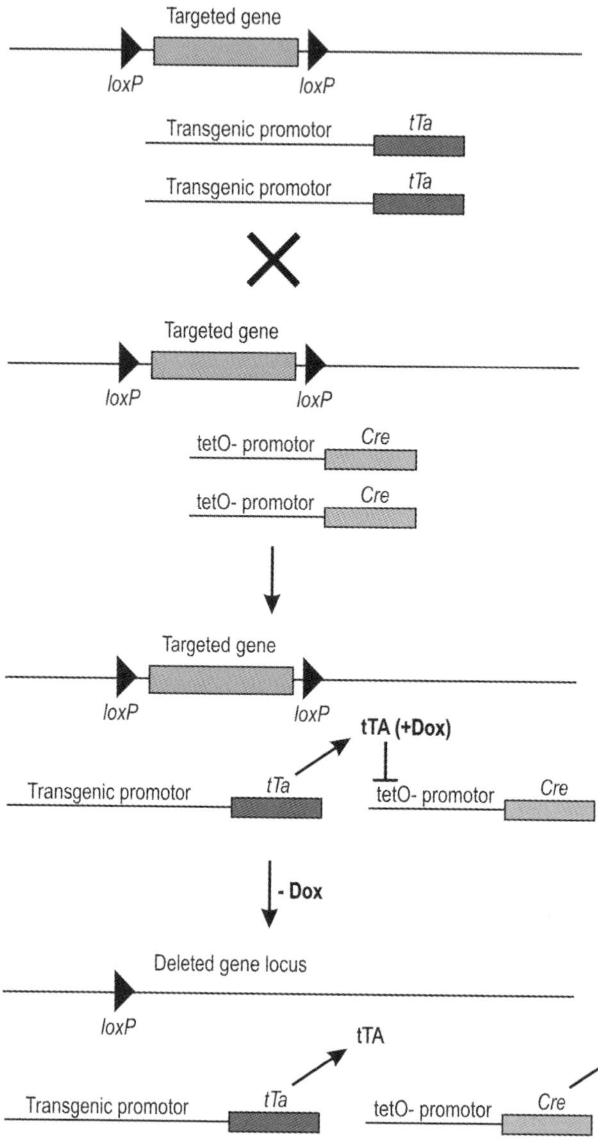

Fig. 2.9 Tet-Off inducible gene targeting. One mouse line is heterozygous for a conditional mutant target locus with LoxP sites flanking the target gene of interest and homozygous for a transgene to express the tetracycline repressor protein. This mouse line is crossed to another that is also heterozygous for the conditional mutant allele and homozygous for a CRE-expressing transgene responsive to tetracycline. The crossing of these lines produce progeny that are heterozygous for both transgenes to maintain the repression of the CRE gene and will carry the condionlly mutant locus and wild-type locus in the expected Mendelian ratios. The removal of doxycycline results in the expression of the CRE gene and the creation of a mutant target locus. The wild-type alleles of the endogenous target locus are not shown.

expressing transgene. To maintain the repression of CRE gene expression and prevent LoxP-mediated recombination of the targeted locus during embryogenesis of the offsprings, doxycycline can be supplied in the drinking water. Upon the removal of doxycycline, CRE gene expression is desinhibited, resulting in LoxP-mediated recombination to disrupt the target gene. A drawback of this system has been the reported leakiness of CRE gene expression in the repressed state.

In the Tet-On version of this system, CRE gene expression from the tetO-minimal promoter is induced rather than repressed by the administration of doxycycline (Fig. 2.9). Hereby, a mutant form of TetR, rtTA, which normally represses the tetO-minimal promoter. Following the addition of doxycycline, rtTA is released from the tetO allowing expression of the CRE gene. Also this systems, as the reverse one described before, lacks a tight repression of the tetO-minimal promoter.

2.4 Transgenic Mice

Whereas a knockout mouse has a gene deleted, transgenic mice may have a new gene added or an extra copy of an existing gene in order to investigate excessive gene expression. Transgenic techniques begin with the development of the fusion gene construct in which the DNA transgene is driven by a specific promoter sequence [27, 28]. The transgene DNA construct is then microinjected into the pronuclei of fertilized mouse oocytes harvested from a hormone treated, super-ovulating donor female mouse (Fig. 2.10). The egg cell is large enough to be seen under the microscope. Through random recombination events, the fusion gene construct becomes integrated into the genome. When the transgene is integrated into the genome before the first cell division, the embryo develops with the foreign gene contained in every somatic cell and germ-line cell. The microinjected eggs are transferred to the oviducts of pseudopregnant, genetically normal female mice (for details see above). The mouse that develops from each microinjected egg is the founder of the mutant line. The transgene is likely to integrate into a different chromosomal location in the genome in each founder mouse. Using tissue specific promoters, the site of incorporation is attempted to be defined. The modified DNA construct inserts at a chromosomal location homologous with the DNA of the promoter. The transgene inserts into only one chromosome. The embryo thus develop as a heterozygote for the transgene. The heterozygotes are then mated with each other to generate homozygous transgenics, heterozygotes, and wildtype littermate controls, following Mendelian distribution. The DNA constructs often contains a reporter gene driven under the same promoter, so that reporter gene positive cells indicate the presence of the transgene in the cell. Concentrations of the reporter gene, the transgene, and the gene product are assayed to determine the overexpression level of the gene in the tissue of interest. Anatomical mapping

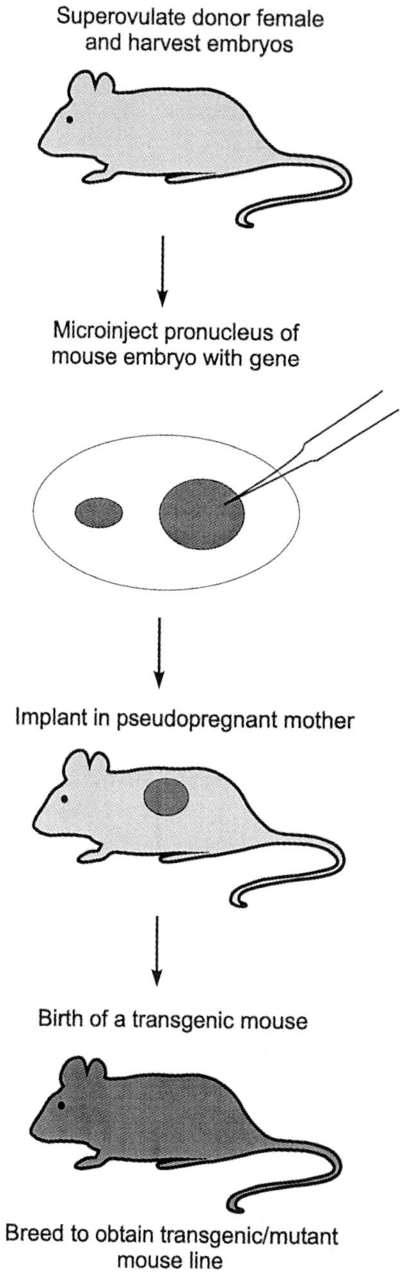

Fig. 2.10 Flowchart scheme of the production of transgenic mice.

of the localization of the reporter gene, the transgene, and the gene product is performed to describe the anatomical distribution of the expression of the transgene during the development of the animal.

2.5 Targeted Gene Disruption in *Drosophila*

The completion of the genome sequence [29] provides unlimited access to all genes of *Drosophila melanogaster.* Nevertheless, despite nearly a century of *Drosophila* genetics, there are many genes for which corresponding mutants are still unavailable. Means to overcome the drawback had been site-selected transposon mutagenesis and RNA-mediated interference. While transposon mutagenesis involves elaborate PCR screening, RNAi generates only gene-specific phenocopies of loss-of-function mutations and does not always cause a true null phenotype. Therefore, methods of targeted gene knockout have been developed. Gene targeting in *Drosophila* is accomplished by an elegant technique that takes advantage of the fly's endogenous homologous recombination machinery in the germline [9]. The technique resembles knockout targeting in mouse ES cells and can target a mutation to any arbitrary locus in the *Drosophila* genome. The method involves three components: i) a transgene that expresses a heat shock inducible-site-specific recombinase (FLP recombinase from yeast), ii) a second transgene that expresses a heat shock-inducible site-specific endonuclease (I-SceI), iii) a transgenic donor vector that carries recognition sites for both enzymes (two 18-base pairs FRT sites and one I-SceI site, respectively) in addition to the wild-type target gene, and the *white* gene as a positive selection marker (allows screening of the offsprings for white eyes). The first step is the random introduction of heat shock-inducible site specific I-SceI endonuclease and FLP recombinase using the P element system. The P-element is a mobile genetic element encoding a transponase which enables the P-element to insert into DNA. The donor vector (Fig. 2.11) is also randomly inserted into the genome by P-element-mediated transformation. For recombination, flies bearing both the heat shock inducible I-SceI and FLP transgenes are crossed to flies that carry the gene to be targeted flanked by FRT recognition sites.

Through heat shock, the site-specific FLP recombinase and the site-specific endonuclease (I-SceI) then excise *in vivo* an extrachromosomal DNA molecule that carries a recombinogenic double-strand break (DSB) within the gene of interest. The presence of this DSB stimulates homologous recombination between the excised donor and the homologous chromosomal target locus (Fig. 2.11). Several classes of recombination at the target locus including allelic substitutions and integration of donor DNA are observable in female gametes. In many cases, the insertional product is a tandem partial duplication of the target gene, with both copies defective because they each lack a portion of the gene (Figure 11). Since the

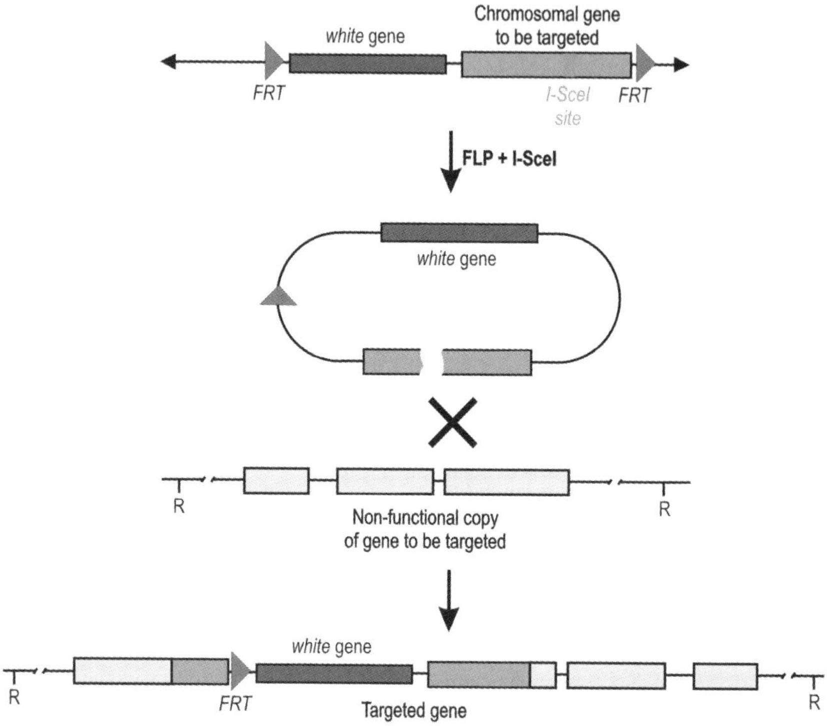

Fig. 2.11 Gene targeting in *Drosophila*. The transgenic targeting vector contains the gene to be targeted in a truncated version with a rare site for the endonuclease I-Sce-I, together with the white gene allowing selection for white eyes in *Drosophila*. The construct is flanked by two FRT sites that are recognized by the FLP recombinase. FLP recombinase-mediated excision and I-SceI endonuclease-mediated cutting releases the extrachromosomal targeting DNA from the transgenic donor vector. This targeting DNA is expected to recombine with the endogenous gene locus to produce a non-functional tandem duplication.

recombination event occurs in the germline, progeny are obtained with the target gene knocked out in every cell.

2.6 Targeted Gene Knockdown in Zebrafish

Recently the genome of the 2 to 4 cm sized zebrafish has been completely sequenced. The haploid genome comprises 1.7×10^9 bp compared to 3×10^9 in human. Comparative mapping shows extensive conservation of blocks of spatially conserved gene order on chromosomes (syntenies) between zebrafish and mammalian genomes. Direct assignment of function of the majority of the 30 000 genes

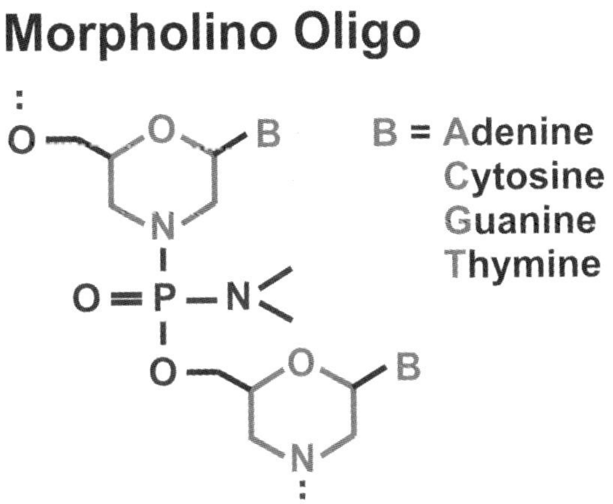

Fig. 2.12 Chemical structure of a morpholino oligonucleotide. The segment of a morpholino oligonucleotide comprising two subunits joined by an intersubunit linkage is shown.

in the zebrafish genome on the basis of this information is facilitated by the development of the rapid, targeted knockdown morpholino technology that can be applied to this model organism. Morpholinos (Fig. 2.12) are modified antisense oligonucleotides which are effective and specific translational inhibitors when injected into macroscopic fertilized zebrafish eggs [30]. They bind to and inactivate selected mRNA sequences. The oligonucleotides are assembled of four subunits, each of which contains one of the four genetic bases adenine, cytosine, guanine and thymine linked to a 6-membered morpholino ring which conveys them a high nuclease resistance. Morpholinos contain 21 to 25 bases in a specific order. The oligonucleotides are designed to bind to the 5′ untranslated regions flanking and including the initiating methionine of a specific mRNA. They act via a steric block of the translational initiation complex, and due to their high affinity and sequence specificity, they yield reliable and reproducible phenotypic results. For single and multiple gene knockdown, morpholinos are injected into the yolk of 1 to 4 cell stage embryos. The embryos develop externally to the mother and therefore, are not resorbed if they die during embryogenesis due to lethality caused by the gene knockdown. After fertilization of the eggs, the basic body plan of the animal including all organs develops within 24 hours (equivalent to about 9 days in the mouse). The embryos are also completely transparent which facilitates observation of cell fates. Embryos were analysed morphologically 1–5 days after injection of morpholinos, i.e. at a developmental stage when the zebrafish already swimms freely (being the case within three days). In contrast to gene targeting in mice, which normally requires months or even years for the determination of gene function, morpholino-based targeted gene inactivation in zebrafish enables the determination of gene function within days. The animal model is particularly suitable for genes critical for vertebrate processes such as somitogenesis and organogenesis.

2.7
Targeted *Caenorhabditis Elegans* Deletion Strains

The genome of this nematode has also been sequenced and consists of six chromosomes with in total 97 Mega base pairs (haploid size) encoding 19 099 genes. The adult nematode is formed by exactly 959 somatic cells including 302 neurons and 95 muscle cells. *C. elegans* has a very short generation time, the length of time needed to develop from a fertilized egg into a mature adult. Three days after the egg is fertilized, the worm can produce offsprings by itself. The gene knockout strategy often involves the use of the random mutagen trimethylpsoralen to mutagenize a very large number of worms [31]. Exposition to the mutagen results in random mutations in germ cells. For example, one sperm could be mutated for a specific gene. Fertilization of an egg by this sperm will result in a heterozygous individual. Because the worm is a self-fertilizing hermaphrodite, it will produce eggs and sperm that bear this mutated gene; one-quarter of its next progeny will be homozygous for the mutation and result in a specific phenotype. Such an animal can be inspected within three days whether the mutant phenotype breeds through.

The screening of worms exposed to a mutagen is performed as follows: the worms are divided into many small subcultures and allowed to have progeny. A portion of each subculture is stored alive in the freezer, and genomic DNA is made from the rest of the culture. This DNA is thus made from the siblings of the frozen worms and carries the same mutations as the frozen worms do. At a very low frequency (about 1 of 200 000 mutagenized genomes) the mutagenesis will produce a small deletion (100–1000 bp) in any gene. If PCR primers flanking an area of the gene of interest are used to amplify from the genomic DNA samples, deletions between the primers can be detected since they will bring the primer sites closer together and thus will generate a PCR amplicon smaller in size than that amplified from wild-type genomic DNA. Thus DNA representing several thousands mutagenized genomes can be screened and a deletion amplicon generated from just one of those genomes will still be detected. Once a DNA pool containing a specific deletion is thus identified, one can work back to identify the subculture of worms in which the deletion occured, the frozen worms from that subculture can be thawed, and live animals carrying the deletion mutation can be identified. Individual mutants are detected and can be recovered and bred further to study gene function. About 10% of the *C. elegans* genes have been defined by mutations (at the end of 2002). One-third of these have a mammalian counterpart, and understanding their functions in *C. elegans* will help us to determine their functions in more complex animals.

References

1 Capecchi MR. Targeted gene replacement. (1994) *Sci Am* **270**, 52–59.

2 Smithies O. Animal models of human genetic diseases. (1993) *Trends Genet* **9**, 112–16.

3 Malakoff D. The Rise of the Mouse, Biomedicine's Model Mammal. (2000) *Science* **288**, 248–53.

4 Metzger D, Chambon P. Site- and time-specific gene targeting in the mouse. (2001) *Methods* **24**, 71–80.

5 Jaenisch R. Transgenic animals. (1988) *Science* **240**, 1468–74.

6 Lankenau DH. Genetics of genetics in Drosophila: P elements serving the study of homologous recombination, gene conversion and targeting. (1995) *Chromosoma* **103**, 659–68.

7 Gloor GB. Gene-targeting in Drosophila validated. (2001) *Trends Genet* **17**, 549–51.

8 Adams MD, Sekelsky JJ. From sequence to phenotype: reverse genetics in Drosophila melanogaster. (2002) *Nat Rev Genet* **3**, 189–98.

9 Rong YS, Titen SW, Xie HB, Golic MM, Bastiani M, Bandyopadhyay P, Olivera BM, Brodsky M, Rubin GM, Golic KG. Targeted mutagenesis by homologous recombination in D. melanogaster. (2002) *Genes Dev* **16**, 1568–81.

10 Brenner S. The genetics of Caenorhabditis elegans. (1974) *Genetics* **77**, 71–94.

11 Horvitz HR, Sulston JE. Joy of the worm (1990) *Genetics* **126**, 287–92.

12 Jorgensen EM, Mango SE. The art and design of genetic screens: caenorhabditis elegans. (2002) *Nat Rev Genet* **3**, 356–69.

13 Nasevicius A, Ekker SC. The zebrafish as a novel system for functional genomics and therapeutic development applications. (2001) *Curr Opin Mol Ther* **3**, 224–8.

14 Summerton J. Morpholino antisense oligomers: the case for an RNase H-independent structural type. (1999) *Biochim Biophys Acta* **489**, 141–58.

15 Udvadia AJ, Linney E. Windows into development: historic, current, and future perspectives on transgenic zebrafish. (2003) *Dev Biol* **256**, 1–17.

16 Blake JA, Richardson JE, Bult CJ, Kadin JA, Eppig JT, and the members of the Mouse Genome Database Group. MGD: The Mouse Genome Database. (2003) *Nucleic Acids Res* **31**, 193–5.

17 Thomas KR, Capecchi MR. Site-directed mutagenesis by gene targeting in mouse embryo-derived stem cells. (1987) *Cell* **51**, 503–12.

18 Nagy A, Rossant J, Nagy R, Abramow-Newerly W, Roder JC. Derivation of completely cell culture-derived mice from early-passage embryonic stem cells. (1993) *Proc Natl Acad Sci USA* **90**, 8424–8.

19 Zhang H, Hasty P, Bradley A. Targeting frequency for deletion vectors in embryonic stem cells. (1994) *Mol Cell Biol* **14**, 2404–10.

20 Thomas KR, Deng C, Capecchi MR. High-fidelity gene targeting in embryonic stem cells by using sequence replacement vectors. (1992) *Mol Cell Biol* **12**, 2919–23.

21 Abremski K, Hoess R. Bacteriophage P1 site-specific recombination. Purification and properties of the Cre recombinase protein. (1984) *J Biol Chem* **259**, 1509–14.

22 Gu H, Zou YR, Rajewsky K. Independent control of immunoglobulin switch recombination at individual switch regions evidenced through Cre-loxP-mediated gene targeting. (1993) *Cell* **73**, 1155–64.

23 Metzger D, Chambon P. Site- and time-specific gene targeting in the mouse. (2001) *Methods* **24**, 71–80.

24 Feil R, Brocard J, Mascrez B, LeMeur M, Metzger D, Chambon P. Ligand-activated site-specific recombination in mice. (1996) *Proc Natl Acad Sci USA* **93**, 10887–90.

25 Kistner A, Gossen M, Zimmermann F, Jerecic J, Ullmer C, Lubbert H,

Bujard H. Doxycycline-mediated quantitative and tissue-specific control of gene expression in transgenic mice. (1996) *Proc Natl Acad Sci USA* **93**, 10933–38.

26 Gossen M, Bujard H. Studying gene function in eukaryotes by conditional gene inactivation. (2002) *Annu Rev Genet* **36**, 153–73.

27 Palmiter RD, Brinster RL. Germ-line transformation of mice. (1986) *Annu Rev Genet* **20**, 465–99.

28 Rossant J, Nagy A. Genome engineering: the new mouse genetics. (1995) *Nat Med* **1**, 592–4.

29 Adams MD, Celniker SE, Holt RA, Evans CA, Gocayne JD, Amanatides PG, Scherer SE, Li PW, Hoskins RA, Galle RF. et al. The genome sequence of Drosophila melanogaster. (2000) *Science* **287**, 2185–95.

30 Heasman J. Morpholino oligos: making sense of antisense? (2002) *Dev Biol* **243**, 209–214.

31 Jansen G, Hazendonk E, Thijssen KL, Plasterk RH. Reverse genetics by chemical mutagenesis in Caenorhabditis elegans. (1997) *Nat Genet* **17**, 119–21.

3
Reporter Gene Assay Systems for the Investigation of G-protein-coupled Receptors

Michaela C. Dinger and Annette G. Beck-Sickinger

3.1
Receptors and Cellular Communication

Higher organisms, e.g. humans, consist of millions of cells that are important for the regulation of physiological functions. Accordingly, cells have to communicate with each other and a dysfunction of cell communication can lead to several diseases. In order to control cellular function, extracellular signaling molecules and a complementary set of receptor proteins have considerable influence [1]. Each cell responds to a specific set of signals that act in various combinations to regulate the behavior of the cell. Because of the unique three-dimensional structure of a receptor protein, only a few signaling molecules are able to bind to a given receptor. Therefore, each cell acts in a characteristic and programmed way. Most extracellular signaling molecules such as adrenaline, serotonin, dopamine and neuropeptide Y are hydrophilic, and can activate receptor proteins only on the surface of the target cell. Accordingly, the receptors transduce signals by converting the extracellular binding event into intracellular signals that alter the behavior of the target cell. There are three major families of cell surface receptors. All of them belong to the large family of transmembrane proteins and each family tranduces extracellular signals in a different way. Ion channel-linked receptors, enzyme-linked cell-surface receptors and G-protein coupled receptors (GPCRs) are reported as transmembrane receptors.

3.1.1
Ion Channel-linked Receptors

One class of cell-surface receptor proteins is the ion channel-linked receptors, also known as ligand-gated ion channels (LGICs). These receptors are very different from GPCRs and are involved in rapid synaptic signaling via the movement of ions through channels gated by neurotransmitters. Typically, ligand-gated ion channels are closed in the resting state, but open in response to an agonist. After activation, the channels undergo spontaneous desensitization. The resulting change in membrane potential represents a signal that can be further processed by the receiving

Molecular Biology in Medicinal Chemistry. Edited by Th. Dingermann, D. Steinhilber, G. Folkers

ISBN: 3-527-30431-2

cell. At present, based on pharmacological and sequence homology criteria, three superfamilies of ligand-gated ion channels are known – the nicotinicoid superfamily, excitatory glutamate receptors and ATP receptors.

The nicotinicoid superfamily consists of several families of receptors including the nicotinic acetylcholine receptors, serotonin 5-hydroxytryptamine $(HT)_3$ receptors, γ-aminobutyric acid $GABA_A$ and $GABA_c$ receptors, and strychnine-sensitive glycine receptors [2]. The most studied and best-understood receptors are the GABA and acetylcholine receptors. They are considered to be pentamers composed of protein subunits and characterized by a large extracellular N-terminal domain.

The second receptor superfamily, the excitatory glutamate receptors, is abundant in most regions of the mammalian central nervous system. Three families – *N*-methyl-D-aspartate (NMDA) receptors, α-amino-3-hydroxy-5-methyl-4-isoxazole propionic acid (AMPA) receptors and kainate receptors – are members of this superfamily [3]. The native receptors consist of four or five subunits that assemble as homomers, doublets or triplets.

The ATP P_{2x} receptor subtype is a member of the third superfamily, and is widely distributed within the central nervous system and the cardiovascular system [4].

3.1.2
Enzyme-linked Cell-surface Receptors

Enzyme-linked receptors are transmembrane proteins with their ligand-binding domain on the outer surface of the plasma membrane. Their cytosolic domain either has an intrinsic enzyme activity or is associated directly with an enzyme. Each subunit of a catalytic receptor has only one transmembrane domain.

The most important receptors of this family are the tyrosine kinase receptors and tyrosine kinase-associated receptors. Many known receptors such as epidermal growth factor (EGF) receptor [5], insulin receptor, platelet-derived growth factor (PDGF) receptor and vascular endothelial cell growth factor (VEGF) receptor belong to the group of receptor tyrosine kinases [6, 7]. These receptors are single transmembrane helices that dimerize after ligand binding. One exception is the insulin receptor, which is already a dimer in the transmembrane. After dimerization, autophosphorylation of the receptor takes place and causes phosphorylation of a variety of signaling molecules on specific tyrosine residues. For example, Ras activity is stimulated, and thus the level of second messenger is increased that modulates transcription factors and other cellular proteins. Receptor tyrosine kinases play an important role in the regulation of cell metabolism, cell proliferation and cell differentiation.

Cytokine receptors belong to the tyrosine kinase-associated receptors. They consist of an extracellular ligand-binding domain, a single transmembrane domain and an intracellular domain. The intracellular domain possesses no catalytic activity. The interleukin receptors are examples for tyrosine kinase-associated receptors. After ligand binding, a tyrosine kinase that is associated to the receptor is activated by phosphorylation. Examples for these tyrosine kinases are Src or Jak kinases. The activation of tyrosine kinases results in the phosphorylation of signal transducers and activators of transcription (STAT) proteins that act as transcription factors.

Furthermore, serine/threonine receptors are also known as enzyme-linked cell-surface receptors, and are involved in many cellular processes like proliferation, migration and differentiation. These receptors consist of a small extracellular region, a single transmembrane domain and a cytoplasmic region that acts as a serine/threonine kinase. The transforming growth factor (TGF)-β receptor represents one member of this family [8, 9].

Another family of enzyme-linked cell-surface receptors is made up of the guanylyl cyclase receptors. The atrial natriuretic peptide receptor is the most popular member of this family and catalyzes the formation of cGMP [10]. Major effects are natriuresis, diuresis and the inhibition of aldosterone synthesis.

3.1.3
GPCRs

GPCRs represent the largest and the most important family of transmembrane receptors characterized by amino acid sequences that contain seven hydrophobic domains. These hydrophobic domains represent the transmembrane-spanning regions of the proteins and gives rise to the second name for this superfamily – 7TM or heptahelix receptors [11]. In mammalians, this family contains more than 600 members. They are involved in a broad spectrum of biological processes by mediating the signals of a wide variety of stimuli such as peptide hormones (glucagon, angiotensin, bradykinin), neurotransmitters (adrenalin, serotonin, dopamine), neuropeptides (neuropeptide Y) as well as light and odorants. Accordingly, they play an important role in physiological regulation and their malfunction can result in many diseases.

In general, ligand binding to the GPCRs results in a conformational change that leads to the association of an intracellular G-proteins [12]. These G-proteins are members of the superfamily of GTPases and are characterized by a heterotrimeric composition – the α, β and γ subunits. Structural and functional classification of G-proteins has been defined by the α subunit that can be divided in the main class $G_{\alpha s}$-, $G_{\alpha i}$-, $G_{\alpha o}$- and $G_{\alpha q}$-proteins (Fig. 3.1). The association of a G-protein to the transmembrane receptor provokes conformational changes in the G-proteins that facilitate GDP release and GTP binding. The GTP binding then causes the α subunit to separate from the β and γ subunits. The $G_{\alpha s}$-protein, known as stimulating G-protein, activates the adenylyl cyclase (AC) that catalyses the production of the second messenger cAMP. In contrast, the $G_{\alpha i}$ protein inhibits the AC leading to a low cAMP level (Fig. 3.1). Activation of the $G_{\alpha q}$-protein is followed by stimulation of phospholipase C that induces the production of the second messengers diacyl glycerol or inositol triphosphate (IP_3) (Fig. 3.1), whereas $G_{\alpha q}$ interacts with K^+ channels. Activation of second messenger results in tremendous signal amplification [13–15]. For example, cAMP can activate the protein kinase A that stimulates the phosphorylation of other proteins (Fig. 3.1). Furthermore the $\beta\gamma$ subunit of the G-protein is known to activate the mitogen-activated protein (MAP)

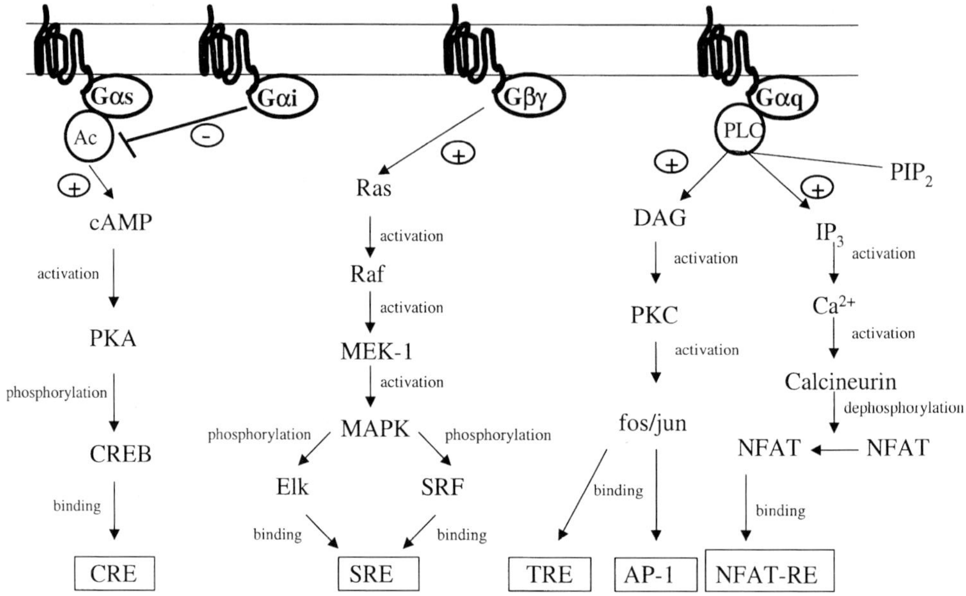

Fig. 3.1 Schematic diagram of GPCR signal transduction pathways showing how, after stimulation of the GPCR and dissociation of G-protein subunits, the major G-protein families signal via different intracellular second messenger pathways to communicate with nuclear promoter elements. (AC, adenylyl cyclase; CRE, cAMP response element; CREB, CRE binding protein; DAG, diacyl glycerol, IP_3, inositol triphosphate; MAPK, mitogen-activated protein kinase; MEK, MAPK kinase; NF-AT, nuclear factor of activated T cells; NF-AT-RE, NF-AT response element; PKA, protein kinase A; PKC, protein kinase C; PLC, phospholipase C; PIP_2, phosphatidylinositol-4,5-bisphosphate; SRE, serum response element; SRF, serum response factor; TPA, 12-*O*-tetradecanoyl-phorbol-13-acetete; TRE, TPA-response element.)

kinase (MAPK) pathway (Fig. 3.1). Recently, in addition to the classical pathways, G-protein-independent pathways have been described, which means GPCRs can activate effector molecules like extracellular signal-related kinase (ERK) 2 or STATs without stimulation of the G-protein [16, 17].

GPCRs have been classified into three main subfamilies: the rhodopsin subfamily (class 1), the calcitonin subfamily (class 2) and the glutamate metabotropic receptor subfamily (class 3) [18]. The class 1 family contains approximately 90% of all GPCRs and is characterized by a N-terminal segment that is highly glycosylated, and possesses a characteristic disulfide bridge between the extracellular loop 1 and extracellular loop 2. Odorants, peptides and other small molecules as well as large glycoproteins bind to receptors of the class 1 family such as adrenaline receptors, serotonin receptors or neuropeptide Y receptors. The most prominent member of this family is rhodopsin, which gives the whole family its name. The crystal structure of rhodopsin was solved in 2000 by Palczewski et al. and is the only available high-resolution structure of a class 1 GPCR [19].

Mainly peptide hormones and neuropeptides such as secretin, glucagon and calcitonin bind to receptors of the class 2 GPCR family. They have a large N-terminal extracellular domain in common that contains six cysteine residues.

The third and, so far, the smallest group of GPCRs (class 3) is characterized by a huge N-terminal segment domain constituted of two lobes that close like a Venus flytrap upon ligand binding. The metabotropic glutamate receptors and $GABA_B$ receptors belong to this receptor subfamily.

Further receptors for fungi and plant ligands have also been characterized and classified into special subtypes.

GPCRs are key proteins in the physiological regulation of cells and their malfunction causes several diseases as elucidated from studies with GPCR knockout mice [20]. At present, GPCRs are the targets of more than 50% of the current therapeutic agents and, as such, the pharmaceutical industry is tremendously interested in the investigation of GPCRs [21]. In terms of the development of new drugs it is important to generate highly specific compounds that bind with a high affinity and high selectivity to one particular GPCR. Only these features guarantee enhanced therapeutic specificity tolerable side-effects.

3.2
Affinity and Activity of GPCR Ligands

As mentioned above, GPCRs are targets for a variety of drugs, and the affinity and the activity of a ligand to its receptor has to be known in order to develop new drugs. The property of affinity describes the avidity or the tenacity with which a ligand will bind to its receptor. Activity, also called intrinsic efficacy, has been viewed as the property of a ligand that can produce a biological response. Potential new drugs should possess high affinity and high selectivity. High selectivity guarantees a reduction of side-effects and high affinity allows a reduction of the dose of drug. The natural ligand of a GPCR is often used as template for drug design and new ligands with structural similarity are generated.

Conventional competition assays with a radiotracer are generally used in order to determine the affinity of a ligand to a receptor. Whole cells or cell membranes expressing GPCRs are incubated with a ligand that is radiolabeled in the presence of the unlabeled, investigated ligand. A constant concentration of radiotracer and different concentrations of unlabeled ligand are used. After reaching the equilibrium, the cells or membranes are separated from the supernatant and the radioactivity from the bound radiolabeled ligand can be measured with a counter. The measured radioactivity is proportional to the concentration of radiolabeled ligand that binds to the receptor. A binding curve can be determined by using different concentrations of unlabeled ligand in the presence of a constant radiotracer concentration and an IC_{50} value can be obtained and transformed to the K_i value according to the Cheng-Prussof equation. The K_i reflects the affinity of a ligand to a GPCR – the lower the K_i, the higher the affinity.

In order to determine the intrinsic efficacy of ligands, their agonistic or antagonistic potential, various assays have to be performed that measure the activation of the G-proteins or the intracellular second messenger upon ligand binding. The investigated second messenger depends on the receptor-activated G-protein. These assays are mostly very time-consuming and expensive. First, it is possible to measure GTP binding after the activation of the receptor through its ligand. Cells or membranes expressing the receptor are incubated in the presence of radiolabeled GTP and the investigated ligand. The radioactivity can be measured after the separation of cells or membranes from the supernatant. Measured radioactivity correlates with the stimulation of the receptor that associates with a G-protein.

Conventional ELISA assays can be used in order to measure second messengers like cAMP, cGMP, diacyl glycerol or IP_3. The concentration of the second messenger cAMP is dependent on the type of the activated G-protein and can result in either an increase ($G_{\alpha s}$) or a decrease ($G_{\alpha i}$) of second messenger.

The determination of the signal on the level of the second messenger is sometimes very difficult as not all cells express the optimal G-proteins or the second messengers are immediately degraded. Therefore, reporter genes in combination with special promoters and enhancer elements are used to generate novel assay systems. After the binding of a ligand to a GPCR, the concentration of second messengers, the activities of enzymes and consequently the binding of transcription factors to the enhancer elements are modulated. Accordingly, reporter gene expression is directly dependent on the activation of a GPCR and its activated signal transduction pathway. The resulting signal is more stable and easier to quantify.

3.3
The Role of Transcription Factors in Gene Expression

The information required for the synthesis of a specific protein is encoded in the sequence of nucleotides in DNA. This information must be transcribed into RNA, the messenger molecule, which is then translated in the cytoplasm by the protein synthesis machinery. Specialized enzymes called RNA polymerases catalyze the process of transcription of DNA into RNA. RNA polymerase II transcribes the majority of the genes; however, many other accessory proteins that interact with the polymerase are essential for the transcription of genes. Numerous of nuclear transcription factors are known and their function is either to activate or to inhibit gene expression [22–24]. They bind to short sequence motifs of double-stranded DNA regions that consist of variable lengths, usually 5–20 bp. These short sequence motifs are located upstream of the core promoter and generally specific to a single transcription factor or to a member of a single transcription factor family. Transcription factors can bind as monomers, homodimers or heterodimers to a specific motif. After binding to their specific binding site, the transcription factors interact with each other and with RNA polymerase II and other basal factors in order to regulate gene expression.

At present, more than 300 transcription factors have been identified in the vertebrate genome. The activity of some transcription factors is modulated by their phosphorylation [25]. Well-understood and investigated transcription factors include cAMP response element binding protein (CREB) [26], serum response element binding factor (SRF), nuclear factor of activated T cells (NF-AT) [27, 28], c-Jun and STAT proteins [29, 30]. The recognition sites of these transcription factors are often used in promoters for the establishment of reporter gene assay systems for the investigation of GPCRs (Fig. 3.1).

3.3.1 CREB

CREB is involved in the cAMP pathway. Upon activation of $G_{\alpha s}$-protein-coupled receptors, that positively regulate AC, the intracellular cAMP level is increased. cAMP itself activates protein kinase A (PKA), enabling the catalytic unit of PKA to enter the nucleus where it phosphorylates CREB at the serine in position 133. The phosphorylated CREB binds to the enhancer element cAMP response element (CRE). The transcription of the linked gene is activated after the binding of CREB-binding protein (CBP) to CREB. Stimulation of $G_{\alpha i}$-coupled GPCRs decreases intracellular cAMP and causes a reduction of CRE-linked reporter gene transcription.

3.3.2 SRF

The MAPK pathway is a very important signaling pathway for receptor tyrosine kinases and is necessary for the phosphorylation of SRF. SRF itself binds to the serum response element, a DNA-binding site present, e.g. in the c-Fos promoter, which leads to the transcription of c-Fos, another transcription factor. Only in combination with the ternary complex factor (TCF/Elk1) can SRF bind to SRE and activate gene transcription. Both, the transcription factors SRF and TCF are activated through MAPK phosphorylation. The $\beta\gamma$ subunits of $G_{\alpha i}$-proteins have been shown to activate this pathway via RAS-dependent MAPK activation. Stimulation of MAPK via $G_{\alpha q}$-coupled receptors results mainly from protein kinase C (PKC) activation.

3.3.3 STAT Proteins

The transcription factors STAT proteins are activated after the stimulation of cytokine receptors and receptor tyrosine kinases. At present, six STAT proteins (STAT1–STAT6) are known. After receptor activation through ligand binding these receptors associate with the tyrosine kinase JAK kinase, which in turn is phosphorylated. Subsequently, activated JAK kinases can activate STAT proteins again through phosphorylation followed by dimer formation. The dimers are translocated into the nucleus and bind to the DNA binding site interferon-stimulated response element (ISRE), and gene transcription is initiated. Recent studies demonstrate the activation and involvement of the Stat3 pathway in $G_{\alpha i}$- and $G_{\alpha o}$-protein-mediated pathways [31].

3.3.4
c-Jun

The transcription factor c-Jun is phosphorylated by a protein kinase, the JNK (c-Jun-N-terminal kinase), which is a new member of the MAPK family. This kinase also phosphorylates the transcription factor ATF2. ATF2 and c-Jun together a heterodimer that binds to the DNA binding site TPA-response element (TRE) in the c-Jun promoter.

3.3.5
NF-AT

NF-AT is found as transcription factor in lymphocytes and is induced by antigen receptor stimulation. NF-AT is phosphorylated and located in the cytoplasm in resting cells. The control of NF-AT-mediated transcription in lymphocytes is dependent on simultaneous signaling through Ca^{2+} and the MAPK that are activated after stimulation of the antigen receptor. The Ca^{2+} pathway activates the phosphatase calcineurin. Calcineurin dephosphorylates NF-AT and leads to a nuclear translocation of NF-AT. NF-AT then assembles in a multimeric complex with AP1 and binds to a major upstream enhancer in the IL-2 cytokine gene. The transcription of genes is now possible. Distinct isoforms of NF-AT are found in human and mice, which are expressed in nonlymphoid cells.

3.4
Reporter Genes

A reporter gene is a sequence of DNA whose product is synthesized in response to the activation of the investigated signaling cascade. The expression of reporter genes can be controlled by enhancer elements, the binding sites for specific transcription factors. The choice of a particular reporter depends on several conditions:

- The cell line used for the experiments. The endogenous activity of each cell is different and it is important to prove that the cell line used does not express the chosen reporter gene.
- The nature of the experiment, because different reporter genes have to be used for the investigation of the dynamic of gene expression and for the investigation of transfection efficiency in the same experiment.
- The choice of the reporter gene also depends on the detection method.

Reporter genes possess special features. They are characterized by a defined nucleotide sequence, low background activity, high sensitivity and the response must be an easily detectable phenotype. In order to perform measurements, the activity of reporter genes has to be quantitative, rapid, easy, reproducible and safe. Measurement methods that are mostly used are radioactivity, fluorescence, luminescence and colorimetric measurements. The reporter gene signal must be significant over background activity, which depends on the reporter chemistry, cell line used, instrumentation sensitivity and interference from chemical activities.

Genetic reporter systems were developed in order to study eukaryotic gene expression and regulation. Furthermore, they are used to study promoter and enhancer sequences, transacting mediators such as transcription factors, mRNA processing and translation. Reporter gene systems are also applied to monitor transfection efficiencies, protein–protein interactions, protein subcellular localization and as indicators of transcriptional activity within cells. Nowadays, reporter gene systems are applied as biological screens for drug discovery [high-throughput screening (HTS)] [32], for the investigation of signaling pathways and gene therapy. They are also used for the investigation of receptor–ligand interactions, especially for the investigation of GPCRs.

Several reporter genes are known, e.g. chloramphenicol acetyltransferase (CAT), β-galactosidase (β-Gal), β-glucuronidase, alkaline phosphatase (AP), secreted AP (SEAP), β-lactamase, luciferase and green fluorescent protein (GFP). Their advantages and disadvantages are listed in Tab. 3.1. Some of these reporter genes are used for the investigation of GPCRs.

3.4.1
CAT

CAT was the first reporter gene used to monitor transcriptional activity in cells [33]. CAT was isolated from *Escherichia coli* and is a trimeric protein that consists of three identical subunits, each with a molecular weight of 25 kDa [34]. The CAT protein is relatively stable in mammalian cells and without transfection there is no endogenous expression. CAT has the requisite properties to be used in transient assays designed to investigate accumulation of protein expression. The natural enzymatic reaction of CAT, the transfer of an acetyl group from acetyl-CoA to the 3-hydroxyl position of chloramphenicol, is used to perform the reporter assays [35]. Radiolabeled acetyl-CoA allows the transfer of the radiolabel to the chloramphenicol substrate. Product formation can be determined by physical separation using thin-layer chromatography (TLC). TLC separates the substrate and product, and the radioactivity can be observed by exposing the TLC plate to X-ray film or by phosphoimaging. The TLC assay allows visual confirmation of the reaction. Quantification is possible by scraping the TLC spots, extracting with an organic solvent and counting the samples in a scintillation counter. Other isotopic assays rely upon direct organic extraction of the more nonpolar products [36]. Alternatively, antibodies against CAT are available and allow quantification by Western blotting or ELISA.

3.4.2
β-Gal

β-Gal is a well-characterized bacterial enzyme and consist of four subunits, each with a molecular weight of 116 kDa. The catalysis of the hydrolysis of β-Gal sugars such as lactose is reported as a functional assay. The enzymatic activity in cell extracts can be assayed with various specialized substrates that allow enzyme activity

Tab. 3.1 Comparison of commonly used reporter genes.

Reporter gene	*Advantages*	*Disadvantages*
Chloramphenicol acetyltransferase (CAT)	no endogenous expression in mammalian cells visual confirmation of enzyme activity	high cost for isotope and TLC systems radioactivity cell lysis additional substrates
β-Galactosidase (β-Gal)	variety of assay formats for use with cell extracts easy to assay	colorimetric assays not very sensitive endogenous activity in some cell types cell lysis
β-Glucuronidase	variety of assay formats	endogenous activity in mammalian cells cell lysis
Alkaline phosphatase (AP)	variety of assay formats easy to assay inexpensive	endogenous activity in mammalian cells colorimetric assay not very sensitive cell lysis
Secreted alkaline phosphatase (SEAP)	secreted protein variety of assay formats low background	addition of substrate
β-Lactamase	secreted form available variety of assay formats sensitive (fluorometric assay)	addition of substrate
Luciferase	fast and easy high sensitivity no endogenous activity	cell lysis addition of substrate relatively labile protein
Green fluorescent protein (GFP)	GFP variants can be detected in the same cell investigation in living cells autofluorescence (no substrates required)	sensitive fluorescence reader

quantification with a spectrophotometer, fluorometer or luminometer. Simple colorimetric assays with o-nitrophenyl-β-D-galactopyranoside (ONPG) or chlorophenol red β-D-galactopyransoide (CPRG) [37] substrates have been developed. Colorimetric assays, which are not very sensitive and have only a narrow dynamic range, have largely been replaced with more sensitive fluorescent- or luminescent-based assays. Substrates such as β-methyl umbelliferyl galactoside and fluorescein digalactoside (FDG) can be used for a fluorescent-based assay [38] and 1,2-dioxetane for the luminescent-based assays [39]. Reporter assays with β-galactosidase are widely used, especially to monitor the transfection efficiency.

3.4.3
β-Glucuronidase

β-Glucuronidase is a tetrameric glycoprotein from *E. coli* composed of four identical subunits of 68 kDa, localized predominantly in the acidic environment of the lysosomes. The enzyme removes terminal *β*-glucuronic acid residues from the nonreducing end of glycosaminoglycans and other glycoconjugates. Colorimetric assays that can be detected with a spectrophotometer were developed in order to measure the activity of this enzyme. Chemiluminescent assays were developed to improve the sensitivity. *β*-Glucuronidase is mostly used for investigations in higher plants [40] because of its endogenous activity in some mammalian cells [41].

3.4.4
AP

AP is a relatively stable protein and dephosphorylates a broad range of substrates at alkaline pH. The standard spectrophotometric assay is based on the hydrolysis of *p*-nitrophenyl phosphate (PNPP) by AP. This assay is inexpensive, rapid and simple, but lacks the sensitivity obtained with other methods. A two-step bioluminescent assay has been developed in order to improve the sensitivity [42]. In this two-step approach, AP first hydrolyses D-luciferin-*O*-phosphate to D-luciferin, which serves as a substrate for luciferase in the second step. Although the sensitivity is much higher, the handling is less convenient. A single-step chemiluminescent assay for AP has been developed as an alternative [43].

3.4.5
SEAP

SEAP is a mutated form of human placental AP, lacking 24 amino acids at the C-terminal end of the protein. Removal of these amino acids prevents the enzyme from anchoring to the plasma membrane, and it is consequently secreted from the cell and can be detected by sampling the culture medium [44]. Cells remain intact and viable for further experimentation. SEAP can be easily and cheaply detected by the hydrolysis of *p*-nitrophenol phosphate that causes a color change. This change can be quantified. Chemiluminescent assays are also used in addition to colorimetric assays. A major advantage of SEAP is that the enzyme is stable to heat and to the phosphatase inhibitor L-homoarginine. Therefore, the cell lysate can be treated with heat or L-homoarginine in order to inactivate the background AP activity within cells that may have entered the culture medium. Thus, the high background observed with the AP reporter system is essentially eliminated with the SEAP system. The combination of secreted reporter protein, low endogenous reporter background, and a wide variety of easy and sensitive assays make SEAP a convenient and versatile reporter system.

3.4.6
β-Lactamase

The 29-kDa monomeric protein *β*-lactamase (penicillin amido-*β*-lactamhydrolase) was isolated from *E. coli*. This enzyme has esterase activity, and its natural function is the cleavage of ampicillin and other antibiotic drugs, e.g. penicillin and cephalosporin. β-Lactamase has been engineered in three forms: secreted, cytosolic and plasma membrane associated [45]. Accordingly, enzyme activity can be assayed from cell extracts or, when secreted, directly from the medium. Colorimetric assays are applied for measurements, e.g. substrates like nitrocefin or [3-(2,4-dinitrostyryl)-(6*R*,7*R*)-7-(2-thienylacetamido)-ceph-3-em-4-carboxylic acid] (PADAC) change their color after reaction with β-lactamase. Enzyme activity with the PADAC substrate is monitored by a decrease in absorbence at 570 nm and with nitrocefin by an increase in absorbence at 570 nm. Recently, a highly sensitive fluorescent substrate for β-lactamase has been developed and as a membrane-permeable ester it can be applied to monitor gene expression in single living cells. This fluorogenic substrate CCF2/AM is a cephalosporin derivate attached to two fluorophores. When the fluorogenic substrate is intact, fluorescence resonance energy transfer (FRET) occurs between the two fluorophores [46]. When the substrate is cleaved by β-lactamase, the two fluorescent parts of the molecule are separated and the FRET effect is disrupted. The resulting wavelength shift is detectable in cell suspension using fluorescence-activated cell sorting, in cell monolayers using conventional plate readers and in individual cells using confocal microscopy. Reporter gene assays with β-lactamase are very sensitive using this fluorogenic substrate, but also expensive.

3.4.7
Luciferase

Luciferase refers to a family of enzymes that catalyze the oxidation of various substrates like luciferin and coelenterazine, which results in light emission. Known luciferases are the bacterial luciferase, the firefly (*Photinus pyralis*) luciferase, and more recently the renilla luciferase from the bioluminescent sea pansy, *Renilla reniformis*. The bacterial luciferases are generally heat-labile dimeric proteins. Accordingly, their application as reporter genes in mammalian cells is limited.

The most popular luciferase is the firefly luciferase. This 60-kDa enzyme, a monooxygenase, catalyses a reaction using D-luciferin and ATP, in the presence of oxygen and Mg^{2+}, resulting in light emission [47]. The total amount of light measured during a given time interval is proportional to the amount of luciferase activity in the sample. The X-ray structure of this luciferase was solved in 1996 [48] and arginine in position 218 was found to be essential as a luciferin-binding site [47]. More recently, reagents have been developed that prolong the half-life of the flash. Prolonged light outputs increase the sensitivity of the assay, and provide a higher half-life of the luminescence and thus a longer measurement period.

Luciferase from *R. reniformis* [49] catalyses the oxidation of coelenteramide and is mostly used in dual-luciferase assay formats. These assays allow the measurement of both firefly and renilla luciferase in the same cell extract. Therefore, firefly luciferase can be used as a reporter gene and renilla luciferase as an internal standard in order to quantify transfection efficiency [50].

Firefly luciferase is mostly used for the study of promoter regulation and regulatory effects of cAMP-elevating agents. At present, luciferase is still the reporter of choice for dynamic measurements of gene expression, because of its high sensitivity and relatively short half-life. It allowed, for example, the investigation of the circadian rhythms in plants [51]. Furthermore, reporter gene assays with firefly luciferase were generated for the investigation of GPCRs, and to identify receptor agonists and antagonists. This system is very popular in the pharmaceutical industry and is used for HTS.

3.4.8 GFP

GFP from the jelly fish *Aequorea victoria* is a 238-amino-acid photoprotein that emits green light with an emission maximum of 509 nm upon excitation at 488 nm [52]. The cDNA was cloned in 1991 [53] and the crystal structure was solved in 1996 [54]. The great benefit of GFP is that it is naturally fluorescent, thus GFP does not require any additional substrates and the fluorescence is not species specific. Furthermore, GFP can be used in living cells and for the generation of fusion proteins. The active chromophore in GFP required for fluorescence is a hexapeptide that contains a cyclized Ser–dehydroTyr–Gly trimer [55]. This chromophore is located in the centre of the protein, whose structure is referred as a β-can structure. Mutations within the active chromophore of the wild-type GFP led to the development of a GFP-variant with enhanced green fluorescence. A cyan fluorescent protein (CFP), a yellow fluorescent protein (YFP) and a blue fluorescent protein (BFP) were also developed as mutations. GFP expression can be directly monitored in living organism such as cells [56]. Other advantages are the stability to heat, extreme pH and chemical denaturants.

In recent years many fusion proteins with GFP have been constructed to investigate protein transport and localization [57, 58]. Imaging of gene expression is easily performed due to the properties of GFP [59]. The development of GFP and its spectral variants allows the simultaneous tracking of different proteins within living cells [60]. Furthermore, the investigation of protein–protein interaction is possible using the FRET technique, because GFP and its variants can be used as FRET pairs [61–63].

Experiments with GFP fusion proteins were performed to investigate GPCR regulation [64]. Recently, GFP was also used to generate a reporter gene assay system for the investigation of GPCRs. The identification of agonists and the determination of ligand affinity was possible [65].

3.5 Reporter Gene Assay Systems for the Investigation of GPCRs

In recent years reporter gene assay systems have been developed for the investigation of GPCRs that offer an alternative to conventional biochemical assays such as competition assays and cAMP assays. Investigation of the signal transduction pathways from receptors at the cell surface to nuclear gene transcription in cells is possible. The change of second messenger after activation of a GPCR can be easily measured by the increase or decrease of the reporter gene expression by using these assay systems. Previously, the level of intracellular second messengers had to be detected by special metabolite assays that were technically difficult, inconvenient, time-consuming and mostly expensive.

The development of reporter gene vectors with a promoter sequence and a reporter gene sequence is essential in order to investigate GPCRs with a reporter gene assay system. The choice of promoter depends on the nature of signaling pathway of the studied GPCR, and dictates the sensitivity and specificity of the reporter. The DNA-binding sites of transcription factors, called enhancer elements, are used in combination with a core promoter that controls the transcription of the reporter genes. Accordingly, the pathway of the studied GPCR has to be known (Fig. 3.1).

The disadvantage of natural promoters such as c-Fos is that they contain binding sites for a number of transcription factors and, thus, the regulation of such promoters depends on several transcription factors. In order to generate reporter gene assays for single signaling cascades, synthetic promoters containing only a single type of transcription factor binding sites have been constructed. The most popular synthetic promoter sequences consist of CRE elements, SRE elements, TRE elements and NF-AT enhancer element; the natural promotor c-Fos, the AP1 binding site and the GAL4 upstream activating sequence are also used for the investigation of GPCRs. Published data demonstrate that luciferase is the most popular reporter gene for the investigation of GPCRs. Furthermore, the application of β-lactamase and SEAP is also possible. Recently, GFP as reporter gene for the investigation of GPCRs has been reported, which has some advantages over the other reporter genes mentioned.

3.5.1 Application of Luciferase as a Reporter Gene

The firefly luciferase (Section 3.4.7) is the most widely used reporter gene for the investigation of GPCRs [66]. The luciferase gene is combined with different promoter elements such as CRE elements, TRE elements, c-Fos promoter and NF-AT enhancer element.

The enhancer element CRE is used for the investigation of $G_{\alpha s}$- and $G_{\alpha i}$-coupled receptors. Therefore, the expression of luciferase is regulated by the increase or decrease of cAMP. After the stimulation of a $G_{\alpha s}$-coupled receptor there is a high level of cAMP. cAMP causes phosphorylation of CREB followed by high expression of luciferase. In order to perform assays with $G_{\alpha i}$-coupled receptors, the assays are usually performed in the presence of forskolin, a stimulator of AC, to raise the cAMP level, so that the inhibition is easier to detect. The stimulation of forskolin itself is measured in parallel (Fig. 3.2).

Himmler et al. used the CRE and luciferase reporter gene system to investigate human dopamine D_1 and D_5 receptors, both $G_{\alpha s}$-coupled receptors [67]. They observed an increase of luciferase expression after stimulation with forskolin or

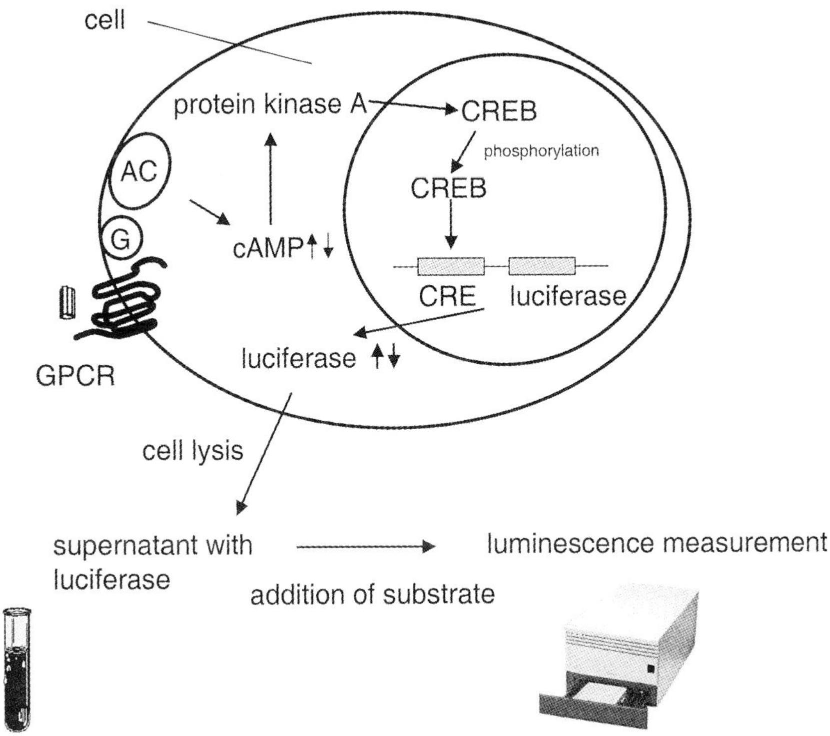

Fig. 3.2 Reporter gene assay system using CRE elements and luciferase for the investigation of GPCRs. Cells coexpressing a GPCR, either $G_{\alpha s}$- or $G_{\alpha i}$-coupled, were incubated with the termed ligand. After the stimulation or inhibition of AC there is an increase or decrease of cAMP, respectively. An increase of the cAMP level leads to the activation of PKA that activates CREB through phosphorylation. The phosphorylated form of CREB now interacts with CRE elements. Subsequently, the transcription of luciferase is initiated. For the investigation of $G_{\alpha i}$-coupled receptors, cells have to be incubated with the ligand in the presence of forskolin. Luciferase is expressed in the cytoplasm, and activity can be measured after cell lysis and addition of a substrate such as luciferin luciferase. A high luciferase activity causes strong luminescence. The luciferase activity can be correlated with the activity of the ligand that activates the GPCR. (AC, adenylyl cyclase; CRE, cAMP-binding element; CREB, CRE binding protein; G, G-protein.)

with dopamine agonists. The identification of agonistic or antagonistic compounds was successful. A similar assay was applied for the investigation of the binding behavior of different ligands to the serotonin, 5-HT_{1B} receptor and the calcitonin receptor C1a [68].

Luciferase in combination with CRE is also popular for use in HTSs in the pharmaceutical industry. The application of this system was reported from different groups, and assays were established for the rapid and sensitive investigation of G_i- and G_s-coupled receptors [32]. Assays for HTSs are mostly performed in microtiter plates.

In order to investigate $G_{\alpha s}$-, $G_{\alpha i}$- and $G_{\alpha q}$-coupled receptors with the same assay system, CRE elements were combined with MRE elements. This system also allows the investigation of $G_{\alpha q}$-coupled receptors [69]. This promoter region can respond to increased intracellular calcium, to stimulate PKC and to change the levels of cAMP. The response can be measured by the increase or decrease of the luciferase activity, and is useful for the pharmacological characterization of both agonists and antagonists.

Furthermore, other studies utilized the application of firefly and renilla luciferase in combination with CREs for the investigation of the human corticotropin releasing factor 1 and 2 receptors ($CRFR_1$ and $CRFR_2$) [50]. The identification of agonists and antagonists was successful.

In order to investigate receptors whose signal transduction pathway changes either intracellular cAMP levels or IP_3/DAG levels, a luciferase gene with the TRE elements as promoter was used as cellular assay system [70]. This system allowed the characterization of the different neurokinin receptors NK1, NK2 and NK3. Furthermore, another system with luciferase and TRE elements was developed for the screening of ligands [71]. The system was also coupled with GFP in order to select cells that stably express the reporter system. These cells were used for the investigation of GPCRs.

Agonists and antagonists of the gonadotropin-releasing hormone receptors (GnRH receptors) were successfully investigated with a luciferase reporter gene assay system in combination with the c-Fos promoter [72]. GnRH receptors are coupled to the inositol-phospholipid pathway and, after activation of the receptor, the transcription of the luciferase gene is induced by the c-Fos promoter.

3.5.2
Application of other Reporter Genes for the Investigation of GPCRs

The application of β-lactamase and SEAP for the investigation of GPCRs has also been reported. Pollok et al. developed a reporter gene assay system with β-lactamase as the reporter gene in combination with the NF-AT promoter [73] and used it to investigate GPCRs that are coupled to $G_{\alpha q}$-proteins. The change in second messenger concentration could be detected through the increase of β-lactamase expression. Pollok et al. also used this system to investigate a $G_{\alpha i}$-protein-coupled receptor. In order to perform these assays the signal of the $G_{\alpha i}$-protein was converted to a $G_{\alpha q}$-protein signal and β-lactamase activity could be measured.

The ligands of the GPCRs DP and EP prostanoid receptors were characterized with a CRE–SEAP assay [74]. Activation of the receptors could be detected by the change of intracellular SEAP levels.

A major disadvantage of the measurement of luciferase and β-lactamase activity is the need to prepare cell extracts and to add a substrate. These cells cannot be used for further investigation, the cell lysate preparation is time-consuming and the addition of a substrate can disturb the sensitive cell system. Accordingly, a reporter gene assay system with CRE elements and GFP as the reporter gene was developed for the investigation of GPCRs in living cells (Fig. 3.3). The advantage of

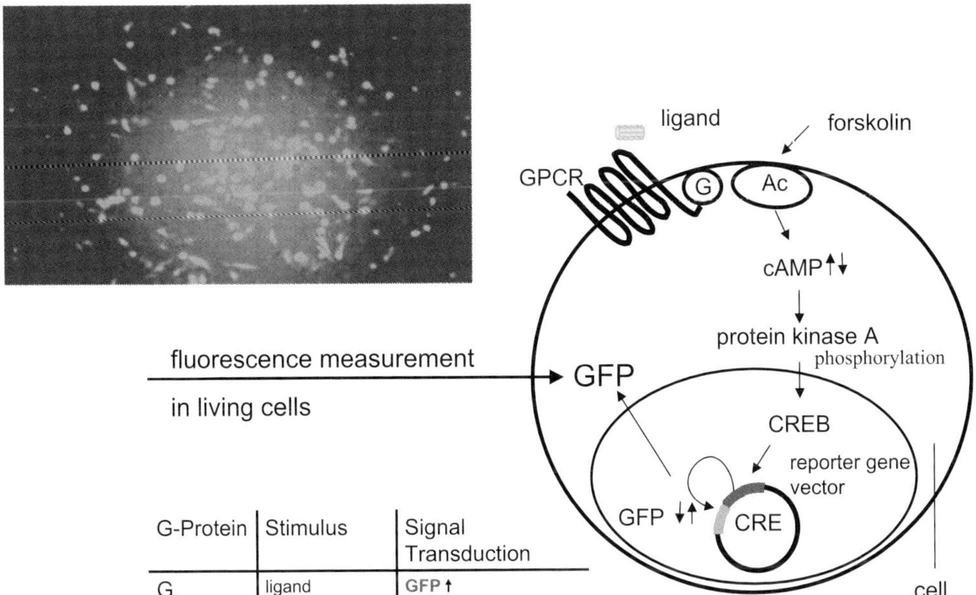

G-Protein	Stimulus	Signal Transduction
G_s	ligand	GFP ↑ green fluorescence ↑
G_i	forskolin	GFP ↑ green fluorescence ↑
	forskolin/ligand	GFP ↓ green fluorescence ↓

Fig. 3.3 Reporter gene assay system using GFP and CREs for the investigation of GPCRs. Cells expressing a GPCR, either $G_{\alpha s}$- or $G_{\alpha i}$-coupled, were transfected with the reporter gene vector consisting of GFP and CRE elements. After ligand binding the stimulation or inhibition of AC is induced and there is an increase or decrease of the cAMP-level, respectively. An increase of the cAMP level leads to the activation of PKA that activates CREB through phosphorylation. The phosphorylated form of CREB now binds to the CRE elements. Accordingly, the transcription of GFP is initiated. For the investigation of $G_{\alpha i}$-coupled receptors, cells have to be incubated with the ligand in the presence of forskolin. GFP is expressed in the cytoplasm and can be measured directly in living cells using a fluorescence reader. The lysis of cells or the addition of a substrate is not necessary. (AC, adenylyl cyclase; CRE, cAMP-binding element; CREB, CRE binding protein; G, G-protein; GFP, green fluorescent protein.)

this system is that the addition of a substrate is not necessary and the lysis of cells is not required.

Different reporter gene vectors with increasing numbers of CRE elements (4, 8 and 12) were cloned to establish this novel system. Studies demonstrated that the sensitivity of the assay was dependent on the numbers of CREs – the vector with 12 CRE elements showed the most significant data. This system can also be applied in living cells. It was used for the investigation of NPY Y_5 receptor, a $G_{\alpha i}$-coupled GPCR. The ligand-activated NPY receptor causes a decrease of the cAMP level and consequently a low expression of GFP (Fig. 3.3). Neuropeptide Y dose-dependent expression of GFP was found and the calculated EC_{50} value was com-

Tab. 3.2 Applications and limitations of reporter gene assays for GPCRs.

Applications of reporter-gene assays	*Problems and limitations*
Identification of GPCR agonists	Sensitive machines for measurements
Pharmacological studies in living cells	Development of a suitable assay system
Identification of second messenger pathways from the receptor to the nucleus	Transfection of cells
High-throughput drug screening	Not all systems are commercially available
Identification of orphan receptors and their ligands	

parable with the EC_{50} value obtained from a competition assay. Thus a system is now available which easily allows us to perform measurements in a single cell. For the first time GFP was used as reporter gene in assays for the investigation of a $G_{\alpha i}$-coupled GPCR. Furthermore, the novel reporter gene assay system has some unique features that make the assay superior to any other reported assays. The data presented show that the assay is at least as sensitive as a luciferase assay. In living cells, analysis is faster and easier without disturbing the sensitive cell system.

Apart from investigating GPCRs, reporter gene assay systems can also be used to investigate signal transduction pathways. Each step of the signal transduction pathway can be disturbed and the effects can be easily measured. Since reporter gene production is the end-point of an entire signaling cascade, the functionality of each step of the signaling pathway can be solved. Furthermore, it is possible to determine which reagent can influence signal transduction and, by using the GFP reporter gene assay system, each step of the signal transduction pathway can easily be detected. After investigation the cells are still alive and can be used for further analysis.

In general, reporter gene assay systems are convenient and easily detectable systems for the investigation of GPCRs (Tab. 3.2), but also for other receptors like tyrosine kinase receptors. Accordingly, they represent opportunities for the fast screening of ligands and for the investigation of receptor–ligand interaction. Furthermore, investigation and characterization of orphan receptors is much easier, and the search for their natural ligands is accelerated.

However, interpretation of the data from reporter gene assays requires some caution. Reporter gene assays require incubation for 4–6 h for the expression of the reporter gene product – exposure of some agonist to GPCRs for this period of time causes receptor desensitization and can cause a shift in the EC_{50} value.

Basal expression of the reporter product can also cause problems. To circumvent this problem cells have to be investigated for their endogenous expression and a suited reporter gene must be selected. As the reporter product is the end-point of an entire signaling cascade, there are also many possibilities for interference with the signaling pathway from other sources. For example, CRE reporter activity can also be activated by serum. The ligand can activate other downstream components of the signaling cascade that can influence the reporter readout. Researchers have to be aware of these problems which can be detected by a carefully designed negative control.

In conclusion, reporter gene assays are magnificent tools for the investigation of GPCRs. The assays are fast, safe, convenient, simple and inexpensive. Determination of ligand activity is possible and these assays are of tremendous help in the continuous search for potential new drugs.

References

1 Uings IJ, Farrow SN. 2000. Cell receptors and cell signaling. *Mol Pathol* **53**, 295–9.
2 Chebib M, Johnston GA. 2000. GABA-activated ligand gated ion channels: medicinal chemistry and molecular biology. *J Med Chem* **43**, 1427–47.
3 Burnashev N, Rozov A. 2000. Genomic control of receptor function. *Cell Mol Life Sci* **57**, 1499–507.
4 Brake AJ, Julius D. 1996. Signaling by extracellular nucleotides. *Annu Rev Cell Dev Biol* **12**, 519–41.
5 Leof EB. 2000. Growth factor receptor signaling: location location location. *Trends Cell Biol* **10**, 343–8.
6 Schlessinger J. 2000. Cell signaling by receptor tyrosine kinases. *Cell* **103**, 211–25.
7 Simon MA. 2000. Receptor tyrosine kinases: specific outcomes from general signals. *Cell* **103**, 13–5.
8 Zhu HJ, Burgess AW. 2001. Regulation of transforming growth factor-beta signaling. *Mol Cell Biol Res Commun* **4**, 321–30.
9 Piek E, Heldin CH, Ten Dijke P. 1999. Specificity diversity and regulation in TGF-beta superfamily signaling. *FASEB J* **13**, 2105–24.
10 Foster DC, Garbers DL. 1998. Dual role for adenine nucleotides in the regulation of the atrial natriuretic peptide receptor guanylyl cyclase-A. *J Biol Chem* **273**, 16311–8.
11 Lefkowitz RJ. 2000. The superfamily of heptahelical receptors. *Nat Cell Biol* **2**, E133–6.
12 Hamm HE, Gilchrist A. 1996. Heterotrimeric G proteins. *Curr Opin Cell Biol* **8**, 189–96.
13 Wess J. 1997. G-protein-coupled receptors: molecular mechanisms involved in receptor activation and selectivity of G-protein recognition. *FASEB J* **11**, 346–54.
14 Marinissen MJ, Gutkind JS. 2001. G-protein-coupled receptors and signaling networks: emerging paradigms. *Trends Pharmacol Sci* **22**, 368–76.
15 Gether U. 2000. Uncovering molecular mechanisms involved in activation of G protein-coupled receptors. *Endocr Rev* **21**, 90–113.
16 Post GR, Brown JH. 1996. G protein-coupled receptors and signaling pathways regulating growth responses. *FASEB J* **10**, 741–9.
17 Brzostowski JA, Kimmel AR. 2001. Signaling at zero G: G-protein-independent functions for 7-TM receptors. *Trends Biochem Sci* **26**, 291–7.
18 Beck-Sickinger AG. 1996. Structural characterization and binding sites of G-protein-coupled receptors. *Drug Discov Today* **1**, 502–13.
19 Palczewski K, Kumasaka T, Hori T, Behnke CA, Motoshima H, Fox BA, Le Trong I, Teller DC, Okada T, Stenkamp RE, Yamamoto M, Miyano M. 2000. Crystal structure of rhodopsin: a G protein-coupled receptor. *Science* **289**, 739–45.
20 Edwards SW, Tan CM, Limbird LE. 2000. Localization of G-protein-coupled receptors in health and disease. *Trends Pharmacol Sci* **21**, 304–8.
21 Sautel M, Milligan G. 2000. Molecular manipulation of G-protein-coupled receptors: a new avenue into

drug discovery. *Curr Med Chem* **7**, 889–96.

22 Brivanlou AH, Darnell JE, Jr. 2002. Signal transduction and the control of gene expression. *Science* **295**, 813–8.

23 Manning AM. 1996. Transcription factors: a new frontier for drug discovery. *Drug Discov Today* **1**, 151–60.

24 Jordan JD, Landau EM, Iyengar R. 2000. Signaling networks: the origins of cellular multitasking. *Cell* **103**, 193–200.

25 Karin M. 1994. Signal transduction from the cell surface to the nucleus through the phosphorylation of transcription factors. *Curr Opin Cell Biol* **6**, 415–24.

26 Habener JF, Miller CP, Vallejo M. 1995. cAMP-dependent regulation of gene transcription by cAMP response element-binding protein and cAMP response element modulator. *Vitam Horm* **51**, 1–57.

27 Boss V, Talpade DJ, Murphy TJ. 1996. Induction of NF-AT-mediated transcription by G_q-coupled receptors in lymphoid and non-lymphoid cells. *J Biol Chem* **271**, 10429–32.

28 Macian F, Garcia-Rodriguez C, Rao A. 2000. Gene expression elicited by NF-AT in the presence or absence of cooperative recruitment of Fos and Jun. *EMBO J* **19**, 4783–95.

29 Treisman R. 1996. Regulation of transcription by MAP kinase cascades. *Curr Opin Cell Biol* **8**, 205–15.

30 Darnell JE, Jr. 1997. STATs and gene regulation. *Science* **277**, 1630–5.

31 Ram PT, Iyengar R. 2001. G protein coupled receptor signaling through the Src and Stat3 pathway: role in proliferation and transformation. *Oncogene* **20**, 1601–6.

32 Goetz AS, Andrews JL, Littleton TR, Ignar DM. 2000. Development of a facile method for high throughput screening with reporter gene assays. *J Biomol Screen* **5**, 377–84.

33 Bronstein I, Fortin J, Stanley PE, Stewart GS, Kricka LJ. 1994. Chemiluminescent and bioluminescent reporter gene assays. *Anal Biochem* **219**, 169–81.

34 Leslie AG, Moody PC, Shaw WV. 1988. Structure of chloramphenicol acetyltransferase at 1.75-A resolution. *Proc Natl Acad Sci USA* **85**, 4133–7.

35 Shaw WV. 1975. Chloramphenicol acetyltransferase from chloramphenicol-resistant bacteria. *Methods Enzymol* **43**, 737–55.

36 Seed B, Sheen JY. 1988. A simple phase-extraction assay for chloramphenicol acyltransferase activity. *Gene* **67**, 271–7.

37 Eustice DC, Feldman PA, Colberg-Poley AM, Buckery RM, Neubauer RH. 1991. A sensitive method for the detection of beta-galactosidase in transfected mammalian cells. *Biotechniques* **11**, 739–40, 742–3.

38 Price J, Turner D, Cepko C. 1987. Lineage analysis in the vertebrate nervous system by retrovirus-mediated gene transfer. *Proc Natl Acad Sci USA* **84**, 156–60.

39 Jain VK, Magrath IT. 1991. A chemiluminescent assay for quantitation of beta-galactosidase in the femtogram range: application to quantitation of beta-galactosidase in *lacZ*-transfected cells. *Anal Biochem* **199**, 119–24.

40 Jefferson RA, Kavanagh TA, Bevan MW. 1987. GUS fusions: beta-glucuronidase as a sensitive and versatile gene fusion marker in higher plants. *EMBO J* **6**, 3901–7.

41 Paigen K. 1989. Mammalian beta-glucuronidase: genetics molecular biology and cell biology. *Prog Nucleic Acid Res Mol Biol* **37**, 155–205.

42 Miska W, Geiger R. 1987. Synthesis and characterization of luciferin derivatives for use in bioluminescence enhanced enzyme immunoassays. New ultrasensitive detection systems for enzyme immunoassays I. *J Clin Chem Clin Biochem* **25**, 23–30.

43 Schaap AP, Akhavan H, Romano LJ. 1989. Chemiluminescent substrates for alkaline phosphatase: application to ultrasensitive enzyme-linked immunoassays and DNA probes. *Clin Chem* **35**, 1863–4.

44 Berger J, Hauber J, Hauber R, Geiger R, Cullen BR. 1988. Secreted

placental alkaline phosphatase: a powerful new quantitative indicator of gene expression in eukaryotic cells. *Gene* **66**, 1–10.

45 Moore JT, Davis ST, Dev IK. 1997. The development of beta-lactamase as a highly versatile genetic reporter for eukaryotic cells. *Anal Biochem* **247**, 203–9.

46 Zlokarnik G, Negulescu PA, Knapp TE, Mere L, Burres N, Feng L, Whitney M, Roemer K, Tsien RY. 1998. Quantitation of transcription and clonal selection of single living cells with beta-lactamase as reporter. *Science* **279**, 84–8.

47 Branchini BR, Magyar RA, Murtiashaw MH, Portier NC. 2001. The role of active site residue arginine 218 in firefly luciferase bioluminescence. *Biochemistry* **40**, 2410–8.

48 Conti E, Franks NP, Brick P. 1996. Crystal structure of firefly luciferase throws light on a superfamily of adenylate-forming enzymes. *Structure* **4**, 287–98.

49 Lorenz WW, Cormier MJ, O'Kane DJ, Hua D, Escher AA, Szalay AA. 1996. Expression of the *Renilla reniformis* luciferase gene in mammalian cells. *J Biolumin Chemilumin* **11**, 31–7.

50 Parsons SJ, Rhodes SA, Connor HE, Rees S, Brown J, Giles H. 2000. Use of a dual firefly and *Renilla* luciferase reporter gene assay to simultaneously determine drug selectivity at human corticotrophin releasing hormone 1 and 2 receptors. *Anal Biochem* **281**, 187–92.

51 Michelet B, Chua N. 1997. Improvement of *Arabidopsis* mutant screens based luciferase imaging in planta. *Plant Mol Biol Rep* **14**, 321–329.

52 Tsien RY. 1998. The green fluorescent protein. *Annu Rev Biochem* **67**, 509–44.

53 Prasher DC, Eckenrode VK, Ward WW, Prendergast FG, Cormier MJ. 1992. Primary structure of the *Aequorea victoria* green-fluorescent protein. *Gene* **111**, 229–33.

54 Ormo M, Cubitt AB, Kallio K, Gross LA, Tsien RY, Remington SJ. 1996. Crystal structure of the *Aequorea victoria* green fluorescent protein. *Science* **273**, 1392–5.

55 Fukuda H, Arai M, Kuwajima K. 2000. Folding of green fluorescent protein and the cycle3 mutant. *Biochemistry* **39**, 12025–32.

56 Pines J. 1995. GFP in mammalian cells. *Trends Genet* **11**, 326–7.

57 Gerdes HH, Kaether C. 1996. Green fluorescent protein: applications in cell biology. *FEBS Lett* **389**, 44–7.

58 Welsh S, Kay SA. 1997. Reporter gene expression for monitoring gene transfer. *Curr Opin Biotechnol* **8**, 617–22.

59 Chalfie M, Tu Y, Euskirchen G, Ward WW, Prasher DC. 1994. Green fluorescent protein as a marker for gene expression. *Science* **263**, 802–5.

60 Ellenberg J, Lippincott-Schwartz J, Presley JF. 1999. Dual-colour imaging with GFP variants. *Trends Cell Biol* **9**, 52–6.

61 Pollok BA, Heim R. 1999. Using GFP in FRET-based applications. *Trends Cell Biol* **9**, 57–60.

62 Sorkin A, McClure M, Huang F, Carter R. 2000. Interaction of EGF receptor and grb2 in living cells visualized by fluorescence resonance energy transfer (FRET) microscopy. *Curr Biol* **10**, 1395–8.

63 Mitra RD, Silva CM, Youvan DC. 1996. Fluorescence resonance energy transfer between blue-emitting and red-shifted excitation derivatives of the green fluorescent protein. *Gene* **173**, 13–7.

64 Milligan G. 1999. Exploring the dynamics of regulation of G protein-coupled receptors using green fluorescent protein. *Br J Pharmacol* **128**, 501–10.

65 Dinger MC, Beck-Sickinger AG. 2002. The first reporter gene assay on living cells: green fluorescent protein as reporter gene for the investigation of Gi-protein coupled receptors. *Mol Biotechnol* **21**, 9–18.

66 Stratowa C, Himmler A, Czernilofsky AP. 1995. Use of a luciferase reporter system for characterizing

G-protein-linked receptors. *Curr Opin Biotechnol* **6**, 574–81.

67 Himmler A, Stratowa C, Czernilofsky AP. 1993. Functional testing of human dopamine D1 and D5 receptors expressed in stable cAMP-responsive luciferase reporter cell lines. *J Recept Res* **13**, 79–94.

68 George SE, Bungay PJ, Naylor LH. 1997. Functional coupling of endogenous serotonin (5-HT_{1B}) and calcitonin (C1a) receptors in CHO cells to a cyclic AMP-responsive luciferase reporter gene. *J Neurochem* **69**, 1278–85.

69 Fitzgerald LR, Mannan IJ, Dytko GM, Wu HL, Nambi P. 1999. Measurement of responses from G_i-, G_s- or G_q-coupled receptors by a multiple response element/cAMP response element-directed reporter assay. *Anal Biochem* **275**, 54–61.

70 Stratowa C, Machat H, Burger E, Himmler A, Schafer R, Spevak W, Weyer U, Wiche-Castanon M, Czernilofsky AP. 1995. Functional characterization of the human neurokinin receptors NK1 NK2 and NK3 based on a cellular assay system. *J Recept Signal Transduct Res* **15**, 617–30.

71 Kotarsky K, Owman C, Olde B. 2001. A chimeric reporter gene allowing for clone selection and high-throughput screening of reporter cell lines expressing G-protein-coupled receptors. *Anal Biochem* **288**, 209–15.

72 Beckers T, Reilander H, Hilgard P. 1997. Characterization of gonadotropin-releasing hormone analogs based on a sensitive cellular luciferase reporter gene assay. *Anal Biochem* **251**, 17–23.

73 Xing H, Tran H, Knapp TE, Negulescu PA, Pollok BA. 2000. A fluorescent reporter assay for the detection of ligands acting through G_i protein-coupled receptors. *J Recept Signal Transduct Res* **20**, 189–210.

74 Durocher Y, Perret S, Thibaudeau E, Gaumond MH, Kamen A, Stocco R, Abramovitz M. 2000. A reporter gene assay for high-throughput screening of G-protein-coupled receptors stably or transiently expressed in HEK293 EBNA cells grown in suspension culture. *Anal Biochem* **284**, 316–26.

4
From the Human Genome to New Drugs: The Potential of Orphan G-protein-coupled Receptors

Remko A. Bakker and Rob Leurs

4.1
Introduction

The initial sequencing of the human genome, completed by the Human Genome Project and Celera [1], has resulted in great excitement in the biomedical community, since it offers tremendous opportunities for understanding human disease and developing innovative drug therapies. Abnormal functioning of certain cells in part of the human body usually results in diseases in humans. Cellular function is principally regulated by a variety of chemical signals that are released by various cell types to regulate human physiology. Upon binding of these chemical messengers to specific receptors expressed on a target cell, the receptor gets activated and will initiate various signals within that cell, resulting in changes in its function. Drugs affect cellular function by interacting with the receptor to mimic or block ligand–receptor binding and thereby regulate the disease process. In this chapter, to understand the importance of the human genome sequence for drug discovery, we will highlight the impact on drug discovery related to guanine nucleotide-binding (G)-protein-coupled receptors (GPCRs).

Drug discovery programs of many pharmaceutical companies focus on the GPCR superfamily. GPCRs have historically proven to be a good class of drug targets and approximately 50% of the currently marketed pharmaceutical products target GPCRs [2]. In many diseases GPCRs are known as "validated targets", i.e. ligands acting at the GPCR are known to affect the disease outcome in man. A few examples of important and frequently used drugs targeting GPCRs are shown in Tab. 4.1.

Bioinformatic analysis of the human genome sequence has revealed several hundred new members of the GPCR family. As GPCRs are one of the most important families of targets in drug discovery in the pharmaceutical industry, it is expected that the various newly identified receptors offer similar potential and will also prove to be good drug targets [2]. Here, we review the various steps that ultimately translate an identified human gene encoding for a GPCR into a validated drug target.

Molecular Biology in Medicinal Chemistry. Edited by Th. Dingermann, D. Steinhilber, G. Folkers

ISBN: 3-527-30431-2

Tab. 4.1 Several important therapeutics that are directed against GPCRs.

GPCR target	*Therapeutic*	*Indication*	*GPCR subtype*
Acetylcholine receptor	Dicyclomine (Bemote)	Irritable bowel syndrome	M1 antagonist
	Ipratropium (Atrovent)	COPD	M3 antagonist
	Tiotropium (Spiriva)	COPD	M1/M3 antagonist
Adrenoreceptor	Atenolol (Tenormin)	Hypertension	β_1 antagonist
	Timolol (Timuptol, Inderal)	Glaucoma/Hypertension	mixed β antagonist
	Salmeterol (Albuterol)	Asthma	β_2 agonist
	Albuterol (Ventolin)	Asthma	β_2 agonist
	Doxazosin (Cardura)	Prostate hypertrophy, antihypertensive	α_1 antagonist
	Carvedilol (Coreg)	Congestive heart failure, antihypertensive	$\beta_1/\beta_2/\alpha_1$
	Clonidine	Antihypertensive	α_2 agonist
	Propranolol (Inderal)	Antiarrhythmic, Antihypertensive, Angina, Anxiety	β antagonist
	Terazosin (Hytrin)	Prostate hypertrophy, antihypertensive	α_1 antagonist
Angiotensin II receptor	Losartan (Cozaar)	Hypertension	AT1 antagonist
	Eprosartan (Teveten)	Hypertension	AT1 antagonist
Histamine receptor	Cetirizine, Loratadine	Allergies	H_1 antagonist
	Cimetidine, Ranitidine	Ulcers	H_2 antagonist
Leukotriene receptor	Pranlukast (Onon)	Asthma, Allergic rhinitis	CysLTR1 antagonist
	Zafirlukast (Accolate)	Asthma	CysLTR1 antagonist
Opioid receptor	Buprenorphine	Narcotic analgesic, pain relief	μ agonist/κ antagonist
	Butorphanol	Narcotic analgesic, pain relief	μ antagonist/κ agonist
	Alfentanil (Alfenta)	Narcotic analgesic, pain relief	μ agonist
	Morphine (Roxanol)	Narcotic analgesic, pain relief	μ agonist
Prostanoid receptor	Epoprostenol (Flolan)	Primary pulmonary hypertension	Prostacyclin receptor agonist
	Misoprostol (Cytotec)	Ulcers	Prostaglandin receptor agonist
Somatostatin	Octreotide (Sandostatin)	abdominal illness, intestinal tumors	sst2 and sst5 agonist
Serotonin receptor	Sumatriptan (Imitrex)	Migraine	$5HT_{1B/1D}$ agonist
	Buspirone (Buspar)	Anxiety	$5HT_{1A}$ agonist
	Olanzapine (Zyprexa)	Psychosis	$5HT_2/D_2$ antagonist
	Ritanserin (Tisterton)	Antidepressant	$5HT_2$ antagonist
	Cisapride (Prepulsic)	Nocturnal heartburn	$5HT_4$ agonist
	Trazodone (Desyrel)	Antidepressant	$5HT_2$ antagonist
	Clozapine (Leponex)	Schizophrenia	mixed
Gonadotrophin-releasing factor receptor	Goserelin (Zoladex)	Cancer	GnRHR/LHRHR agonist
Dopamine receptor	Ropinerole (Requip)	Parkinson's disease	D_2, D_3, D_4 agonist

4.2
GPCRs and the Human Genome

4.2.1
GPCR Architecture, Signaling and Drug Action

GPCRs form the fourth protein family in the human genome [1], and play an important role in many physiological and pathophysiological processes. The family comprises a large number (more than 1000) of transmembrane proteins sharing a highly similar architecture: an extracellular N-terminus, three extracellular and three intracellular loops, and seven transmembrane α-helices that connect the extra- and intracellular loops. Due to the presence of the seven transmembrane domains, these receptors have been labeled "serpentine" because of their resemblance to a coiled snake. They contain an extracellular ligand-binding site, critical to their differential activation, as well as intracellular domains for coupling to proteins such as G-proteins. Although GPCRs form a family of related transmembrane receptors, they tend to be selectively activated by a highly diverse set of ligands, including biogenic amines, peptides, lipid analogs, nucleotides, amino acid derivatives, Ca^{2+}, and sensory stimuli such as light (photons), taste and odor (Fig. 4.1). Accordingly, the superfamily of GPCRs can be subdivided into several subgroups such as the rhodopsin family (family A), the secretin receptor-like family (family B) and the metabotropic glutamate receptor-like family (family C). The various GPCR families reflect various structural elements that enable the receptors to bind the diverse ligands.

This high specificity in ligand recognition is also exploited to target exogenous ligands (drugs) to specific GPCR proteins, explaining why GPCRs are so successful as drug targets. Moreover, GPCRs are relatively easily amenable to drug develop-

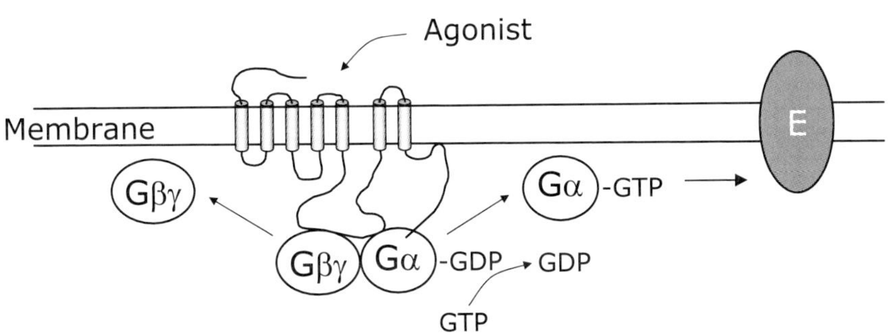

Fig. 4.1 Schematic representation of GPCR signaling. The GPCR consists of an extracellular N-terminal domain, and three extra- and three intracellular loops, which are connected by seven α-helical transmembrane domains. The GPCR is located in the plasma membrane and can be activated by extracellular ligands. Activation of the receptor results in the intracellular activation of heterotrimeric G-proteins by promoting the GDP–GTP exchange on the G_α subunit. Upon activation of the heterotrimeric G-protein, the G-protein dissociates into the G_α and $G_{\beta\gamma}$ subunits, both of which can activate effector proteins (E) in the cell to generate second messengers.

ment due to their transmembrane localization, which alleviates the need for drugs to enter the cells in order to be effective.

Upon recognition of an extracellular signaling molecule, a conformational change triggers intracellular heterotrimeric G-proteins to modulate a variety of downstream biochemical pathways [3]. A GPCR exists in equilibrium between two (or more) different states: an inactive state and an active state [4]. In an active state, the GPCR is able to couple to a G-protein and to produce a biological response. In an inactive state, the receptor is unable to produce a biological response. Binding of molecules that activate the GPCR (an *agonist*) shifts the equilibrium as the GPCR is stabilized in the active state. Many of today's drugs acting at GPCRs function by preventing the action of the endogenous signaling molecule at the targeted GPCR. These ligands, previously known as antagonists, can now be classified as "inverse agonists", which, stabilize the GPCR in an inactive conformation, and "neutral antagonists", which do not shift the conformational equilibrium themselves, but prevent the interaction with the endogenous agonist by binding to the GPCR.

4.2.2
Identification of GPCRs

Their common, seven transmembrane structural domains are thought to result in highly conserved structures. During the 1980s, recombinant DNA technology was introduced to the neurosciences, and boosted the discovery of both receptors and gene-encoded peptide/protein ligands. The purification and subsequent cloning of the hamster β_2 receptor gene in 1986 [5] can be seen as one of the hallmarks in this field. The work of Strader et al. revealed a common, seven transmembrane GPCR structure and a DNA sequence similarity in the transmembrane regions with bovine rhodopsin, the first GPCR that was cloned [6]. The sequence homology between both GPCRs allowed the subsequent cloning of many other GPCRs in the 1980s–1990s on the basis of homology. Either low-stringency hybridization screening or degenerate polymerase chain reaction (PCR) was used to identify novel, yet related, GPCR sequences. This approach resulted in a large number of GPCRs for which the native ligands were initially unknown. However, relatively simple sequence alignments allowed reasonable predictions of ligands for a number of the earliest cloned orphan receptors. Initially, the identification of novel drug targets mainly focused on the testing of known endogenous ligands against highly related receptors. This approach has led to the identification of numerous receptor subtypes. The first "deorphanized" GPCR was the 5-hydroxytryptamine $(HT)_{1A}$ receptor that was cloned based on sequence homology to the related β_2 receptor [7] and appeared to respond to serotonin. However, the approach is not always successful and greatly depends on the amino acid identity with already identified receptors. For many years various academic and industrial groups have, for instance, attempted to clone the pharmacologically defined histamine H_3 receptor. Despite the availability of the genes encoding the human H_1 and H_2 receptor since the early 1990s [8], it was not until 1999 that Lovenberg et al. finally succeeded in cloning the human histamine H_3 receptor gene [9] using information from the Incyte expressed sequence tag (EST) database (see Section 4.3). Phylogenetic analysis indicates that the gene encoding the human H_3 receptor is not closely related to those that encode human H_1 or H_2 receptors (overall homology around 22%); in fact, the H_3 receptor gene resembles more those encoding the α_2-adrenoceptors. These three different histamine receptor genes may have evolved from different "ancestors" and apparently all three acquired crucial elements for the recognition of histamine. The poor resemblance of the H_3 receptor gene with the genes encoding the two other histamine receptor subtypes explains why the various homology-based approaches to clone the H_3 receptor gene were unsuccessful [10].

The successful cloning of the H_3 receptor gene was promptly followed by the cloning of the histamine receptor gene encoding the H_4 receptor, for which only sparse pharmacological evidence of its existence was available [11]. The H_4 receptor gene was cloned by several groups concurrently based on the sequence of the H_3 receptor that had become available [12, 14]. The new histamine receptor shows highly restricted expression in the immune system and is currently regarded as a potentially interesting drug target for inflammatory conditions.

4.2.3
GPCRs in the Postgenomic Era: Orphan Receptors

With the available information on the human genome, a systemic exploitation of the human GPCR genome by a functional genomic approach may now be applied to discover novel drug targets. To find new GPCRs in the genome, highly conserved sequence motifs within the seven transmembrane domains that are common to all GPCRs are used to query the entire genome. Such use of bioinformatics has resulted in the *in silico* identification of many DNA sequences encoding putative human GPCRs. Although the complete genome has been sequenced, the prediction of the total number of GPCRs is not yet fully established. There is considerable debate concerning the total number of GPCRs in the human genome and estimates vary widely from 1000 up to 2000 [1]. Recent progress in genome research unveiled a total of around 300 GPCRs that are considered to have pharmaceutical value as potential targets for drug development [15]. Of these, 160 GPCRs are activated by 97 known transmitters, including biogenic amines, amino acids, peptides, nucleotides, fatty acid derivatives and polypeptides. However, approximately 140 of these novel GPCRs do not bind any known endogenous ligand and this offers interesting new opportunities for drug discovery aimed at these so-called "orphan GPCRs" (oGPCRs) – GPCRs whose function and ligand are still unknown [15].

In addition to these GPCRs of human origin, pathogens may harbor GPCR-encoding genes that may be expressed in the host [16]. Karposi-associated human herpes virus (HHV-8), for instance, encodes ORF74, a highly constitutively active receptor thought to play an important role in the development of disease [17]. Likewise, human cytomegalovirus (CMV) encodes four GPCRs in its genome. One of the human CMV-encoded GPCRs, US28, is highly constitutively active [18], plays an important role in development of atherosclerosis [19] and, with the recent discovery of nonpeptidergic inverse agonists for this receptor [20], appears promising as a potential drug target. In addition, native ligands have recently been identified for a variety of insect GPCRs [21] – insect receptors may ultimately be targeted to selectively control insects.

In view of the large number of "untargeted" GPCRs, many pharmaceutical/biotechnology companies and academic laboratories are currently putting great effort into identifying ligands for existing and newly discovered putative receptors. Many company's drug discovery programs include a program that attempts to exploit the Human Genome Project to gain access to new GPCR families, including oGPCRs, to obtain novel drug leads. To "deorphanize" oGPCRs entails the identification of the natural endogenous, or native, ligands for these receptors. The functions of these novel receptors are being elucidated using functional genomics approaches. The search for ligands for orphan receptors may involve classical screening methods such as the monitoring of changes in the levels of well-known second messengers such as cAMP and Ca^{2+} or may involve highly sophisticated technologies that are largely based upon a profound understanding of the biology that is involved in multiple levels of the receptor-activation process and beyond, the fundamental basis of which is made possible by recent advances in molecular biology, computational power and bioinformatics, microfluidics, nanotechnology, etc. As such, the deorphanizing of receptors has become a truly multidisciplinary discipline, and in the following sections we will indicate how fundamental knowledge of bioinformatics, receptor expression and signal transduction systems is exploited for the identification of ligands for oGPCRs.

4.3
Ligand Hunting

4.3.1
Reverse Pharmacology Approaches to oGPCRs

Historically, the concept of receptors was developed on the basis of the observed biological effects of exogenously administered drugs (reviewed in [22]). Many receptors and receptor subtypes for known endogenously chemical messengers were discovered by investigating the action of biologically active molecules. The development of the field of molecular pharmacology on the basis of the early receptor concepts by Ariens and many of his fellow scientists has paved the road to the rational design of drugs on the basis of a proven interaction with a known receptor system. Drug action revealed the existence of a relevant receptor system which was "chemically exploited". This process led to many of today's important drugs and is still a useful approach for known receptor systems.

However, with oGPCRs the situation is very different. Knowledge of the endogenous signaling molecules is lacking. Moreover, it is very difficult to predict beforehand the potential of specific ligands acting at the oGPCR of choice. Therefore, the process of identification of the endogenous ligand for an orphan receptor is also called "reverse pharmacology"; one first has to establish the endogenous ligand for the oGPCR, then one needs to investigate the function of the oGPCR–ligand pair in physiological processes, where after potential drug targeting the oGPCR could be developed. Although reverse pharmacology is a demanding and high-risk approach, the final innovation for the drug discovery process is usually extremely advantageous. Moreover, in recent years, technological advances and increased insights in GPCR function have greatly contributed to the development of successful strategies.

4.3.2
In Silico Approaches

Databases can be employed to obtain information on phylogenetic relationships, tissue expression, existence of "polymorphic variants" and their potential linkage with a specific phenotype to identify an oGPCR *in silico*. Phylogenetic analysis of GPCR sequences can reveal the relation of the oGPCR to other already known receptors [23] and may be used as a tool to predict the class of orphan receptor ligands [24]. As exemplified by the histamine H_4 receptor, a phylogenetic analysis of the H_4 receptor sequence clearly matches it with the related histamine H_3 receptor (Fig. 4.2).

Currently, more and more gene array data are becoming available, and careful analysis of the available data will allow us to deduce important information about gene expression data *in silico*. In order to evaluate the potential of an oGPCR for therapeutic applications, detailed knowledge of the tissue expression is of utmost importance. If one is interested to develop drugs against central nervous system disorders, one is of course not interested in oGPCRs that are only expressed in the periphery. As we will discuss later, information on tissue expression is also very important for the search of the endogenous ligand.

The availability of the sequence of the human genome and of other species such as the mouse [25] allows genome-wide analysis of, for instance, single-nucleotide polymorphisms (SNPs) that may be linked to a specific phenotype, and thus to the function of a particular gene and the encoded protein. Analysis at the level of the genome is anticipated to bring unequalled advances in the understanding of gene function and their causative role in disease. Such a functional genomics approach has been successfully applied to the identification of the PTC gene encoding a phenylthiocarbamide-sensitive GPCR that is responsible for the taste sensitivity to phenylthiocarbamide. Several SNPs in this gene result in the loss of taste sensitivity [26].

Alternatively, the generation of mice in which the expression of a particular GPCR has been eliminated by genetic manipulation (a receptor "knockout") can provide clues to the potential physiological role of the receptor. If a particular ligand is encoded by the genome, as is the case for peptidergic ligands, knockouts may also be generated that lack a particular transmitter. The potential phenotype of the knockout may therefore harbor clues to the function of the receptor and assist

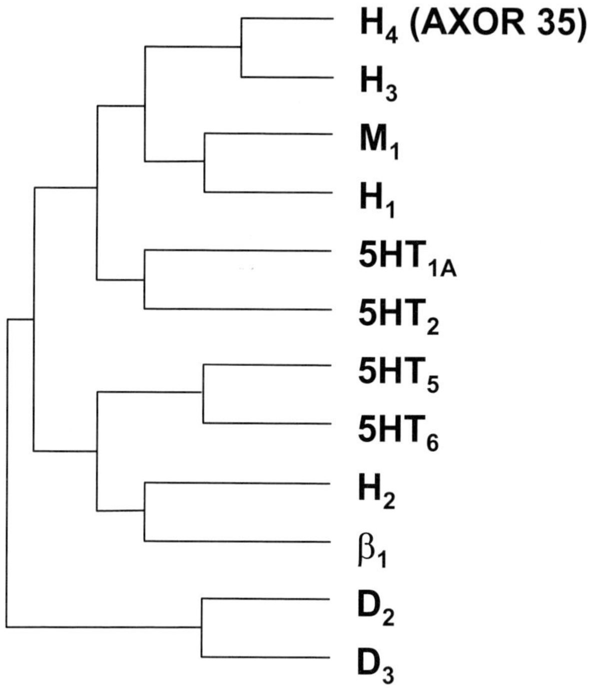

Fig. 4.2 Phylogenetic tree analysis of several aminergic GPCRs places AXOR 35 close to the histamine H_3 receptor. Pharmacological analysis subsequently confirmed AXOR 35 to represent a novel human histamine receptor, the H_4 receptor. Figure kindly provided by Dr. S. Rees (GSK, Stevenage, UK).

in the validation of a GPCR as a potential drug target [27]. A knockout of the intimal thickness-related receptor, for instance, has shown this oGPCR to be involved in vascular remodeling [28]. Lexicon Genetics and Deltagen currently offer a high-throughput *in vivo* mammalian knockout technology to generate various knockout mice (including various GPCR knockouts) and analyze their phenotypes [27, 29]. Access to the relevant (commercial) database allows one to obtain information on the potential role of the oGPCR in physiology.

4.3.3 Tissue Expression

As mentioned in the previous section, knowledge of the tissue expression of the oGPCR is one of the first steps in order to decide if the oGPCR is really of therapeutic interest. In addition to the *in silico* approaches, there are various other means to obtain data on tissue expression of the oGPCR. The most popular technique for tissue profiling has been the analysis of mRNA expression by Northern blot analysis, reverse transcription-PCR (RT-PCR) or *in situ* hybridization.

In Northern blot analysis, mRNA from various tissues is size-separated by gel electrophoresis, blotted and immobilized on nitrocellulose or nylon membranes, and incubated with a radioactive or fluorescently labeled oligonucleotide or cDNA probe, which is based on the DNA sequence of the oGPCR of interest. By means of the known base-pairing specificity of DNA–RNA molecules, hybridization of the probe will only occur with the oGPCR mRNA molecules. Analysis of the radioactive or fluorescent signal that remains on the sheet upon appropriate washings allows us to deduce tissue expression. Alternatively, by performing less stringent washings (low-stringency hybridization screening), the probe may hybridize to RNA molecules coding for novel GPCRs that exhibit sufficient sequence homology to the probe. Currently, blots with various human tissues can be obtained commercially.

Tissue mRNA expression can also be obtained by means of a variation of the PCR. Using the enzyme reverse transcriptase, the tissue mRNA is transcribed into cDNA, where after the oGPCR cDNA molecules are amplified via a PCR reaction with two oGPCR-specific oligonucleotides.

The most detailed mRNA expression data can be obtained via *in situ* hybridization. Tissue slices are incubated under appropriate conditions with a labeled (radioactive or fluorescent) oligonucleotide or cDNA probe, which is based on the DNA sequence of the oGPCR of interest. Upon washing, the readout gives highly detailed information on the variation in expression levels in complex tissues, such as the brain.

As mRNA expression does not necessarily reflect actual protein expression, one should perhaps try to measure protein expression levels instead of mRNA expression. However, for protein expression profiling one preferably needs a specific antibody for Western blot analysis or immunohistochemistry. Although the generation of antibodies to oGPCRs is not impossible, it often requires substantial resources and is therefore not very attractive.

As knowledge of GPCR expression is fundamental for the whole process of reverse pharmacology, large-scale molecular anatomical approaches aimed at mapping the expression level (expression profiling) of all GPCRs known, including orphan receptors, allows an early prioritization of those GPCRs that may prove useful in drug discovery.

4.3.4 Expression of an oGPCR of Interest

To identify the endogenous ligand for an interesting oGPCR, we must establish cell lines that express the orphan receptors and use these cell lines for screening for potential ligands. Many mammalian cell lines are nowadays available for so-called "heterologous expression" of oGPCRs. To express an oGPCR of interest, the oGPCR cDNA sequence is first cloned from a relevant tissue or genomic DNA, an *epitope tag* such as a FLAG, c-*myc*, V5 or hemagglutinin is often incorporated in the N-terminal domain of the receptor to allow detection of the receptor protein at the cell surface using antibodies directed against the epitope tag. The oGPCR cDNA is

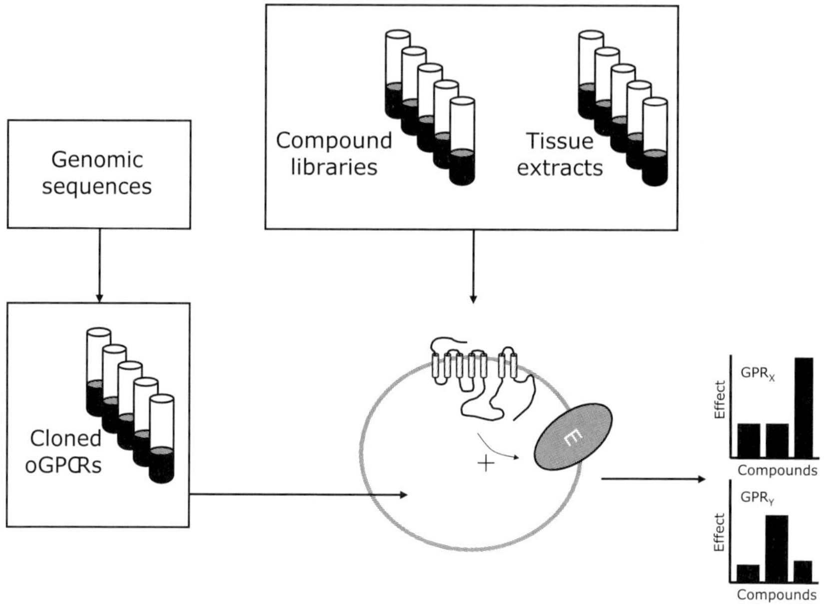

Fig. 4.3 cDNA encoding GPCRs derived from the genome are cloned into appropriate expression vectors in order to achieve adequate expression of the receptor in the cells used for the ligand identification process. When the cells express the receptor of interest, the cells may be challenged with potential ligands that may be derived from compound libraries or tissue extracts and the cells are then monitored for instance for changes in cellular activity by performing functional assays (see Section 4.3.5). By repeating this strategy, GPCRs may reveal specific responses for particular GPCR–ligand pairs.

inserted in specialized DNA-vectors, which are transferred ("transfected") into mammalian cells (Fig. 4.3). A variety of techniques (chemical transfection, electroporation, lipofection and infection techniques using, for instance, baculovirus, adenovirus or Semliki Forest Virus) are currently available to effectively transfer DNA into a variety of cells. Examples of commonly used cell lines for the expression of GPCRs include human embryonic kidney (HEK) 293 and Chinese hamster ovary (CHO) cells. Transfection of GPCR genes in these cells usually results in relatively high expression levels, which facilitates the whole process of reverse pharmacology. Caution has to be taken in relation to potential interference with the screening system of endogenously expressed GPCRs of the host cells. Conversely, it may well be that these expression systems do not provide the appropriate cellular environment in terms of G-protein expression [30] or, perhaps more strikingly, are examples of the association of receptor activity-modifying proteins (RAMPs) with a GPCR and the formation of heterodimeric GPCRs which are crucial for the formation of a proper functional GPCR [31, 32].

In addition to the large variety of these mammalian cell lines, several nonmammalian cell lines including insect cells, *Xenopus laevis* melanophores [33] and yeasts are also used for the heterologous expression and characterization of GPCRs

[34], including the identification of potential native [35–37] and surrogate [38] ligands for oGPCRs. An important advantage of the yeast expression system is the limited set of native G-proteins and GPCRs in yeast.

4.3.5
Screening Approaches

There are two starting points for the identification of native ligands for orphan receptors. One uses native tissues in which a receptor is known to be expressed. The presence of the receptor results in a particular ligand-binding site in certain tissues and, upon ligand binding, functional responses are induced that are not mediated through any of the known GPCRs that bind related ligands, suggesting additional GPCRs exist. Thus, an identified ligand is binding to an unidentified receptor. This was the case for nocistatin, a peptide ligand derived from the same precursor as nociceptin, the endogenous ligand for the opioid receptor-like 1 receptor [39]. Another example of an oGPCR that was deorphanized in this manner is the HIV coreceptor Bonzo as a CXCL16 chemokine receptor (CXCR6) [40]. By first searching human EST databases, CXCL16 was identified as a potential novel membrane-bound chemokine receptor ligand. Subsequent studies revealed specific cell types to bind to CXCL16-expressing cells. An expression cloning approach was then successfully used to identify a receptor for CXCL16. A cDNA expression library with mRNA was made from CXCL16-binding cells and transfected into HEK293 cells, which were then evaluated for CXCL16 binding by FACS analysis. Ultimately, individual cDNA clones were identified that, when expressed in HEK293 cells, showed strong specific binding to CXCL16. Complete sequencing of the cDNA revealed the identity of CXCR6 [40].

Usually, however, when trying to identify a ligand for an oGPCR, the researchers establish cell lines expressing the oGPCR and then use this artificial system to screen for potential ligands. In such a chemical genetics methodology, oGPCR-expressing cells are utilized as vehicles for small molecule target identification.

Cells expressing the receptor of interest can then be used for either radioligand-binding assays (non-functional assays) or functional assays that involve the measurement of cellular responses that may occur upon receptor activation. Due to the requirement of a radioactively labeled ligand, the radioligand-binding approach is less useful for deorphanizing receptors, unless there is good chance that the orphan receptor is highly homologous to an existing GPCR for which high-affinity ligands are available. This approach has been successful for the deorphanizing of, for instance, the α-laurotoxin receptors [41, 42] and can assist the deorphanizing of receptors by confirming the binding of an identified ligand to the receptor [9, 12, 14]. Functional assays, however, are often essential for the identification of native ligands for oGPCRs.

For the identification of the native ligand for an orphan receptor using a functional assay, diverse *libraries* consisting of large collections of potential endogenous ligands are panned for their *biological responses* in cells expressing the receptor. Such compound libraries may comprise many of the known agonists that are available, including novel peptides that may be identified on the basis of bioinformatics analysis of the genes encoding putative peptides. If no hits are obtained from these known or putative GPCR agonists, a search for novel ligands may then be initiated by screening against biological extracts of tissues in which the oGPCR is known to be expressed, as well as fluids and cell supernatants.

4.4 Screening for oGPCR Ligands using Functional Assays

The physiological response that is observed following GPCR activation is mostly governed by the coupling to and activation of G-proteins by the activated receptor. GPCRs couple to second-messenger signaling cascade mechanisms via heterotrimeric G-proteins. Most classical functional assays used to measure GPCR activity are usually based on the measurement of either G-protein or effector activation. Screening for oGPCRs ligands is usually performed by means of exploitation of the biology of receptor systems, and may involve classical screening methods such as monitoring changes in the levels of well-known second messengers such as cAMP and Ca^{2+} or may involve highly sophisticated technologies that are largely based upon a profound understanding of the biology that is involved in the receptor-mediated activation of signal transduction cascades. These novel high-throughput screening (HTS) assays utilize automation and miniaturization, and can be performed more quickly and efficiently than traditional drug discovery screening techniques. Comprehensive analysis techniques and the use of convenient assay systems for ligand identification for oGPCRs are coming to fruition. Table 4.2 lists a number of oGPCRs that have been paired with their native ligand in recent years. Once such receptor–ligand pairs are identified, these HTS assays can be focused on finding potential therapeutics for these new targets. In this section we highlight the various experimental approaches that have been developed for functional screening of oGPCRs. If applicable, the methodology will be highlighted by the successful deorphanization of specific oGPCRs.

4.4.1 GTPγS-binding Assays

GTPγS-binding assays monitor the activation of the receptor-associating G-proteins and, although the assay is particularly suitable for G_i- and G_o-coupled receptors, it was recently demonstrated that the assay is also applicable to GPCRs coupling to the G_q family of G-proteins following an immunoprecipitation procedure [43]. By substituting GTP for [^{35}S]GTPγS, G_α-protein activation results in the incorporation of the nonhydrolyzable and radioactively labeled [^{35}S]GTPγS, allowing the quantification of activated G_α-proteins.

Although there are examples of oGPCRs that have been deorphanized using this assay, including the identification of eicosanoids exhibiting agonistic activities against the oGPCR TG1019 [44], the assay is seldom used as the primary screen-

Tab. 4.2 Recently deorphanized receptors and their identified ligands.

Ligand	*Receptor*	*Assays used*	*Year identified*	*Reference*
5-HT	5-HT_{1A}	radioligand binding	1988	7
Adenosine	A_{2A}	cAMP	1990	106
Adenosine	A_1	cAMP	1991	107
5-HT	5-HT_{1D}	cAMP	1991	108
Bombesin-like peptides	bombesin	oocytes	1993	109
Nociceptin/orphanin FQ	ORL-1	cAMP	1995	83
Neuropeptide Y	Y2	radioligand binding	1995	110
AZ3B	C3a	$[Ca^{2+}]_i$, oocytes	1996	111
Stromal cell-derived factor-1	CXCR4 (LESTR, fusin)	$[Ca^{2+}]_i$	1996	112
CIRL	α-laurotoxin	radioligand binding	1997	41
Calcitonin gene-related peptide (CGRP)	CGRP receptor; CRLR	cAMP	1996, 1998	32, 113
Ghrelin	GHS-R	$[Ca^{2+}]_i$	1996, 1999	91, 114
Leukotriene B_4	BLTR	radioligand binding	1997	115
Hypocreting/orexins	orexin-A and -B	$[Ca^{2+}]_i$	1998	47, 85
Thyrotropin-releasing hormone	THR_2	radioligand binding	1998	116
Prolactin releasing peptide	GPR10 (hGR3)	arachidonic acid	1998	88
Apelin	APJ	ext. pH (cytosensor)	1998	92
CRLR	CGRP	oocytes	1998	32
B-lymphocyte chemoattractant	BLR1	Chemotactic activity	1998	117
Sphingosine 1-phosphate	EDG 1, 3, 5, 6 and 8	$[Ca^{2+}]_i$, cAMP, oocytes	1998–2000	118
Lysophosphatic acid	EDG 2, 4 and 7	$[Ca^{2+}]_i$, cAMP, yeast	1998–2000	119
CIRL-2	α-laurotoxin	radioligand binding	1999	42
Histamine	H_3 (GPCR97)	cAMP, radioligand binding	1999	9
Melanin concentrating hormone	SLC-1 (MCH1, GPR24)	$[Ca^{2+}]_i$	1999	120
Urotensin II	UII (GPR14, SENR)	$[Ca^{2+}]_i$	1999	90
Motilin	motilin (GPR38)	$[Ca^{2+}]_i$ (aequorin)	1999	121
Leukotriene D4	Cys-LT1R (HG55, HMTMF81)	$[Ca^{2+}]_i$	1999	122
Leukotriene B_4	BLT2 (GPR16)	cAMP, radioligand binding	2000	123
Neuromedin U	NMU1R (FM-3, GPR66)	$[Ca^{2+}]_i$	2000	124, 125
Neuromedin U	NMU2R (FM-4, TGR-1)	$[Ca^{2+}]_i$	2000	124, 126
Prostaglandin D2	PGD2 (CRTH2)	$[Ca^{2+}]_i$	2000	127
Neuropeptide FF and AF	NPFF2 (HLWAR77)	$[Ca^{2+}]_i$	2000	128
Neuropeptide FF	NPFF1 (OT7T022), NPFF2	Oocytes	2000	129
Sphingosylphosphorylcholine	SPC (OGR-1)	$[Ca^{2+}]_i$	2000	130
Histamine	H_4 (AXOR35)	cAMP, radioligand binding	2000	12, 14
UDP-glucose	$P2Y_{14}$ (KIAA0001, GPR105)	Yeast	2000	36
Leukotriene C4 and D4	Cys-LT2R (PSECO146, AXOR54)	$[Ca^{2+}]_i$, oocytes	2000	131
Eskine	CCR10 (GPR2)	$[Ca^{2+}]_i$	2000	132

Tab. 4.2 *(continued)*

Ligand	*Receptor*	*Assays used*	*Year identified*	*Reference*
CXCL16	BONZO, STRL33, TYMSTR (CXCR6)	flow cytometric analysis, $[Ca^{2+}]_i$	2000	40
β-PEA and tryptamine	TA_1, TA_2	oocytes, cAMP, radio-ligand binding	2001	133
Melanin concentrating hormone	MCH2 (AXOR21)	$[Ca^{2+}]_i$	2001	134
ADP	P2Y12 (SP1999)	oocytes	2001	135
Trace amines	TA_1, TA_2	oocytes	2001	136
Psychosine	TDAG-8	cAMP	2001	137
Lysophosphatidylcholine	G2A	$[Ca^{2+}]_i$	2001	138
KiSS-1	metastatin (GPR54, hOT7T175, AXOR12)	$[Ca^{2+}]_i$	2001	139
Sphingosylphosphorylcholine and lysophosphatidyl-choline	GPR4	radioligand binding, reporter-gene assay	2001	53
NPAF	mrgA4	$[Ca^{2+}]_i$	2001	96
NPFF	mrgA1	$[Ca^{2+}]_i$	2001	96
Relaxin	LGR7	cAMP	2001	45
Relaxin	LGR8	cAMP	2001	45
Short-chain carboxylic acid anions	GPR41, GPR43	yeast	2002	37
Complement fragments	C5L2	radioligand binding	2002	140
Neuropeptide W	GPR7 and GPR8	GTPγS	2002	100
Bile acids	M-BAR (BG37)	cAMP	2002	141
5,8,11-Eicosatriynoic acid	GPR40	$[Ca^{2+}]_i$, reporter gene assay	2002	49
Adenine	rat adenine receptor	$[Ca^{2+}]_i$	2002	98
Proenkephalin A gene products	sensory neuron-specific GPCRs (SNSRs)	$[Ca^{2+}]_i$	2002	99
Nicotinic acid	HM74 (GPR109)	$[Ca^{2+}]_i$ (aequorin), cAMP, GTPγS	2003	79, 80
Nicotinic acid	HM74A	GTPγS	2003	80
Sphingosylphosphorylcholine	GPR12	$[Ca^{2+}]_i$ (aequorin)	2003	142

ing assay. However, it may be used for the confirmation of the coupling to and the stimulation of G_i-proteins of an activated oGPCR once potential ligands have been identified.

4.4.2 Measurements of cAMP

As cAMP is one of the most important secondary messengers for controlling various metabolic pathways, a number of different assays have been developed over the years to measure the formation of cAMP from AMP by adenylyl cyclase (AC) upon receptor stimulation. The AC effector enzyme is activated by G_s-proteins and inhibited by G_i-proteins, and therefore these assays can be used to monitor the activation of a wide variety of GPCRs.

The development of sensitive and reproducible radioimmunoassays and fluorescence-based immunoassays have largely replaced the older techniques that relied on labeled ATP, followed by the extraction and determination of the amount of newly formed labeled cAMP or the disappearance of the precursor.

Assays for the measurement of cAMP have been widely used to assist the deorphanization of oGPCRs (Tab. 4.2), including, for instance, the receptors for relaxin [45] and histamine H_3 [9].

4.4.3 Ca^{2+} Measurements

Various GPCRs elevate intracellular Ca^{2+} by triggering both Ca^{2+} release from intracellular stores and extracellular Ca^{2+} influx. A typical GPCR-mediated Ca^{2+} response is characterized by a rapid transient rise of the intracellular Ca^{2+} concentration, which is followed by a sustained elevation of the Ca^{2+} concentration. It is widely accepted that activation of Ca^{2+} mobilization through GPCRs is associated with the phospholipase C (PLC)-catalyzed hydrolysis of membrane inositide phospholipids. G_q-coupled receptors may activate G-proteins that belong to the family of G_q-proteins to activate PLC-mediated hydrolysis of phosphatidyl-4,5-biphosphate [$PI(4,5)P_2$] resulting in the formation of inositol-1,4,5-trisphosphate [$Ins(1,4,5)P_3$] and 1,2-diacylglycerol (DAG). $Ins(1,4,5)P_3$ receptors in the endoplasmic reticulum (ER) mediate the subsequent release of Ca^{2+} from the ER and thereby raise $[Ca^{2+}]_i$, whereas the released DAG may activate other effectors such as isoforms of protein kinase C (PKC) and protein kinase D (PKD), and affect $[Ca^{2+}]_i$ influx by activating transient receptor potential channel (TRPC) channels [46].

By monitoring the changes in the levels of intracellular Ca^{2+} in cells, Sakurai et al. have identified the orexins, a family of hypothalamic neuropeptides that were isolated from tissue extracts, as the natural ligands for previously orphan receptors, which are now known as the orexin receptors [47]. In their study they used Fura-2, one of many commercially available fluorophores that change their fluorescence characteristics upon binding to Ca^{2+}. In the last few years many more methods have become available that can be used for the measurement of changes in $[Ca^{2+}]_i$ [48]. One of the most known plate readers for the measurements of $[Ca^{2+}]_i$ is the Fluorometric Imaging Plate Reader (FLIPR™) which has also been used extensively for the deorphanization of oGPCRs. The FLIPR system monitors the response of cell populations in well plates to potential drug candidates. One of the latest oGPCRs to be deorphanized using the FLIPR is GPR40, a GPCR for long-chain fatty acids such as eicosatriynoic acid [49]. Another assay system for monitoring $[Ca^{2+}]_i$ is based on the aequorin photoprotein from the jelly fish *Aequorea victoria*. Binding of Ca^{2+} to aequorin results in the oxidation of the coelenterazine substrate to produce photons that can be detected by a luminometer (Fig. 4.4).

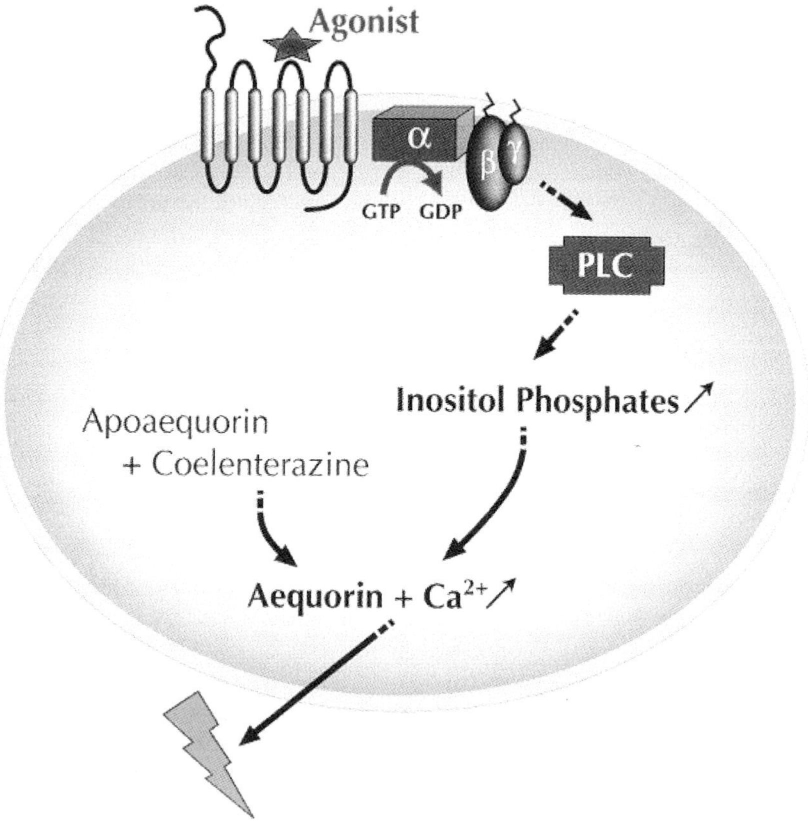

Fig. 4.4 Measuring receptor mediated changes in intracellular Ca^{2+} via the aequorin system. Receptor-stimulation mediated activation of PLC via either $G_{\alpha q/11}$ or, for instance, chimeric $G_{\alpha q5i}$ results in the formation of inositolphosphates and a subsequent rise in intracellular Ca^{2+} (see Sections 4.4.3 and 4.4.6). Interaction of Ca^{2+} with aequorin, a photoprotein that is coexpressed together with the receptor of interest, results in the oxidation of coelenterazine, with which the cells are loaded prior to receptor stimulation, to produce photons which can be detected by a luminometer. Figure kindly provided by Dr. P. Meoni (Euroscreen, Brussels, Belgium).

4.4.4
The Cytosensor Microphysiometer

In contrast to measuring changes in intracellular calcium or cAMP, more generic assays can also be used to measure receptor activation. The "cytosensor", for instance, is a microphysiometer that utilizes the changes in the extracellular pH that are accompanied by changes in cellular activity upon, for instance, GPCR activation, and has been successfully used to identify ligands for the CCR5 receptor [50] and for the pharmacological characterization of a variety of GPCRs [51]. The assay has a rather low throughput, a reason why many other assays have gained popularity over this particular assay system. However, the assay may be useful for deorphanizing GPCRs as there is no need for prior knowledge of the signal transduction pathways that may be activated by these oGPCRs.

4.4.5 Reporter Gene Assays

Changes in the levels of intracellular second messengers may also lead to alteration in expression of various genes due to the modulation of the activity of transcription factors. These transcription factors may bind to specific transcription factors present in the promotor regions of genes to enhance or repress gene expression. Reporter gene assays for monitoring the activation of various transcription factors include CRE, TRE, SRE, NF-κB and AP-1, mediated transcription of, for instance, luciferase and β-galactosidase, for which the enzyme activity can be assayed in the cell lysate, and green fluorescent protein (GFP). Reporter gene technology is now widely used to monitor the cellular events associated with signal transduction and gene expression. The principal advantage of these assays is their high sensitivity, reliability, convenience and adaptability to large-scale measurements [52]. Reporter gene assays have been used to assist the deorphanization of various oGPCRs, including, for instance, the receptor for sphingosylphosphorylcholine that promotes SRE-mediated gene transcription [53], as well as the histamine H_4 receptor [14].

In addition, reporter genes may be used as marker genes to quantify cell proliferative responses or the formation of cAMP. In these cases the reporter gene is not under direct transcriptional control as is the case for reporter gene assays [54]. A cell-based technology platform for the functional evaluation of drugable gene families, including GPCRs, is called Receptor Selection and Amplification Technology (R SAT™) and makes use of such a marker gene as a functional readout [55]. This proprietary assay has recently been used for the development of non-peptidergic ligands for the urotensin II receptor [56] and has shown that the application of cell-based assays may yield ligands with receptor-binding sites that are distinct from the binding site that is used by the native receptor ligand [57].

4.4.6 GPCR–G-protein Coupling

To increase the likeliness of identification of a native ligand for an orphan receptor, the functional screen that is used needs to be adaptable to a wide range of GPCRs. Thus, activation of GPCRs that couple to distinct family members of heterotrimeric G-proteins needs to result in a uniform response.

Receptors are *promiscuous* in their coupling specificity towards their binding to and activation of G-proteins, and as a result a receptor may activate G-proteins that belong to various G-protein families. Although some GPCRs couple to members from all families of heterotrimeric G-proteins [58], most GPCRs couple with some specificity to various G-proteins. The cellular context, however, will also determine which G-proteins are ultimately activated upon activation of a particular GPCR: apart from the affinity of particular GPCRs for the various G-proteins, the specific amounts of the G-proteins also expressed play a role. Based on this principle, specific G-proteins may be coexpressed together with GPCRs to promote the activation of these coexpressed G-proteins in order to funnel signal transduction through a common pathway. This technique has proven to work especially well with $G_{\alpha 16}$ and its murine homologue $G_{\alpha 15}$ [59, 60], G-proteins that display a very restricted expression pattern of endogenous expression [61]. These promiscuous G-proteins belong to the G_q family of G-proteins, activation of which results in the activation of PLC, resulting in the production of diacylglycerol and inositolphosphates, and a concomitant raise in intracellular levels of Ca^{2+} and its downstream effectors, and therefore appear ideal for screening purposes (see Section 4.4.3) [62].

Given that the expression of both the receptor and the G-protein are adequate, receptor promiscuity should in principle allow virtually any receptors to signal through specific G-proteins to activate a particular signal transduction cascade. Nonetheless, not all receptors appear to interact with G-proteins with the same affinity and for some receptors it is not very evident to set up a functional screening assay.

The C-terminus of the G_α-protein is a key domain in the GPCR–G-protein interaction [63]. Chimeric G-proteins, in which the last few amino acids have been replaced by corresponding amino acids that are present in other families of G-proteins, have been widely used in screening assays to facilitate a functional coupling of receptors, allowing the functional screening for ligands modifying the receptor activity [60, 64]. Probably the most commonly used chimeric G-proteins are those in which the last five amino acids of $G_{\alpha q}$ have been replaced by those of, for example, $G_{\alpha i}$, $G_{\alpha o}$ or $G_{\alpha z}$, and are commonly referred to as $G_{\alpha qi5}$, $G_{\alpha qo5}$ and $G_{\alpha qz5}$. These chimeric $G_{\alpha q}$-proteins have an increased promiscuity towards $G_{\alpha i/o}$-coupled receptors, allowing $G_{\alpha i/o}$-coupled receptors to couple to the rise in intracellular Ca^{2+} levels, and are routinely used in screening assays using the FLIPR (see Section 4.4.3) [65].

Alternatively, a cocktail of G-proteins, including chimeric G-proteins, can be coexpressed together with the oGPCR of interest to increase the likeliness that the oGPCR couples to appropriate G-proteins to produce a biological response and thereby increase the likeliness of identification of a native ligand for the oGPCR.

Receptor promiscuity can be extended by forcing the receptor to couple to a designated G-protein that physically links to the GPCR protein by means of a *fusion protein*. Such fusion proteins of receptors and G-proteins can be used as a tool for ligand screening and have now been successfully described for receptors fused to either G_s, G_i or G_q [66]. The agonist-bound receptors can interact with and activate the associated G-proteins in these fusion proteins. Such GPCR–G-protein fusions are used for screening purposes, as exemplified by the oGPCR TG1019 that was fused to a $G_{\alpha i1}$-protein and expressed in insect cells for the functional evaluation of TG10119 using GTPγS-binding assays, which resulted in the identification of several eicosanoids exhibiting agonistic activities against TG1019 [44].

4.4.7
Ligand-independent GPCR Activity

Heterologous expression of GPCRs may result in relatively high expression levels and often allows for the detection of agonist-independent or "constitutive receptor activity". This property of GPCRs is also found for several oGPCRs [67], including various viral GPCRs [68]. Recent studies have demonstrated that many antagonists, of a wide variety of different receptor types, actually possess the intrinsic ability to decrease constitutive receptor activity. Many antagonists have thus recently been reclassified as inverse agonists based on the application of a variety of functional assays that, unlike radioligand-binding techniques, can differentiate competitive antagonists from inverse agonists. Functional assays are successfully used to identify drug candidates that reduce cellular responses resulting from constitutive receptor activity, such as inverse agonists, and are the preferred drugs for treating diseases in which ligand-independent receptor activity may be important. One should keep in mind, however, that the observed apparent agonist-independent biological response may be the consequence of a permanent stimulation by a ubiquitous ligand that is produced by the cell line used to express the receptor.

Constitutive receptor activity is used as an approach to identify ligands for receptors. As not all wild-type GPCRs exhibit a high degree of constitutive activity, coexpression of these receptors with appropriate G-proteins may increase basal signaling properties and thus allow for better opportunities to identify compounds that reduce the intrinsic receptor activity of such receptors [69]. Alternatively, the receptors can be engineered such that "constitutively active mutant" (CAM) receptors are created by an amino acid substitution-induced alteration in receptor conformation. The CAM receptors are usually created based on an analogy of the effects on receptor activity that is found upon mutation of specific residues in other GPCRs. A sequence that is often considered when a CAM receptor is desired is the GPCR signature triplet amino acid sequence DRY that is found downstream of the third transmembrane domain (TM3) of most GPCRs and is involved in the interaction of the receptor with G-proteins. Another domain that can be considered to generate a CAM is TM6. The relative orientation of TM3 and TM6 is thought to be important for receptor activity – the change from an inactive to an active GPCR is thought to be accompanied by a change in the relative orientation of these two TM domains with a concomitant rotation of TM6.

The creation of a CAM receptor makes the (natural) agonist ligand obsolete to induce a functional receptor response in a cellular assay, which is the basis of constitutively activating receptor technology (CART) [70]. When such CAM receptors are created and used for functional screens, the principle is to identify ligands by their capability to reduce the intrinsic receptor activity; in other words, to identify inverse agonists for the mutated receptor. The identified ligand and/or close analogs thereof may then be used to study their potential interaction with the wild-type receptor.

4.4.8 Novel Screening Strategies

In addition to the screening methods describe above, several new technologies are becoming available for GPCR characterization and, thus, also for oGPCR screening.

GPCRs transduce most of their signals to heterotrimeric G-proteins. After exchange of GDP bound to the G_α subunit for GTP, the activated heterotrimeric G-proteins dissociate from the receptor into a G_α subunit bound to GTP and a $G_{\beta\gamma}$ heterotrimer. Fluorescence resonance energy transfer (FRET)-based assays utilize the mechanism of resonance energy transfer from a donor to an acceptor moiety, which is a function of the distance between these two moieties. Therefore, FRET-based assays can be used to monitor the distance between fluorophores. Recently, such a FRET-based assay has been adapted to the dissociation of the heterotrimeric G-proteins upon receptor activation [71], resulting in a functional assay that, in principle, could be used for any GPCR, provided that the fluorescent G-proteins interact with the GPCRs in the same manner as the wild-type G-proteins.

One of the latest technological additions to the field of measuring biological activities at the cellular level is high-content screening. The principle of high-content screening is based upon the spatial rearrangement of proteins that occurs during the initiation of signal transduction processes and therefore these kinds of assays are also known as protein translocation assays. Protein translocations can be monitored in time upon, for instance, receptor stimulation using fluorescent fusion proteins, e.g. with GFP from the jelly fish *A. victoria* [72]. The strength of the assay is the automated spatial recognition and quantification of the fluorophores within the cell by specifically designed algorithms, which can theoretically be done for several fluorophores simultaneously, thus allowing the automated monitoring of the activation of various signal transduction cascades in the same cell at the same time.

As many proteins may participate in the signal transduction cascades activated by a GPCR, a whole range of proteins can be considered for fluorescent labeling for high-content screening purposes, including GPCRs, G-proteins [73], arrestins [74] and transcription factors such as NF-κB [75].

GPCRs are normally translocated from the cellular surface to intracellular acidic endosomes upon prolonged stimulation by agonists. Therefore, fluorescent labeling of GPCRs, e.g. using GFP, as is also often used to confirm the correct targeting of the expressed GPCR to the plasma membrane and may in time allow the non-invasive detection of the tagged receptor upon agonist stimulation [76] (Fig. 4.5). Tagging the receptor, however, may alter the pharmacological properties of the receptor. Alternatively, fluorescently tagged proteins that play a role in receptor desensitization may also be used. Members of the arrestins are most widely used [74].

A novel method to determine ligand specificity for GPCRs is the immobilization of functional GPCRs on, for instance, a glass surface [77]. In this manner, many GPCRs can be assessed simultaneously for their ligand- or G-protein-binding po-

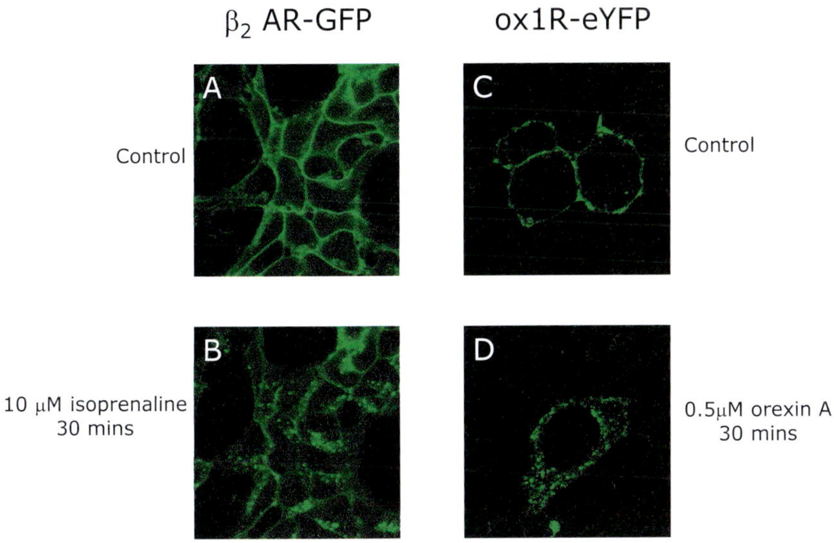

Fig. 4.5 Tagging of GPCRs at their intracellular C-terminal tail with a fluorescent protein allows tracking of the receptor proteins in time upon stimulation. GFP fusion proteins of the β_2-adrenergic (A) or orexin 1 receptor (C) results in a clear membrane-associated localization of the receptors. Stimulation of the β_2-adrenergic receptor with the β_2 receptor agonist isoprenaline (B) or the orexin 1 receptor with the agonist orexin A (D) results in the translocation of the fluorescently tagged receptors from the plasma membrane to intracellularly located vesicles. Figure provided by and printed with permission from Dr. G. Milligan (Glasgow University, UK).

tential using either radioactive or fluorescently labeled ligands or using surface plasmon resonance [78]. These techniques are still under development; however, the generation of arrays of functional GPCRs for novel HTS strategies will offer the potential for deorphanizing GPCRs by screening them against known ligands.

4.5 Future Prospects

The completion of the human genome sequence allows the development of potential therapeutic molecules directed against novel targets derived from the use of the genome data. The promise of genomics is the identification of novel drug targets, including GPCRs [1]; however, genomics is insufficient to determine whether a particular gene corresponds to an optimal drug target. The genomic revolution has propelled a flurry of technological advances, which promises to result in a rapid target validation process for disease etiology. In the absence of viable drug targets, the traditional target identification has been circumvented by initially searching for efficacious molecules in model systems for disease. As an illustration, although nicotinic acid has been used for decades for the treatment of dyslipidemia producing a desirable normalization of cardiovascular risk factors, the mechanism of action was unknown. Recently, two oGPCRs have been identified to be high- and low-affinity GPCRs for nicotinic acid, and provide the opportunity for the development of new therapeutics [79, 80]. Similarly, receptors for the hormone relaxin, one of the first reproductive hormones to be identified [81], have only recently been identified as LGR7 and LGR8 [45], two receptors that belong to the family of leucine-rich repeat-containing GPCRs. In terms of drug discovery opportunities, the orphan receptors provide incredible potential as drug discovery targets. Many drug discovery projects aim at an integrated approach enabling a systemic exploitation of the human GPCR genome by using a functional genomic approach to discover novel drug targets. The anticipated bridging of functional genomics-based target validation and high-throughput ligand identification processes will define the pipeline for pharmacological exploitation of genome data.

The search for native ligands for orphan receptors was thought to lead to the discovery of many new neuropeptides that are proposed to become the focus of may drug discovery programmes [82], since neuropeptides are thought to become the counterpart of the current pharmaceutical research that is mostly founded upon amino acid derivatives and biogenic amine systems. Indeed, recent orphan receptor programmes have, for example, identified the orphanin FQ/nociceptin [83, 84], the orexins/hypocretins [85, 86], the prolactin-releasing peptide [87, 88], the cyclic peptide urotensin II [89, 90], ghrelin [91] and apelin [92] as neuropeptide ligands for heretofore orphan receptors. The identification of the ligands for these receptors has boosted neuropeptide research, and these receptors are now thought to be implicated in a wide variety of physiological and pathophysiological conditions [93], including regulation of appetite [94], atherosclerosis [95] and pain transmission [39].

Recently, on the basis of sequence homology of an oGPCR, a whole family of *mas*-related genes encoding oGPCRs was identified that are only expressed in a specific subset of sensory neurons that are known to detect painful stimuli [96]. MrgA1 and mrgA4 were subsequently identified as potential NPFF and NPAF GPCRs, respectively [96]. MrgA1 and MrgC11 have also been shown to be specifically activated by surrogate agonists, suggesting endogenous ligands of Mrg receptors are likely to be RF(Y)G and/or RF(Y) amide-related peptides [97]. Interestingly, however, an oGPCR related to the MrgA1 has recently been identified as an adenine receptor [98]. Furthermore, SNSR1, a receptor that belongs to the family of sensory neuron-specific GPCRs, is closely related to MrgC11 and has been shown to be activated by proenkephalin A gene products [99].

Once an activating ligand has been found, the final, and unequivocally most challenging, step involves the use of the identified ligand to evaluate the physiological role of the receptor and its potential as a therapeutic target for drug discovery. Therefore, the challenge still remains to establish relevant biological cellular and physiological systems in which to test these molecules for efficacy. It is important to have comprehensive and detailed information on receptor and ligand expression profiles in both normal and diseased tissues at this stage in order to direct these studies most efficiently. In addition, the availability of transgenic or gene knockout animals can also be useful in evaluating receptor function. Interestingly, GPR8, recently identified as a neuropeptide W receptor [100], is absent in rodents [101], making target validation for such a receptor more difficult.

Ultimately, most, if not all, orphan receptors will be deorphanized. As yet, native ligands have not been identified for all oGPCRs. Although deorphanizing oGPCRs may seem rather straightforward, it is often not that simple. There are, for instance, oGPCRs for which we know synthetic ligands to bind and even activate, although as yet the natural ligands for the receptors remain elusive. An example of such a receptor for which synthetic agonist ligands have been identified is the bombesin 3 receptor [102]. Although the availability of these agonists has allowed receptor pharmacology studies, including the evaluation of the signaling pathways that are activated by this receptor, the natural ligands for these GPCRs remains unknown.

Some of the remaining oGPCRs have been attributed important physiological roles. For some of the oGPCRs expression is induced by particular stimuli, providing potential clues to the receptor function. Leukocyte-specific signal transducers and activators of transcription (STAT)-induced GPCR (LSSIG), for example, is a novel murine oGPCR with the highest homology to human GPR43, and these receptors are induced upon stimulation of STAT3, indicating these receptors might play pivotal roles in differentiation and immune response of monocytes and granulocytes [103]. Another example of an oGPCR with a potentially important physiological role is GPR30 [104], a GPCR that is associated with estrogen receptor expression in breast cancer and which appears to inhibit cell growth [105].

These oGPCRs provide incredible potential as drug targets for the development of valuable therapeutics. It is unlikely that the current interest in GPCRs as drug targets will diminish once more oGPCRs are paired with their natural ligand. Their involvement in such a multitude of cell processes makes them undeniable important. Further technological advances will only quicken the pace of drug discovery. Time will tell whether deorphanized GPCRs will provide the anticipated drug targets of the future.

References

1 JC Venter et al., *Science* **291**, 1304–51 (2001); TGIS Consortium, *Nature* **409**, 860–921 (2001).

2 P Ma, R Zemmel, *Nat Rev Drug Discov* **1**, 571–2 (2002).

3 AD Jensen et al., *J Biol Chem* **276**, 9279–90 (2001); P Ghanouni et al., *Proc Natl Acad Sci USA* **98**, 5997–6002 (2001); T Okada et al., *Trends Biochem Sci* **26**, 318–24 (2001); JA Ballesteros et al., *Mol Pharmacol* **60**, 1–19 (2001).

4 T Kenakin, *Annu Rev Pharmacol Toxicol* **42**, 349–79 (2002); A Christopoulos, T Kenakin, *Pharmacol Rev* **54**, 323–74 (2002); T Kenakin, *Nat Rev Drug Discov* **1**, 103–10 (2002).

5 RA Dixon et al., *Nature* **321**, 75–9 (1986).

6 J Nathans, DS Hogness, *Cell* **34**, 807–14 (1983).

7 A Fargin et al., *Nature* **335**, 358–60 (1988).

8 MD De Backer et al., *Biochem Biophys Res Commun* **197**, 1601–8 (1993); H Fukui et al., *Biochem Biophys Res Commun* **201**, 894–901 (1994); N Moguilevsky et al., *Eur J Biochem* **224**, 489–95 (1994); I Gantz et al.,

Biochem Biophys Res Commun **178**, 1386–92 (1991); BA Chowdhury et al., *Soc Neurosci Abstr* **19**, 84 (1993).
9 TW Lovenberg et al., *Mol Pharmacol* **55**, 1101–7 (1999).
10 R Leurs et al., *Trends Pharmacol Sci* **21**, 11–2 (2000).
11 DG Raible et al., *Am J Respir Crit Care Med* **149**, 1506–11 (1994).
12 T Nakamura et al., *Biochem Biophys Res Commun* **279**, 615–20 (2000); Y Zhu et al., *Mol Pharmacol* **59**, 434–41 (2001); KL Morse et al., *J Pharmacol Exp Ther* **296**, 1058–66 (2001); T Oda et al., *J Biol Chem* **275**, 36781–6 (2000).
13 T Nguyen et al., *Mol Pharmacol* **59**, 427–33 (2001); LB Hough, *Mol Pharmacol* **59**, 415–9 (2001).
14 C Liu et al., *Mol Pharmacol* **59**, 420–6 (2001).
15 O Civelli et al., *Trends Neurosci* **24**, 230–7 (2001).
16 MJ Smit et al., *Pharmac Acta Helv* **74**, 299–304 (2000).
17 E Cesarman et al., *J Exp Med* **191**, 417–22 (2000).
18 P Casarosa et al., *J Biol Chem* **276**, 1133–7 (2001).
19 DN Streblow et al., *Cell* **99**, 511–20 (1999).
20 P Casarosa et al., *J Biol Chem* **278**, 5172–8 (2003).
21 F Staubli et al., *Proc Natl Acad Sci USA* **99**, 3446–51 (2002); Y Park et al., *Proc Natl Acad Sci USA* **99**, 11423–8 (2002); H-J Kreienkamp et al., *J Biol Chem* **277**, 39937–43 (2002); G Cazzamali et al., *Biochem Biophys Res Commun* **298**, 31–6 (2002); G Cazzamali et al., *Proc Natl Acad Sci USA* **99**, 12073–8 (2002); S Nishi et al., *Endocrinology* **141**, 4081–90 (2000); N Birgul et al., *EMBO J* **18**, 5892–900 (1999).
22 MR Bennett, *Neuropharmacology* **39**, 523–46 (2000); AH Maehle et al., *Nat Rev Drug Discov* **1**, 637–41 (2002); DJ Triggle, *Pharmac Acta Helv* **74**, 79–84 (2000).
23 P Vernier et al., *Trends Pharmacol Sci* **16**, 375–81 (1995).
24 P Joost et al., *Genome Biol* **3**, RESEARCH0063. 1–16 (2002).
25 RH Waterston et al., *Nature* **420**, 520–62 (2002).
26 U-K Kim et al., *Science* **299**, 1221–5 (2003).
27 BP Zambrowicz et al., *Nat Rev Drug Discov* **2**, 38–51 (2003).
28 S Tsukada et al., *Circulation* **107**, 313–9 (2003).
29 A Abuin et al., *Trends Biotechnol* **20**, 36–42 (2002).
30 F Quan, L Thomas, M Forte, *Proc Natl Acad Sci USA* **88**, 1898–902 (1991).
31 PM Sexton et al., *Cell Signal* **13**, 73–83 (2001); SM Foord et al., *Trends Pharmacol Sci* **20**, 184–7 (1999).
32 LM McLatchie et al., *Nature* **393**, 333–9 (1998).
33 ME Nuttall et al., *J Biomol Screen* **4**, 269–78 (1999); G Chen et al., *Mol Pharmacol* **57**, 125–34 (2000).
34 MH Pausch, *Trends Biotechnol* **15**, 487–94 (1997).
35 S Wilson et al., *Br J Pharmacol* **125**, 1387–92 (1998); GC Zondag et al., *Biochem J* **330**, 605–9 (1998).
36 JK Chambers et al., *J Biol Chem* **275**, 10767–71 (2000).
37 AJ Brown et al., *J Biol Chem* **19**, 19 (2002).
38 C Klein et al., *Nat Biotechnol* **16**, 1334–7 (1998).
39 E Okuda-Ashitaka et al., *Nature* **392**, 286–9 (1998).
40 M Matloubian et al., *Nat Immunol* **1**, 298–304 (2000).
41 VG Krasnoperov et al., *Neuron* **18**, 925–37 (1997).
42 K Ichtchenko et al., *J Biol Chem* **274**, 5491–8 (1999).
43 JJ Carrillo et al., *J Pharmacol Exp Ther* **302**, 1080–8 (2002).
44 T Hosoi et al., *J Biol Chem* **277**, 31459–65 (2002).
45 SY Hsu et al., *Science* **295**, 671–4 (2001).
46 T Hofmann et al., *Nature* **397**, 259–63 (1999); L Birnbaumer et al., *Proc Natl Acad Sci USA* **93**, 15195–202 (1996); MD Bootman et al., *Cell* **83**, 675–8 (1995); MK Topham et al., *J Biol Chem* **274**, 11447–50 (1999); K Kiselyov et al., *Trends Neurosci* **22**, 334–7 (1999).

47 T Sakurai et al., *Cell* **92**, 573–85 (1998).

48 DA Zacharias et al., *Curr Opin Neurobiol* **10**, 416–21 (2000).

49 CP Briscoe et al., *J Biol Chem*, M211495200 (2002).

50 M Samson et al., *Biochemistry* **35**, 3362–7 (1996).

51 YL Wu et al., *Cell Res* **7**, 69–77 (1997); MM Berglund et al., *Peptides* **20**, 1043–53 (1999); S Ahmad et al., *Ann NY Acad Sci* **863**, 108–19 (1998).

52 SJ Hill et al., *Curr Opin Pharmacol* **1**, 526–32 (2001); LH Naylor, *Biochem Pharmacol* **58**, 749–57 (1999).

53 K Zhu et al., *J Biol Chem* **276**, 41325–35 (2001).

54 MJ Smit et al., *Methods Enzymol* **343**, 430–47.

55 GE Croston, *Trends Biotechnol* **20**, 110–5 (2002).

56 GE Croston et al., *J Med Chem* **45**, 4950–3 (2002).

57 TA Spalding et al., *Mol Pharmacol* **61**, 1297–302 (2002).

58 KL Laugwitz et al., *Proc Natl Acad Sci USA* **93**, 116–20 (1996).

59 G Milligan et al., *Trends Pharmacol Sci* **17**, 235–7 (1996).

60 E Kostenis, *Trends Pharmacol Sci* **22**, 560–4 (2001).

61 TT Amatruda, 3rd, et al., *Proc Natl Acad Sci USA* **88**, 5587–91 (1991).

62 S Offermanns et al., *J Biol Chem* **270**, 15175–80 (1995).

63 HE Hamm, *Proc Natl Acad Sci USA* **98**, 4819–21 (2001); HR Bourne, *Curr Opin Cell Biol* **9**, 134–42 (1997).

64 G Milligan, S Rees, *Trends Pharmacol Sci* **20**, 118–24 (1999).

65 P Coward et al., *Anal Biochem* **270**, 242–8 (1999); SM Mody et al., *Mol Pharmacol* **57**, 13–23 (2000).

66 G Milligan, *Trends Pharmacol Sci* **21**, 24–8 (2000); R Seifert et al., *Trends Pharmacol Sci* **20**, 383–9 (1999); B Bertin et al., *Proc Natl Acad Sci USA* **91**, 8827–31 (1994); Z-D Guo et al., *Life Sci* **68**, 2319–27 (2001).

67 D Eggerickx et al., *Biochem J* **309**, 837–43 (1995).

68 PM Murphy, *Nat Immunol* **2**, 116–22 (2001).

69 ES Burstein et al., *Mol Pharmacol* **51**, 312–9 (1997); ES Burstein et al., *FEBS Lett* **363**, 261–3 (1995).

70 D Chalmers et al., *Nat Rev Drug Discov* **1**, 599–608 (2002).

71 C Janetopoulos et al., *Science* **291**, 2408–11 (2001).

72 RN Ghosh et al., *Biotechniques* **29**, 170–5 (2000); BR Conway et al., *J Biomol Screen* **4**, 75–86 (1999).

73 J-Z Yu, MM Rasenick, *Mol Pharmacol* **61**, 352–9 (2002).

74 DA Groarke et al., *J Biol Chem* **274**, 23263–9 (1999).

75 KL Pierce et al., *Nat Rev Neurosci* **2**, 727–33 (2001).

76 SSG Ferguson, *Pharmacol Rev* **53**, 1–24 (2001); L Kallal et al., *Trends Pharmacol Sci* **21**, 175–80 (2000); L Kallal et al., *Methods Enzymol* **343**, 492–506 (2002); G Milligan, *Br J Pharmacol* **128**, 501–10 (1999).

77 Y Fang et al., *Chembiochem* **3**, 987–91 (2002); Y Fang et al., *J Am Chem Soc* **124**, 2394–5 (2002); L Neumann et al., *Chembiochem* **3**, 993–8 (2002).

78 MA Cooper, *Nat Rev Drug Discov* **1**, 515–28 (2002); C Bieri et al., *Nat Biotechnol* **17**, 1105–8 (1999).

79 S Tunaru et al., *Nat Med* **3**, 3 (2003).

80 A Wise et al., *J Biol Chem*, M210695200 (2003).

81 FL Hisaw, *Proc Soc Exp Biol Med* **23**, 661 (1926).

82 O Civelli et al., *Brain Res* **848**, 63–5 (1999).

83 JC Meunier et al., *Nature* **377**, 532–5 (1995); RK Reinscheid et al., *Science* **270**, 792–4 (1995).

84 E Okuda-Ashitaka et al., *Brain Res Mol Brain Res* **43**, 96–104 (1996).

85 L de Lecea et al., *Proc Natl Acad Sci USA* **95**, 322–7 (1998).

86 T Sakurai et al., *Cell* **92**, 573–85 (1998).

87 S Hinuma et al., *J Mol Med* **77**, 495–504 (1999).

88 S Hinuma et al., *Nature* **393**, 272–6 (1998).

89 Y Coulouarn et al., *FEBS Lett* **457**, 28–32 (1999).

90 RS Ames et al., *Nature* **401**, 282–6 (1999); M Mori et al., *Biochem Biophys Res Commun* **265**, 123–9 (1999); HP

Nothacker et al., *Nat Cell Biol* **1**, 383–5 (1999); Q Liu et al., *Biochem Biophys Res Commun* **266**, 174–8 (1999).
91 M Kojima et al., *Nature* **402**, 656–60 (1999).
92 K Tatemoto et al., *Biochem Biophys Res Commun* **251**, 471–6 (1998).
93 L de Lecea et al., *Cell Mol Life Sci* **56**, 473–80 (1999).
94 M Kojima et al., *Curr Opin Pharmacol* **2**, 665–8 (2002).
95 SD Katugampola et al., *Clin Sci (Lond)* **103** (Suppl 48), 171S–5S (2002).
96 X Dong et al., *Cell* **106**, 619–32 (2001).
97 SK Han et al., *Proc Natl Acad Sci USA* **99**, 14740–5 (2002).
98 E Bender et al., *Proc Natl Acad Sci USA* **99**, 8573–8 (2002).
99 PM Lembo et al., *Nat Neurosci* **19**, 19 (2002).
100 Y Shimomura et al., *J Biol Chem* **277**, 35826–32 (2002).
101 DK Lee et al., *Mol Brain Res* **71**, 96–103 (1999).
102 D Weber et al., *J Peptide Sci* **8**, 461–75 (2002); SA Mantey et al., *J Biol Chem* **272**, 26062–71 (1997); TK Pradhan et al., *Eur J Pharmacol* **343**, 275–87 (1998).
103 T Senga et al., *Blood* **101**, 1185–7 (2003).
104 C Carmeci et al., *Genomics* **45**, 607–17 (1997).
105 TM Ahola et al., *Endocrinology* **143**, 3376–84 (2002).
106 C Maenhaut et al., *Biochem Biophys Res Commun* **173**, 1169–78 (1990).
107 F Libert et al., *EMBO J* **10**, 1677–82 (1991).
108 C Maenhaut et al., *Biochem Biophys Res Commun* **180**, 1460–8 (1991).
109 Z Fathi et al., *J Biol Chem* **268**, 5979–84 (1993).
110 D Gerald et al., *J Biol Chem* **270**, 26758–61 (1995).
111 RS Ames et al., *J Biol Chem* **271**, 20231–4 (1996).
112 CC Bleul et al., *Nature* **382**, 829–33 (1996).
113 N Aiyar et al., *J Biol Chem* **271**, 11325–9 (1996).
114 AD Howard et al., *Science* **273**, 974–7 (1996).
115 T Yokomizo et al., *Nature* **387**, 620–4 (1997).
116 H Itadani et al., *Biochem Biophys Res Commun* **250**, 68–71 (1998); J Cao et al., *J Biol Chem* **273**, 32281–7 (1998).
117 MD Gunn et al., *Nature* **391**, 799–802 (1998).
118 MJ Lee et al., *Science* **279**, 1552–5 (1998); JR Van Brocklyn et al., *Blood* **95**, 2624–9 (2000); JR Van Brocklyn et al., *J Biol Chem* **274**, 4626–32 (1999); Y Yamazaki et al., *Biochem Biophys Res Commun* **268**, 583–9 (2000); DS Im et al., *J Biol Chem* **275**, 14281–6 (2000); K Sato et al., *Biochem Biophys Res Commun* **253**, 253–6 (1998); K Sato et al., *FEBS Lett* **443**, 25–30 (1999); N Ancellin, T Hla, *J Biol Chem* **274**, 18997–9002 (1999).
119 DS Im et al., *Mol Pharmacol* **57**, 753–9 (2000); JR Erickson et al., *J Biol Chem* **273**, 1506–10 (1998); S An et al., *J Biol Chem* **273**, 7906–10 (1998); S An et al., *Biochem Biophys Res Commun* **231**, 619–22 (1997).
120 PM Lembo et al., *Nat Cell Biol* **1**, 267–71 (1999); Y Shimomura et al., *Biochem Biophys Res Commun* **261**, 622–6 (1999); D Bachner et al., *FEBS Lett* **457**, 522–4 (1999); Y Saito et al., *Nature* **400**, 265–9 (1999); J Chambers et al., *Nature* **400**, 261–5 (1999).
121 SD Feighner et al., *Science* **284**, 2184–8 (1999).
122 HM Sarau et al., *Mol Pharmacol* **56**, 657–63 (1999); KR Lynch et al., *Nature* **399**, 789–93 (1999).
123 M Kamohara et al., *J Biol Chem* **275**, 27000–4 (2000).
124 AD Howard et al., *Nature* **406**, 70–4 (2000); R Raddatz et al., *J Biol Chem* **275**, 32452–9 (2000).
125 JA Hedrick et al., *Mol Pharmacol* **58**, 870–5 (2000); PG Szekeres et al., *J Biol Chem* **275**, 20247–50 (2000); M Kojima et al., *Biochem Biophys Res Commun* **276**, 435–8 (2000); R Fujii et al., *J Biol Chem* **275**, 21068–74 (2000).
126 M Hosoya et al., *J Biol Chem* **275**, 29528–32 (2000); L Shan et al., *J Biol Chem* **275**, 39482–6 (2000).

127 H Hirai et al., *J Exp Med* **193**, 255–61 (2001).
128 NA Elshourbagy et al., *J Biol Chem* **275**, 25965–71 (2000).
129 S Hinuma et al., *Nat Cell Biol* **2**, 703–8 (2000); JA Bonini et al., *J Biol Chem* **275**, 39324–31 (2000).
130 Y Xu et al., *Nat Cell Biol* **2**, 261–7 (2000).
131 J Takasaki et al., *Biochem Biophys Res Commun* **274**, 316–22 (2000); CE Heise et al., *J Biol Chem* **275**, 30531–6 (2000); HP Nothacker et al., *Mol Pharmacol* **58**, 1601–8 (2000).
132 DI Jarmin et al., *J Immunol* **164**, 3460–4 (2000).
133 B Borowsky et al., *Proc Natl Acad Sci USA* **98**, 8966–71 (2001).
134 J Hill et al., *J Biol Chem* **276**, 20125–9 (2001).
135 FL Zhang et al., *J Biol Chem* **276**, 8608–15 (2001).
136 B Borowsky et al., *Proc Natl Acad Sci USA* **17**, 17 (2001).
137 DS Im et al., *J Cell Biol* **153**, 429–34 (2001).
138 JH Kabarowski et al., *Science* **293**, 702–5 (2001).
139 T Ohtaki et al., *Nature* **411**, 613–7 (2001); AI Muir et al., *J Biol Chem* **276**, 28969–75 (2001).
140 D Kalant et al., *J Biol Chem* **278**, 11123–9 (2003); SA Cain, PN Monk, *J Biol Chem* **277**, 7165–9 (2002).
141 T Maruyama et al., *Biochem Biophys Res Commun* **298**, 714–9 (2002).
142 A Ignatov et al., *J Neurosci* **23**, 907–14 (2003).

Part II
Synthesis

Molecular Biology in Medicinal Chemistry. Edited by Th. Dingermann, D. Steinhilber, G. Folkers

ISBN: 3-527-30431-2

5
Stereoselective Synthesis with the Help of Recombinant Enzymes

Nagaraj N. Rao and Andreas Liese

5.1
Stereoselective Synthesis before the Advent of Genetic Engineering

One of the earliest recorded stereoselective synthetic processes using naturally occurring microorganisms was the manufacture of optically active lactic acid **1** by fermentation in 1880 (Fig. 5.1).

In 1858, Pasteur had earlier demonstrated the microbial resolution of racemic tartaric acid. The fermentation of the ammonium salt of racemic tartaric acid with *Penicillium glaucum* yielded (*S*,*S*)-tartaric acid **2** [1].

As the discovery of new drugs, both natural and synthetic in origin, progressed and their structures were elucidated, they were obtained either by extraction from natural sources or by classical organic synthetic procedures. In the latter case, this generally led to the formation of racemic mixtures when the molecules had optically active centers. Table 5.1 lists a few drugs that were available in the 1980s as racemic mixtures.

It was soon recognized that the different enantiomers of an optically active drug molecule did not necessarily exhibit the same pharmacological actions. On the contrary, in several cases, they exhibited totally different areas of pharmacological activity and were sometimes even toxic. Table 5.2 gives examples of a few drugs whose enantiomers display dissimilar pharmacological activities.

Research into drug molecule structure–activity relationships has greatly helped the understanding of the pharmacological actions of drug molecules. The demand for chiral drugs is based on the following:

- The cell-surface receptors are biological molecules which are often chiral by themselves.
- Efficient drug molecules must match the receptor's symmetry.
- Compliance with regulatory requirements. The US Food and Drug Administration, for example, now instructs pharmaceutical research companies to rigidly evaluate whether or not a novel drug molecule can be produced as a single isomer [2].

Molecular Biology in Medicinal Chemistry. Edited by Th. Dingermann, D. Steinhilber, G. Folkers

ISBN: 3-527-30431-2

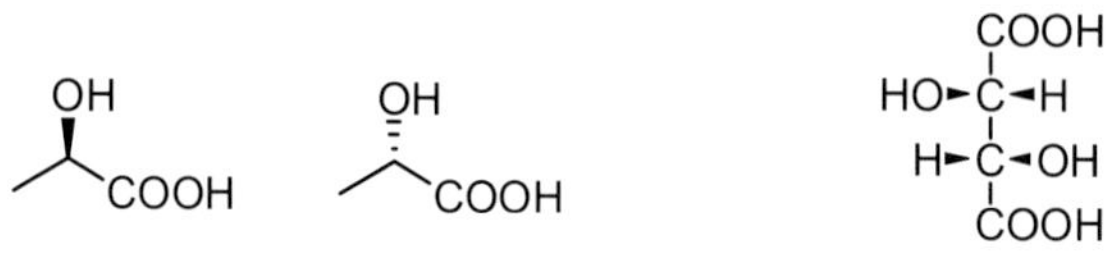

(R)-1 **(S)-1** **2**

Fig. 5.1 (*R*)- and (*S*)-lactic acid **1**; (*S*,*S*)-tartaric acid (**2**).

It was well known that fermentation processes involved stereoselective steps, e.g. in the case of the penicillins, and the ability of enzymes to catalyze chemical reactions in a stereoselective manner was recognized quite early. Whole-cell fermentation and biotransformation with isolated enzymes thus laid the foundations for stereoselective synthesis. As early as 1921, Neuberg and Hirsch discovered that the condensation of benzaldehyde **10** and acetaldehyde in the presence of yeast gives optically active 1-hydroxy-1-phenyl-2-propanone **12** (Fig. 5.2) with the formation of pyruvic acid **11** as an intermediate [3]. The enzyme pyruvate decarboxylase catalyzes the reaction. The German pharmaceutical firm Knoll AG used this compound to chemically synthesize L-(−)-ephedrine **13** by a process patented in 1930 [4].

Another milestone in stereoselective biotransformations before the advent of genetic engineering was the discovery of the stereo- and regioselective hydroxylation by *Rhizopus arrhuis* of progesterone **14** to 11-α-hydroxy progesterone **15**, a valuable intermediate for the synthesis of cortisone **16** (Fig. 5.3). The chemical synthesis of cortisone is extremely tedious and economically unviable. The discovery of the microbial route for hydroxylation dramatically reduced the price of cortisone to 3% of the then prevailing prices [5, 6].

The variety of substrates and the types of reactions being stereoselectively catalyzed by biocatalysts are increasing steadily. Examples of typical stereoselective reactions being carried out with the help of natural biocatalysts in the field of medicinal chemistry are listed in Table 5.3.

5.2
Classical Methods of Strain Improvement for Stereoselective Synthesis

The traditional methods to identify new enzymes and microorganisms are based on screening techniques, e.g. of soil samples, or strain collection by enrichment cultures. Generally, the microorganisms isolated from nature produce the desired enzyme at levels that are too low to offer a cost-effective production process. Therefore, their cloning and overexpression would be highly desirable for process development. Not all the microorganisms can be cultured using common fermentation technology. The percentage of accessible microorganisms from those present in a sample is estimated to be between 0.001 and 1%, depending on the sample's origin. The problem is compounded by the fact that only a few of the enzymes found in nature are suitable for a certain synthetic or scaling-up problem – characteristics

Tab. 5.1 Examples of drugs available in the 1980s as racemic mixtures.

Name	***Chemical structure***	***Remarks***
Dobutamine	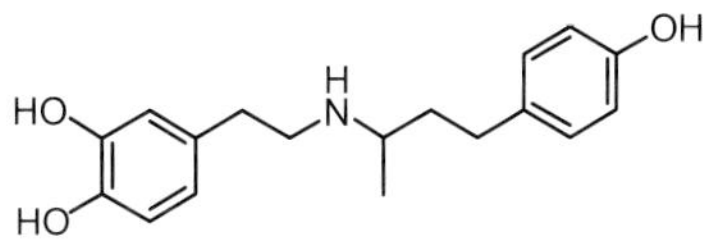 **3**	cardiotonic
Verapamil	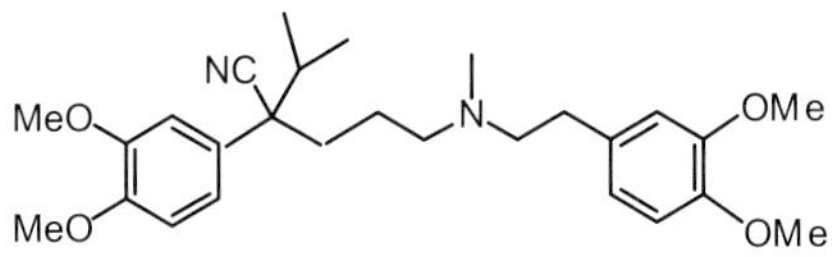 **4**	calcium antagonist, coronary dilator
Norfenefrine	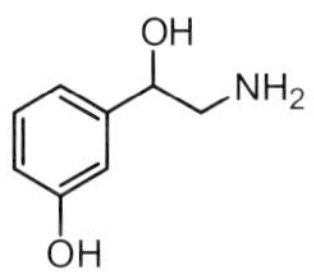 **5**	adrenergic
Fenfluramine	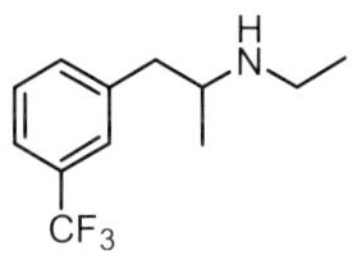 **6**	anorexic
Ibuprofen	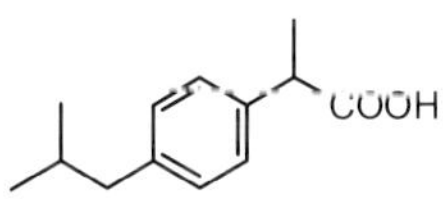 **7**	anti-inflammatory
Atenolol	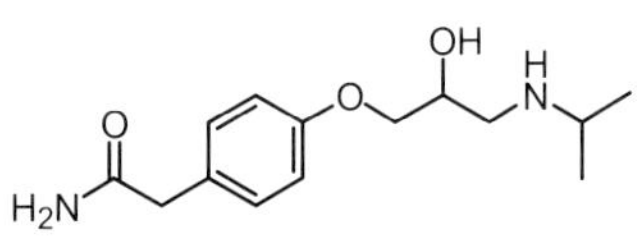 **8**	β-adrenergic blocker
Oxazepam	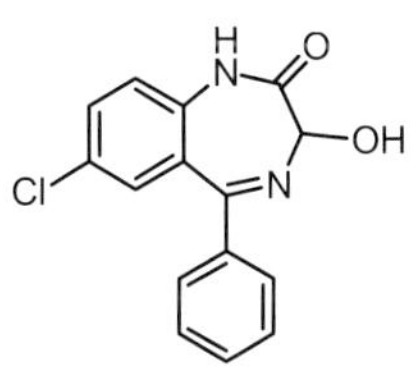 **9**	sedative

Tab. 5.2 Influence of stereochemistry on pharmacological activity of some drugs.

Name	*D-Enantiomer*	*L-Enantiomer*
Methyldopa	not active	anti-hypertensive
Penicillamine	anti-arthritic	highly toxic
Propoxyphene	analgesic	anti-tussive
Methorphan	anti-tussive	analgesic
Isoproterenol	high bronchodilatory activity	low bronchodilatory activity

such as activity, stability, substrate specificity and enantioselectivity may leave much to be desired [7].

Until recently, these limitations were overcome by methods such as screening for alternative biocatalysts, changes of the reaction system (e.g. from hydrolysis to esterification), modification of reaction parameters, etc. For this reason, the modification of the microorganism would be highly desirable for process development. Strain improvement by direct evolution, i.e. improvement by mutation and selection, has been successfully used in many industrial microbial applications for many years [8]. Success with the method of random genetic mutation and recombination caused by *in vitro* directed evolution of enzymes has been significant. Subsequent screening or selection for desired properties leads to the identification of the desired enzyme. When evolution processes are carried out in the laboratory, this so-called microevolution occurring in bacterial cultures grown in the chemostat gives rise to altered enzyme specificity. The rate of production of existing enzymes and the expression of previously dormant genes are also typically affected by

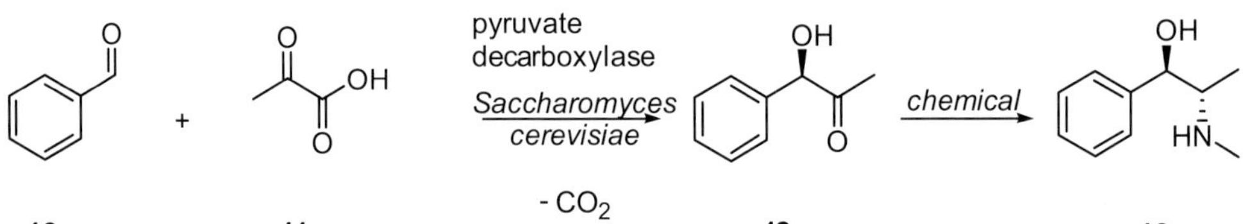

10 11 12 13

Fig. 5.2 L-Ephedrine production using microbial, stereoselective biotransformation.

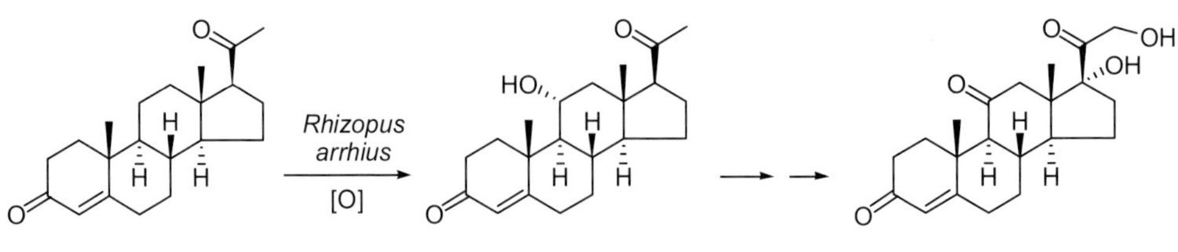

14 15 16

Fig. 5.3 Microbial stereo- and regioselective reduction of progesterone to 11-α-hydroxy progesterone.

Tab. 5.3 Typical stereoselective reactions carried out with the help of natural biocatalysts.

Reaction type	*Biocatalyst*	*Substrate*	*Product*
Oxidation	oxygenase	simvastatin	6-β-hydroxymethylsimvastatin
Oxidative deamination	D-amino acid oxidase	cephalosporin C	α-keto-adipinyl-7-ACA
Hydroxylation	benzoate dioxygenase	benzene	1,2-*cis*-dihydroxycatechol
Reduction	pyruvate decarboxylase	benzaldehyde	(1*R*)-phenylacetylcarbinol
Reductive amination	leucine dehydrogenase	trimethylpyruvic acid	L-*tert*-leucine
Esterification	lipase	3-benzylthio-2-methylpropanoic acid	zofenopril side-chain
Hydrolysis	lipase	(*S*)-ibuprofen methoxy-ethyl ester	ibuprofen
Isomerization	isomerase	glucose	D-fructose
Aldol condensation	aldolase	*N*-acetylmannosamine	*N*-acetylneuraminic acid

microevolution [9]. Classical methods of strain improvement also relied heavily on the use of chemical mutagens and creative selection techniques to identify superior strains for achieving a certain objective. These methods have been successfully used in the industrial production of antibiotics, vitamins and amino acids. However, the genetic and metabolic profiles of mutant strains were poorly characterized and mutagenesis remained a random process [10].

The classical methods of strain improvement can be summarized as follows:

- Screening techniques
- Strain collection by enrichment cultures
- Random genetic mutation
- *In vitro* directed evolution
- Chemical mutagenesis

Traditional mutagenesis does not have the ability to endow a cell with genes from other microorganisms nor can it impart their functions. Such possibilities have been opened up with the advent of recombinant DNA technology [11].

5.3 Genetic Engineering and the Advent of Recombinant Enzymes for Stereoselective Synthesis

New technologies for enzyme discovery and tailoring are the key drivers for a renaissance in the industrial application of enzymes. Stereoselectivity is one of the major advantages of biocatalysis routes for the production of chiral compounds.

Genetic engineering together with the identification and amplification of relevant structural genes have become viable alternatives to mutagenesis and random screening procedures. The introduction of genes into organisms by employing recombinant DNA techniques is a powerful method for the construction of strains with desired genotypes. The net effect is to place the transferred gene into a new host to accomplish a specific biotechnological purpose. The opportunity to introduce heterologous genes and regulatory elements enables the construction of metabolic configurations with novel beneficial characteristics. Furthermore, this approach avoids the complication of uncharacterized mutations that are frequently obtained with classical mutagenesis of whole cells. The improvement of cellular activities by manipulation of enzymatic, transport and regulatory functions of the cell using recombinant DNA technology leads to industrially relevant strains of microorganisms [12, 13].

During the past two decades, gene clusters for the production of secondary metabolites, especially antibiotics, have been identified and discovered for many industrial microorganisms by employing recombinant DNA techniques [14]. In the course of the last decade, progress in biochemistry, protein chemistry, molecular cloning, random and site-directed mutagenesis, and fermentation technology has opened up unlimited access to a variety of enzymes and microbial cultures as tools in organic synthesis. This process has been further accelerated by combinatorial chemistry, high-throughput biological screening and structure-based drug design. However, success has been modest, partly because of the difficulty involved in predicting the exact protein structure required for the stereoselective reaction of a given substrate [15].

There has been significant progress in the establishment of mutation and selection methodology. DNA shuffling technology that mimics natural evolution is employed for artificial DNA recombination and a phage-displayed combinatorial peptide library offers rapid selection for proteins expressed for mutated genes.

The rapid progress in genetic engineering and the development of many new recombinant DNA techniques has provided us with the tools to design, modify and engineer naturally occurring biomolecules. For industrial applications, the engineered biomolecules have to be stable and active for prolonged use under unusual, often harsh, bioprocess conditions within wide ranges of temperature, pH and reaction environment. Thus, sustained activity and stability may also be required in the presence of strong solvents, highly reactive substrates, high salt concentrations or extremes of pH. For this reason, efforts are directed at engineering of cellular physiology in order to achieve process improvement. The ultimate aim of metabolic engineering is the design and creation of suitable biocatalysts, which are capable of leading to maximum productivity and yields of desired products. The elimination or reduction of byproduct formation is also an important goal. Genetic selection allows us to alter the selectivity of biocatalysts, increase their substrate range and thermostability, and confer new functions to existing proteins. It also enables us to optimize multistep processes, and identify novel peptide ligands for biological receptors and characterize ligand–receptor interactions *in vivo* [16].

To address industrially relevant problems in enzyme processes, attempts have been made to design enzymes that will confer or enhance useful properties by using biomolecular engineering techniques. Although a rational design approach based on structure–function relationships is widely used, the directed evolution approach is generally favored for many industrial enzymes due to the difficulties of relating the desired application to the required properties. Functional genomics, proteomics, fluxomics and physiomics are expected to contribute significantly to metabolic pathway engineering [17].

Bacteria are ideally suited to serve as hosts for recombinant DNA. They grow rapidly in nutrient broth and readily accept foreign DNA to be copied along with their own genes. Many bacteria possess extrachromosomal DNA (plasmids) that replicate separately from the bacterial genome. Until the mid-1990s, *Escherichia coli* was the dominant host for the production of polypeptide and protein pharmaceuticals, in terms of economic value. By using a novel tRNA codon pair, an aminoacyl-tRNA synthetase and an amino acid, the genetic code of *E. coli* has been expanded to incorporate unnatural amino acids "with a fidelity rivaling that of natural amino acids" [18]. Mammalian cell production has overtaken *E. coli* in recent years in terms of economic value [19].

The common modern methods of strain improvement are summarized below:

- Identification and amplification of relevant structural genes
- Recombinant DNA and recombinant RNA techniques to manipulate enzymatic, transport and regulatory functions
- Site-directed mutagenesis, particularly of active-site residues
- Random mutagenesis by protein truncation with selection
- Molecular cloning
- Combinatorial chemistry
- High-throughput biological screening
- DNA shuffling techniques
- Altering protein topology by quaternization, by using monomeric mutases or directed evolution

Once genetic engineering techniques began to yield microorganisms and enzymes with the desired levels of activities and other properties for industrial exploitation, several processes for stereoselective syntheses have been developed or improved upon. This, in turn, has had a profound influence on the manufacturing methods for several pharmacologically active compounds, amino acids, vitamins and other compounds of medicinal interest. Genetic engineering techniques have enabled the large-scale supply of many enzymes at affordable prices [20]. In fact, biocatalysts have been commercially employed for the synthesis of complex drug intermediates, specialty chemicals, and also commodity chemicals for use by the food, pharmaceutical and agrochemical industries [21].

5.4
β-Lactam Antibiotics

Biosynthetic studies have revealed that antibiotics are derived from primary metabolic precursors. Examples of such precursors include the short-chain fatty acids (acetate and propionate) and amino acids. The structural genes of several of these antibiotic-synthesizing enzymes have been successfully cloned and characterized. In the case of the *β*-lactam antibiotics, this includes the structural genes of L-*δ*-(aminoadipoyl)-l-cysteinyl-D-valine (ACV) synthase. ACV is the tripeptide intermediate in the biosynthetic pathway of the *β*-lactam antibiotics.

5.4.1
Penicillins

Penicillin acylase from *E. coli* has been widely studied with regard to the synthesis of semisynthetic antibiotics. Strategies employed for improving the expression of penicillin acylase include changing the carbon source, coexpression of penicillin acylase with chaperone proteins, increasing the transcriptional and translational efficiency, and using more suitable combinations of host/vector systems [22].

The need to systematically develop an optimal expression strategy for the overproduction of penicillin acylase in *E. coli* has become extremely relevant. This is because the product, 6-aminopenicillanic acid (6-APA) **18**, formed by the hydrolysis of penicillin G **17**, is a crucial starting compound for the synthesis of a large number of semisynthetic *β*-lactam antibiotics **19** (Fig. 5.4). *Vohra* et al. cloned the penicillin G acylase (PGA) gene (*pac*) into a stable *asd*+ vector (pYA292) and expressed it in *E. coli*. This recombinant strain produced 1000 U PGA/g cell dry weight, 23 times more than that produced by parental *E. coli* ATCC 11105. The

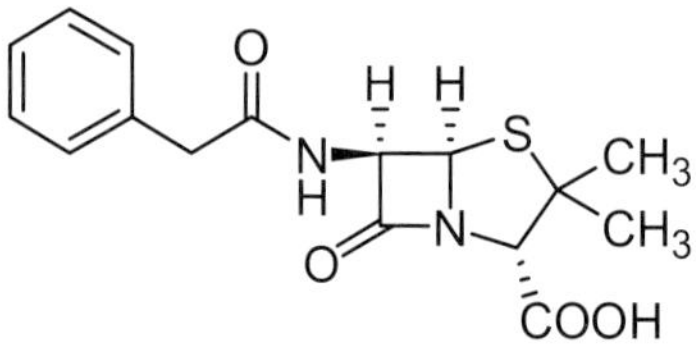

17

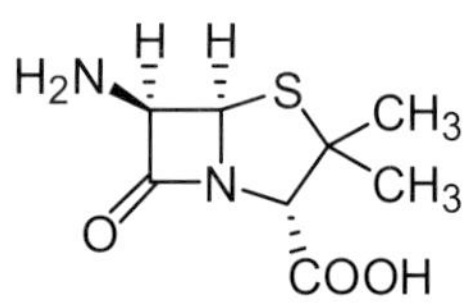

18

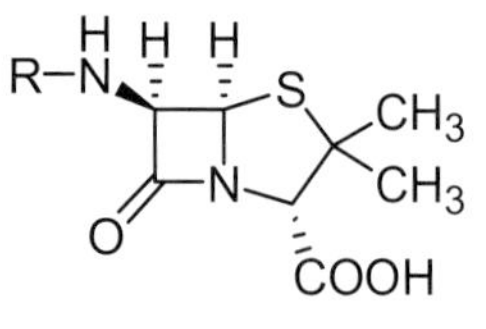

R = acyl substituent

19

Fig. 5.4 Structures of penicillin G **17**, 6-APA **18** and semisynthetic penicillins **19**.

significant increase in the *pac* production was accompanied by absence of glucose catabolite repression, nonrequirement of inducers like phenylacetic acid and isopropyl-β-D-thiogalactopyranoside, 100% stability of pRT4 plasmid in the absence of any selective marker in the growth medium, and absence of inclusion bodies formation. Such properties are highly desirable for industrial applications [23].

PGA of *E. coli* is an important industrial enzyme used for the production of semisynthetic antibiotics from the key β-lactam precursors 6-APA and 7-aminodeacetoxymethylcephalosporanic acid (7-ADCA) **20**. Different high-expression recombinant systems have been developed in which the cloned PGA gene is under the control of a strong promoter and PGA is therefore overproduced. The expression of the chromosomal gene is repressed in a medium not supplemented with the inducer and its expression requires the presence of phenylacetic acid (PAA) in the medium. PAA also inhibits the intracellular proteolysis of the pre-pro enzyme (ppPA) in the cytoplasm. *Maresova* et al. improved a recombinant *E. coli* strain RE 3 (pKA18) overproducing PGA from plasmid-borne gene PGA by using the technique of continuous culture in a chemostat. The chemostat culture of the recombinant bacterium yields components for assembling a new production strain. In this particular case, the new recombinant strain with modified traits was constructed by means of retransformation of the evolved host ERE3 with ancestral plasmid pKA18 [24].

The penicillin amidase (acylase) was originally found in *Penicillium chrysogenum* Q176 in 1950 [25]. The discovery of this enzyme has vastly contributed to the industrial manufacture of β-lactam antibiotics. Today, *E. coli* and *Bacillus megaterium* are the strains mainly used for sourcing the enzyme. Genetic engineering techniques were applied as early as 1979 to obtain a cosmid hybrid *E. coli* 5K strain, which had a 10-fold higher activity with regard to enzyme production [26].

The lysine pathway in *P. chrysogenum* has been genetically engineered for β-lactam overproduction. This example demonstrates the fact that gene disruption of diverging branches of a biosynthetic pathway leads to channeling of the intermediates towards the desired end-products. On the other hand, amplification of a specific gene encoding a putative bottleneck enzyme in a biosynthetic pathway does not necessarily result in overproducing strains since other bottlenecks may still occur after removal of the first one [27].

The penicillin acylase gene (*pac*) from *E. coli* ATCC 11105 genomic DNA was cloned into pUC19 vector using conventional techniques by Gumusel et al. The *E. coli* strain JM 109 transformed by this construct was shown to possess high plasmid stability. Glucose exhibited a repressor effect on penicillin acylase production. The optimum parameters for the isolated and purified enzyme *pac* were established [28]. *Lin* et al. observed that the production of recombinant penicillin acylase could be enhanced in *E. coli* by increasing the intracellular concentration of the periplasmic protease DegP. Periplasmic processing and inclusion bodies normally limited the overproduction of *pac*. The amount of these periplasmic inclusion bodies was significantly reduced and *pac* activity was significantly increased upon coexpression of DegP. The stability of recombinant *pac* was not affected by the expression of DegP. The strategy of using protease for reducing the amount of inclusion bodies and improving recombinant protein production has been used here for the first time [29]. The same group also expressed heterologously the *Providencia rettgeri pac* gene in *E. coli* for the high-temperature-oriented production of bacterial penicillin acylase. It was possible to produce the enzyme at a temperature of 37°C, provided the pH of the culture was almost neutral [30].

Antibiotic-resistant activity has been unexpectedly discovered in functionally unrelated DNA fragments. The hyperthermophilic archaeon *Pyrococcus furiosus* does not show any *β*-lactamase activity. However, a DNA fragment conferring very low ampicillin-resistance activity on *E. coli* has been isolated from its genomic DNA library. After carrying out 50 rounds of DNA shuffling and screening at steadily increasing concentrations of ampicillin, the ampicillin-resistance activity of this clone on *E. coli* was significantly enhanced [31].

5.4.2 Cephalosporins

A genetically engineered strain of *Pseudomonas cepacia* BY 21 (KCTC 8922F), which is capable of producing a large amount of glutaryl 7-aminocephalosporanic acid (GL-7-ACA) acylase, has been developed and patented [32]. This acylase catalyzes the reaction producing 7-aminocephalosporanic acid (7-ACA) **21**, which is the starting material for the semisynthetic cephalosporin antibiotics.

Kim et al. cloned and expressed the gene coding for glutaryl 7-ACA acylase from *Pseudomonas diminuta* KAC-1 in *E. coli.* The *E. coli* BL 21 carrying pET2B, the plasmid construct for high expression of the glutaryl 7-ACA acylase gene, produced this enzyme at about 30% of the total proteins with 3.2 U/mg protein. Growth at a temperature less than 31°C and deletion of signal peptide played a major role in the high expression of the *P. dimunita* KAC-1 acylase gene 333 in the recombinant *E. coli* cells. Cephalosporin acylases are an industrially important group of enzymes which hydrolyze cephalosporin C (CPC) **22** and/or GL-7-ACA **23** to **21** (Fig. 5.5) [33]. Zheng et al. obtained two newly engineered bacteria which could secretorily express the gene encoding GL-7-ACA acylase from *Pseudomonas* sp. 130 with high activity. The highest specific activity of GL-7-ACA acylase was in the intact cell as compared with that of transformants constructed by the authors. The specific activity of GL-7-ACA acylase remained constant in the 50th generation of mutants transferred on agar plates [34].

The *Pseudomonas* strain SY-77-1 producing GL-7-ACA **23** when subjected to mutation yielded a strain, called GK-16, whose ability to produce GL-7-ACA was 20-fold higher than that of the parent strain. It was found to be suitable for the industrial production of 7-ACA [35]. Later, the GL-7-ACA amidase gene was subcloned on multicopy plasmids in order to enhance the activity further 6-fold [36].

The production processes of cephalosporins 7-ADCA and 7-ACA by *P. chrysogenum* expressing heterologous expandases have been studied for several years and provide typical examples of metabolic engineering for extending the product spectrum of biocatalysts [37–40]. DSM produces penicillin G/V by fermentation

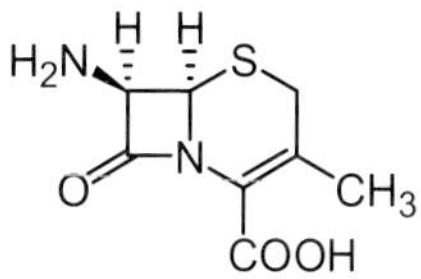

20

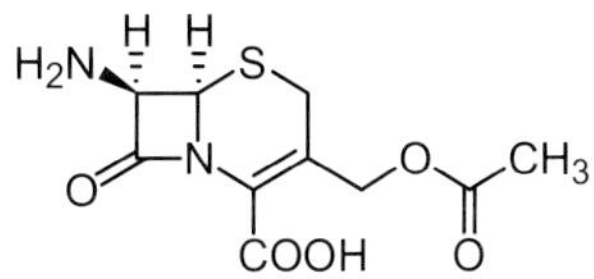

21

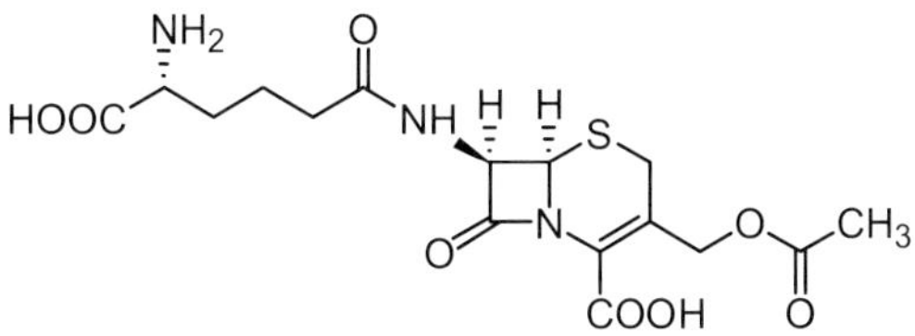

22

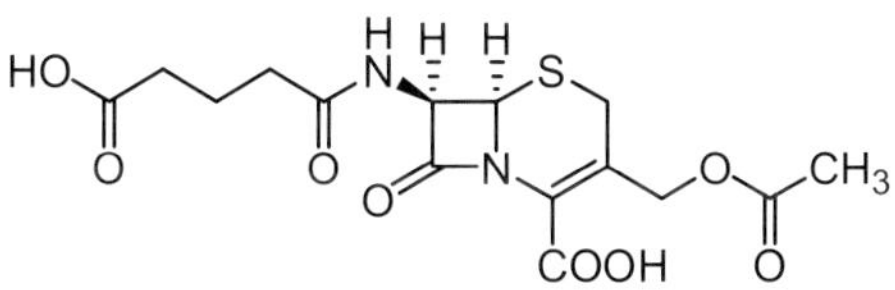

23

Fig. 5.5 7-ADCA **20**, 7-ACA **21**, CPC **22** and GL-7-ACA **23**.

using *P. chrysogenum* strains that have been improved both by classical methods as well as by genetic engineering [41].

The *in vitro* conversion of penicillin N analogs to cephalosporins was first demonstrated in resting cells and cell-free extracts of *Streptomyces clavuligerus* NP1. Sim et al. optimized the conversion rate of a highly expressed soluble source of recombinant *S. clavuligerus* NRRL 3585 deacetoxycephalosporin C synthase (ScDAOCS) through detailed characterization of its cofactor requirements. The conversions of penicillin G and ampicillin were enhanced by 21- and 35-fold, respectively. The exclusion of dithiothreitol (DTT) and ascorbate from the expandase reaction had a beneficial effect on the product yields. Magnesium sulfate and potassium chloride were also found to be not essential [42].

Yield improvements through metabolic engineering have been demonstrated for a number of other systems. Skatrud et al. increased the production of cephalosporin C by *Cephalosporium acremonium* by 15% by overexpressing the *cefEF* gene [43].

One of the first large-scale examples to demonstrate the power of metabolic pathway engineering is that of cephalexin production. By the year 2000, cephalexin

Step 1:	Conversion of benzaldehyde into D-phenylglycine amide
Step 2:	Fermentation using sugar and adipic acid to get 7-ADCA adipic acid amide
Step 3:	Enzymatic hydrolysis of 7-ADCA adipic acid amide into 7-ADCA and adipic acid
Step 4:	Enzymatic condensation of 7-ADCA with D-phenylglycine amide to give cephalexin

Fig. 5.6 Advancement in synthetic methods for cephalexin.

synthesis consisted of an organic synthesis step (*Step 1*), a fermentation step (*Step 2*) and two enzymatic steps (*Steps 3* and *4*, Fig. 5.6).

The process has been commercialized by DSM and is based on the high productivity of the *Penicillium* strain. The expandase enzyme genes from a *Cephalosporium* strain were implanted into the *Penicillium* strain. This engineered strain was able to convert sugar directly to the 7-ADCA moiety. This process, which is more efficient, makes the chemical route redundant. As a result, toxic chemicals such as peracetic acid, dichloromethane, phosphorous pentachloride, trimethylsilyl chloride, trimethylamine hydrochloride, bis(trimethyulsilyl)urea, pyridine hydrobromide, D-phenylglycine acid chloride hydrochloride, phenylacetic acid and *n*-butyl alcohol, which were necessary for the chemical synthesis, are not being used any more. The two-step process for 7-ADCA led to a reduction in solvent consumption by about 90%, auxiliary reagents by 100%, energy consumption by about 40% and waste generated by about 15% [44]. Recently, Bristol-Myers Squibb has patented a process for the direct production of desacetylcephalosporin C by culturing a strain of *Acremonium chrysogenum* containing recombinant nucleic acid encoding *Rhodosporidium toruloides* cephalosporin esterase [45].

The production of 7-ACA has been facilitated by the use of genetic engineering techniques. D-amino acid oxidase (DAAO) is a highly stereoselective flavoenzyme, which catalyzes the oxidative deamination of D-amino acids to produce the corresponding α-keto acids and ammonia. The enzyme is widely distributed in mammals and eukaryotic organisms. Enzymes form only a few of them, such as those from pig kidney and the yeasts *Rhodotorula gracilis* and *Trigonopsis variabilis*, have given homogeneous preparations. The coenzyme, reduced FAD, is reoxidized spontaneously by molecular oxygen to give hydrogen peroxide.

Detailed studies on the properties of the *R. gracilis* DAAO (RgDAAO) have shown that this oxidase exhibits certain properties favorable for industrial applications. The most important industrial application is the bioconversion of cephalosporin C in the first reaction of the two-step route to 7-ACA, whereas the second reaction is the deacylation of the cephalosporin C side-chain by GL-7-ACA acylase.

The use of this flavozyme is often accompanied by problems arising from the costly and tedious procedure of purification, a low enzyme turnover and the loss of the coenzyme [46]. The same group has engineered a His-tagged RgDAAO starting from a chimeric recombinant protein with six additional amino acid residues at the N-terminus. The His-tagged protein was successfully expressed in *E. coli* as a sol-

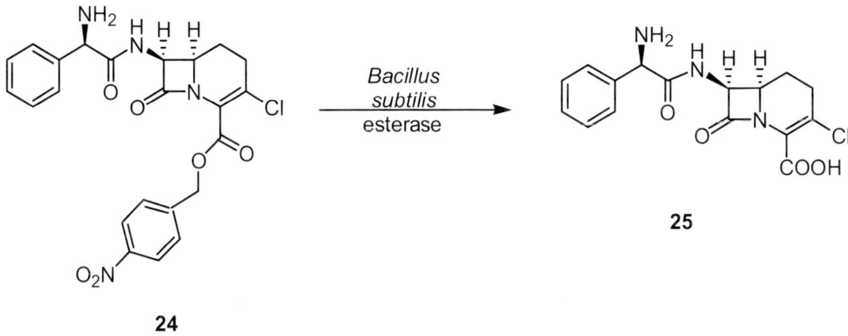

Fig. 5.7 Stereoselective biotransformation in synthesis of Loracarbef **25**.

uble protein and a fully active holoenzyme. No side activities, such as β-lactamase activity or catalase activity, were observed in significant amounts. The enzyme was active and had increased thermal stability after covalent matrix immobilization. The enzyme could also be repeatedly used after simple recovery from the reaction medium. The His-RgDAAO has been purified by single-step metal-chelate chromatography to a high specific activity and stability. No side reactions are catalyzed [47].

As in the case of the manufacture of cephalexin, the manufacture of 7-ACA from cephalosporin C by the use of DAAO has proved to be a "green chemistry" process. The use of chlorinated solvents, toxic additives and extremely low temperatures is avoided. Downstream processing is simplified because of the purity of the product formed. The fact that over 50 patents have been filed relating to this biotransformation with DAAO indicates the industrial significance attached to it [46].

Genetic engineering techniques have been used successfully to modify enzyme characteristics for the production of drugs. The *p*-nitrobenzyl ester **24** of the cephalosporin antibiotic Loracarbef® **25** is poorly soluble in water and dimethyl formamide is used as a cosolvent. However, the esterase from *Bacillus subtilis*, which hydrolyses the ester to Loracarbef® **25**, is inhibited by DMF (Fig. 5.7).

The combination of epPCR and DNA shuffling led to the generation of a variant that had 150 times higher activity compared to the wild-type in 15% DMF. The thermostability of this enzyme has been increased further by about 14°C by directed evolution [48, 49].

5.5 Polyketide Antibiotics

Polyketide antibiotics are naturally occurring compounds produced by eukaryotic and prokaryotic microorganisms and plants. They are partly composed of a carbocyclic or alicyclic carbon chain derived from the condensation of carboxylic acids. They are assembled from simple building blocks of specific structures containing two to five carbon atoms. Their biosynthesis by polyketide synthases is similar to

Tab. 5.4 Genetically engineered microorganisms for the synthesis of polyketides.

Genetically engineered strain	*Product, Remarks*
Recombinant *Streptomyces peucetius* SGF-107	increased production of daunomycin
Recombinant *Saccharopolyspora erythrea*	improved biosynthesis of erythromycin
Streptomyces avermitilis mutant ATCC 31780 and L9	increased biosynthesis of avermectin
Streptomyces coelicolar, cloned and expressed with gene cluster of *Sorangium cellulosum*	efficient biosynthesis of anticancer agent epothilone
Streptomyces galilaeus, cloned and expressed with actinorhodin biosynthesis genes	biosynthesis of anthraquinones aloesaponarin II and desoxy-erythrolaccin

the biosynthesis of fatty acids, with repetitive condensations of acyl thioesters. The way in which these building blocks are assembled by a specific enzyme subunit complex determines the final structure of the compound produced. They exhibit a broad-spectrum biological activity, including anti-infective (erythromycin), immunosuppressant (rapamycin) and carcinogenic (aflatoxin) activities.

The polyketides represented about 3% of the total world sales in 1998 for prescription pharmaceuticals, with an estimated value of US$8.5 billion [50]. Table 5.4 gives a summary of the various genetically engineered microorganisms used for the production of pharmacologically interesting polyketides.

The earliest example of the generation of novel polyketides through the genetic manipulation of polyketide synthase (PKS) genes was reported in 1990 by Strohl's group. Cloning of actinorhodin biosynthesis genes in the aclarubicin-producing strain *Streptomyces galilaeus* led to the biosynthesis of the anthraquinones aloesaponarin II and desoxyerythrolaccin [51]. The laboratories of Hopwood and Khosla developed and successfully used a combinatorial approach towards the generation of novel type II PKS products by mixing and matching various genes of the type II PKSs [52–54].

Semisynthetic polyketides are created by the modification of bioactive, natural polyketides [50, 51]. Since microorganisms gradually develop resistance to polyketide antibiotics over a period of time, genetic engineering provides the tools to create new diversity and new efficacy [55].

Anthracycline polyketide antibiotics such as daunorubicin **26** (daunomycin) and doxorubicin **27** (adriamycin) (Fig. 5.8) are important antibiotics with potent antitumor activity. However, they are very expensive because of low titers and the formation of a complex mixture of products by the producing bacteria, *Streptomyces lividans*, leading to the use of expensive downstream processing techniques [56]. Recently, Rosabal et al. cloned and sequenced a chromosomal DNA fragment from the daunomycin producer *Streptomyces peucetius* SGF-107, which activates the biosynthesis of two polyketide antibiotics, daunomycin and actinorhodin. Partial DNA sequencing of the activating fragment revealed that there existed two open reading frames, whose deduced products exhibited similarities to that of other known transcriptional regulators of the MarR and ArsR family, respectively. The produc-

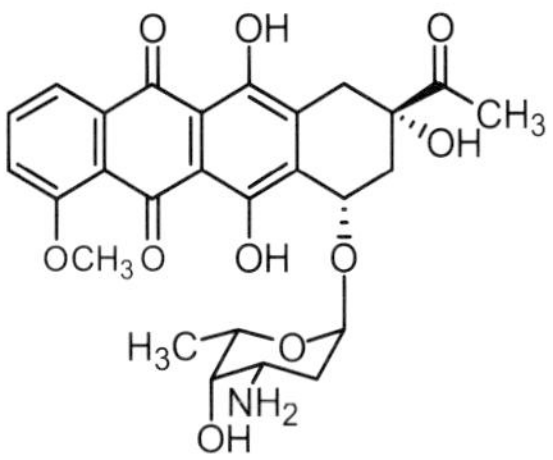

26

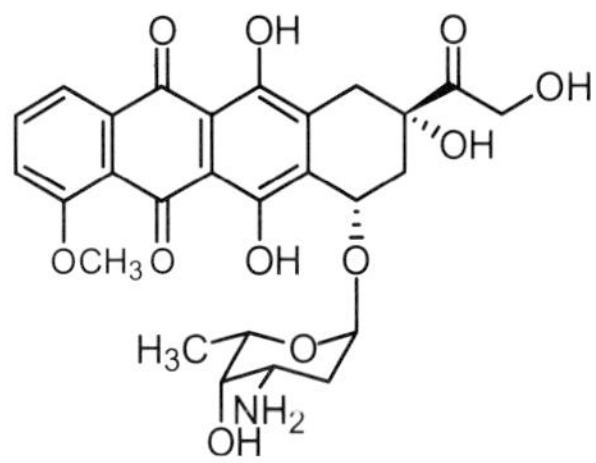

27

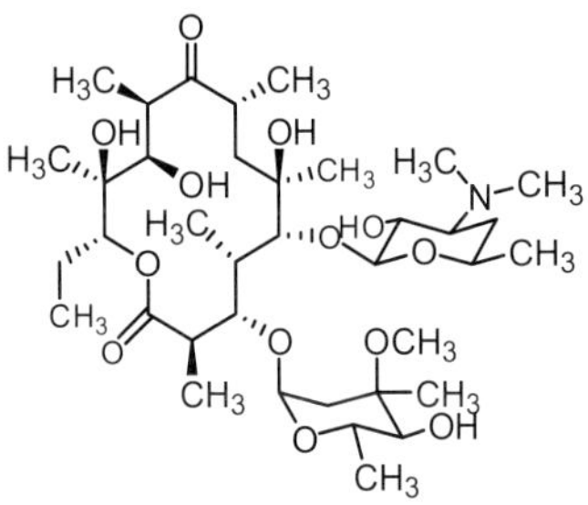

28

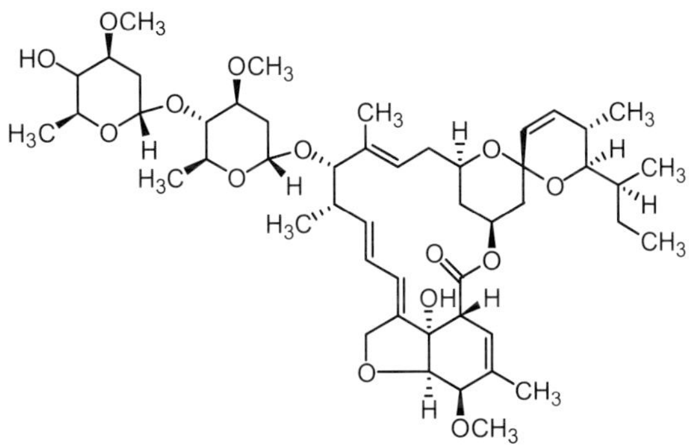

29

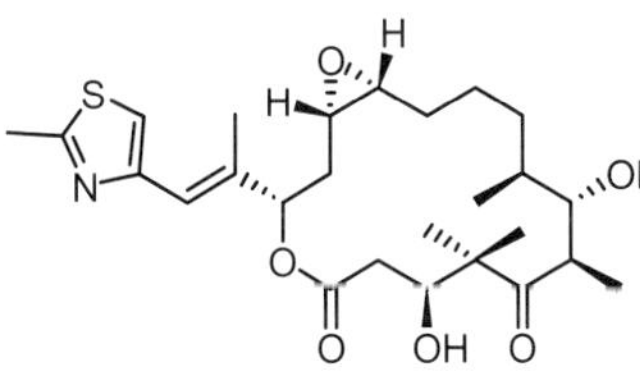

30

Fig. 5.8 Structures of daunorubicin **26**, doxorubicin **27**, erythromycin A **28**, avermectin A1a **29** and epothilone A **30**.

tion of daunomycin increased by as much as 8 times in the recombinant strain. The presence of the recombinant plasmid was tested during and after the fermentation process. It was found to be present at the end of the fermentation process. The activation was observed in several strains [57].

Most of the work in the field of engineering polyketide biosynthesis has concentrated on the engineering of the biosynthesis of erythromycin **28**, which is naturally produced by *Saccharopolyspora erythrea*. New insight has been gained into the production of erythromycin in *S. erythrea*. The biosynthesis involves only three giant genes, each of which codes for a protein of over 300 kDa. Each protein is in turn made up of two inexact repeats that can be divided into six modules. In fact, novel molecules can be generated by using different combinations and permutations of these basic modules, as well as by introducing point mutations within functional domains [52, 53, 58]. The work being carried out with PKSs illustrate how many different compounds can be produced in a common host. It is now realized that investment in optimization of a general host system is beneficial for the production of several different compounds [40].

The actinomycetes, in general, and the *Streptomyces* species, in particular, have been considered to be among the most important industrial microorganisms due to their ability to synthesize a wide range of valuable antibiotics and other secondary metabolites. Of all the secondary metabolites containing antihelminthic activity, avermectin **29** is considered as the most important metabolite because of its superior commercial potency. Avermectin has replaced several existing drugs in the conventional market because of its excellent antihelminthic activities. Several molecular biological studies have been carried out to understand the biosynthetic pathway and regulation. Hwang et al. have studied the optimal pH conditions for efficient transformation of protoplasts and intact cells in the high avermectin producing *Streptomyces avermitilis* mutant ATCC 31780 and L9. They found that the protoplast buffer pH was the most important factor influencing transformation efficiency. They also developed and optimized a process of electroporation of intact cells without lysozyme treatment in order to compensate for the lowering of productivities as a result of the protoplasting process [59].

The polyketide anticancer agent epothilone **30** stabilizes microtubules in a manner similar to Taxol. Its biosynthetic pathway could be transplanted into a host with better production properties. Therefore it appears that these natural products can be transferred to other hosts to increase yield and productivity. While the epothilone producer *Sorangium cellulosum* yields only about 20 mg/l of the polyketides with a long doubling time of 16 h, cloning of the gene cluster and expression of the genes in the actinomycete *Streptomyces coelicolar* produced epothilones A and B with a 10-fold faster replication [60].

5.6 Vitamins

Genetic engineering techniques have influenced greatly the production methods for various vitamins.

5.6.1 L-Ascorbic Acid (Vitamin C)

Microorganisms have always played a crucial role in the industrial manufacture of vitamin C **38**. As early as 1923, it was discovered that the bacterium *Acetobacter suboxydans* had the ability to carry out the regiospecific oxidation of D-sorbitol **32**, which was obtained by the chemical reduction of D-glucose **31**, into L-sorbose **33** (Fig. 5.9). This biotransformation is also carried out by *Gluconobacter oxydans*. L-

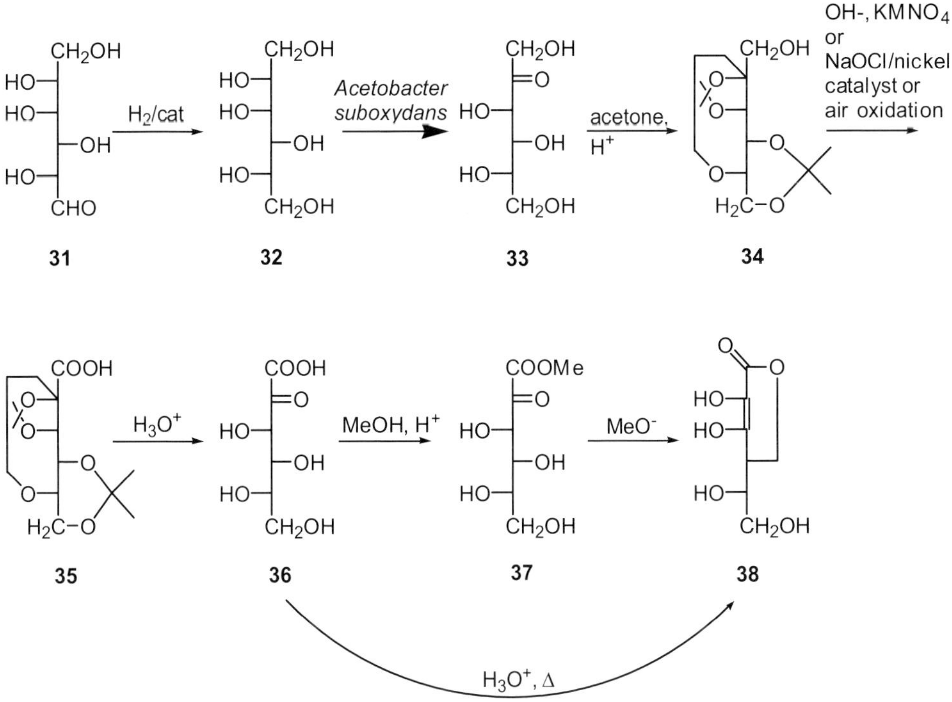

Fig. 5.9 Reichstein–Gruessner process for synthesis of L-ascorbic acid **38**.

Sorbose is an important intermediate for the manufacture of vitamin C by this process, which is also known as the Reichstein–Gruessner process [61, 62].

The Reichstein–Gruessner process is very time-consuming, and has environmental and economical limitations. For this reason, much effort has been directed at finding and developing alternative technologies with the help of genetic engineering (see Tab. 5.5). Sonoyama et al. have described a tandem fermentation process in which the microbial oxidation of D-glucose to 2,5-diketo-D-gluconate (2,5-DKG) is carried out by *Erwinia* species. The subsequent reduction of 2,5-DKG to the key intermediate 2-keto-L-gulonate (2-KLG) is catalyzed by *Corynebacterium* species [63]. In an effort to change this into a one-stage process, the *Erwinia herbicola* species has been genetically transformed with the *Corynebacterium* gene, which encodes 2,5-DKG reductase (DKGR), the enzyme that catalyzes the conversion of 2,5-DKG to 2-KLG. After optimizing the culture conditions, these recombinant strains of *Erwinia* produced about 120 g/l of 2-KLG within 120 h and with a molar yield from glucose of about 60%, compared to around 50% by the classical route. Follow up studies illustrated the potential economic viability of the metabolically engineered strain for vitamin C production and have led to a number of US patents [64–66].

Another approach is to convert D-sorbitol or L-sorbose into 2-KLG via L-sorbosone. *Gluconobacter melanogenus* and *Gluconobacter oxydans* UV10 have the

Tab. 5.5 Genetically engineered microorganisms for the synthesis of L-ascorbic acid.

Microorganism	*Reaction catalyzed*	*Remarks*
Acetobacter suboxydans *Gluconobacter oxydans*	D-sorbitol to L-sorbose	oxidation
Erwinia species	D-glucose to 2,5-DKG	two-stage, tandem fermentation
Corynebacterium species	2,5-DKG to 2-KLG	
Erwinia herbicola with *Corynebacterium* gene encoding 2,5-DKG reductase	D-glucose to 2-KLG	single-stage fermentation
Gluconobacter oxydans T-100 cloned with genes from *Gluconobacter melanogenus* or *gluconobacter oxydans* UV10	D-sorbitol to 2-KLG via L-sorbosone	improved yields of 2-KLG
Mixed culture of *Flavimonas oryzihabitans* OA2 and *Pseudomonas* substrates	D-glucose to 2,5-DKG via D-gluconate and 2-KDG	microbial transformations with more convenient cepacia 3/27

necessary enzymes, i.e. a membrane-located sorbose-FAD dehydrogenase (SDH) and a cytoplasmic NAD(P)-L-sorbosone DH (SNDH). The genes coding for SDH and SNDH have been cloned into 2-KLG producing *G. oxydans* T-100 strain. The L-sorbose was completely consumed. The recombinant strains were able to improve the yields of 2-KLG by 230% from a 15% D-sorbitol-based medium [67].

Sulo et al. converted D-glucose to 2,5-DKG in 92% yield within 150 h via fermentation using a mixed culture of two Gram-negative bacteria isolated from soil. The first strain, *Flavimonas oryzihabitans* OA2, produces 2-KDG from D-glucose via D-gluconate. The second strain, *Pseudomonas cepacia* 3/27, converts 2-KDG to 2,5-DKG. This approach has opened up a new possibility to produce ascorbic acid by microbial transformation with more convenient substrates and provides important leads for genetic engineering [68].

5.6.2
Biotin (Vitamin B_8, Vitamin H)

An essential nutrient for many microorganisms and animals, biotin **39** is largely manufactured by a complicated and expensive chemical synthesis method – the so-called Goldberg and Sternbach process. Studies connected to the metabolic pathway for biotin synthesis from pimelic-CoA were first carried out in *E. coli*. Subsequently, all the enzymes involved in biotin synthesis from pimelic acid in *Bacillus sphaericus* were identified. It was discovered that mutants of *B. sphaericus* and *Serratia marcescens*, which were resistant to the biotin analogs acidomycin and 5-(2-thienyl)-*n*-valeric acid, secreted significant quantities of biotin pathway intermediates. This led to the isolation of the desired genes from these organisms. The genes involved in biotin synthesis are organized in two clusters, bioXUF and bio-

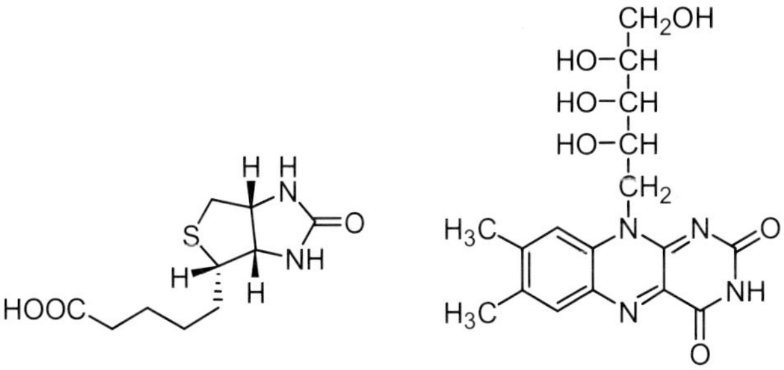

39 40

Fig. 5.10 Structures of biotin **39**, riboflavin **40** and nicotinamide **42**.

DAYB, and have been cloned on *E. coli* vectors [69]. *E. coli* transformed with these genes produced relatively high concentrations of biotin and its intermediates [70].

Similarly, in order to increase further the biotin productivity of the *S. marcescens* mutants, their biotin operons were cloned on several plasmids. These hybrid plasmids harbored the 7.2-kb DNA fragments, coding for the five genes of biotin synthesis. Reintroduction into *S. marcescens* mutants led to the selection of a stable recombinant strain exhibiting a high D-biotin production level [71].

5.6.3 Riboflavin (Vitamin B_2)

Riboflavin **40** (see Fig. 5.10), which is used in human nutrition and therapy, is produced both by synthetic and by fermentation processes. Although bacteria (*Clostridium* species) and yeasts (*Candida* species) are also good producers, two closely related ascomyceti fungi, *Ashbya gossypii* and *Eremothecium ashbyii*, are considered to be excellent riboflavin producers. The biosynthesis pathway of riboflavin involves a key step of conversion of ribulose-5-phosphate into 3,4-dihydroxy-2-butanone-4-phosphate. This step is catalyzed by the enzyme 3,4-dihydroxy-2-butanone-4-phosphate synthase. This enzyme has been purified from the flavinogenic yeast *Candida guilliermondii* as well as from *E. coli*. The *E. coli* synthase gene has been cloned, sequenced and expressed. The gene codes for a protein of 24 kDa, which is also the size of the yeast enzyme. Fermentation processes are expected to dominate in the future [71].

5.6.4 Nicotinamide (Vitamin B_3 or PP)

A microbial process for the conversion of 3-cyanopyridine **41** into nicotinamide **42** has gained industrial relevance in recent years. The chemical alkaline hydrolysis of 3-cyanopyridine to nicotinamide (Fig. 5.11) is accompanied by a 4% yield of nicotinic acid – the product of further hydrolysis. The microorganism used for the biotransformation is a *Rhodococcus rhodochrous* J1 strain expressing 3-cyanopyri-

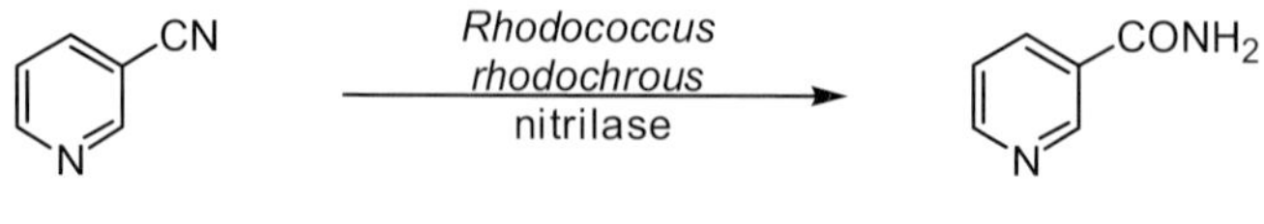

41 **42**

Fig. 5.11 Hydrolysis of 3-cyanopyridine **41** to nicotinamide **42**.

dinase (a nitrilase) activity. When the enzyme is induced with benzonitrile, the liberated ammonia is utilized as a source of nitrogen for growth. Nicotinic acid is not metabolized further. The process has been commercialized by Lonza. This microorganism has also gained industrial importance for the manufacture of acrylamide by hydrolysis of acrylonitrile [72].

5.6.5 Vitamin B_{12}

The primary natural source of vitamin B_{12} is bacterial metabolic activity. In industry, *Propionibacterium shermanii* and *Pseudomonas denitrificans* have been used because of the high yields and also due to their ability to grow rapidly. Mutagenic treatments, such as UV light, ethyleneimine, nitrosomethylurethane, etc., have led to improved productivity, but the addition of cobalt ions and, frequently, 5,6-dimethylbenzimidazole (5,6-DBI) cannot be dispensed with. Potential precursors such as glycine, threonine and δ-aminolevulinic acid have a beneficial effect on the synthesis of vitamin B_{12}. It has been discovered that betaine and choline have a strong stimulatory effect on bacterial vitamin B_{12} production. Moreover, porphyrine-less mutants and catalase-negative mutants are generally overproducers of vitamin B_{12}. The industrial fermentation is carried out in two stages: an anaerobic stage initially to promote growth and cobamide biosynthesis, and a subsequent shift to an aerobic process, which promotes 5,6-DBI biosynthesis and the conversion of cobamide to cobalamine. The growth of *P. denitrificans* parallels cobalamine synthesis under aerobic conditions when 5,6-DBI and cobalt salts are directly added to the culture. By mutation and selection procedures, strains have been obtained that produce more than 150 mg of vitamin B_{12} per liter. Moreover, several vitamin B_{12} derivatives, such as hydroxycobalamine and coenzyme B11, can now be produced by direct fermentation [71]. Several genetically engineered *P. denitrificans* strains have been systematically constructed in the laboratories of Aventis, the world's leading manufacturer of vitamin B_{12} [73].

5.6.6 D-Pantothenic Acid (Vitamin B_5)

In the industrial manufacture of D-(−)-pantothenic acid **45**, microbial, stereospecific reduction of ketopantoyl lactone **43** to D-(−)-pantoyllactone **44** with washed cells of *Rhodotorula minuta* or *Candida parapsilosis* is carried out (Fig. 5.12).

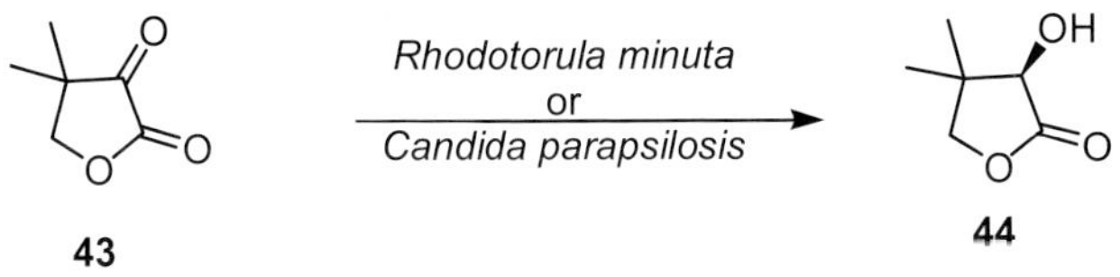

Fig. 5.12 Stereospecific microbial reduction of ketopantoyl lactone **43** to D-(−)-pantoyllactone **44**.

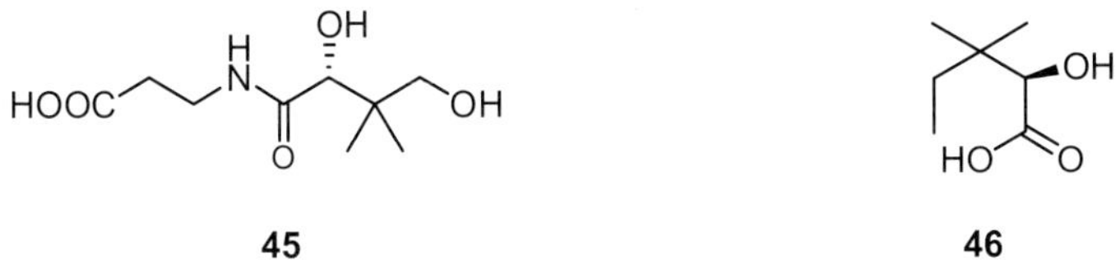

Fig. 5.13 D-Pantothenic acid **45** and D-pantoic acid **46**.

Subsequent condensation with *β*-alanine then gives pantothenic acid. Takeda Chemical Industries have developed a direct, high-yielding fermentation process for D-pantoic acid **46** and D-pantothenic acid **45** (Fig. 5.13).

Several *Enterobacteriaceae* genera (e.g. *Citrobacter, Klebsiella, Enterobacter* and *Escherichia*), having resistance to salicylic acid and/or *α*-ketoisovaleric acid, *α*-ketobutyric acid, *α*-aminobutyric acid, *β*-hydroxy-aspartic acid and *O*-methyl threonine, were selected as high producer strains. *E. coli* mutants and recombinant strains, which were transformed with plasmids carrying genes involved in the biosynthesis of pantothenic acid, gave high yields of D-pantothenic acid from glucose, with *β*-alanine as a precursor [71]. Degussa has recently patented a process for the preparation of D-pantothenic acid by the fermentation of *Coryneform* bacteria in which the nucleotide sequence coding for the Zwa1 gene product is over-expressed [74]. The process includes the following steps:

- Fermentation of the D-pantothenic acid-producing bacteria in which at least the gene that codes for Zwa1 gene product is enhanced.
- Concentration of the D-pantothenic acid in the medium or in the cells of the bacteria.
- Isolation of the D-pantothenic acid produced.

5.6.7
Vitamin A

Another example of the application of metabolic engineering in order to convert native metabolic intermediates to desirable end-products is the production of the *β*-carotene precursor for vitamin A. Genetic transformation of *Zymomonas mobilis* and *Agrobacterium tumefaciens* with four of the *β*-carotene genes led to the formation of yellow colonies on agar plates. These colonies contained significant quantities of the precursor [75].

5.6.8 Vitamin K (Menaquinone)

An antihemorrhagic factor widely used in medicine for various types of bleeding symptoms, vitamin K has largely been manufactured by chemical synthesis. Vitamin K_2 is found only in bacteria and not in fungi or yeasts. Facultative anaerobic bacteria contain particularly Vitamin K_2 and its homologs. *Flavobacterium meningosepticum* has been identified as a high producer of this metabolite. A mutant of this microorganism, which was resistant to the vitamin K_2 biosynthesis inhibitor 1-hydroxy-2-naphthoate (HNA), was selected. This mutant overproduced vitamin K_2 intracellularly on a medium containing L-tyrosine and *iso*-pentanol. Microbial fermentation processes for specific synthesis of vitamin K analogs may become commercially viable in the near future [71].

5.7 Steroids

The size of the world market for steroids exceeds 1000 tons per year. Mutants of *Mycobacterium* species, which are devoid of steroid-ring degradation activities, have been in use at Schering for the production of androsten-dione and androsta-dien-dione on a scale of 200 m^3. These intermediates are used for subsequent synthesis of steroid drugs using chemical and biotechnological routes. The biotechnological routes that have attained commercial significance include hydroxylations (e.g. at the 11α or 11β positions with *Curvularia* species), dehydrogenations ($\delta 1$ position; hydrocortisone to prednisolone) and reductions (17-keto reduction). Mediator proteins have also been used [76]. Using genetic engineering techniques, new yeast strains are being developed for use in synthesis of steroid compounds. Dumas et al. have developed strains that have reduced 20α-hydroxysteroid dehydrogenase activity. As a result, the yield and selectivity of steroid biotransformation processes are improved. Moreover, the product obtained is of higher purity [77].

5.8 Other Drugs

5.8.1 L-Dihydroxyphenyl Alanine (L-DOPA)

L-DOPA is an optically active amino acid used in the treatment of Parkinson's disease. The traditional method of production of L-DOPA starts from the dihydroxyphenyl compound and involves either homogeneous catalysis or biocatalysis.

Since L-tyrosine has one phenolic hydroxy group less than L-DOPA, efforts have been made to hydroxylate it using biocatalysts. Until recently, this was not possible due to the fact that subsequent hydroxylation led to the formation of unwanted

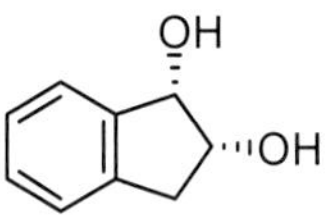

47

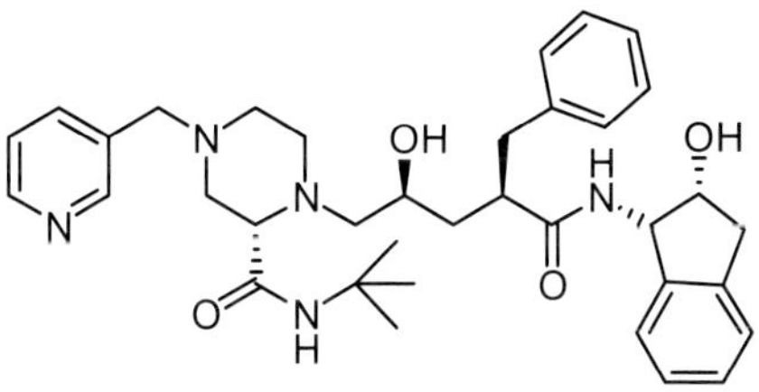

48

Fig. 5.14 *Cis*-(1*S*,2*R*)-indandiol **47** and indinavir **48**.

trihydroxylated contaminants. Krämer et al. cloned both the genes constituting 4-hydroxyphenylacetate-3-hydroxylase, i.e. a monooxgenase and a $FADH_2$-NADH oxidoreductase, in an expression vector for *E. coli*. The corresponding genes *hpaB* and *hpaC* were functionally expressed, and direct synthesis of L-DOPA from L-tyrosine occurred [78].

5.8.2 Crixivan®

The compound *cis*-(1*S*,2*R*)-indandiol **47** is a precursor in an engineered biosynthetic pathway for the production of the human immunodeficiency virus protease inhibitor Crixivan® , which is the sulfate form of indinavir **48** (Fig. 5.14). With the help of random mutagenesis of the gene encoding toluene dioxygenase from *Pseudomonas putida*, *Zheng* et al. have improved the biotransformation yields of **47** [79].

5.8.3 N-Acetyl Neuraminic Acid (NANA)

NANA **51**, also known as sialic acid, is the substrate of the influenza virus neuraminidase. This enzyme was the first virus-associated enzyme to be discovered. Several analogs of NANA have been developed and some, such as Zanamivir®, are efficient inhibitors of influenza A virus. The enzyme Neu5Ac (NANA) aldolase (synthase) catalyzes the formation of NANA from *N*-acetyl-D-mannosamine **50**, an epimer of *N*-acetyl-D-glucosamine **49** (Fig. 5.15).

Genetic engineering techniques have been employed to obtain better expression of the enzyme. Mahmoudian et al. have developed an immobilized enzyme process for NANA production using Neu5ac aldolase from an overexpressing recombinant strain of *E. coli*. The strain *E. coli* TG1 (pMexAld) was constructed and gave comparatively larger quantities of the Neu5Ac aldolase [80]. The NANA synthase of *Neisseria meningitidis* serogroup B has been manufactured by expression of the cloned gene. The *siaC* gene of a serogroup B *N. meningitidis* was cloned by PCR using primers derived from a commonly available sequence. The substrate re-

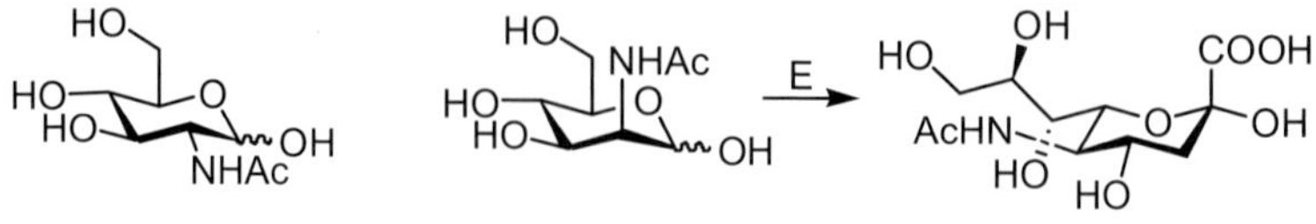

49 50 51

Fig. 5.15 Biotransformation of *N*-acetyl-D-glucosamine **49** to NANA **51** via *N*-acetyl-D-mannosamine **50**.

quirements of the synthase indicated that it has a number of advantages over other aldolases, particularly with respect to its use of phosphoenolpyruvate as a methyl group donor [81, 82]. Genetic engineering techniques have been introduced into the enzymatic process developed for NANA by Kragl et al. [83]. Tabata et al. have coupled bacteria expressing *N*-acetyl-D-glucosamine 2-epimerase and *N*-acetyl-D-neuraminic acid synthetase in order to produce NANA. The *srl1975*gene of *Synechocystis* sp. PCC6803, a phototrophic cyanobacterium, was cloned by PCR and expressed in *E. coli*, and the recombinant *E. coli* showed GlcNAc 2-epimerase activity. This is the first example of the cloning of the gene for GlcNAc 2-epimerase from prokaryotes [84].

5.8.4
Cholesterin Biosynthesis Inhibitors (HMG-CoA Reductase Inhibitors)

The chiral compound (*S*)-4-chloro-3-hydroxybutanoate (CHBE) **53** is an important chiral building block in the synthesis of 4-(2-(arylethyl)hydroxyphosphinyl)-3-hydroxy-butanoic acids, a new class of cell-selective inhibitors of cholesterol biosynthesis. CHBE has been manufactured by the reduction of the corresponding keto ester, 4-chloro-3-oxo-butanoic acid ethyl ester (COBE) **52** with the help of an NADPH-dependent dehydrogenase from *Geotrichum candidum* SC 5469 (Fig. 5.16). With the help of molecular cloning and overexpression of the gene encoding an NADPH-dependent carbonyl reductase from *Candida magnoliae*, the same biotransformation has been carried out to give a product with 100% enantiomeric excess. The S1 gene was overexpressed in *E. coli*, and the enzyme thus expressed was purified to homogeneity and had the same catalytic properties as the enzyme

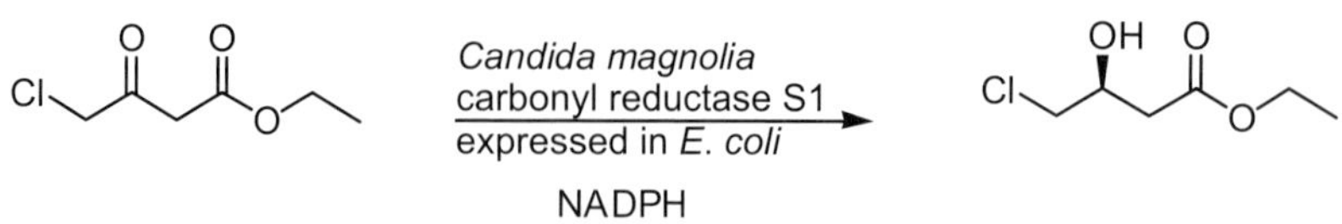

52 53

Fig. 5.16 Stereoselective reduction of (*R*)-4-chloro-3-oxo-butanoic acid ethyl ester (COBE) **52** to (*S*)-4-chloro-3-hydroxybutanoate (CHBE) **53**.

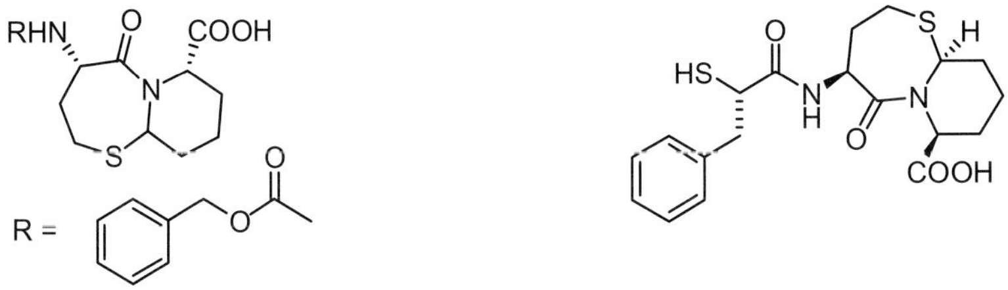

54 55

Fig. 5.17 BMS-199541-01 **54** and omapatrilat **55**.

from *C. magnoliae*. The biotransformation was carried out in an organic solvent two-phase system using *n*-butyl acetate [85].

5.8.5
Omapatrilat

Omapatrilat (BMS-186716) **55** is a new vasopeptidase inhibitor under development, in which the compound BMS-199541-01 **54** is a key chiral intermediate. In order to synthesize BMS-199541-01, a strain of *Sphingomonas paucimobilis* SC 16113, which had a novel L-lysine ε-aminotransferase, was isolated. The corresponding gene (*lat*) was cloned and overexpressed in *Escherichia coli*. A reaction yield of 65–70% was obtained in the biotransformation using this biocatalyst (Fig. 5.17) [86].

5.8.6
Hydromorphone

The synthesis of the semisynthetic opiate drug hydromorphone involves the microbial oxidation of morphine **56** to morphinone **57** by a dehydrogenase enzyme, followed by the reduction of the double bond by a transhydrogenase to yield hydromorphone **58**. The soluble pyridine nucleotide transhydrogenase (STH) of *P. fluorescenes* was overexpressed in *E. coli*, and applied for the regeneration of both NADH and NADP in the production of hydromorphone (Fig. 5.18) [87].

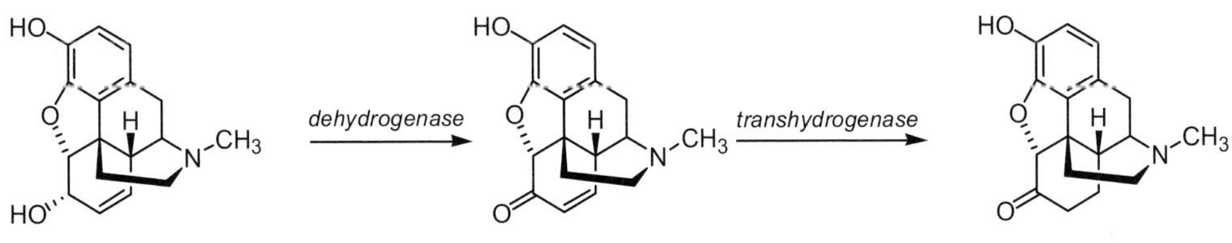

56 57 58

Fig. 5.18 Stereo- and regioselective biotransformation of morphine **56** to hydromorphone **58** via morphinone **57**.

5.9
Concluding Remarks

The recognition of the importance of chirality in drug molecules towards pharmacological action and the concurrent development of newer and more effective tools for genetic engineering have led to newer approaches for the synthesis of chiral drugs. Intensive research in several areas of molecular biology is creating a better understanding of nature's methods of biotransformation, in addition to throwing more light on the causes of several diseases. Stereoselective syntheses with the help of either recombinant microorganisms or biocatalysts isolated from them have provided excellent support for the efforts of medicinal chemists to synthesize chiral drug molecules. In 2001, about 36% of the worldwide sales of formulated pharmaceutical products of US$410 billion was made-up of single-enantiomer drugs [88]. The market share of single-enantiomer drugs is reported to be steadily increasing – a situation conducive to more intensive research and development in the field of stereoselective syntheses.

References

1 L Pasteur, *C R Acad Sci (Paris)* **46**, 615–8, 1858.
2 K-E Jaeger, T Eggert, A Eippes, MT Reetz, *Appl Microbiol Biotechnol* **55**, 519–30, 2001.
3 C Neuberg, J Hirsch, *Biochem Z* **115**, 282–310, 1921.
4 G Hildebrandt, W Klavehn, *German Patent DE 548459*, 1930.
5 DH Peterson, HC Murray, SH Epstein, LM Reineke, A Weintraub, PD Meister, HM Leigh, *J Am Chem Soc* **74**, 5933–6, 1952.
6 OK Sebek, D Perlman, in: *Microbial Technology*, 2nd edn, vol. 1, 484–488, Academic Press, New York, 1979.
7 UT Bornscheuer, *Biocatal Biotransform* **19**, 85–97, 2001.
8 FH Arnold, JC Morre, *Adv Biochem Eng Biotechnol* **58**, 2–14, 1997.
9 R Borris, in: JF Kennedy, ed., *Biotechnology, vol 7a, Enzyme Technology*, 35–62, VCH, Weinheim, 1987.
10 GN Stephanopoulus, AA Aristidou, J Nielsen, *Metabolic Engineering – Principles and Methodologies*, Academic Press, San Diego, CA 1998.
11 JE Bailey, in: SY Lee, ET Papoutsakis, eds, *Metabolic Engineering*, xi–xvi, Marcel Dekker, New York, 1999.
12 H Sahm, L Eggeling, in: J Lederberg, editor-in-chief, *Encyclopedia of Microbiology*, 2nd edn, vol. 1, 152–161, Academic Press, San Diego, CA, 2000.
13 H Zhao, K Chockalingam, Z Chen, *Curr Opinion Biotechnol* **13**, 104–10, 2002.
14 DY Ryu, D-H Nam, *Biotechnol Prog* **16**, 2–16, 2000.
15 KM Koeller, C-H Wong, *Nature* **409**, 232–40, 2001.
16 SV Taylor, P Kast, D Hilvert, *Angew Chem Int Ed* **40**, 3310–35, 2001.
17 G Chotani et al., *Biochim Biophys Acta* 434–55, 2000.
18 L Wang, PG Schultz, *Chem Commun* 1–11, 2002.
19 JR Schwartz, *Curr Opin Biotechnol* **12**, 195–201, 2001.
20 UT Bornscheuer, M Pohl, *Curr Opinion Chem Biol* **5**, 137–43, 2001.
21 A Liese, K Seelbach, A Buchholz, J Haberland, in: A Liese, K Seelbach, C Wandrey, eds, *Industrial Biotransformations*, 95–392, Wiley-VCH, Weinheim, 2000.
22 WBL Alkema, EJ de Vries, CMH Hensgens, JJ Polderman-Tijmes, BW Dijkstra, DB Janssen, in: A Bruggink, ed., *Synthesis of β-Lactam*

Antibiotics, 250–79, Kluwer, Dordrecht, 2001.

23 PK Vohra, R Sharma, DR Kashyap, R Tewari, *Biotechnol Lett* **23**, 531–5, 2001.

24 H Maresova, V Stepanek, P Kyslik, *Biotechnol Bioeng* **75**, 46–52, 2001.

25 K Sakaguchi, S Murao, *Bull Agric Chem Soc Jpn* **23**, 411, 1950.

26 K Matsumoto, in: A Tanaka, T Tosa, T Kobayashi, eds, *Industrial Application of Immobilized Biocatalysts*, 67–88, Marcel Dekker, New York, 1993.

27 J Casqueiro, O Banuelos, S Gutierrez, JF Martin, *Focus Biotechnol* **1**, 147–59, 2001.

28 F Gumusel, SI Ozturk, NK Korkut, C Gelegen, E Bermek, *Enzyme Microb Technol* **29**, 499–505, 2001.

29 W-J Lin, S-W Huang, CP Chou, *Biotechnol Bioeng* **73**, 484–92, 2001.

30 S-W Huang, Y-H Lin, H-L Chin, W-C Wang, B-Y Kuo, CP Chou, *Biotechnol Prog* **18**, 668–71, 2002.

31 H Zhau, K Chockalingam, Z Chen, *Curr Opinion Biotechnol* **13**, 104–10, 2002.

32 YH Khang, *Rep Korea Patent KR 52257*, 2000; *Chem Abstr* **136**: P261916p.

33 D-W Kim, K-H Yoon, *Biotechnol Lett* **23**, 1067–71, 2001.

34 Y-G Zheng, Y Li, J-F Chen, W-H Jiang, G-P Zhao, E-D Wang, *Biotechnol Lett* **23**, 1781–7, 2001.

35 S Ichikawa, Y Murai, S Yamamoto, Y Shibuya, T Fujii, K Komatsu, R Kodaira, *Agric Biol Chem* **45**, 2225–9, 1981.

36 A Matsuda, K Komatsu, *J Bacteriol* **163**, 1222–8, 1985.

37 C Cantwell, R Beckmann, P Whiteman, SW Queener, EP Abraham, *Proc R Soc Lond B* **248**, 283–9, 1992.

38 L Crawford, AM Stepan, PC McAda, JA Rambosek, MJ Condes, VA Vinvi, CD Reeves, *Biotechnology* **13**, 58–62, 1995.

39 T Isogai, M Fukagawa, I Aramori, M Iwami, H Kojo, T Ono, Y Ueda, M Kohsaka, H Imanaka, *Biotechnology* **9**, 188–91, 1991.

40 J Nielsen, *Appl Microbial Biotechnol* **55**, 263–83, 2001.

41 A Schmid, JS Dordick, B Hauer, A Kiener, M Wubbolts, B Witholt, *Nature* **409**, 258–67, 2001.

42 J Sim, T-S Sim, *Enzyme Microb Technol* **29**, 240–45, 2001.

43 PL Skatrud, AJ Tietz, TD Ingolia, CA Cantwell, DL Fisher, JL Chapman, SW Queene, *Bio/Technology* **7**, 477–85, 1989.

44 R Schoevaart, T Kieboom, *Chem Innovation* **12**, 33–8, 2001.

45 S-JD Chiang, BD Basch, *PCT WO 01 66767*, 2001; *Chem Abstr* **135**, P241034a.

46 MS Pilone, L Pollegioni, *Biocat Biotrans* **20**, 145–59, 2002.

47 S Fantinato, L Pollegioni, MS Pilone, *Enzyme Microbial Technol* **29**, 407–12, 2001.

48 FH Arnold, JC Moore, *Adv Biochem Eng/Biotechnol* **58**, 1–14, 1997.

49 L Giver, A Gershenson, P-O Freskgard, FH Arnold, *Proc Natl Acad Sci USA* **95**, 12809–13, 1998.

50 AS Anderson, Z An, WR Strohl, in: J Lederberg, editor-in-chief, *Encyclopedia of Microbiology*, 2nd edn, vol. 3, 773–86, Academic Press, San Diego, CA, 2000.

51 WR Strohl, ed., *Biotechnology of Antibiotics*, Dekker, New York, 1997.

52 DA Hopwood, *Chem Rev* **97**, 2465–97, 1997.

53 DA Hopwood, C Khosla, *Ciba Found Symp* **171**, 88–106, 1992.

54 C Khosla, RJX Zawada, *Trends Biotechnol* **14**, 335–41, 1996.

55 G Chotani, T Dodge, A Hsu, M Kumar, R LaDuca, D Trimbur, W Weyler, K Sanford, *Biochim Biophys Acta* **1543**, 434–55, 2000.

56 A Grein, *Adv Appl Microbiol* **32**, 203–4, 1987.

57 G Rosabal, C Rodriguez, R del Sol, S Planas, C Vallin, F Malpartida, *Biotechnol Lett* **23**, 289–93, 2001.

58 C Khosla, PB Harbury, *Nature* **409**, 247–52, 2001.

59 Y-S Hwang, J-Y Lee, E-S Kim, C-Y Choi, *Biotechnol Lett* **23**, 457–62, 2001.

60 L Tang, S Shah, J Chung, J Carney, L Katz, C Khosla, B Julien, *Science* **287**, 640–2, 2000.

61 AJ Kluyver, FJ de Leeuw, *Tijdschr Verg Geneesk* **10**, 170, 1924.

62 T Reichstein, H Gruessner, *Helv Chim Acta* **17**, 311–28, 1934.

63 T Sonoyama, H Tani, K Matsuda, B Kageyama, M Tanimoto, K Kobyashi, S Yagi, H Kyotani, K Mitsushima, *Appl Environ Microbiol* **43**, 1064–9, 1982.

64 S Anderson, CB Marks, R Lazarus, J Miller, K Stafford, J Seymour, W Light, WH Rastetter, DA Estell, *Science* **230**, 144–9, 1985.

65 S Anderson, R Lazarus, J Miller, *US Patent 5032514*, 1991.

66 KG Hardy, H van de Pol, JF Grindley, MA Payton, *US Patent 4945052*, 1990.

67 Y Saito, Y Ishii, M Hayashi, Y Imao, T Akashi, K Yoshikawa, Y Noguchi, S Soeda, M Yoshida, M Niwa, J Hosoda, K Shimonura, *Appl Environ Microbiol* **63**, 454–60, 1997.

68 P Sulo, D Hudecova, A Properova, I Basnak, I Sedlacek, *Biotechnol Lett* **23**, 693–6, 2001.

69 R Gloecker, I Ohsawa, D Speck, C Ledoux, S Bernand, M Zinsius, D Villeval, Kisou, T Kamagowa, Y Lemoire, *Gene* **87**, 63–70, 1990.

70 J Sabatie, D Speck, J Reymund, C Herbert, L Caussin, D Weltin, K Gloecker, Y Lemoine, SW Brown, *J Biotechnol* **20**, 29–50, 1991.

71 S de Baets, S Vandedrinck, EJ Vandamme, in: J Lederberg, editor-in-chief, *Encyclopedia of Microbiology*, 2nd edn, vol. 4, 837–53, Academic Press, San Diego, CA, 2000.

72 M Petersen, A Kiener, *Green Chem* **4**, 99–106, 1999.

73 J-H Martens, H Barg, MJ Warren, D Jahn, *Appl Microbiol Biotechnol* **58**, 275–85, 2002.

74 N Dusch, G Thierbach, *PCT WO 02 24936*, 2002; *Chem Abstr* **136**, P261914m.

75 N Misawa, M Nakagawa, K Kobayashi, S Yamano, Y Izawa, K Nakamura, K Harashima, *J Bacteriol* **172**, 6704–12, 1990.

76 V Reipa, MP Mayhew, VL Vilker, *Proc Natl Acad Sci USA* **94**, 13554–8, 1997.

77 B Dumas, G Cauet, T Achstaetter, E Degryse, *PCT WO 02 12536*, 2002; *Chem Abstr* **136**, P166158j.

78 M Krämer, S Mäschen, M Wubbolts, Poster presentation at *Biotechnology Conference BioTrans 2001*, Darmstadt, 2001.

79 N Zheng, BG Stewart, JC Moore, RL Greasham, DK Robinson, BC Buckland, C Lee, *Metab Eng* **2**, 339–348, 2000.

80 M Mahmoudian, D Noble, CS Drake, RF Middleton, DS Montgomery, JE Piercey, D Ramlakhan, M Todd, MJ Dawson, *Enzyme Microb Technol* **20**, 393–400, 1997.

81 RS Blacklow, L Warren, *J Biol Chem* **237**, 3520–6, 1962.

82 W-D Fessner, M Knorst, 2002, *German Patent DE 10034586*; *Chem Abstr* **136**, 166160d.

83 U Kragl, D Gygax, O Ghisalba, C Wandrey, *Angew Chem Int Ed Engl* **30**, 827–28, 1991.

84 K Tabata, S Koizumi, T Endo, A Ozaki, *Enzyme Microb Technol* **30**, 327–33, 2002.

85 Y Yasohara, N Kizaki, J Hasegawa, M Wada, M Kataoka, S Shimizu, *Biosci Biotechnol Biochem* **64**, 1430–6, 2000.

86 RN Patel, A Banerjee, VB Nanduri, SL Goldberg, RM Johnston, RL Hanson, CG McNamee, DB Brzozowski, TP Tully, RY Ko, TL LaPorte, DL Cazzuline, S Swaminathan, C-K Chen, LW Parker, JJ Venit, *Enzyme Microb Technol* 376–89, 2000.

87 B Boonstra, DA Rathbone, CE French, EH Walker, NC Bruce, *Appl Environ Microbiol* **66**, 5161–6, 2000.

88 AM Rouhi, *Chem Eng News* **June 10**, 43–50, 2002.

6 Nucleic Acid Drugs

Joachim W. Engels and Jörg Parsch

6.1 Introduction

Nucleic acids are the memory of all the information of life of every plant and every animal. Nucleic acids are part of every cell of every living thing. Deoxyribonucleic acid (DNA) consists of long double-stranded chains of nucleotides. The nucleotides themselves consist of a sugar moiety [ribose for ribonucleic acid (RNA) and 2-deoxyribose for DNA], a phosphate group and a nucleobase. Generally, only four different nucleobases can be found in natural DNA and RNA. These are adenine, guanine, cytosine and thymine in DNA, and uracil instead of thymine in RNA. The chains of nucleotides, called oligonucleotides are double stranded in the nucleus. These double-stranded chains are mainly held together by hydrogen bonds between the nucleobases. The base pairs are always guanine–cytosine and adenine–thymine (uracil) (Fig. 6.1) [1].

The information saved in the DNA must be translated into proteins. Therefore the DNA must first be transcribed into messenger RNA (mRNA), which leaves the nucleus. The mRNA will be translated into proteins at the ribosomes. All these processes are possible points of attack of nucleic acid drugs. Several different concepts or mechanisms of action of nucleic acid drugs are now under investigation. The most important is the antisense concept (see Section 6.4.1). Other concepts under investigation are the triple helix (triplex) concept, the RNA interference (RNAi) concept and ribozymes as drugs.

6.2 Chemical Synthesis of Oligonucleotides

Oligonucleotides are long-chain molecules built in any nucleotide sequence order. Chemical synthesis results in condensation of single nucleotides until the required sequence and length is reached. It is not important for the synthesis if the nucleotides are DNA, RNA or modified building blocks. The reaction conditions only differ in the reaction times and in the choice of activating agents. Oligonucleotide

Molecular Biology in Medicinal Chemistry. Edited by Th. Dingermann, D. Steinhilber, G. Folkers

ISBN: 3-527-30431-2

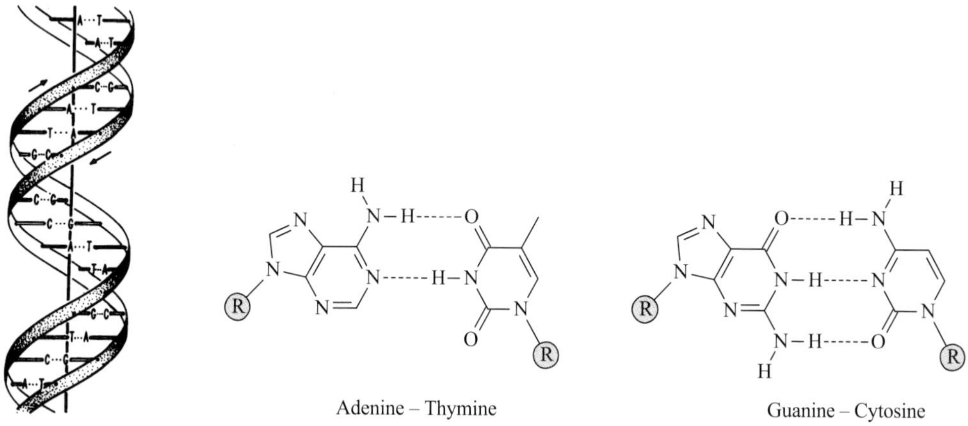

Fig. 6.1 Watson–Crick base pairing.

synthesis on the solid phase was developed by Letsinger in 1975 [2, 3]. The main advantage of solid-phase synthesis of oligonucleotides is that purification is only necessary after the last step of synthesis and not after every reaction. Furthermore, it is possible to synthesize more than one oligonucleotide at the same time at one oligonucleotide synthesizer. Several methods of oligonucleotide synthesis have been developed over the last 25 years. Building blocks of nucleotides with different protecting groups are required for each method. These methods are the so-called phosphordiester method, phosphortriester method, H-phosphonate method [4] and phosphoramidite method [5]. Nowadays only the phosphoramidite method plays an important role in the solid-phase synthesis of oligonucleotides. Figure 6.2 shows the synthesis cycle of the phosphoramidite method. The first nucleotide, fixed at the solid support, is deprotected at the 5′ position with trichloroacetic acid (TCA). Then the next building block is added, activated with an activator and coupled to the first nucleotide. Tetrazole or a derivative is used as an activator in most cases. The unreacted hydroxyl groups are capped with acetic anhydride to stop the building chain at this position. The capping avoids the synthesis of undesired sequences. The coupled dimer is oxidized by a mixture of iodine and water. Synthesis can be stopped after this step or the final nucleotide of the chain can be deprotected to start a new coupling cycle.

All protecting groups are cleaved off after the end of the synthesis of the desired oligonucleotide. The purification can be performed by high-performance liquid chromatography (HPLC) or gel electrophoresis.

Another possibility to obtain oligonucleotides is enzymatic synthesis, especially for longer oligonucleotides. Enzymatic synthesis utilizing modified building units is a valuable alternative to chemical synthesis. Enzymatic incorporation of natural or suitably modified nucleoside triphosphates by phage RNA polymerases (e.g. T7) has been successfully used; however, this approach is limited by the acceptance of the polymerase for sugar, backbone or base modifications.

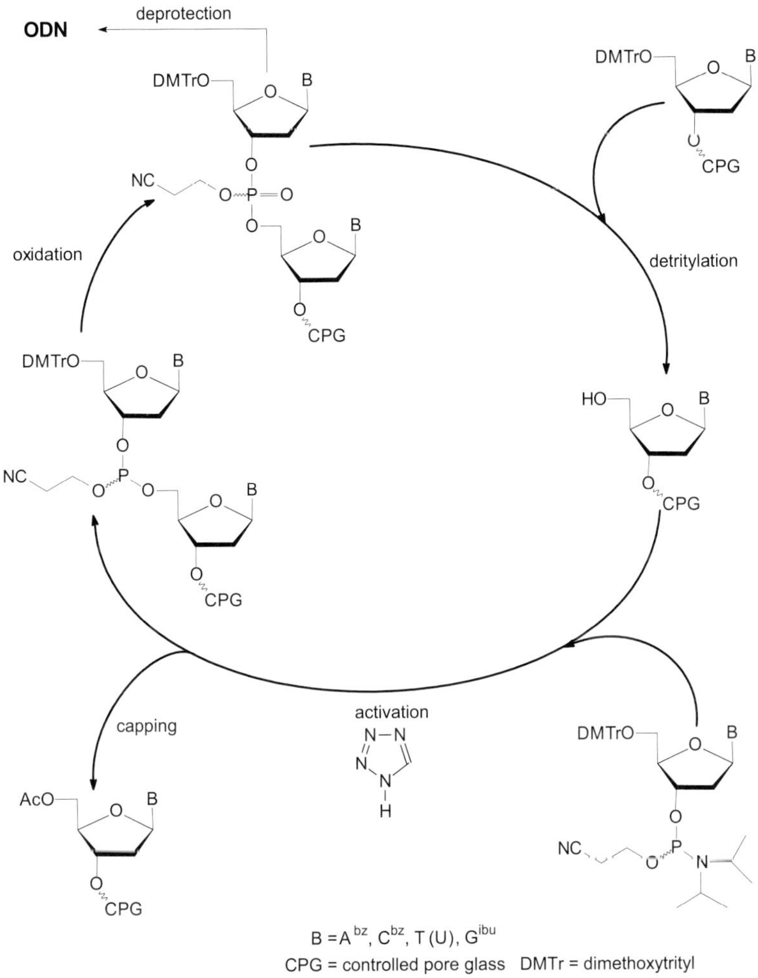

Fig. 6.2 Solid-phase synthesis of oligonucleotides.

6.3 Chemical Modifications of Oligonucleotides

In this section we outline the structural chemistry and properties of oligonucleotide derivatives. Chemical variation of natural oligonucleotide structure is necessary to render these compounds useful in biological systems. The following prerequisites must be fulfilled for good nucleic acid drugs:

- They must be sufficiently stable against nucleases in serum and within cells.
- They should enter the various organs of the body. After distribution to the desired tissue, they must be able to penetrate cellular membranes to reach their site of action.

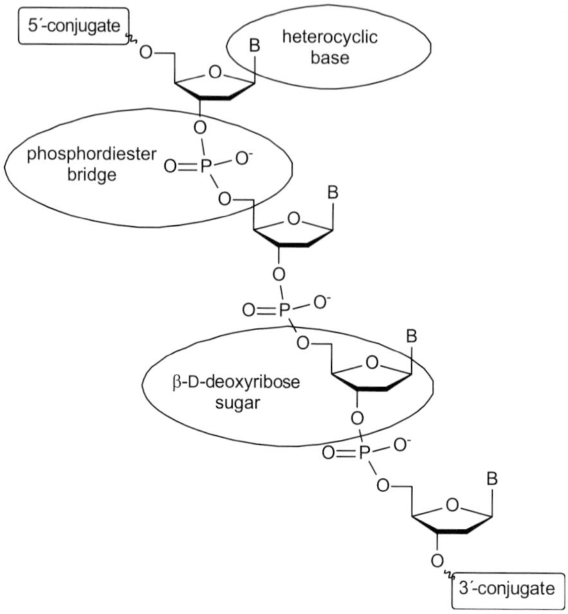

Fig. 6.3 Modification of oligonucleotides.

- They must form stable Watson–Crick or Hoogsteen complexes with complementary target sequences under physiological conditions.

Figure 6.3 shows the possibilities for modification for oligonucleotides. Modifications of the backbone through sugar modifications or substitutions at the phosphate moiety were performed as well as the incorporation of substituted or completely changed nucleobases. Also the incorporation of conjugates at the 3′ or 5′ end should be mentioned as well as completely backbone modified nucleic acids like polypeptide nucleic acids (PNAs) or locked nucleic acids (LNAs). All these modified nucleic acids have been synthesized and tested in recent years.

Unmodified oligonucleotides are widely used as tools in molecular biology. However, in cellular and animal experiments it was observed that oligonucleotides with a natural phosphodiester internucleoside linkage are degraded in serum within a few hours, mainly by the action of fast cleaving 3′-exonucleases that are accompanied by slower cleaving endonucleases. Significant 5′-exonuclease activity has also been observed in certain tissues. Chemically modified oligonucleotides will be described in the following sections.

6.3.1 2′-Modifications

2′-*O*-Modified oligonucleotides (Fig. 6.4) can be considered as analogs of oligoribonucleotides. 2′-*O*-Methyl ether can be found in naturally occurring, RNA e.g. at

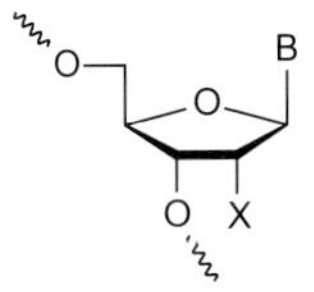

X = O-alkyl, F, NH_2, etc.

Fig. 6.4 2′-Modifications of oligonucleotides.

certain positions in tRNAs, rRNAs and snRNAs. The stability of (RNA)·(2′-*O*-alkyl-RNA) heteroduplexes [6] clearly depends on the nature of the 2′-*O*-alkyl group and increases in the order 2′-*O*-dimethylallyl < 2′-*O*-butyl < 2′-OH < 2′-*O*-allyl < 2′-*O*-methyl < 2′-methoxyethoxy. As early as in 1987 it was reported that 2′-*O*-methyl RNA forms a more stable duplex with a complementary RNA strand than unmodified DNA or even RNA [7]. The 2′-*O*-alkylribonucleotides prefer the C(3′)-endo conformation in a duplex with RNA. This sugar pucker has been found as a key structural element in RNA–RNA duplexes that are generally more stable than DNA–DNA duplexes of the same sequence. It is worthwhile mentioning that the 2′-*O*-methoxyethoxy modification does not only result in enhanced binding affinity to RNA, but at the same time renders the oligomer more stable towards nuclease degradation. RNase H activity is not supported by 2′-*O*-modifications so that the general policy is to locate them at the flanks of antisense oligonucleotides mainly consisting of phosphorothioate windows with modified ones to reduce negative effects of phosphorothioates (see Section 6.3.3). In many cases, particularly for application in animal models, even the flanks contain the phosphorothioates to further improve their stability. Apparently the cytotoxic side-effects of phosphorothioates are not so severe in conjunction with a 2′-*O*-alkyl modification.

In addition to 2′-*O*-alkyl modifications, 2′-substituted nucleotides such as 2′-fluoro or 2′-amino are known. A uniformly modified 2′-deoxy-2′-fluoro oligonucleotide [8] exhibits a considerably increased binding affinity for RNA as compared to the DNA oligonucleotide without compromising base pair specificity, whereas the 2′-amino modification [9] destabilizes the duplex with RNA. However, additional modifications, such as phosphorothioate bridges, are necessary to render these oligonucleotide derivatives sufficiently stable to nucleases. The introduction of uniform 2′-deoxy-2′-fluoro sugars leads to loss of RNase H activation.

6.3.2 Alkyl- and Arylphosphonates

Many different alkyl- and arylphosphonates (Fig. 6.5) have been investigated in recent years. The main representatives of these groups are methyl- and phenylphosphonates. Apart from the phosphorothioates, the methylphosphonates are probably the second best investigated class of oligonucleotide derivatives with a modification on phosphorus and they have been used very early for specific antisense inhibition of gene expression [10]. In methylphosphonates the negatively

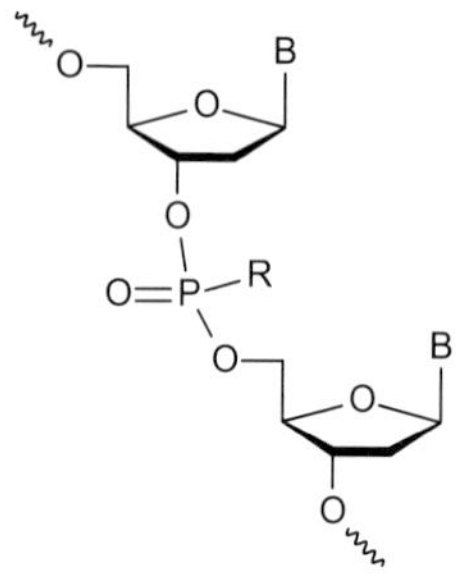

R = alkyl, aryl

Fig. 6.5 Aryl- and alkylphosphonate oligonucleotides.

charged phosphate oxygen is replaced by a neutral methyl group. The methylphosphonate linkage is highly stable to degradation by nucleases. However, a major problem is posed by the chirality of the methylphosphonate bridge, which can have either the R_P or the S_P configuration. Therefore, as with phosphorothioates, methylphosphonate oligonucleotides resulting from standard synthesis usually consist of a mixture of 2^n diastereomers, where n is the number of such linkages. It has been shown for short oligonucleotides with a methylphosphonate backbone that the all-R_P backbone shows a significantly higher melting point (T_m) than a mixture of diastereomers or even the corresponding phosphodiester oligonucleotide when hybridized to complementary nucleic acids.

Uniformly methylphosphonate modified oligonucleotides do not form duplexes with RNA that induce RNase H cleavage. This may be a limitation to their use as antisense oligonucleotides and could explain the relatively high concentration required for effective translation arrest in antisense experiments. Chimeric oligonucleotides containing both methylphosphonate and a window of at least five to seven phosphodiester linkages retain RNase H activity, and the poor solubility of uniformly modified compounds in biological systems can be overcome.

Alkylphosphonates are available through automated oligonucleotide synthesis using methylphosphonamidites as synthons [11]. Since the alkylphosphonate bridge is more base labile than the natural phosphodiester linkage, much milder conditions are necessary for cleavage from the solid support and for deprotection. Similarly, phenylphosphonate- and phenylphosphonothioate-containing oligonucleotides can be prepared from the corresponding nucleoside-3′-phenylphosphonamidites [12]. The binding affinity of the arylphosphonate-containing oligomers depends strongly on the sequence, so that some duplexes are destabilized by −0.3 to −1.3 K, whereas others are stabilized by +0.2 to +0.5 K relative to their natural congeners. The synthesis of oligodeoxynucleotide pentadecamers containing two octylphosphonate linkages with stereoregular or stereorandom chirality has also been described [13]. The difference in T_m was −3.4 K per modification of the stereorandom and about −2.3 to −4.0 K per modification for the stereoregular configured oligonucleotides.

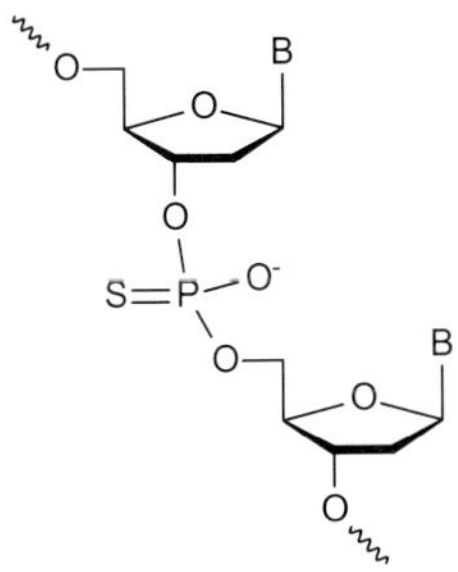

Fig. 6.6 Phosphorothioate oligonucleotides.

6.3.3 Phosphorothioates

Nucleases degrade oligonucleotides by a nucleophilic attack at the phosphodiester linkage. The replacement of a nonbonding phosphate oxygen by other atoms is an easy way to make oligonucleotides resistant against nucleases.

The phosphorothioate modification (Fig. 6.6) provides significant stabilization against degradation by nucleases. The phosphorothioate bridge also represents a center of chirality. It must be mentioned, however, that nuclease resistance depends strongly on the configuration at phosphorus. The S_P diastereomers are substrates of nucleases S1 or P1, while the R_P diastereomers are cleaved by snake venom phosphodiesterase. Cellular uptake of phosphorothioate oligonucleotides is similar to phosphodiester oligonucleotides. An important property of phosphorothioates is their ability to mediate RNase H degradation of the RNA after hybridization.

This modification enhances the stability of the oligonucleotides against enzymatic degradation and, at the same time, is compatible with activation of the RNase H. In spite of all these promising characteristics, the phosphorothioates show some disadvantages. The thermal stability of the complex formed with the target RNA is not as stable as that formed with an unmodified oligonucleotide. The hybridization properties of diastereoisomeric phosphorothioates are sufficiently strong to use them under *in vivo* conditions, although there is an average loss of −0.5 K per phosphorothioate linkage in the T_m for the racemic mixture. Moreover, for reasons which are not well understood, the phosphorothioate oligonucleotides can have a much higher affinity to certain proteins such as those responsible for polyanion binding. This can result in undesired side-effects often associated with cytotoxicity.

A phosphorothioate also represents a molecule which bears a further chirality center in comparison with unmodified nucleotides. The S_P- and R_P-phosphorothioates showed different physiochemical properties. Heterodimers formed between oligoribonucleotides and all-R_P-phosphorothioates showed a T_m as compared with the less stable heterodimers formed with all-S_P-phosphorothioates or the random mixture of diastereomers [14]. The DNA–RNA complex contain-

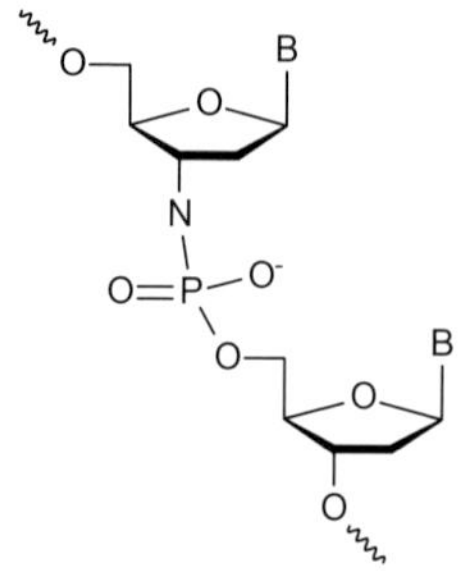

Fig. 6.7 3′ Phosphoramidate oligonucleotides.

ing the phosphorothioate of all-R_P configuration was found to be more susceptible to RNase H-dependent degradation as compared with hybrids having either all-S_P counterparts or the random mixture of diastereomers. Interestingly, 3′-exonucleases present in human plasma appear to degrade phosphorothioates of the R_P configuration, but not those of the S_P configuration [15].

6.3.4 N3′-P5′-phosphoramidates

N3′-P5′-phosphoramidate oligonucleotides (Fig. 6.7) show remarkable resistance to nucleases and strong hybridization, but do not activate RNase H. Extensive studies on the inhibition of gene expression are still awaited.

6.3.5 Morpholino Oligonucleotides

Morpholino oligonucleotides are very stable to nucleases and they are RNase H insensitive. They have found acceptance mainly in developmental studies in the zebra fish and *Xenopus*, both systems where the compounds can be administered (injected) to the embryo.

6.3.6 PNAs

In a PNA the sugar-phosphate backbone is replaced by an *N*-aminoethylglycine-based polyamide structure (Fig. 6.8) [16]. An advantage of PNAs is that they bind with higher affinity to complementary nucleic acids than their natural counterparts following the Watson–Crick base-pairing rules [17–19]. The N-terminus of a PNA corresponds to the 5′ end and the C-terminus to the 3′ end of a DNA. In case of the antiparallel PNA–DNA or PNA–RNA duplexes, respectively, the T_m is increased by approximately 1 or 1.5 K/base, respectively, as compared with the corresponding DNA–DNA or DNA–RNA duplex. A further advantage of PNA is that base mismatches often give rise to a significantly larger reduction in T_m compared to DNA.

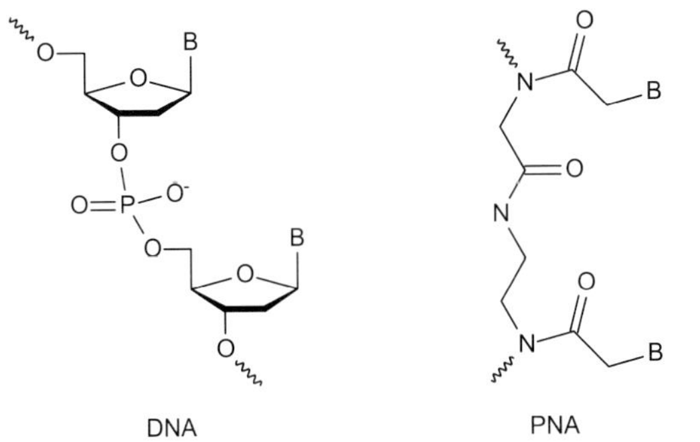

Fig. 6.8 DNA and PNA oligonucleotides.

PNA is extremely stable to nucleases and peptidases. One limitation of PNA, however, is that it cannot stimulate RNase H cleavage. Consequently, it has been found that PNAs are less effective antisense agents than phosphodiester [20] and phosphorothioate [21] oligonucleotides.

In contrast, DNA/PNA chimeras [19] with more than four nucleotides are able to stimulate the cleavage of RNA by RNase H on formation of a chimera–RNA duplex. RNA cleavage occurs at the ribonucleotides which base-pair with the DNA part of the chimera. PNA/DNA chimeras also obey the Watson–Crick rules on binding to complementary DNA and RNA [22, 23], whereby T_m strongly depends on the PNA:DNA ratio in the chimeras. The T_m of 5′-DNA/3′-PNA chimeras, in which the PNA and DNA parts are of equal length, lies roughly between those observed for the corresponding pure PNA and DNA. In contrast to pure PNA, 5′-DNA/PNA chimeras bind exclusively in the antiparallel orientation to DNA and RNA under physiological conditions. The chimeras show also much better cellular uptake than pure PNA.

6.3.7
Sterically LNAs

Nucleic acid analogs with conformationally restricted sugar–phosphate backbones based on (3′*S*,5′*R*)-2′-deoxy-3′5′-ethano-*β*-D-ribofuranosyladenine and -thymine (bicyclo-DNA) were reported to form more stable duplexes than the corresponding natural congeners (Fig. 6.9) [24]. The bicyclo-DNA (A_{10}) formed more stable triplexes with d(T_{10}) of the pyrimidine purine-pyrimidine motif than natural d(A_{10}). Recently, a new DNA analog, "bicyclo-[3.2.1]-DNA", has been described [25] which has a rigid phosphodiester backbone, emulating a B-DNA-type conformation, and to which the nucleobases are attached via a flexible open-chain linker. Although bicyclo-[3.2.1]-DNA forms less stable duplexes with complementary DNA than natural DNA, base-mismatch discrimination is slightly enhanced compared to pure DNA duplexes. Importantly, bicyclo-[3.2.1]-DNA oligomers are resistant to 3′-exonuclease degradation.

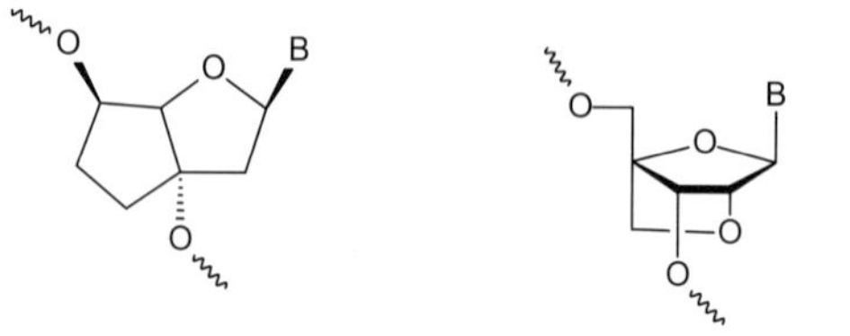

Fig. 6.9 Sterically locked oligonucleotides.

A novel modification with unusually high binding affinity is LNA, in which the 2′ oxygen is linked by a methylene bridge to the 4′ carbon forming a methylene-linked bicyclic ribofuranosyl nucleoside. The sugar conformation of this derivative is locked in the N-type (3′-endo) conformation [26]. An unprecedented increase of +3 to +8 K per modification in the thermal stability of duplexes towards both DNA and RNA was reported when evaluating mixed sequences of partly or fully modified LNA.

6.3.8 Oligonucleotide Conjugates

The covalent attachment of nonnucleosidic molecules to either the 3′ or 5′ ends of oligonucleotides can also modulate the properties of antisense oligonucleotides. This type of derivatization is chemically simple and allows modulation of nuclease stability, cellular uptake and organ distribution of oligonucleotides.

6.3.8.1 5′-End Conjugates

Conjugation of molecules to the 5′ end of oligonucleotides can be made straightforward by coupling a phosphoramidite or H-phosphonate derivative of the desired molecule to the 5′-hydroxy group of the oligomer following chain elongation by solid-phase synthesis. A broad range of phosphoramidite derivatives of ligands, such as fluorescein, biotin, cholesterol, dinitrophenyl, acridine and psoralen derivatives, is commercially available. Alternatively, the 5′-terminal hydroxy group of the oligonucleotide is reacted with an aminoalkyl linker phosphoramidite, which after deprotection results in a free aminoalkyl function. The amino function of the oligonucleotide can then be reacted post-synthesis in solution with suitably activated conjugate molecule derivatives, such as active esters, isothiocyanates or iodoacetamides.

6.3.8.2 3′-End Conjugates

The conjugation of molecules to the 3′ end of oligonucleotides is conveniently achieved by using correspondingly functionalized solid supports, which in addition bear a hydroxyl function from which the oligonucleotide chain is extended during solid-phase synthesis. After oligonucleotide synthesis is complete, the oligonucleotide conjugate is cleaved from the solid support and deprotected by ammonia treatment. Using this method, relatively exotic derivatives, such as the anionophoric moiety of pamamycin [19], can be easily introduced. Similarly to the 5′-end conjugation, appropriate 3′-amino-modifier solid supports are commercially available which allow coupling of suitably activated derivatives of the molecule to be conjugated with the 3′-amino alkyl group in a post-synthetic solution-phase step.

Conjugation to the 3′ end of oligonucleotides usually results in strong stabilization against 3′-exonucleases, which are the predominant nucleases in human serum.

6.4 Mechanism of Action

Since the discovery of the DNA structure by Watson and Crick [27] in 1953 many different ideas of mechanisms of action of nucleic acid drugs have been developed. The most promising are shown in Fig. 6.10. They are the antisense, the anti-gene (triple helix) and the RNAi concept as well as the use of ribonucleic acids as ribozymes. The mechanism of action of these concepts are discussed in this section. At present, only nucleic acid drugs following the antisense concept and ribozymes are on the market or in clinical test phases (*cf.* Section 6.7).

6.4.1 The Antisense Concept

The antisense concept follows the most important way of action to modulate the transfer of genetic information to proteins. Some mechanisms of action by which an oligonucleotide can induce a biological effect are complex. Antisense oligonucleotides can be classified into two main classes. On the basis of their mechanism of action – RNase H-dependent oligonucleotides, which induce the degradation of mRNA by RNase H, and steric blocking oligonucleotides, which physically prevent or inhibit the splicing or the translation machinery. Most of the investigated antisense oligonucleotides work via the RNase H-dependent mechanism.

6.4.1.1 Steric Blocking

The genetic information of live is saved in the DNA double helix and the DNA must be translated into proteins in order to obtain this information, i.e. the DNA will be transcribed into mRNA which will be translated to proteins. Antisense oligonucleotides interact with these mRNA strands to block the translation of mRNA to proteins at the ribosomes. Therefore the antisense oligonucleotides specifically form hydrogen bonds with their nucleobases to the complementary bases in the Watson–Crick manner. The binding is sequence specific so that one mismatch reduces the stability of the formed duplex significantly. In most cases, under physio-

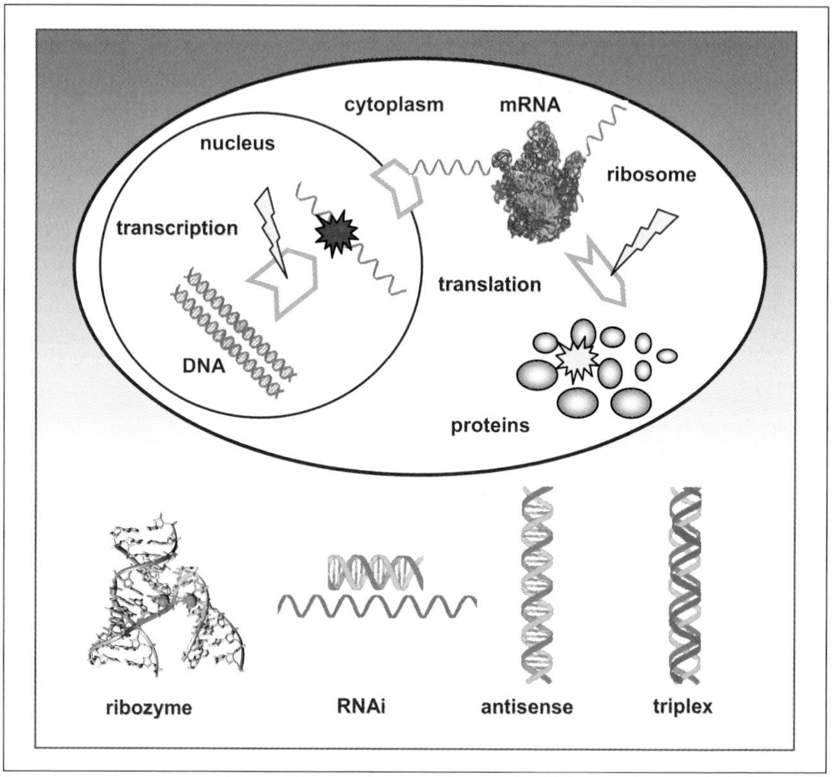

Fig. 6.10 Oligonucleotide interference.

logical conditions, a duplex with one or more mismatches will be cleaved quickly. Thus, it is necessary to know the sequence of the mRNA which should be blocked exactly. Many sequences are known, and since the whole human genome was sequenced more and more information about the effects produced by the corresponding mRNA will be available. This knowledge makes the synthesis of specific antisense oligonucleotides possible.

An antisense oligonucleotide can bind to the corresponding mRNA through Watson–Crick hydrogen bonding, and the antisense oligonucleotide and mRNA form a new duplex. The translation of the mRNA to proteins is blocked through the binding to the mRNA and the formation of a double helix makes it impossible for the translation enzymes to translate a double-stranded helix. Because of the only temporary stability of these double helices, modifications of nucleotides are being developed to obtain more stable helices under physiological conditions to improve the steric blocking. The objective of this development will be to form a double helix which will have a longer life-time than the corresponding mRNA alone in order to optimize the mechanism of action.

6.4.1.2 RNase H Activation

The second and most important mechanism of action of antisense oligonucleotides after steric blocking is the activation of RNase H [28, 29]. RNase H normally degrades RNA primers during DNA replication or selectively excises wrongly incorporated RNA nucleotides in DNA strands. In DNA–RNA heteroduplexes, the RNase H selectively hydrolyses the RNA strand. The cleaved mRNA will be immediately degraded by exonucleases while intact mRNA is protected by specific 5′ and 3′ structures against degradation of exo- and endonucleases. The RNase H mechanism also explains why antisense RNA shows no effect on inhibition of translation. The RNA-cleaving activity of RNase H will not be induced by RNA–RNA duplexes [30].

Chemically modified antisense oligonucleotides should also activate the RNase H. While phosphorothioates activate RNase H, others like methylphosphonates or PNA and LNA nucleotides do not activate this enzyme. The activation of RNase H in phosphorothiates is dependent on the stereochemistry at the phosphorous atom. In partly modified antisense oligonucleotides this disadvantage of the nonspecific action of phosphodiester or phosphorothioate oligonucleotides is turned to an advantage. The incorporation of five or six modified nucleotides in an unmodified strand is enough to activate the RNase H [30]. However, in view of the lability of unmodified oligonucleotides against nucleases and the weak antisense effects observed for most derivatives which do not activate RNase H, the development of mixed-backbone oligonucleotides or chimeric oligomers, in which the advantages of the individual structural elements are combined, appears to be the approach of choice at present.

6.4.2 The Triplex Concept

The first publications about triple helixes appeared only 4 years after the discovery of the DNA double-helix structure by Watson and Crick [31]. The triple helices consist of one polypurine and two polypyrimidine strands. While T×A·T triple helixes occurred rarely in cells C^+×G·C triple helixes are not stable under physiological conditions (Fig. 6.11). The cytosine in the third strand must be protonated to form a stable triple helix. To protonate a cytosine a pH of 6.0 was necessary. The binding of the third stand to the Watson–Crick double helix resulted in Hoogsteen or reverse Hoogsteen hydrogen bonding (Figs. 6.11 and 6.12) [1, 32, 33].

Short oligonucleotides can bind base-specifically, forming Hoogsteen or reverse Hoogsteen hydrogen bonds in the major groove of the DNA double helix in a sequence specific manner. T×A·T and C^+×G·C base triads were formed in the parallel mode and G×G·C, A×A·T and T×A·T base triads were formed in the antiparallel mode (× = Hoogsteen and · = Watson–Crick hydrogen bonds) (Fig. 6.12).

The application of triple-helical nucleic acids is one possibility to regulate gene expression *in vivo*. In the triple-helix concept (anti-gene concept), the third nucleic acid strand should hybridize with the DNA double helix in the nucleus and inhibit

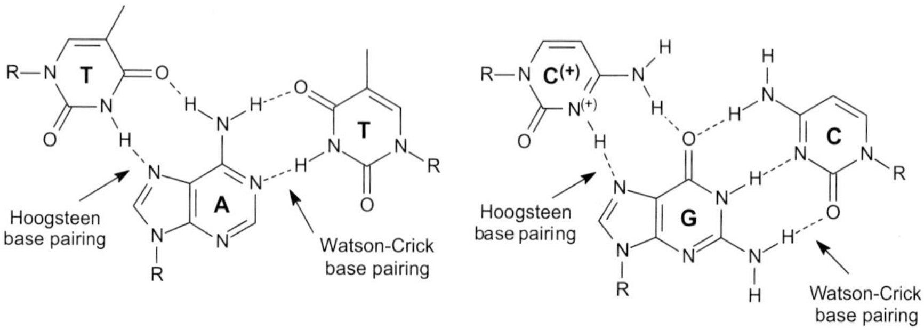

Fig. 6.11 Watson–Crick and Hoogsteen base pairing.

the translation of the DNA to the corresponding RNA. A major advantage of the anti-gene concept compared with the antisense concept is the fact that every cell has only a diploid set of chromosomes, so that only two equal DNA sequences per cell and not hundreds or thousands of mRNA strands must be inhibited.

There are four different possible mechanisms of action of triple-helical nucleic acids:

- The triplex-forming oligonucleotide can interact with a binding position of one component of the transcription or replication sequence.
- The change in conformation caused by forming the triple helix can disturb the binding of an enzyme imported for translation.

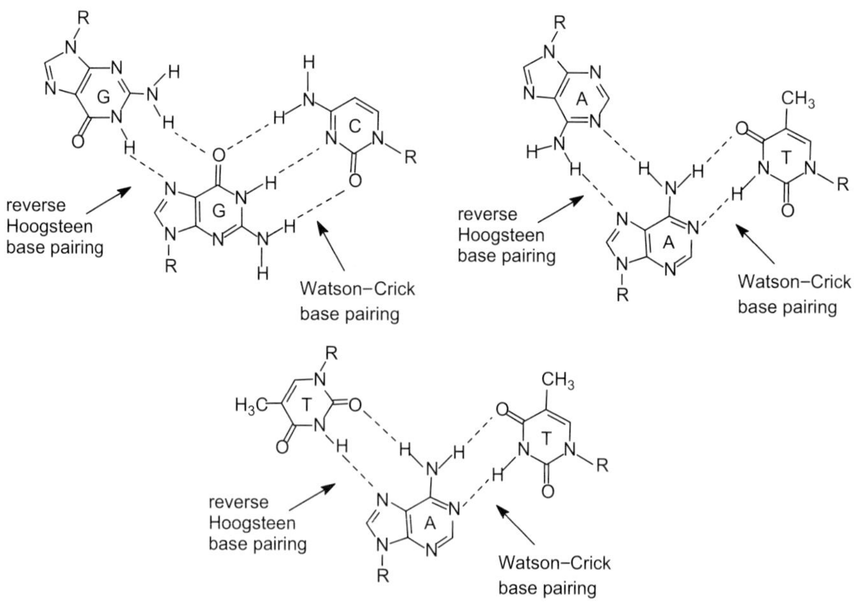

Fig. 6.12 Possible triple helix motifs.

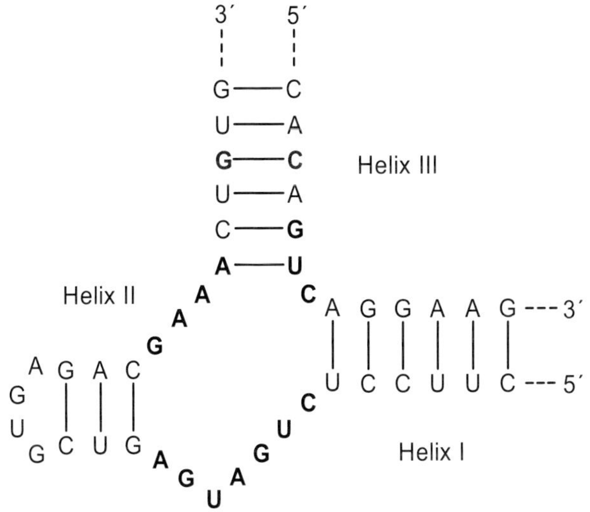

Fig. 6.13 Hammerhead ribozyme.

- A triplex-forming oligonucleotide bound downstream of a RNA polymerase promoter region could inhibit initiation of transcription.
- A triplex-forming oligonucleotide bound to the DNA could act as a barrier to transcription or replication enzymes.

The problems of *in vivo* application of triplex-forming oligonucleotides are similar to those for antisense oligonucleotides, i.e. stability in biological media, absorption to the nucleus and pharmacokinetics.

6.4.3
Ribozymes

Ribozymes are catalytically competent RNAs that occur either in nature or have been obtained by *in vitro* selection [34, 35]. Naturally occurring ribozymes mostly catalyze RNA cleavage and ligation reactions.

Here, we concentrate on the hammerhead and the hairpin ribozyme that belong to the class of small ribozymes (Figs. 6.13–6.15) [34]. They have attracted considerable attention because of their possible straightforward chemical synthesis and, in particular, for incorporating nucleotide analogues. They have also been applied for the inhibition of gene expression on the RNA level. These ribozymes were originally discovered as structural motifs in certain plant pathogen viroids for the self-cleavage of catenated RNA [36]. This cleavage is an intramolecular trans-esterification reaction subsequently devised for intermolecular reactions by placing the substrate and ribozyme part on separate RNA molecules [37, 38]. Thus it became possible to analyze the ribozyme kinetically like a conventional enzyme and this aided the understanding of the structure–function relationship. Furthermore,

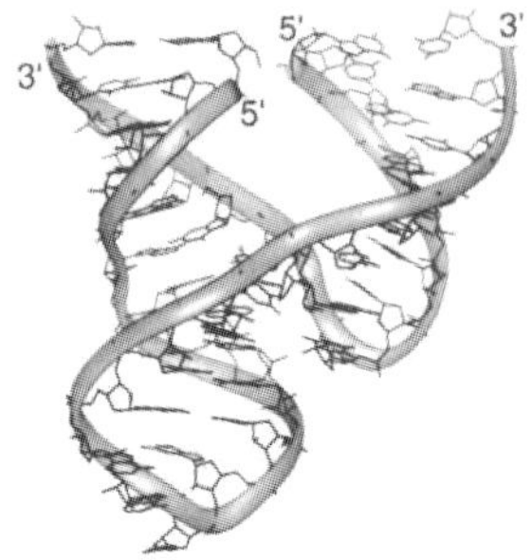

Fig. 6.14 X-ray structure of a hammerhead ribozyme.

the separation of the substrate from the ribozyme made the ribozyme suitable for cleaving any given RNA and suitable for the sequence-specific inhibition of gene expression [39, 40].

6.4.3.1 Structure and Reaction Mechanism

The structure of the hammerhead ribozyme has been determined by X-ray crystallography in the ground state as well as for a construct approaching the transition state of the reaction [42–44]. A two-dimensional representation of the hammerhead ribozyme complexed with a substrate is given in Fig. 6.14.

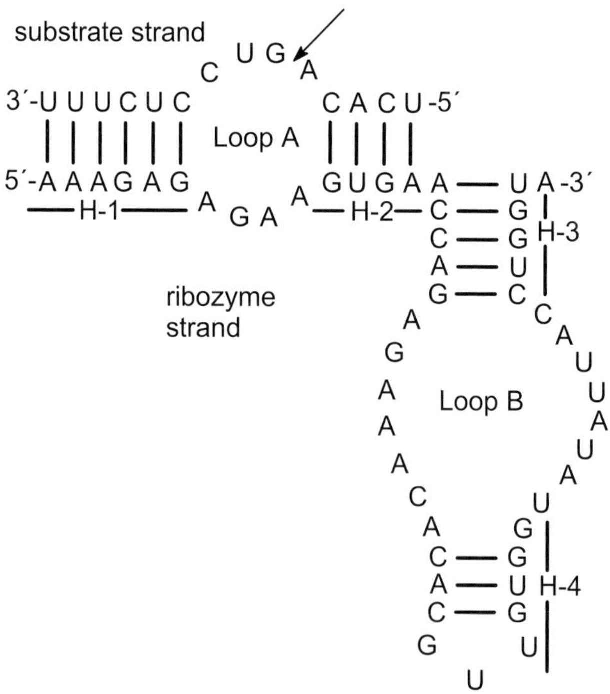

Fig. 6.15 Hairpin ribozyme.

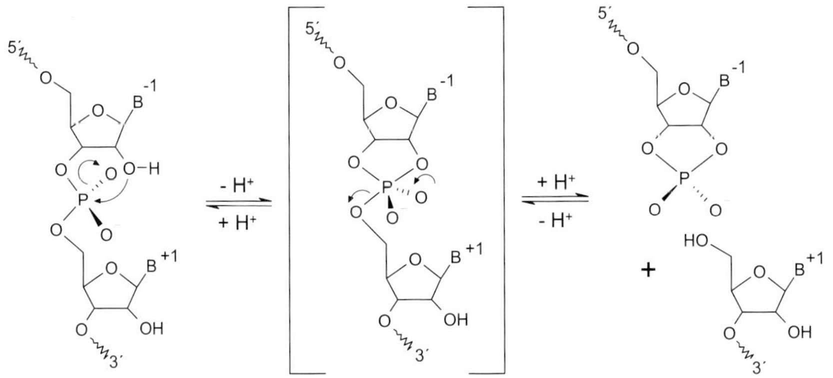

Fig. 6.16 Mechanism of ribozyme cleavage.

For most of such studies the ribozyme is prepared by chemical synthesis as it permits the introduction of modified nucleotides at specific positions. The ribozyme consists of a central core with invariant nucleotides, a helix II which stabilizes this core, and two binding arms to form helices I and III with the substrate. Mg^{2+} is required for optimal activity and one of the roles of the metal ion is to induce conformational changes to reach the catalytically competent structure.

The ribozyme catalyses a phosphoryl transesterification reaction of an internucleotidic 3′,5′ phosphate of the substrate to a 2′,3′ nucleoside cyclic phosphate which results in cleavage of the internucleotidic bond (Fig. 6.16) [45, 46].

The reaction is sequence specific, in that cleavage occurs 3′ to the triplet of the general formula NUH, where N is any nucleotide, U is uridine and H is any nucleotide except guanosine. The targeting of a particular NUH triplet for cleavage in an RNA is determined by the sequence of the ribozyme-binding arms that has to be complementary to the upstream and downstream sequences of the triplet to form the ribozyme–substrate complex.

The ribozyme is cleaved with an inversion of configuration on phosphorus, thus indicating an in-line mechanism for the transesterification reaction [47]. The favored mechanism for the hammerhead is in the presence of a metal ion with an active role of the metal ion in catalysis. However, there is also the possibility that a functional group of one of the nucleobases is involved – the mechanism favored for the hairpin ribozyme. Thus nucleobase functional groups can in principle act as general acid–base catalysts.

6.4.3.2 **Triplet Specificity**

The natural ribozymes cleave 3′ to NUH triplets. The X-ray structure has been solved not only in the ground but also close to the transition state. This emphasizes the dynamics of the system that must adopt the active structure by as-yet unknown conformational cleavage. However, efforts to obtain hammerhead ribozymes with altered cleavage specificity were more successful. A ribozyme could be selected cleaving 3′ to NUR where R can be a guanosine or an adenosine [48]. This indicates that this particular sequence space for NUG cleavage is very limited. An interesting feature of this ribozyme is its ability to cleave triplets where the central U can also be either a C or a G, thus differing considerably from the conventional hammerhead ribozyme with its strict requirement for a central U. Nevertheless, this new ribozyme expands the spectrum of cleavage sites in an RNA, which is important for the inhibition of gene expression.

6.4.3.3 Stabilization of Ribozymes

Ribozymes are readily hydrolyzed by RNases that are particularly active in serum. When ribozymes are applied exogenously for the inhibition of gene expression they will come into contact with serum and therefore their stabilization against such degradation is an important prerequisite for successful development as a drug. A simple protection against RNases is removal of the 2′-OH group, but retaining it at those positions which are essential for the catalytic activity of the ribozyme. Indeed, changing the ribose of the pyrimidine nucleosides to 2′-fluoro-2′-deoxyribose increased the half-life considerably with only a small reduction in catalytic power [49]. This was improved by replacing uridines at positions 4 and 7 by 2′-amino-2′-deoxyuridine [50]. Subsequent work showed that the amino group can be incorporated at all pyrimidine nucleoside positions with the desired effect [51, 52]. Alkylation of the 2′-OH group also protects against degradation and can be incorporated at all positions (except for the purine nucleosides in the core region) without impairment of activity [53, 54]. Degradation by exonucleases can be prevented either by introduction of phosphorothioates or an inverted nucleotide with a 3′,3′ linkage. Combination of these modifications increases the half-life of the ribozyme from a few minutes to days without serious impairment of catalytic competence. In conclusion, the problem of stabilization of the ribozyme against nuclease degradation can be considered solved.

6.4.3.4 Inhibition of Gene Expression

As reported for suppressing gene expression, the hammerhead and, to a lesser extent, the hairpin ribozyme have been investigated, and many examples of successful applications have been described [39, 55–57]. These approaches can be divided into two classes depending on the mode of delivery of the ribozyme to cells. One, described as endogenous application, consists of cloning the sequence for the ribozyme behind a suitable promoter into a vector, either a plasmid or a retroviral vector. After transfection or transduction, the ribozyme is transcribed in the cell to find its mRNA target. This approach has been adopted by several laboratories and shown to be successful, e.g. in interfering with HIV replication, as discussed in several reviews [39, 55–57]. One advantage of this mode of delivery is that the gene can be stably integrated into the host DNA, as in a gene therapy protocol. Also, once integrated, the supply of ribozyme is permanent. However, the choice of vector is crucial, particularly for application to humans where vector safety is of considerable concern. In the other mode of application, exogenous delivery, the ribozymes are prepared either by chemical synthesis or by transcription and are added to the cells from the outside.

6.4.3.5 Colocalization

An important aspect for the efficient interference with gene expression is the colocalization of the ribozyme with its target RNA. In this respect, endogenous delivery has the advantage in that the ribozyme can be directed to the nucleus or to the cytoplasm either by the choice of promoter or by combining it with subcellular localization signals [58, 59]. Colocalization of ribozyme and target in the nucleolus demonstrates this point very convincingly [60, 61]. At present synthetic ribozymes designed for exogenous delivery miss such signals and localization cannot be directed. However, it is hoped that the attachment of peptide signals, e.g. those important for nucleus import, might improve this situation. Chemical modification can apparently influence localization somewhat. Thus, phosphorothioate groups, introduced to achieve nuclease stability, direct the ribozyme preferentially to the nucleus, although interference with gene expression seems to take place in the cytoplasm [62].

6.4.4 RNAi

A newly developing approach for targeting mRNA is called post-transcriptional gene silencing or RNAi [63–65] (Fig. 6.17).

RNAi is the process by which double-stranded RNA targets mRNA for destruction in a sequence dependent manner. The mechanism of RNAi initially involves processing of long (around 500–1000 nucleotides) double-stranded RNA into 21- to 25-bp "trigger" fragments [66] by a member of the RNase III family of nucleases called DICER [67–69]. When incorporated into a larger, multicomponent nuclease complex named RISC (RNA-induced silencing complex), the processed trigger strands form a "guide sequence" that targets the RISC to the desired mRNA sequence and promotes its destruction [68]. RNAi has been used successfully for gene silencing in various experimental systems, including petunias, tobacco plants, neurospora, *Caenorhabditis elegans*, insects and zebra fish. The use of long double-stranded RNA to silence expression in mammalian cells has been tried, largely without success [70]. More recent reports using short interfering RNA (siRNA; see below) seem to be more promising [71]. It has been suggested that mature, as opposed to embryonic, mammalian cells recognize these long double-stranded RNA sequences as invading pathogens. This triggers a complex host-defense reaction that effectively shuts down all protein synthesis in the cell through an interferon-inducible serine/threonine kinase enzyme called protein kinase R (PKR). PKR phosphorylates the α subunit of eukaryotic initiation factor 2 (EIF 2), which glob ally inhibits mRNA translation. The long double-stranded RNA also activates 2′,5′-oligoadenylate synthetase, which in turn activates RNase L. RNase L indiscriminately cleaves mRNA. Cell death is the understandable result of these processes. Recently, a number of reports have suggested that siRNAs (RNA double strands of around 21–22 nucleotides in length) do not trigger this host-defense response, and therefore might to be able to silence expression in mammalian somatic cells if appropriately modified to contain 3′-hydroxy and 5′-phosphate

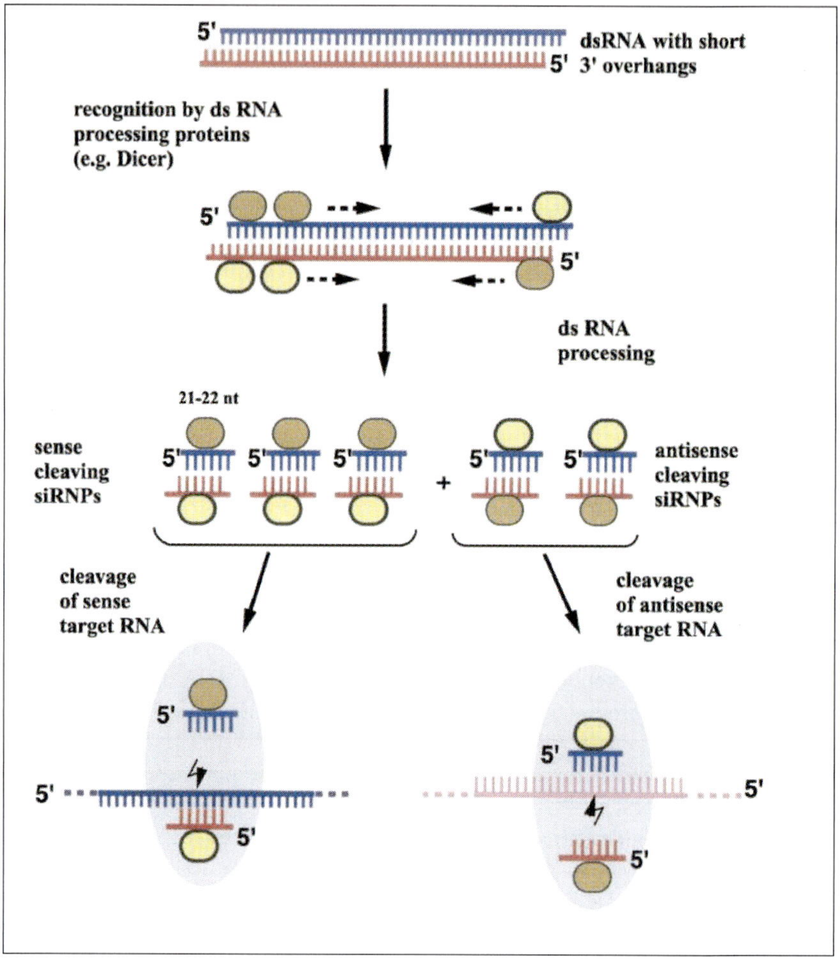

Fig. 6.17 RNAi mechanism.

groups [72–74]. The universality of this approach and the types of gene that can be modified using this strategy in mammalian cells are under intense studies at this time.

6.5 Positioning Identification of Ribozyme Accessible Sites on Target RNA

Trans-acting hammerhead ribozymes can potentially cleave any RNA as long as it contains any of the cleavable triplets. The presence of such triplets is usually not limiting, but the identification of ribozyme-accessible regions on the target RNA is an important issue for the success of the inhibition of gene expression. RNAs have extensive secondary structures and it is to be expected that regions that are open for docking of an oligonucleotide such as antisense oligodeoxynucleotides or ribozymes will be limited. Until recently such sites have been identified mostly by trial and error. However, experimental and theoretical approaches have been developed to facilitate the detection of accessible RNA sites. One such approach is scanning the labeled target RNA by hybridization to a library of oligonucleotides of known sequence on an array [75]. Another is the annealing of a randomized oligodeoxynucleotide to a labeled transcript of the RNA and subsequent cleavage of this complex by RNase H. The sites of cleavage can then be scanned for triplets susceptible for cleavage by ribozymes [76]. The analysis of transcripts is not entirely satisfactory as the secondary structure of the mRNA in the cell might be quite different from that of the transcript. Additionally, part of the mRNA might be covered by proteins preventing oligonucleotide annealing. Thus, extension of such analyses to native mRNA is most desirable even though experimentally this is very challenging because of the low stability of the cleavage products. Indeed, the analysis of murine DNA methyltransferase mRNA in cell extracts has been successfully conducted with oligonucleotides and RNase H [77]. Sites thus selected were equally effective for inhibition by antisense oligodeoxynucleotides and ribozymes. Computational methods have also been developed to identify such sites [78, 79]. One direct comparison demonstrated that sites identified by the computational and by the RNase H methods were consistent [80]. More recently, ribozymes with randomized substrate-binding arms have been employed for cleavage of transcripts by the hairpin ribozyme [81, 82].

6.6 Delivery

The exogenous delivery of ribozymes circumvents the problem of vector choice, but is faced with the problems of nuclease degradation of the supplied ribozymes in the serum and the hurdle of efficient cellular uptake. The problem of degradation of exogenously delivered ribozymes by serum nucleases has essentially been solved by chemical modification as discussed above. Cellular uptake in cell culture is in general achieved by the use of cationic lipids as carriers, where the positive charge is localized on the outside of the lipid for electrostatic interaction with the negative charge of the oligonucleotide. Various compositions of such lipids are available, some more suited for certain cell lines than others. In the future it is hoped that the derivatization of ribozymes with peptides of known transport properties might alleviate this situation. Such a strategy might also pave the way for cell type-directed delivery. However, there are cell types in culture that take up ribozymes quite readily without such aids as long as they are chemically modified [83]. It is assumed that these cells, Chinese hamster ovary cells in this example, have a very active membrane that facilitates oligonucleotide transport. Most interestingly, chemically stabilized ribozymes are taken up by tissue upon local application in animal models without carrier [84–87]. This agrees with the *in vivo* application of antisense oligodeoxynucleotides that is also achieved without carriers [88, 89]. Thus, uptake of oligonucleotides differs in cell culture and in animals, and our understanding of these processes is still very unsatisfactory.

Targets for chemically modified ribozymes in cell culture have been c-*myb* [90], N-*ras* [91], luciferase [92], vascular endothelial growth factor receptor [87], multiple drug resistance gene [83, 93] and Hepatitis C virus 5′-long terminal repeat [94]. *In vivo* model studies with synthetic ribozymes have been reported for inhibition of expression of amelogenin [84], protein kinase C [51], stromelysin mRNA [85] and the dopamine D2 receptor [86].

Only cell culture experiments will be considered here in more detail. The degrees of inhibition surprisingly do not vary very much in these studies – 51–77% inhibition (Tab. 6.1). However, the results are difficult to compare as they were obtained with differences in cell lines, mRNA targets, concentrations of ribozymes, the types of chemical modifications and carriers, and often with no screen for the best ribozyme annealing or cleavage site. Factors affecting the efficiency of ribozyme inhibition also include the lifetime of the protein and the mRNA. Obviously, the longer lived the protein is, the more difficult it will be to achieve inhibition.

Tab. 6.1 Antisense drug pipeline.

Target	*Lead indication*	*Company*	*Product*	*Phase*
CMV	CMV retinitis	ISIS	Vitravene	drug
PKC-α	nonsmall cell lung cancer	ISIS/Lilly	Affinitac	III
C-RAF	ovarian cancer	ISIS	ISIS 5132	II
H-ras	pancreatic cancer	ISIS	ISIS 2503	II
ICAM-1	Crohn's disease	ISIS	ISIS 2302/ Alicaforsen	III
ICAM-1	topical psoriasis	ISIS	ISIS 2302/ Alicaforsen	II
ICAM-1	ulcerative colitis	ISIS	ISIS 2302/ Alicaforsen	II
HCV	hepatitis C	ISIS	ISIS 14803	II
Tumor necrosis factor-α	Crohn's disease, topical psoriasis	ISIS	ISIS 104838	II
c-*myc*	cancer, restenosis	AVI-Biopharm	Resten-NG	III
bcl	cancer	Genta/Aventis	Genasense	III
HIV	AIDS	Enzo Biochem	HGTV43	II
Adenosine A1	respiratory disease	EpiGenesis	EPI2010	II
Protein kinase	cancer	Hybridon	Gem 231	II
Ribonucleotide reductase	cancer	Lorus Therapeutics	GTI2040	II
Transforming growth factor-β2	cancer	Antisense Pharma	AP 12009	I/II
Vascular endothelial growth factor receptor-1	cancer	Ribozyme Pharmaceuticals	Angiozyme	II
HCV	hepatitis C	Ribozyme Pharmaceuticals	Heptazyme	I
HER2	cancer	Ribozyme Pharmaceuticals	Herzyme	I

Conversely, RNAs with a long half-life are considered better substrates for ribozymes [95].

Interestingly, most authors observe a certain degree of inhibition of protein expression even by the inactive ribozyme. The simplest explanation put forward by most investigators is a translation arrest. An additional mechanism to explain the inhibition could be cleavage of the target RNA by a double-stranded RNase that has been identified in human cells [96] or even by an RNAi mechanism.

6.7 Application

The literature on the application of oligonucleotides for the inhibition of gene expression is vast. Table 6.1 summarizes those examples where the technique has advanced to clinical phase trials, although this collection is probably not complete. Not surprisingly, considering the enormous expense involved, these investigations are almost exclusively conducted by companies. The only notable exception is the study by Gewirtz's laboratory on chronic myelogenous leukemia for *ex vivo* treatment [93]. Not all the examples listed can be discussed here so that some comments must suffice. Common to all oligonucleotides used in the clinical trials is one aspect of their chemical nature in that they all are phosphorothioate-containing oligonucleotides. The first approved antisense drug was developed by ISIS, Fomivirsen [94] or Vitravene®, a tradename of Novartis, certainly a fantastic achievement. However, the number of patients with cytomegalovirus (CMV) retinitis has grown less and less because of the successful treatment of AIDS, at least in the Western hemisphere. Intracellular adhesion molecule (ICAM)-1 from the same company, as a medication for Crohn's disease, has not yet been approved, but is being pursued further under the tradename Alicaforsen™. The development of the Genta anti-*BCL2* oligonucleotide Genasense™ for the treatment of malignant melanoma in conjunction with chemotherapy is well advanced [95]. Genasense is in phase III clinical trial developed together with Aventis.

Affinitac™ is a potent inhibitor of protein kinase C in nonsmall cell lung cancer that shows almost double survival times (16 instead of 8 months) in combination with carboplatin/taxol. The main objective is to obtain a good synergistic profile by the combination with chemotherapeutic agents (as is the goal with Genasense), with taxol in this case. Affinitac is also being tested in a single-agent (phase II) trial against nonHodgkin's lymphoma. Here, a codevelopment together with Eli Lilly is under way [96].

6.8 Conclusion

Nucleic acids, particularly antisense oligonucleotides, are now being considered as drug candidates. One compound, Vitravene, is currently on the market and several promising compounds are now in phase III clinical trials. Their success will define the potential impact of antisense oligonucleotides. In addition, the ribozyme approach also seems to result in applicable drugs, although some hurdles still have to be overcome. The latest concept, RNAi, is in its infancy, but has a very high potential since it is a natural concept. Much of what we have learned with antisense and ribozymes can be adapted here. The most serious limitation so far is the poor cellular uptake and delivery, which will decide about the future of oligonucleotides as drugs.

References

1 W Saenger, *Principles of Nucleic Acid Structure*, Springer Verlag, New York, 1984.

2 RL Letsinger, JL Finnan, GA Heaver, WB Lunsford, *J Am Chem Soc* 1975, **97**, 3278–9.

3 RL Letsinger, WB Lunsford, *J Am Chem Soc* 1976, **98**, 3655–61.

4 PJ Garegg, I Lindh, T Regberg, J Stawinski, R Strömberg, *Tetrahedron Lett* 1986, **27**, 4055–8.

5 SL Beaucage, MH Caruthers, *Tetrahedron Lett* 1981, **22**, 1859–62.

6 SM Freier, KH Altmann, *Nucleic Acids Res* 1997, **25**, 4429–43.

7 H Inoue, Y Hayase, A Imura, S Iwai, K Miura, E Ohtsuka, *Nucleic Acids Res* 1987, **15**, 6131–48.

8 AM Kawasaki, MD Casper, SM Freier, EA Lesnik, MC Zounes, LL Cummins, C Gonzalez, PD Cook, *J Med Chem* 1993, **36**, 831–41.

9 H Aurup, T Tuschl, F Benseler, J Ludwig, F Eckstein, *Nucleic Acids Res* 1994, **22**, 20–4.

10 PS Miller, CH Agris, L Aurelian, KR Blake, A Murakami, MP Reddy, SA Spitz, PO Ts'O, *Biochimie* 1985, **67**, 769–76.

11 J Engels, A Jaeger, *Angew Chem* 1982, **94**, 931.

12 M Mag, J Muth, K Jahn, A Peyman, G Kretzschmar, JW Engels, E Uhlmann, *Bioorg Med Chem* 1997, **5**, 2213–20.

13 M Mag, K Jahn, G Kretzschmar, A Peyman, E Uhlmann, *Tetrahedron* 1996, **52**, 10011–24.

14 M Koziolkiewicz, A Krakowiak, M Kwinkowski, M Boczkowska, WJ Stec, *Nucleic Acids Res* 1995, **23**, 5000–5.

15 M Koziolkiewicz, M Wojcik, A Kobylanska, B Karwowski, B Rebowska, P Guga, WJ Stec, *Antisense Nucleic Acid Drug Dev* 1997, **7**, 43–8.

16 PE Nielsen, M Egholm, RH Berg, O Buchardt, *Science* 1991, **254**, 1497–500.

17 M Egholm, O Buchardt, L Christensen, C Behrens, SM Freier, DA Driver, RH Berg, SK Kim, B Norden, PE Nielsen, *Nature* 1993, **365**, 566–8.

18 B Hyrup, PE Nielsen, *Bioorg Med Chem* 1996, **4**, 5–23.

19 E Uhlmann, *Biol Chem* 1998, **379**, 1045–52.

20 AC van der Laan, P Havenaar, RS Oosting, E Kuyl-Yeheskiely, E Uhlmann, JH van Boom, *Med Chem Lett* 1998, **8**, 663–8.

21 MA Bonham, S Brown, AL Boyd, PH Brown, DA Bruckenstein, JC Hanvey, SA Thomson, A Pipe, F Hassman, *Nucleic Acids Res* 1995, **23**, 1197–203.

22 E Uhlmann, DW Will, G Breipohl, D Langner, A Ryte, *Angew Chem Int Ed Engl* 1996, **35**, 2632–5.

23 AC van der Laan, R Brill, RG Kuimelis, E Kuyl-Yeheskiely, JH van Boom, A Andrus, R Vinayak, *Tetrahedron Lett* 1997, **38**, 2249–52.

24 M Tarkov, M Bolli, B Schweizer, C Leumann, *Helv Chim Acta* 1993, **76**, 481–510.

25 C Epple, C Leumann, *Chem Biol* 1998, **5**, 209–16.

26 AA Koshkin, SK Singh, P Nielsen, VK Rajwanshi, R Kumar, M Meldgaard, CE Olson, J Wengel, *Tetrahedron* 1998, **54**, 3607–30.

27 JD Watson, FHC Crick, *Nature* 1953, **171**, 737–8.

28 J Minshull, T Hunt, *Nucleic Acid Res* 1986, **14**, 6433–51.

29 RY Walder, JA Walder, *Proc Natl Acad Sci USA* 1988, **85**, 5011–5.

30 E Uhlmann, A Peyman, *Chem Rev* 1990, **90**, 543–84.

31 G Felsenfeld, DR Davies, A Rich, *J Am Chem Soc* 1957, **79**, 2023–4.

32 K Hoogsteen, *Acta Crystallogr* 1959, **12**, 822–3.

33 K Hoogsteen, *Acta Crystallogr* 1963, **16**, 907–16.

34 STh Sigurdsson, JB Thomson, F Eckstein, in: RW Simons, M Grunberg-Manago, eds, *RNA Structure and Function*, 339–76, Cold Spring Harbor Laboratory Press, Cold Spring Harbor, NY, 1998,

35 C Carola, F Eckstein, *Curr Opin Chem Biol* 1999, **3**, 274–83.

36 RH Symons, *Annu Rev Biochem* 1992, **61**, 641–71.

37 OC Uhlenbeck, *Nature* 1987, **328**, 596–600.

38 J Haseloff, WL Gerlach, *Nature* 1988, **334**, 585–91.

39 B Bramlage, E Luzi, F Eckstein, *Trends Biotechnol* 1998, **16**, 434–8.

40 J Rossi, *J Chem Biol* 1999, **6**, R33–7.

41 HW Pley, KM Flaherty, DB McKay, *Nature* 1994, **372**, 68–74.

42 WG Scott, JT Finch, A Klug, *Cell* 1995, **81**, 991–1002.

43 WG Scott, JB Murray, JRP Arnold, BL Stoddard, A Klug, *Science* 1996, **274**, 2065–9.

44 JB Murray, DP Terwey, L Maloney, A Karpeisky, N Usman, L Beigelman, WG Scott, *Cell* 1998, **92**, 665–73.

45 RG Kuimelis, LW McLaughlin, *Chem Rev* 1998, **28**, 1027–44.

46 DM Zhou, K Taira, *Chem Rev* 1998, **28**, 991–1026.

47 K Yoshinari, K Taira, *Nucleic Acids Res* 2000, **28**, 1730–42.

48 NK Vaish, PA Heaton, O Fedorova, F Eckstein, *Proc Natl Acad Sci USA* 1998, **95**, 2158–62.

49 WA Pieken, DB Olsen, H Aurup, F Eckstein, *Science* 1991, **253**, 314–7.

50 O Heidenreich, F Benseler, A Fahrenholz, F Eckstein, *J Biol Chem* 1994, **269**, 2131–8.

51 M Sioud, DR Sorensen, *Nat Biotechnol* 1998, **16**, 556–61.

52 A Beaudry, J DeFoe, S Zinnen, A Burgin, L Beigelman, *Chem Biol* 1999, **7**, 323–34.

53 G Paolella, BS Sproat, AI Lamond, *EMBO J* 1992, **11**, 1913–9.

54 TC Jarvis, FE Wincott, LJ Alby, JA McSwiggen, L Beigelman, J Gustofson, A Direnzo, K Levy, M Arthur, J Maculic-Adamic, A Karpeisky, C Gonzalez, TM Woolf, N Usman, DT Stinchcomb, *J Biol Chem* 1996, **271**, 29107–12.

55 HA James, I Gibson, *Blood* 1998, **91**, 371–82.

56 AR Muotri, L Veiga Pereira, LR Reis Vasques, CFM Menck, *Gene* 1999, **237**, 303–10.

57 C Klebba, OG Ottmann, M Scherr, M Pape, JW Engels, M Grez, D Hoelzer, SA Klein, *Gene Therapy* 2000, **7**, 408–16; D Castanotto, M Scherr, J Rossi, *Methods Enzymol* 2000, **313**, 401–20.

58 E Bertrand, D Castanotto, C Zhou, C Carbonnelle, NS Lee, P Good, S Chatterjee, T Grange, R Pictet, D Kohn, D Engelke, JJ Rossi, *RNA* 1997, **3**, 74–88.

59 NS Lee, E Bertrand, JJ Rossi, *RNA* 1999, **5**, 1200–9.

60 DA Samarsky, G Ferbeyre, E Bertrand, RH Singer, R Cedergren, MJ Fournier, *Proc Natl Acad Sci USA* 1999, **96**, 6609–14.

61 A Michienzi, L Cagnon, I Bahner, JJ Rossi, *Proc Natl Acad Sci USA* 2000, **97**, 8955–60.

62 B Bramlage, S Alefelder, P Marschall, F Eckstein, *Nucleic Acids Res* 1999, **27**, 3159–67.

63 A Fire, S Xu, MK Montgomery, SA Kostas, SE Driver, CC Mello, *Nature* 1998, **391**, 806–11.

64 K Nishikura, *Cell* 2001, **107**, 415–8.

65 T Tuschl, *Nat Biotechnol* 2002, **20**, 446–8.

66 SM Elbashir, W Lendeckel, T Tuschl, *Genes Dev* 2001, **15**, 188–200.
67 E Bernstein, AA Caudy, SM Hammond, GJ Hannon, *Nature* 2001, **409**, 363–6.
68 SM Hammond, S Boettcher, AA Caudy, R Hobayashi, GJ Hannon, *Science* 2001, **293**, 1146–50.
69 RF Ketting, SE Fischer, E Bernstein, T Sijen, GJ Hannon, RH Plasterk, *Genes Dev* 2001, **15**, 2654–9.
70 S Yang, S Tutton, E Pierce, K Yoon, *Mol Cell Biol* 2001, **21**, 7807–16.
71 PJ Paddison, AA Caudy, GJ Hannon, *Proc Natl Acad Sci* USA 2002, **99**, 1443–8.
72 D Yang, H Lu, JW Erickson, *Curr Biol* 2000, **10**, 1191–200.
73 PD Zamore, T Tuschl, PA Sharp, D PBartel, *Cell* 2000, **101**, 25–33.
74 SM Elbashir, J Martinez, A Patkaniowska, W Lendeckel, T Tuschl, *EMBO J* 2001, **20**, 6877–88.
75 M Sohail, EM Southern, *Curr Opin Mol Therapeut* 2000, **2**, 264–71.
76 KR Birikh, YA Berlin, H Soreq, F Eckstein, *RNA* 1997, **3**, 429–37.
77 M Scherr, M Reed, C-F Huang, AD Riggs, JJ Rossi, *Mol Therapy* 2000, **2**, 26–38.
78 V Patzel, U Steidl, R Kronenwett, R Haas, G Sczakiel, *Nucleic Acids Res* 1999, **27**, 4328–34.
79 DH Mathews, ME Burkard, SM Freier, JR Wyatt, DH Turner, *RNA* 1999, **5**, 1458–69.
80 M Scherr, JJ Rossi, G Sczakiel, V Patzel, *Nucleic Acids Res* 2000, **28**, 2455–61.
81 Q Yu, DB Pecchia, SL Kingsley, JE Heckman, JM Burke, *J Biol Chem* 1998, **273**, 23524–33.
82 J zu Putlitz, Q Yu, JM Burke, JR Wands, *J Virology* 1999, **73**, 5381–7.
83 L Mariani, L Citti, S Nevischi, F Eckstein, G Rainaldi, *Cancer Gene Therapy* 2000, **7**, 905–9.
84 SP Lyngstadaas, S Risnes, BS Sproat, PS Thrane, HP Prydz, *EMBO J* 1995, **14**, 5224–9.
85 CM Flory, PA Pavco, TC Jarvis, ME Lesch, FE Wincott, L Beigelman, SW Hunt III, DJ Schrier, *Proc Natl Acad Sci USA* 1996, **93**, 745–58.
86 P Salmi, BS Sproat, J Ludwig, R Hale, N Avery, J Kela, C Wahlestedt, *Eur J Pharmacol* 2000, **388**, R1–2.
87 TJ Parry, C Cushman, AM Gallegos, AB Agrawal, M Richardson, LE Andrews, L Maloney, VR Mokler, FE Wincott, PA Pavco, *Nucleic Acids Res* 1999, **27**, 2569–77.
88 ST Crooke, *Methods Enzymol* 2000, **313**, 3–45.
89 S Agrawal, ER Kandimalla, *Mol Med Today* 2000, **6**, 72–81.
90 TC Jarvis, LJ Alby, AA Beaudry, FE Wincott, L Beigelman, JA McSwiggen, N Usman, DT Stinchcomb, *RNA* 1996, **2**, 419–28.
91 M Scherr, M Grez, A Ganser, JW Engels, *J Biol Chem* 1997, **272**, 14304–13.
92 B Bramlage, S Alefelder, P Marschall, F Eckstein, *Nucleic Acids Res* 1999, **27**, 3159–67.
93 SM Luger, SG O O'Brien, J Ratajczak, MZ Ratajczak, R Mick, EA Stadtmauer, PC Novell, JM Goldman, AM Gerwitz, *Blood* 2002, **99**, 1150–8; JB Opalinska, AM Gewirtz, *Nat Rev Drug Discov* 2002, **7**, 503–13.
94 RL Grillone, R Lanz, *Drugs of Today* 2001, **37**, 245–55.
95 B Jensen, V Wacheck, E Heere-Ress, H Schlagbauer-Wadl, C Hoeller, T Lucas, M Hoermann, U Hollenstein, K Wolff, H Pehamberger, *Lancet* 2000, **356**, 1728–33.
96 A Yuen, J Halsey, G Fisher, R Advani, M Moore, M Saleh, P Ritch, G Harker, F Ahmed, C Jone, P Polikoff, W Keiser, J Hwoh, J Holmlund, A Dorr, B Sikic, Presented at *37th Annual Meeting of the American Society of Clinical Oncology*, San Francisco, CA, 2001.

Part III
Analysis

Molecular Biology in Medicinal Chemistry. Edited by Th. Dingermann, D. Steinhilber, G. Folkers

ISBN: 3-527-30431-2

7 Recent Trends in Enantioseparation of Chiral Drugs

Bezhan Chankvetadze

7.1 Introduction

The field of enantioseparation attracts attention from two viewpoints: (1) preparation of chiral biologically active compounds in enantiomerically pure form, and (2) enantioselective analysis of chiral drugs in the process of their production, formulation, storage and use. In the production fields, chromatographic enantioseparation techniques have strong competitors such as diastereomeric crystallization, kinetic resolution, chiral pools, catalytic asymmetric synthesis, membrane-based technologies, etc. However, there is no strong alternative to separation-based techniques in the analytical field. Chiroptical techniques such as traditional polarimetry, optical rotation dispersion or circular dichroism spectrometry, as well as nuclear magnetic resonance spectroscopy, cannot compete with separation-based techniques such as gas chromatography (GC), high-performance liquid chromatography (HPLC), supercritical fluid chromatography (SFC), capillary electrophoresis (CE) and capillary electrochromatography (CEC) in terms of accuracy, sensitivity, simplicity, analysis time, costs, etc. The scope of this chapter is to discuss the current trends and to give an insight on the future developments in the field of enantioseparation.

7.2 Current Status in the Development and use of Chiral Drugs

***Driven by the needs of the drug industry and fueled by the ingenuity of chemists, sales of single-enantiomer chiral compounds keep accelerating.* Stephen C. Stinson, *C & E News* 2001, 79(20), 45–57.**

According to a recent prognosis, the demand for chiral raw materials, intermediates and active ingredients will grow by 9.4% annually between 2000 and 2005. The drug industry is considered to be the driving force behind this strong growth. Thus, from the expected market of $15.1 billion, drug manufacturers will be re-

Molecular Biology in Medicinal Chemistry. Edited by Th. Dingermann, D. Steinhilber, G. Folkers

ISBN: 3-527-30431-2

sponsible for $11.5 billion (76%) [1]. What makes chirality such an attractive issue for drug developers and manufacturers? There are several major driving forces that favor the development of the chiral business, especially in the pharmaceutical industry.

The first and the most important issue representing the basis of all the activity in the field is the significant difference observed between the pharmacological and toxicological properties of enantiomers of biologically active chiral compounds. There are many examples known, and numerous books, book chapters and overviews have been written on this topic [2–7]; this topic is not discussed in further detail here.

The second driving force is a protective mechanism that allows pharmaceutical companies to prolong their proprietary rights of chiral drugs, which have been previously marketed as racemates, and will be redeveloped and patented as single-enantiomer entities [7]. Recent examples of this kind include the single *S*-enantiomer of the highly potent gastric acid secretion inhibitor omeprazole introduced to the European market in 2001 by AstraZeneca [8, 9], the *S*-enantiomer of the chiral antidepressant drug citalopram by Forest Laboratories [1], the *R*-enantiomer of the chiral antiallergic drug cetirizine introduced in 2001 by UCB Pharm [10, 11], and (*S*)-ibuprofen that has now been used in Austria and Switzerland for several years [7], and has recently been introduced to the German market.

The third issue that favors the topic of chirality is that of regulation. For instance, the act by the European Agency for the Evaluation of Medicinal Products (EMA) effective since 1 January 1998, notes: "If a new racemate appears promising, both enantiomers should be studied separately as early as possible to assess the relevance of stereoisomerism for effect and fate *in vivo*" [12]. Regulatory issues like this facilitate the fact that many of the recently developed chiral drugs enter the market as single enantiomers. Examples include the cholesterol-lowering drug atorvastatin (Lipitor®) approved in 1996 and, with sales of US$4 billion, being one of the best-selling drugs in the United States in 1999 [13]. A strong increase in sales over a very short time was also observed for other single-enantiomer chiral drugs such as amprenavir, which was approved in April 1999 by the US Food and Drug Administration (FDA) for the treatment of HIV infections, and for the first representative of a new class of antibiotics, Linezolid, which was approved by the FDA in April 2000 [13]. This list also includes the chiral respiratory drug montelucast, the antireumatoidal (former gastrointestinal) infliximab, the prostate hyperplasia agent tamsulosin and the ophthalmic drug for the treatment of glaucoma latanaprost [14].

There are several strategies currently used by different companies to enter the chiral business market. Some of the companies are re-developing their drugs or already known chiral drugs from other companies as new single-enantiomer entities. Thus, for example, a list of international patent applications by the company Sepracor in 1999 included the enantiomers of the following drugs: (*S*)- and (*R*)-terodiline, (−)- and (+)-pantoprazole, (−)- and (+)-zileuton, (−)- and (+)-ketoconazole, (+)- and (−)-doxazosin, (+)- and (−)-cetirizine, (+)- and (−)-sibutramine, (−)- and (+)-pirbuterol, (*R*)- and (*S*)-ondasetron, (+)- and (−)-cisapride, and (+)-

and (−)-zopiclone [7]. Other companies are developing pharmacologically active chiral metabolites of chiral or prochiral drugs as new drug entities in a single enantiomeric form. Examples include (*R*)- and (*S*)-norfluoxetine developed by Eli Lilly [7], and the active metabolite of zopiclone.

Thus, as can be seen from the few examples mentioned above, chirality is an important topic in the pharmaceutical industry, as well as in clinical practice in order to establish better correlations between drug concentrations and therapeutic effects, to avoid adverse effects of drug actions, and to better address the effects of age, genotypes, diet, drug interactions, etc., on the efficacy of therapy with chiral drugs.

As mentioned above, separation techniques are not the only available tools for the production of chiral drugs in enantiomerically pure form. Strategies based on the chiral pool, catalytic asymmetric synthesis [15–17], etc., represent a valuable alternative. However, whenever one considers a chiral drug, analytical enantioseparation techniques are required in order to control the enantiomeric purity of stages of production (irrespective of the production technology) and storage, as well as to follow the enantioselective effects in its action and biotransformations.

7.3 The Role of Separation Techniques in Chiral Drug Development, Investigation and Use

***The chiral technology revolution was founded on the ability to measure enantiomeric purity.* [1]**

The impact of instrumental enatioseparation techniques on chiral drug development and use is immense [1]. In order to evaluate the role of enantiospecific effects in chiral drug action the enantiomers must be prepared at least in amounts sufficient for basic pharmacological and toxicological studies. Together with classical diastereomeric crystallization, chromatographic separation techniques have played an important part in the production of enantiomerically pure chiral drugs [14, 18, 19]. Actually, at the early stage of drug development, low-pressure column liquid chromatography in its simplest batch mode was applied for the enantioseparation of those chiral drugs that were impossible to synthesize or resolve by existing techniques due to lack of suitable functional groups or required structural properties. In the early 1970s, chromatographic techniques made it possible to obtain the enantiomers of chiral drugs such as mephenythoin, methyprylon, glutethimid, hexobarbital, chlorthalidone, oxazepam, biglumide, etc. [18]. The enantiomers of thalidomide, which had been previously synthesized based on a very expensive intermediate [20], became easily available for pharmacological studies for the first time via direct chromatographic enantioseparation [21].

In the early 1980s, the first column for analytical-scale enantioseparation using HPLC was commercialized. This fact and the following extension of the market of chiral HPLC columns was a key event in promoting the broad application of enantioseparation techniques in pharmaceutical and biomedical analysis. Chiral GC, although having a longer history than HPLC, does not play a primary role in chiral drug development, and its use is limited due to the polar nature and low volatility of most drugs. A few excellent studies must be mentioned as an exception from the aforementioned status [22–24]. As shown in these studies, chiral GC may offer a great potential not only for analytical-, but also for preparative-scale separation of some chiral gases [23, 24].

Chiral SFC possesses a certain potential for analytical- [25, 26] and, especially, preparative-scale enantioseparation [25]. However, this technique did not gain the status that it apparently deserves in this field. One of the reasons for this appears to be the lack of support from instrument companies as well as relatively low research activity by academic laboratories in the field of enantioseparation using SFC.

The first publication that described enantioseparation using CE appeared in the same year [27] as the publication describing chiral SFC. After several problematic years, the breakthrough for chiral CE came in 1992 [28]. Chiral CE still enjoys rapid development and increasingly penetrates to the industrial environment [29, 30]. Early applications of another electromigration technique, CEC, to enantioseparation appeared in 1992–1993 [31, 32]. Despite some impressive achievements, this technique is still struggling for its place in the field of enantioseparation [33, 34]. Future needs will show which techniques will survive better from the wide range of presently available techniques for preparative and analytical enantioseparation.

7.4 Preparation of Enantiomerically Pure Drugs

There are three primary sources of pure enantiomers: racemates, the rich diversity of chiral molecules (carbohydrates, terpenes, alkaloids, peptides) that occur naturally as pure enantiomers and prochiral compounds. The principal methods of obtaining pure enantiomers based on the aforementioned sources are summarized in Fig. 7.1. The strategy discussed in this section is based on the use of racemates as a source for pure enantiomers. The way of transforming racemates to pure

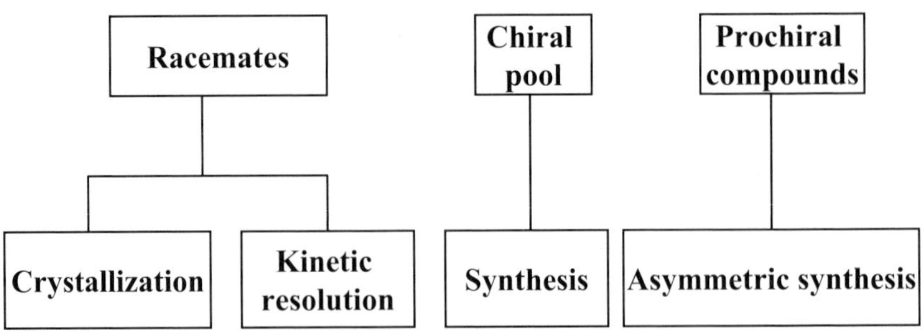

Fig. 7.1 The principal methods of obtaining of pure enantiomers.

enantiomers is based on kinetic resolution or crystallization. Each of these strategies can be performed in different variations that are very briefly discussed below.

7.4.1 Resolution of Racemates

The resolution of racemates is the oldest technique and still remains the major method for industrial synthesis of enantiomerically pure biologically active molecules. Enantiomerically enriched or pure chiral compounds from the racemates might be obtained based on preferential crystallization, diastereomeric crystallization, kinetic resolution or membrane-based separation.

Preferential crystallization is technically feasible only for racemates that are so-called conglomerates. Less than 20% of all racemates are conglomerates. This technique is used in the industrial scale for production of chloramphenicol [35] and α-methyl-L-dopa [36]. In a few cases it is also possible to perform direct crystallization of salts with achiral acids or bases. For instance, the resolution of DL-lysine can be performed by preferential crystallization of its *p*-aminobeneze-sulfonate salt. Naproxen can be resolved in the same way as its salt with ethylamine [37].

Derivatization of a racemate by reaction with an optically pure compound yields a mixture of diastereomers with different physical properties. Such mixtures may be resolved by physical methods such as crystallization. This method is often referred to as classical resolution. A comprehensive discussion of the principles and techniques of diastereomeric crystallization is beyond the scope of this chapter. However, it must be noted that this technique may appear very useful for obtaining enantiomerically pure chiral drugs for early experimental and clinical studies. In addition, this technique is still used for the production of enantiomerically pure chiral drugs such as amoxycillin, captopril, *cis*-diltiazem, naproxen, cefalexin, cefadroxyl, timolol, dextromethorphan, etambutol, etc. [38]. The industrial route to (*S*,*S*)-ethambutol is shown as an example in Fig. 7.2 [38].

Diastereomeric crystallization suffers from the general disadvantage that applies to all enantioseparation processes. In particular, the yield of the desired enantiomer never exceeds 50%. If the unwanted enantiomer cannot be racemized or applied to other purposes, 50% unwanted byproduct accompanies the production of the desired enantiomer. This may be a severe problem in the large-scale manufacture of enantiomers, but does not apply for the production of the enantiomers on the gram or few kilogram scale. Another bottleneck of diastereomeric crystallization is the fact that it is impossible to rationalize this technology based on current knowledge. The crystallization methods have to be developed by trial and error based on intuition and experience. The technical problem of the availability of some natural chiral resolving agents in large quantities also needs to be addressed in certain cases.

Kinetic resolution is gaining in importance for obtaining enantiomerically pure chiral compounds. This method can be defined as a process in which one of the enantiomers of a racemic mixture is more readily transformed compared to the other enantiomer. Kinetic resolution can be performed with biological or chemi-

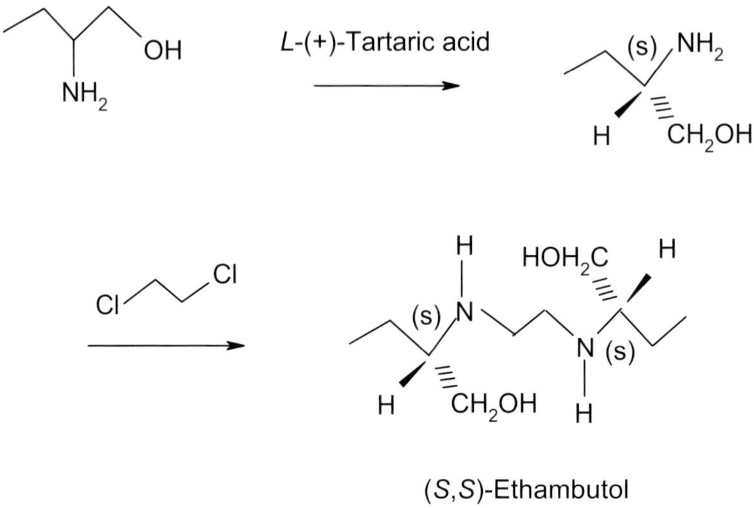

Fig. 7.2 Industrial route to (*S*,*S*)-ethambutol by diastereomeric crystallization. (Reproduced from [38] with permission.)

cal catalysts. This process may lead to 50% maximum enantiomeric yield, similar to diastereomeric crystallization. Thus, a combination of any preparative- and production-scale enantioseparation technique with racemization of the unwanted enantiomer is highly desirable.

The advantage of enzymes as catalysts includes their high efficiency, substrate selectivity, chemoselectivity, regioselectivity and stereoselectivity. In addition, enzymes work under mild conditions and do not require organic solvents. The representatives of all major classes of enzymes such as oxidoreductases, transferases, hydrolases, lyases, isomerases and ligases can be used in the production of enantiomerically pure drugs. Commercially viable techniques for the manufacture of (*S*)-naproxen [38], *cis*-diltiazem [39, 40] and several other drugs have been developed.

Membrane-based technologies are also considered to be promising for the production of enantiomerically pure chiral compounds. There are in fact two alternative ways for applying membranes for enantioselective transport [41] as well as membrane bioreactors [42]. The former technique has not yet been widely applied for the manufacture of enantiomerically pure drugs or drug intermediates, whereas membrane bioreactors are well established for the manufacture of L-α-amino acids [42].

7.4.2
Chiral Pool

Enantiomerically pure compounds available in nature or large-scale byproducts of fermentation processes represent valuable raw materials for the manufacture of enantiomerically pure drugs. Many of these materials, such as carbohydrates, amino acids, terpenes, alkaloids, etc., have prices comparable to bulk petrochemicals, and find a variety of industrial applications as chiral substrates, chiral auxiliaries and as a source of chiral ligands for asymmetric catalysts.

Carbohydrates in the intact form, as well as their transformation products (basically reduction and oxidation), are widely used as chiral intermediates in the pharmaceutical and fine chemical industry. Among the hydroxy acids both D- and L-lactic acids represent interesting starting materials for obtaining enantiomerically pure herbicides, e.g. (*R*)-fluazifop-butyl [43], (*R*)-mecoprop [44] and (*R*)-flamprop-isopropyl [45]. (*R*,*R*)-tartaric acid, as well as being a widely used resolving agent, also represents an interesting chiral starting material. For example, it is used in the Zambon synthesis of (*S*)-naproxen [46]. (*S*,*S*)-tartaric acid is also a natural product that is rather rare. However, it can be obtained synthetically and is used in the production of fosfomycin [47]. With regard to other hydroxy acids, β-hydroxybutyric acid represents a certain interest for the pharmaceutical industry. (*R*)-β-hydroxybutyric acid can be used for the synthesis of a carbapenem intermediate [48] and captopril [49], whereas (*S*)-β-hydroxybutyric acid is useful for the construction of the side-chain of vitamin E [50].

Natural amino acids that are readily available in bulk quantities represent the most important class of compounds for the chiral pool. For example, L-proline, available from fermentation, is the key raw material for a variety of acetylcholine esterase (ACE) inhibitor drugs such as captopril, enalapril, lisinopril, etc. [51]. Other readily available amino acids, e.g. L-glutamic acid [53], L-aspartic acid [54] and L-threonine [55], have been used as chiral synthons for the synthesis of azthreonam [52] and carbapenem antibiotics, such as thienamycin. Together with the aforementioned compounds, some unnatural amino acids manufactured on a large scale, terpenes, alkaloids, etc., can be used as raw materials in the chiral pool strategy [56].

7.4.3
Catalytic Asymmetric Synthesis

Catalytic asymmetric synthesis is gaining importance for obtaining enantiomerically pure materials [15–17]. Asymmetric synthesis can be performed with biological catalysts as it was briefly mentioned in Section 7.4.1. Here, chemocatalysts such as heterogeneous metal catalysts, homogeneous complexes, and soluble chiral acids and bases are briefly discussed.

Early studies in the field of heterogeneous asymmetric catalysis were performed in order to better understand the origin of chirality. For this purpose, naturally available chiral solids such as quartz, silk, cellulose, etc. were used as a support for metallic catalysts. The L-dopa process developed by Knowles and Sabacky at Monsanto in 1966–1968 [15, 57] was the first process to be realized for the production of an enantiomerically pure pharmaceutical drug on an industrial scale by asymmetric catalysis. In 1971, Kagan and Dang described an asymmetric hydrogenation catalyst containing a chiral diphosphine, derived from L-tartaric acid [58]. In the

1980s, Noyori et al. introduced the so-called BINAP series of chiral ligands based on the derivatives of the axially chiral binaphthyl group [16, 59]. Today, catalytic asymmetric hydrogenation based on the aforementioned and related catalysts represent one of the popular ways for the manufacture of enantiomerically pure chiral drugs. The examples include (*S*)-naproxen, the side-chain of vitamin E, isochinolin alkaloids including morphine, benzomorphanes and morpinanes such as the antitusive drug dextromethorphan [61], carbapenem antibiotics [16], (*R*)-1,2-propanediole as the intermediate of levofloxacin [16], carnitine [16], statins [62], phosphmycin [63], anthracycline intermediates [64], denopamine [16], fluoxetine [16], duloxetine [16], orphenadrin [16], neobenodine [16], prostaglandins [65], *β*-blockers [66], the antihelminthic (and anti-inflammatory) drug levamisol [67], etc.

Asymmetric oxidations, including epoxidation, dihydroxylation and aminoxidation, represent important ways for obtaining enantiomerically pure pharmaceuticals on a commercial scale [17]. Sharpless epoxidation is used for the manufacture of (*R*)- and (S)-glycidol, which are important intermediates for enantiomerically pure *β*-blockers. Asymmetric dihydroxylation is used in the synthesis of the chiral calcium antagonist *cis*-diltiazem [68].

Catalytic asymmetric sulfoxidation may gain in importance in the future due to an increasing market for chiral gastric acid secretion inhibitors having a sulfoxide structure such as omeprazole, pantoprazole, rabeprazole, peprazole and lansoprazole. The first enantiomerically pure drug of this class, (*S*)-omeprazole (esomeprazole), is already on the market [8, 9]. This drug is obtained via asymmetric catalytic sulfoxidation according the scheme shown in Fig. 7.3 [69].

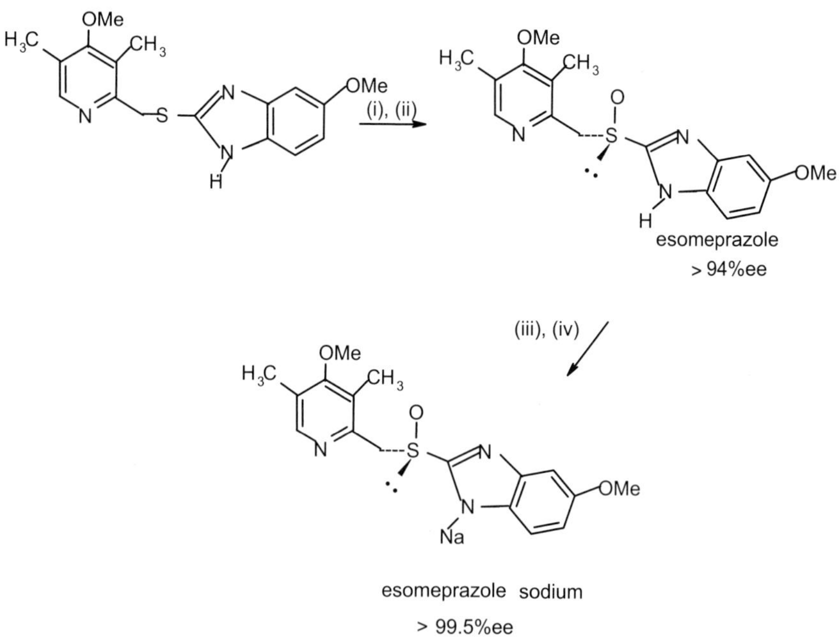

Fig. 7.3 Synthetic schema for obtaining of (*S*)-omeprazole on a production scale. (Reproduced from [69] with permission.)

In addition to the few reaction types mentioned in this subsection, a wide variety of catalytic asymmetric transformations such as hydrogenation of the C=N bond, hydrosilylation, isomerization, hydroformylation, cyclopropanation, addition, hydrocyanation, condensation, etc., can be used for the synthesis of enantiomerically pure drugs. Comprehensive overviews of this topic can be found elsewhere [38, 41, 42].

7.4.4 Chromatographic Techniques

As mentioned above, column chromatography in its simplest batch mode played an important part in obtaining relatively small amounts of enantiomerically pure chiral pharmaceuticals [18, 19]. However, due to the small capacity (low productivity), high dilution of the products and large consumption of organic solvents, this technique did not establish itself as an economically viable method for large-scale production of drug enantiomers. Among the chromatographic techniques, simulated moving bed (SMB) chromatography is primarily considered for the preparative- and product-scale preparation of enantiomerically pure drugs. This technique is discussed in detail below. In addition, a few recent applications of GC-SMB, countercurrent chromatography and micropreparative enantioseparation using electromigration techniques are summarized very briefly in Section 7.4.4.2.

7.4.4.1 SMB

SMB is a continuous chromatographic process that simulates a countercurrent movement of the stationary phase and the mobile phase. A continuous operating mode is achieved by continuously feeding the eluent and the compound mixture to be separated into the system, and by continuous withdrawal of the separated compounds (raffinate and extract) (Fig. 7.4) [70]. Feeding of the eluent and a mixture of enantiomers as well as the withdrawal of the resolved fractions is achieved by four independent pumps. A fifth pump is used for recycling of the main flow within the system. This experimental set-up allows us to save a substantial amount of the eluent, because only small amounts of fresh mobile phase have to be added into the system to compensate for the loss of the mobile phase caused by the withdrawal of the resolved enantiomers. Another significant advantage of SMB technology is that it is a continuous process, whereas all other chromatographic techniques are discontinuous.

The countercurrent movement of the mobile and stationary phases is simulated in the following way: the stationary phase is divided in several separate chromatographic columns that are connected in a cyclic series. Each column head is equipped with valves that allow the addition of eluent and feed, and removal of the raffinate and extract from both component lines. The inlet and outlet lines will be shifted after a given time from one column to the subsequent column in the direction of the mobile phase, thus simulating the movement of the stationary phase in the opposite direction. After a complete cycle, the four lines reach their initial position. The technological parameters for running a SMB system under optimal conditions can be calculated by available software based on analytical runs [71, 72].

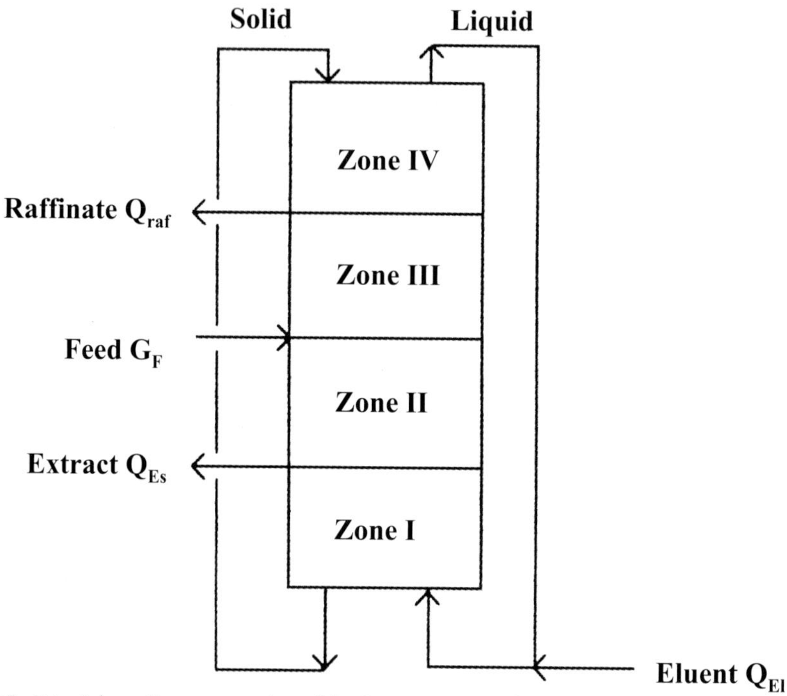

Fig. 7.4 Schematic representation of the SMB unit. (Reproduced from [70] with permission.)

Some examples of preparative enantioseparation of chiral drugs using the SMB technique are summarized in Tab. 7.1 [74–92]. Overall, SMB chromatography is a powerful tool for the production of enantiomerically pure compounds on a large scale within a short phase of development. The availability of software that allows scaling up an analytical enantioseparation to SMB technology is a crucial advantage. The characteristics of SMB chromatography, i.e. providing pure products (both enantiomers) in a predictable way and reducing the risk of failure, shorten the time necessary for development and indicate great promise of this technique. The applicability of SMB technology to a wide variety of compounds may favor the transformation of chromatography from a method of last resort to the method of choice for fast and predictable process development.

SMB allows a drastic reduction of the costs of enantioseparation, mainly due to the reduction in the amounts of the chiral stationary phase (CSP) (50–60% lower than in HPLC) and in eluent consumption (up to 10 times compared with the batch chromatographic process). It allows production scales of 10–100 tons per year, with separation costs as low as $30/kg of pure enantiomer [74]. The coupling of SMB with racemization and/or enantioselective crystallization techniques is even more promising.

Several pharmaceutical and fine-chemical companies have installed SMB units of various capacities. Among the first was UCB Pharm (Belgium), which in 1997

Tab. 7.1 Selected examples of applications of SMB chromatography for the production of enantiomerically pure compounds. (Reproduced from [14] with permission.)

Chiral compound	*CSP*	*Capacity*	*Consumption of the mobile phase (l/g)*	*Reference*
Aminoglutethimide	Chiralcel OJ	7.5 g/h·kg CSP	0.740 ml	74
1,1′-Binaphthyl-2,2′-diol	Pirkle-type 3,5-DNBPG	2000 g was resolved per day		75
Chiral drug candidate (potent partial agonist at muscarinic receptors)	Chiralpak AD	~30–60 g/h·kg	~0.190–0.380 ml	76
CGS 26 214	Chiralcel OJ	47 g/day·kg	4.561	77
Cycloalkanone	Chiralcel OC	1082 g/day·kg	0.280	78
DOLE	Chiralcel OF	272 g/day·kg	0.440	79
EMD 53 986	polyacrylamide	319 g/day·kg	2.540	80
EMD 53 986	Chiralpak AD	432 g/day·kg	2.600	80
EMD 77 697	Chiralcel OD	451 g/day·kg	1.640	81
EMD 122 347	Chiralpak AD	311 g/day·kg	0.590	82
Formoterol	Chiralcel OJ	1.2 g/h·kg	0.515	74
Hetrazepine	cellulose triacetate (CTA)	119 g/h·kg	0.094	83
Guaifenesin	Chiralcel OD	10.0 g/h·kg	0.380	74
Oxo-oxazolidine	L-Chiraspher	6.8 g/h·kg	0.900	84
Phenylethylalcohol	Chiralcel OD	0.98 g/l	5.300	85
Praziquantel	CTA	4.7 g/h·kg	0.056	86
Propranolol	Chiralcel OD	3.83 g/h·kg	0.200	87
Sandoz-epoxide	CTA	59 g/day·kg	0.800	88
Sandoz-epoxide	Chiralcel OD	37 g/day·kg	0.500	89
1a,2,7,7a-Tetrahydro-3-methoxy-napht(2,3–6) oxirane	CTA	1.45 g/h·kg	0.400	90
Threonine	Chirosolve L-proline	30 g was resolved		91
Tramadol	Chiralpak AD	50 g/h·kg	0.500	92

adopted a large-scale SMB unit for the multiton production of an enantiomerically pure drug. The biggest SMB unit is currently installed by Aeroject Fine Chemicals (Rancho Cordova, CA) with a capability 50 metric tons of enantiomer per year. Together with the above-mentioned companies, Bayer (Leverkusen, Germany), Universal Pharma Technologies (Lexington, MA), Daicel (Himeji, Japan) and several other companies possess SMB units of various capacities. Recently, the Danish drug company H. Lundbeck started an SMB unit with an annual capacity in the "double-digit" tons. The unit will be used in the production of the *S*-enantiomer of the company's best-selling drug, the antidepressant citalopram, which was commercialized instead of the racemic citalopram in 2002 [14].

As noted above in the SMB process, the lines are shifted synchronously. As a consequence, the number of columns in each zone remains the same at each moment, regardless of the location of the inlet/outlet lines. Recently, an asynchronous

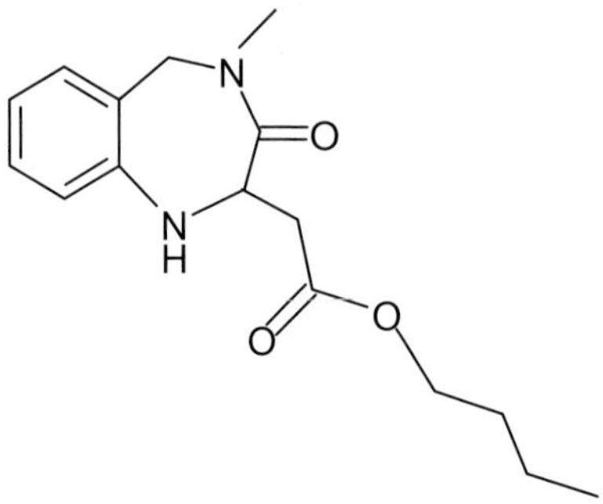

Fig. 7.5 Structure of SB-553261. (Reproduced from [94] with permission.)

analogue of SMB named VARICOL was proposed [93]. It was illustrated by numerous calculations and experiments [93, 94] that this technique offers certain advantages over the well-established SMB technology. The numerical calculations indicated that a five-column VARICOL permits the same purities of the products to be reached as a six-column SMB with the same productivity. When providing products with the same purity, the productivity of the VARICOL process was 18.5% higher compared to the SMB process with an equal number of columns [93].

The number of the columns in the VARICOL process can vary in each zone. This makes a VARICOL machine more complex than an SMB unit, but offers a number of advantages: If properly designed the feed flux does not pollute the extract or raffinate stream when the number of columns is temporarily zero in zones II and III, respectively. The eluent flux will also not unnecessarily dilute the extract or raffinate flux when the number of column is equal to zero in zones I and IV [93].

The enantiomers of the GlaxoSmithKline research product SB-553261 (Fig. 7.5) were resolved using the VARICOL process on amylose tris(3,5-dimethylphenylcarbamate) (Chiralpak AD) [94]. The SMB separation was performed with six columns, and VARICOL process was realized with six, five and four columns. The results of this comparison shown in Tab. 7.2 indicated that, for a six-column system (the same amount of CSP), VARICOL performs better than SMB, in terms

Tab. 7.2 Comparison between SMB and VARICOL processes (Reproduced from Ref. 94 with permission).

	Productivity ($kg_{prod}/kg_{CSP}/day$)	Eluent constitution ($m^3{}_{eluent}/kg_{prod}$)
Six-column SMB (1/2/2/1)	0.604	0.922
Six-column VARICOL ({1}/{2.25}/{2}/{0.75})	0.664	0.888
Five-column VARICOL ({0.95}/{1.85}/{1.5}/{0.7})	0.725	1.050
Four-column VARICOL ({0.85}/{1.5}/{1.15}/{0.5})	0.906	1.392

of both increased specific productivity and reduced eluent consumption. By accurately distributing columns among zones, VARICOL makes it possible to reduce the number of columns in the system. However, this comes at the expense of higher eluent consumption [94].

7.4.4.2 Other Chromatographic and Electrophoretic Techniques for the Preparation of Enantiomers

Low-pressure column chromatography, closed-loop recycling chromatography and peak-shaving techniques are not discussed here. These techniques are summarized in previous overviews on this topic [19, 95–97], and are currently not very often used due to small capacities, high consumption of eluent and high costs. Two techniques, counter-current chromatography (CCC) and micropreparative-scale electrophoretic enantioseparation, are briefly addressed below.

Summarizing the state of the art of enantioseparation in CCC and centrifugal partition chromatography (CPC), one of the experts in the field noted recently that "a total of 18 papers in 18 years constitutes a modest output compared with the literature for other chiral separation techniques" [98]. The author, mentioning the potential of CCC and CPC as powerful preparative techniques due to their high capacity, low cost of stationary phases (i.e. solvent mixture) and low solvent consumption (10 times less than for HPLC), stressed the rather poor efficiency (in the range of 1000 theoretical plates) as a basic problem for enantioseparation using these techniques [98]. In order to counterbalance the low efficiency of CCC and CPC, CSPs with high enantioselectivity are required. Recent developments in the design of CSPs made it possible to achieve enantioselectives in the range of several tens [99–105]. With selectivities like these, CCC and CPC may became economically viable techniques for the production of enantiomerically pure compounds [98, 106, 107]. In addition to the research articles summarized elsewhere [98, 106], a promising application of cinchona alkaloid derivatives as chiral selectors in CCC was published recently [107].

A few reports on the preparative-scale enantioseparation of chiral drugs using GC and GC-SMB indicate that these techniques may appear useful for some special applications [23, 24].

The first study to explore the potential of classical slab-gel electrophoresis for micropreparative-scale enantioseparation was published in 1998 by Stalcup et al. [108]. Using the example of the short-acting β_2-adrenergic agonist terbutaline, it was shown that classical gel electrophoresis is indeed a viable method for the separation of milligram quantities of chiral compounds. In the same year it was shown by Thormann's group [109] and Stalcup et al. [110] that commercially available free-flow electrophoresis devices such as RF 3 and MiniPhor from Protein Technologies (Tucson, AZ) and Mini Prep Cell from Bio-Rad (Hercules, CA) can be used for the preparative collection of milligram amounts of enantiomers. The fractions of *R*-(−)- and *S*-(+)-methadone collected in a study [109] were significantly enriched at the front and reverse sides, but it was impossible to obtain enantiomerically pure fractions. CE analysis of the fractions of piperoxan collected in amounts of 0.5 mg in a run time of 9–10 h indicated that the fractions were almost enantiomerically pure [110]. Glukhovsky and Vigh described the micropreparative enantioseparation of DNS-Phe based on the isoelectric focusing principle and using the commercially available free-flow electrophoretic unit Octopus [111].

Together with slab gel electrophoresis and free-flow electrophoretic units, capillary electrophoresis can also be used for micropreparative separation of enantiomers [112, 113]. Based on theoretical calculations, Kaniansky et al. have shown that the sample capacity may be several orders higher when performing a separation in the isotachophoretic mode compared to the capillary zone electrophoresis (CZE) mode. Further, in the experimental set-up using a combination of 1.0 mm I.D. and 0.8 mm I.D. fluorinated ethylene–propylene copolymer capillaries, microgram amounts of racemic 2,4-dinitrophenyl-DL-norleucine were separated with high enantiomeric purity and a recovery of about 75% in 20 min [113].

Chankvetadze et al. [112], dealing with separation selectivity enhancement in capillary electrophoresis based on counterflow principles, have shown that continuous micropreparative enantioseparation can be performed in the capillary format when using counterpressure, hydrodynamic pressure, electro-osmotic flow (EOF) or the carrier ability of a chiral selector as counterbalancing mobility against the effective electrophoretic mobility of the analyte enantiomers. As it has been shown experimentally in their study, the counterpressure can be adjusted in the way that two enantiomers will migrate towards the opposite electrodes or only one enantiomer will migrate from the inlet vial towards the outlet vial [112].

Glukhovsky and Vigh have applied the counterflow migration principle based on the carrier ability of a single isomer of heptakis-(6-sulfato)-β-cyclodextrin in combination with the aforementioned free-flow electrophoretic unit Octopus and reported the enantioseparation of terbutaline with production rates of 0.1 mg/h (for each enantiomer) at an enantiomeric purity of 99.99% [114, 115]. Recently, Schneiderman et al. reported the micropreparative electrophoretic enantioseparation of the chiral drug Ritalin [116].

7.5 Bioanalysis of Chiral Drugs

Separation techniques (GC, HPLC, SFC, CE and CEC) represent major tools for enantioselective bioanalysis. These techniques are widely accepted in academic and industrial laboratories, and are finding their place in national and international pharmacopoeias, substituting for the traditional polarimetric technique. All the aforementioned instrumental techniques offer the advantages of high accuracy, low sample consumption and high sensitivity (lower detection limit of the minor enantiomeric impurity). The coverage of analytes, flexibility, robustness, method development and application requirements, costs, and environmental issues vary from method to method, and selection of the method has to be made based on the specific problem. However, there are some general trends in the use of each method that are discussed in detail below.

7.5.1 GC

If one does not consider the few early studies on column and planar chromatography, GC represents the first instrumental technique that made enantioseparation feasible [117]. Relatively high peak efficiency in GC enables a baseline enantioseparation even when the enantioseparation factor is rather low (around 1.1). GC was quite widely used for the enantioselective analysis of many analytes of biomedical relevance, and some drugs in the 1970s and early 1980s.

Chiral GC is especially useful for chiral gases that are impossible to be analyzed by liquid-phase techniques. The simultaneous enantioseparation of chiral inhalation anesthetics desflurane, isoflurane and enflurane is shown as an example in Fig. 7.6 [22]. Although the enantioseparation was achieved in a rather short time (within 2.5 min), the authors could further shorten the analysis time of enflurane below 10 s (Fig. 7.7) [22]. Such ultrafast enantioseparations are at least very difficult to perform in HPLC, if possible at all. Very short analysis times may be ad-

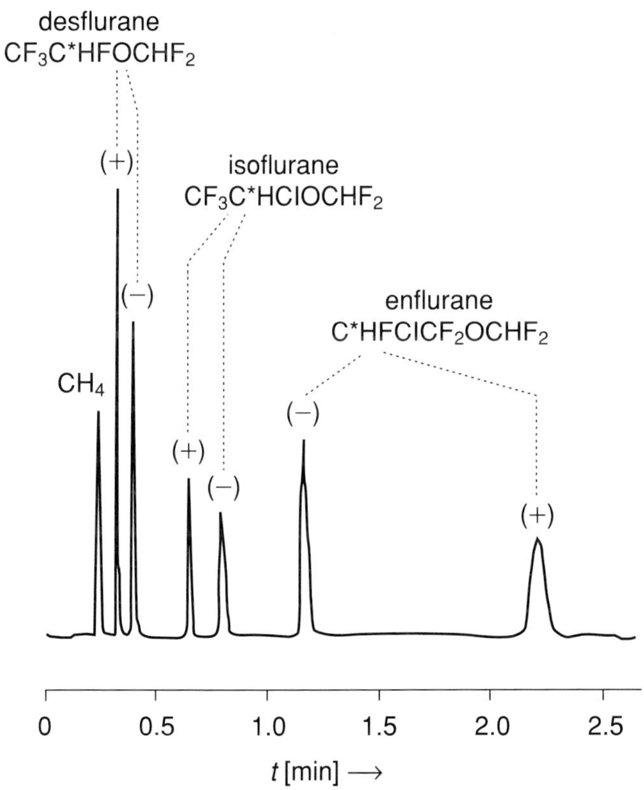

Fig. 7.6 Simultaneous GC enantioseparation of chiral inhalation anesthetics. (Reproduced from [22] with permission.)

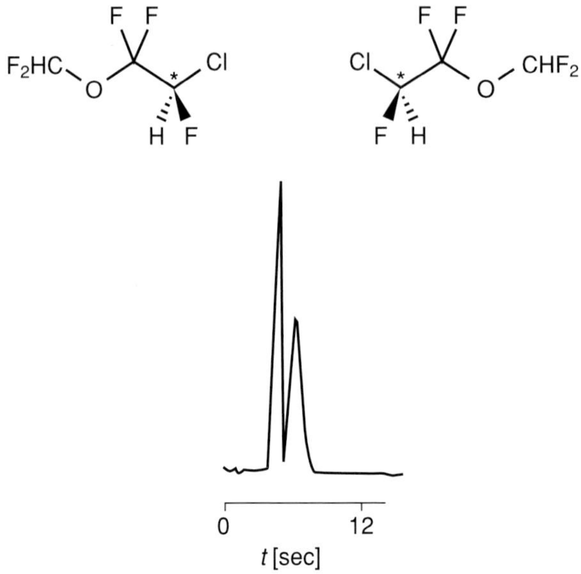

Fig. 7.7 Enantioseparation of enflurane with a very short analysis time. (Reproduced from [22] with permission.)

vantageous when monitoring fast pharmacokinetic processes or pharmacokinetic–pharmacodynamic relationships in living organisms.

The quantitative determination of the inhalation anesthetic isoflurane in blood samples during and after surgery has been performed by enantioselective head-space GC/mass spectrometry (MS) [118].

A relatively new trend in the field of enantioselective GC is the enantiomerization study by dynamic gas chromatography (DGC) or stopped flow gas chromatography (sfGC) [119]. This topic is important because many regulatory guidelines require that "the stability protocol for enantiomeric drug substances and drug products should include a method capable of assessing the stereochemical integrity of the drug substance and drug product" [120]. The results of the determination of the enantiomerization barrier of the chiral drug thalidomide are shown in Tab. 7.3 [121]. Other examples of this kind include chiral diazepines [122]. Both of

Tab. 7.3 Comparison of the enantiomerization barriers of thalidomide obtained by dynamic chromatography and computer simulation with ChromWin and by sfGC. (Reproduced from [121] with permission.)

T (°C)	*$\Delta G^{\neq}$ (DGC) (kJ/mol)*	*$\Delta G^{\neq}$ (sfGC) (kJ/mol)*
200	154 ± 2	150 ± 3
210	157 ± 2	154 ± 3
220	159 ± 2	155 ± 3

the aforementioned techniques, DGC and sfGC, require only minute amounts of samples [118]. In addition, in sfGC the possible catalytic effect due to the presence of a CSP can be circumvented by a multidimensional approach whereby the separation column is isolated from the reactor column [119]. Only a few drugs are gaseous and volatile compounds. Derivatization techniques are available for some analytes that do not belong to this group. However, a derivatization is associated with additional costs, it is time consuming, may not always proceed quantitatively and may affect the enantiomeric ratio in the original sample. Due to these reasons GC covers only a narrow niche in the bioanalysis of chiral drugs at present, whereas liquid-phase techniques such as HPLC and CE dominate.

7.5.2 HPLC

HPLC represents the major technique for enantioselective bioanalysis in industrial laboratories at present. The advantages of this technique include the suitability for nonvolatile and thermolabile compounds, availability of a wide range of CSPs and chiral columns, good compatibility with biological samples, and availability of a wide variety of accurate and sensitive detectors.

Both normal- and reversed-phase HPLC can be applied to bioanalysis. However, the reversed-phase mode appears to be advantageous due to simpler sample preparation. Biological samples can be injected directly on the reversed-phase column after the precipitation of proteins. Solid-phase extraction (SPE) and sample preconcentration are also possible if required. In the case of normal phase separations, the biological samples need to be extracted into organic solvents before their injection onto the separation column. This may appear more laborious, time consuming and expensive. In addition, some problems may also occur from the viewpoint of sample recovery from the biological matrix. Almost all major types of CSPs previously used only in the normal-phase mode can also be adapted to reversed-phase applications. Together with the aforementioned separation modes, the so-called polar-organic mode has become increasingly popular over the last decade [123–129]. In this mode a single polar-organic solvent (basically methanol, ethanol, 2-propanol or acetonitrile) or mixtures are used as mobile phases. The advantages of this mode include the extension of the list of suitable CSPs, in some cases unique selectivities not seen either in normal phase or with aqueous-organic eluents, very high separation factors observed for selected analytes [99, 100], and good compatibility with various types of detectors, especially with mass spectrometers. Low-molecular-weight alcohols (as a mobile phase) are also more environmentally friendly compared with acetonitrile and chlorinated organic solvents.

The chiral column is the most important part of a HPLC enantioseparation system. The number of CSPs described in the literature is approaching 200. On the one hand, this illustrates the power of the technique, but, on the other hand, makes the selection of the optimal CSP extremely difficult. There is no theory enabling column selection based on the structure of the analyte. However, based on available databases and statistical/chemometric approaches, it is feasible to design a separation system quite well.

Among the CSPs that are currently commercially available, the CSPs most widely used in bioanalysis include polysaccharide-based materials such as Chiralcel OD, Chiralpak AD, Chiralcel OJ and Chiralpak AS [130–132]. CSPs based on macrocyclic antibiotics such as vancomycin [133] and teicoplanin [134], and the modified Pirkle-type chiral column Welk-1 [135] are also well suited for bioanalytical studies.

Polysaccharide-based CSPs have been developed for normal-phase applications [130]. However, as recent studies show, these materials are very versatile and also useful under reversed-phase conditions [136–139].

A few examples of the application of cyclodextrin-based CSPs for enantioselective bioanalysis can be still found in the current literature [140], whereas the application of protein-based materials is decreasing. However, the latter CSPs appear to be very useful for the investigation of the stereoselective aspects of protein–drug interactions [141].

Together with the aforementioned chiral phases, the recently commercialized cinchona-based CSPs appear to be attractive for enantioselective bioanalytical studies [142, 143].

The most recent trend in the field of polysaccharide-based materials is their covalent immobilization in order to broaden the range of applicable mobile phases for analytical- and preparative-scale applications [146–150]. Miniaturization has become a major trend in HPLC in few last years, and it is intensively promoted by the most recent developments in the fields of proteomics, genomics, metabonomics, combinatorial synthesis and chip-based technologies. This trend is not strong enough in chiral HPLC. Just a few groups working in the field of chiral CEC also perform research in capillary LC [34]. This direction deserves more attention because the most recent developments in the fields of proteome research and combinatorial synthesis will pose the same requirements of being fast, economical, efficient, highly sensitive and environmentally friendly for chiral separations because many of the drug candidates generated with combinatorial approaches, as well as compounds of biological origin, are chiral. In combination with the state of the art nanospray interfaces, capillary-based separations may appear more sensitive than those with the common size columns.

Current trends in enantioselective bioanalysis include combinations of HPLC with MS [151], highly sensitive laser-based fluorescence detectors [152] and molecular imprinted polymers for on-line sample preparation/preconcentration [153, 154].

7.5.3
CE

CE represents a powerful technique for analytical enantioseparation [28–30]. The applications of this technique have been summarized in recent review papers [155–158]. Here, the characteristics of CE and chromatographic enantioseparation techniques are addressed using some recent applications of chiral CE in biomedical and pharmaceutical analysis as examples. Criteria such as suitability of the samples of interest, success rate (separation power, sensitivity), reliability of results (accuracy, reproducibility), simplicity to learn and use the technique, speed (to learn a technique and method development), environmental pollution (hazardous materials used and their scale), costs (of equipments, required consumables such as chiral selectors, buffers, accessories, etc.), and acceptance in academic and industrial environment will be discussed.

Chiral CE may be used in the early stage of drug development (synthesis of a chiral drug candidate), production processes of a chiral drug entity, formulation studies, preclinical and clinical phase studies, and drug storage and use.

In the early drug development process (synthesis) the enantiomers are determined as pure chiral compounds or as a mixture together with other synthetic intermediates and side products. For this kind of analyte CE offers advantages compared to GC (except the few aforementioned cases of anesthetic gases) and is at least, not less suitable than HPLC. Requirements such as high separation efficiency, high selectivity, automation, etc., are even better met by CE than by HPLC.

Solid (as well as some liquid) drug formulations commonly contain a matrix (cellulose derivatives, starch, lower-molecular-weight oligosaccharides, etc.) that may create certain problems due to adsorption on the inner capillary wall and affect the EOF. Indeed, high-molecular-weight compounds may also create problems in HPLC (contamination of the column, irreversible adsorption on the stationary phase, precipitation on the filter, etc.). A possible solution to these problems in CE may be precoating the capillary wall. Alternatively, the matrix ingredients can be removed from the sample by various sample pretreatment processes (membrane filtration, liquid–liquid extraction, etc.).

In preclinical and clinical phase studies, a chiral drug candidate and its possible metabolites (phase I and phase II) have to be determined in biological media such as plasma, serum, urine, saliva, cerebrospinal fluid and tissue homogenates. Most of these sample matrixes (especially urine) are more compatible with CE than with chromatographic techniques. High-molecular-mass ingredients of plasma may cause problems similar to the aforementioned polysaccharides due to their adsorption on the capillary inner wall.

The parent drug may be polar or may undergo metabolic transformations to polar compounds. This relates especially to phase II metabolites such as glucoronides, sulfates, mercapturates, etc. CE is the technique of choice for the analysis of charged compounds.

It is certainly unimaginable in GC and very difficult in chiral HPLC to achieve the enantioseparation of a chiral drug and its phase I and phase II metabolites in a single run. However, this is possible in chiral CE. Figure 7.8 shows the simultaneous enantioseparation of phase I and II metabolites of the chiral antihistaminic drug dimethindene as an example [159]. Thus, as this example illustrates, CE may work better for a chiral drug and its metabolites possessing different polarities. The success rate of a method is one of the most important criteria for the selection of an analytical technique. As mentioned in the previous section, the number of CSPs

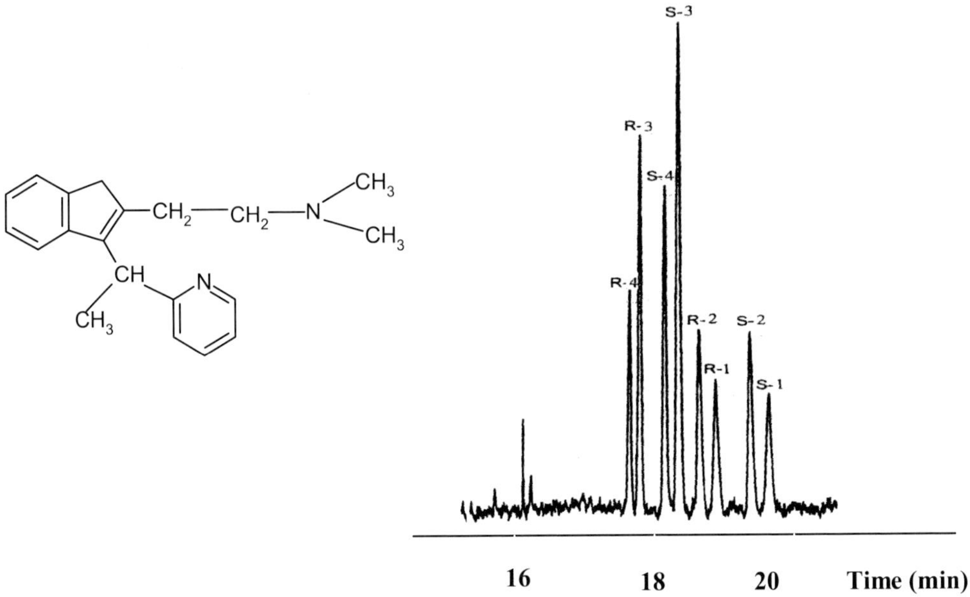

Fig. 7.8 Simultaneous CE enantioseparation of phase I and II metabolites of dimethindene: 1,6-hydroxydimethindine; 2,6-hydroxy-*N*-demethyldimethindine; 3,6-hydroxydimethindine-glucoronide; 4,6-hydroxy-*N*-demethyldimethindine-glucoronide. (Reproduced from [159] with permission.)

for HPLC enantioseparations described in the literature is approaching 200. The number of effective CE chiral selectors is lower. Although CE enables a much faster and cost-effective screening of the chiral selectors compared to HPLC, and the method development time in CE may be much shorter, in general terms and just for two enantiomers the success rate for both of these techniques may appear comparable. However, the superiority of chiral CE compared to HPLC becomes evident as soon as one needs to separate and simultaneously enantioseparate multicomponent mixtures containing the parent chiral drug, its phase I and phase II metabolites, synthetic intermediates, and/or degradation products. The example of the simultaneous enantioseparation of the phase I and phase II metabolites of the antihistaminic drug dimethindene is shown in Fig. 7.8 [159]. One additional example of rather complex enantioseparation is shown in Fig. 7.9 [160].

The enantioseparation of thalidomide is not a problem either in HPLC [21, 161–164], CE [165–167] or CEC [168–170]. Rather more difficult is the simultaneous enantioseparation of thalidomide and its biologically relevant metabolites that were recently detected in incubation mixtures of racemic thalidomide with fraction S9 from human liver and plasma samples from male volunteers who had received racemic thalidomide orally [171]. The teratogenicity of thalidomide caused its withdrawal in 1961. However, this compounds exhibits remarkable effects against

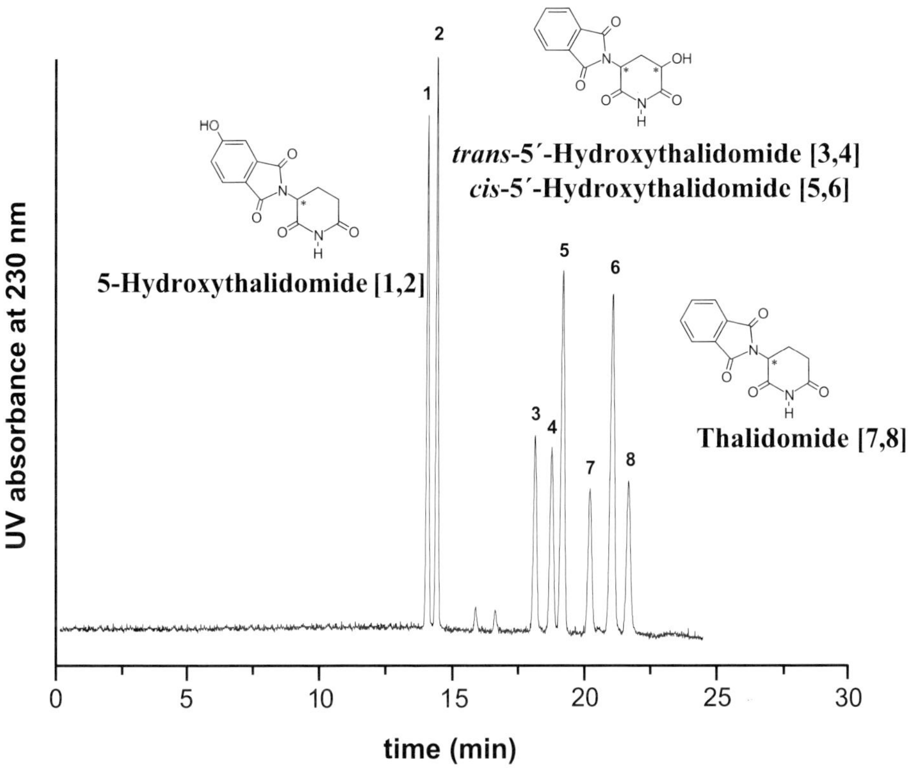

Fig. 7.9 Simultaneous CE enantioseparation of thalidomide and its phase I metabolites. (Reproduced from [160] with permission.)

leprosy [172], inhibits HIV-1 virus [173], suppresses the release of tumor necrosis factor-α [174], etc. Recently, thalidomide was approved by the FDA for the treatment of erythema nodosum leprosum [175]. Thus, investigation of the teratogenicity mechanisms of this drug is very topical. The metabolites of thalidomide are suspected to be potential teratogenic agents and they are formed stereoselectively [176]. This means that some of the metabolites are formed primarily by metabolic transformation of (*R*)- and others of (*S*)-thalidomide. In order to follow this phenomenon, a method for the simultaneous separation and enantioseparation of thalidomide and its metabolites is required. All experiments to separate and enantioseparate thalidomide and its three major metabolites in a single run using either HPLC with chiral columns or CE with a single chiral selector in normal polarity mode were unsuccessful. However, all of them were resolved in a single run using carrier mode CE with a combination of two chiral selectors (Fig. 7.9) [160]. A combination of two chiral selectors based on column switching is also possible in chromatographic techniques. However, this is associated with several technical problems (high back pressure, increased dead volumes, excess peak dispersion, etc.). In addition, chiral selectors can be mixed in CE in any desired ratio (only limited by the solubility). This is difficult and at least very time-consuming in HPLC and GC.

If one compares chiral selectors exhibiting the same thermodynamic enantioselectivity of recognition in an ideal experiment, then it is clear that the rate of success of an enantioseparation is much higher in CE than in HPLC. This advantage stems from the fact that peak efficiency is much higher in CE and the resolution factor increases proportional to $N^{1/2}$. In addition, selector–selectand interactions may be enormously intensified in CE by increasing the concentration of chiral selector. This may be limited by the solubility of a chiral selector, high viscosity of background electrolyte or high current in the case of charged chiral selectors. However, available frames are quite wide in order to optimize a separation. Thus, the separation power of CE is definitely higher and the tools available for the adjustment of a chiral separation are multivariate compared to chromatographic techniques.

Reproducibility of migration times and separation as a whole has been a matter of uncertainty in CE for a long time. However, as most recent developments indicate, this is not a critical issue any more. Several efficient techniques became available for the elimination of the contribution of the capillary wall to the mobility of an analyte either by adjusting the mobility of the EOF or by its suppression. In addition, the chiral selectors used in CE are becoming increasingly uniform and better characterized. Both of the aforementioned trends facilitate the development of well-reproducible chiral CE methods.

Sensitivity has been considered as a major problem in CE for a long time. In one overview several years ago it was noted that, "It is unlikely that chiral CE will be used in drug bioanalysis because enantiomers must be determined at very low levels, often ng/ml or lower" [177]. The most recent developments in chiral CE are basically opposing the aforementioned opinion and chiral CE is very successfully applied in the field of bioanalysis [155–158]. There are several possibilities of sample preconcentration such as on capillary solid-phase (micro)extraction, membrane filtration/preconcentration, sample stacking and sample sweeping [179–181]. In addition, one also has to consider a wide variety of detection cells with an extended optical path length for CE. In principle, one may have even the same path length in CE as in HPLC, providing that the separation factor (which is rather easily adjustable in chiral CE) allows this. Further, comparing the limit of detection (LOD) in CE and HPLC, one has to consider that a sample zone elutes as a much sharper peak in a significantly shorter period of time and is less diluted in CE than in HPLC. However, the most important argument for a bright future of chiral CE in drug bioanalysis is that this technique is compatible with such a sensitive, specific and universal detection method as MS [182–185].

Another aspect of sensitivity that cannot remain outside the scope of this discussion is the following. In the majority of cases it is important to detect an impurity of the minor enantiomer in the presence of the major one. Therefore, it may happen that the detection limit of the minor enantiomeric impurity in the presence of the major one will become a more important issue than the overall LOD.

In this case the separation factor will appear as important for the final success as the overall sensitivity of the system. This happens because the method must allow a sufficient difference between the migration times of two enantiomers in order to detect the minor enantiomer without overlapping by the major one. The adjustment of the enantiomer migration order can also be very useful in this case [186 188]. CE offers certain advantages compared to HPLC for both the optimization of a separation and the design of the enantiomer migration order. Thus, the challenges of chiral CE compared to HPLC from the viewpoint of detection sensitivity do not look as critical as one may assume based on the comparison of the standard size of the separation chamber and the path lengths of detection cells.

The fascination of CE is the remarkable simplicity of this technique. No pump, injector valves and separate detection cells are required in this technique. This not only simplifies the experiment markedly, but also eliminates the sources of additional peak broadening due to sample injection and detection. A single CE equipment can successfully be used not only to perform various CE separation modes (capillary zone electrophoresis, micellar electrokinetic capillary chromatography, capillary isotachophoreis, capillary isoelectric focusing, capillary gel electrophoresis), but also capillary HPLC and CEC separations [170].

CE is more rapid compared to chromatographic techniques from the viewpoint of method development. Changing and conditioning a column is very time-consuming in chromatography. However, changing a capillary and/or chiral selector takes only a few minutes in CE.

CE as microanalytical technique requires minute amounts of solvents. This makes the technique especially environmentally friendly and inexpensive. Further saving costs stems from extremely low amount of chiral selectors and buffers. This allows us to study chiral recognition properties of rather expensive and/or exotic materials that are available only in small amounts. Fused-silica capillaries used in CE are rather cheap as well as most of the accessories. The basic CE equipment of medium quality does not cost more than a HPLC or GC instrument. However, the costs from the use of the latter two exceed those of CE.

At present CE does not have any acceptance problems in academic laboratories. In the industry, HPLC still remains the dominant technique for chiral separation. However, taking into account the experience in academic institutions, it seems highly probably that in the nearest future CE will become the technique of choice for analytical enantioseparation in the pharmaceutical, food, chemical and agrochemical industries as well as in bioanalytical and clinical laboratories.

7.5.4 CEC

Several principal advantages of CE for enantioseparation compared to HPLC and GC were emphasized in subsection 7.5.3. Together with the electrokinetic separation mechanism, the presence of a mobile chiral selector instead of stationary bed in chromatographic techniques is responsible for the aforementioned advantages of CE. On the other hand, the mobile chiral selector is not ideal as (1) it must be replaced after every analysis, which leads to higher consumption of a chiral selector and problems with the run-to-run reproducibility of a given separation, and (2) the presence of a chiral selector in a separation capillary creates significant problems for on-line coupling of CE with a mass spectrometer and renders the use of chiroptical (polarimetric and circular dichroism) detectors practically impossible in CE. Thus, a hybrid technique between chiral HPLC and CE has been developed since the early 1990s called "CEC" [31–34]. CEC is expected to combine the high peak efficiency characteristic of electrically driven separations with the high separation selectivity of multivariate CSPs available in HPLC. In addition, the stationary bed allows us to avoid the aforementioned problems related to the application of the mobile chiral selector in CE. Recent developments in chiral CEC have been summarized in several review papers [33–34].

Although chiral CEC has been successfully applied to enantioseparation of chiral analytes of different structures [33, 34], only a few applications to real problems in pharmaceutical, biomedical, environmental, etc., analyses can be found in the literature. Only one publication describes a not very successful attempt of the simultaneous enantioseparation of thalidomide and its phase I hydroxy metabolites [189]. The CSP in this study was a rather sophisticated mixture of two polysaccharide derivatives. In addition to that example, a validated chiral CEC assay for the enantiomers of the β-blocker metoprolol using a vancomycin-containing CSP in the polar-organic mode was published [190]. Other work describes the enantioseparation of the antidepressant drug venlafaxine and its main metabolite *O*-desmethylvenlafaxine, also with a vancomycin-based CSP. The method was also applied to real plasma samples of patients under clinical treatment with venlafaxine [191]. Together with pure drugs and biomedical samples, CEC may be applied to enantiomeric purity control of chiral drug formulations. This was illustrated for tablets of the contraceptive drug Trigoa containing levonorgestrel and ethinylestradiol as the active ingredients [192].

More applications of chiral CEC to real problems of pharmaceutical and biomedical importance will certainly appear in the near future. This optimistic conclusion is supported by the numerous applications of this technique to standard racemic mixture of enantiomers summarized in review articles [33, 34], on impressive plate numbers and short analysis times already described for selected chiral analytes [193], as well as on the continuously better understanding of the underlying mechanisms of (chiral) CEC.

7.6 Future Trends

The analysis of the most recent developments in the field of enantioseparation indicates a continuous increase of its importance for the development of new drugs as well as for the most efficient application of chiral drugs that are already in use [194]. For analytical purposes the miniaturized techniques such as CE, capillary LC and CEC, performing the job at least as effectively as HPLC, will gain in importance. The miniaturized techniques are more flexible, less expensive and more environmentally friendly. For preparative-scale applications and production of enantiomerically pure chiral drugs, SMB chromatography and its more efficient modifications may appear advantageous in certain cases over diastereomeric crystallization, enzymatic resolution, chiral pool and catalytic asymmetric synthesis. However, the later decision needs to be made on the case-to-case basis considering time constrains, drug development steps, economic issues, scales, etc.

References

1 SC Stinson, *C & E News* 2002, **80**, 43–50.
2 R Crossley, *Chirality and Biological Activity of Drugs*, CRC Press, Boca Raton, FL, 1995.
3 HY Aboul-Enein, IW Wainer, eds, *The Impact of Stereochemistry on Drug Development and Use (Chemical Analysis 142)*, Wiley, New York, 1997.
4 M Eichelbaum, AS Gross, *Adv Drug Res* 1996, **28**, 1–64.
5 M Vakily, R Mehvar, D Brocks, *Ann Pharmacother* 2002, **36**, 693–701.
6 R Mehvar, D Brocks, *J Pharm Pharm Sci* 2001, **4**, 185–200.
7 I Agranat, *Drug Dev Today* 1999, **4**, 313–21.
8 W Creutzfeldt, *Z Gastroenterol* 2000, **38**, 893–97.
9 M Robinson, *Eur J Gastroenterol Hepatol* 2001, **13**, S43–7.
10 GF Clough, P Boutsiaki, MK Church, *Allergy* 2001, **56**, 985–88.
11 DY Wang, F Hanotte, C De-Vos, P Clement, *Allergy* 2001, **56**, 339–43.
12 *Investigation of Chiral Active Substances*, Note for Guidance by The European Agency for the Evaluation of Medicinal Products, 1998.
13 JV Schaus, *C & E News* 2000, **78**, 55–78.
14 B Chankvetadze, *J Sep Sci* 2001, **24**, 691–705.
15 WS Knowles, *Angew Chem* 2002, **114**, 2096–107.
16 R Noyori, *Angew Chem* 2002, **114**, 2108–23.
17 KB Sharpless, *Angew Chem* 2002, **114**, 2126–35.
18 G Blaschke, *Angew Chem* 1980, **92**, 14–25.
19 G Blaschke, B Chankvetadze, in: F Gualtieri, ed., *New Trends in Synthetic Medicinal Chemistry*, 139–73, Wiley-VCH, Weinheim, 2000.
20 YF Shealy, CE Oplinger, JA Montgomery, *J Pharm Sci* 1968, **57**, 757–64.
21 G Blaschke, HP Kraft, K Fickentscher, F Köhler, *Arzneim Forsch* 1979, **29**, 1640–42.
22 V Schurig, H Grosenick, M Juza, *Recl Trav Chim Pays-Bas* 1995, **114**, 211–9.
23 M Juza, E Brawn, V Schurig, *J Chromatogr A* 1997, **769**, 119–27.
24 M Juza, O Di Giovanni, G Biressi, V Schurig, M Mazzoti, M Morbidelli, *J Chromatogr A* 1998, **813**, 333–47.
25 G Terfloth, *J Chromatogr A* 2001, **906**, 301–7.
26 KW Phinney, *Anal Chem* 2002, **72**, 204A–11A.
27 E Gassman, JE Kuo, RN Zare, *Science* 1985, **230**, 813–5.
28 S Fanali, M Cristalli, R Vespalec, P Bocek, *Adv Electrophor* 1994, **7**, 1–86.
29 B Chankvetadze, *Capillary Electrophoresis in Chiral Analysis*, Wiley, Chichester, 1997.
30 B Chankvetadze, G Blaschke, *J Chromatogr A* 2001, **906**, 309–63.
31 S Mayer, V Schurig, *J High Resolut Chromatogr* 1922, **15**, 129–31.
32 S Li, DK Lloyd, *Anal Chem* 1993, **65**, 3684–90.
33 M Lämmerhofer, F Svec, JMJ Frechet, W Lindner, *Trends Anal Chem* 2002, **19**, 676–98.
34 S Fanali, P Catarcini, G Blaschke, B Chankvetadze, *Electrophoresis* 2001, **22**, 3131–51.

35 DF Reinhold, RA Firestone, WA Gaines, JM Chemerda, M Sletzinger, *J Org Chem* 1968, **33**, 1209–13.

36 G Amirad, *Experientia* 1959, **15**, 1–7.

37 Industria Chimica Profarmaco, *European Patent 298385*, 1988.

38 RA Sheldon, *Chirotechnology, Industrial Synthesis of Optically Active Compounds*, Marcel Dekker, New York, 1993.

39 JW Young, RL Bratzler, in: *Proc Chiral 90 Symp*, 23–8, Spring Innovations, Stockport, 1990.

40 LA Hulsof, JH Roskam, *European Patent 034714*, 1989.

41 LJ Brice, WH Pirkle, in: S Ahuja, ed., *Chiral Separations: Applications and Technology*, 309–34, Oxford University Press, New York, 1996.

42 AS Bormarius, K Drauz, U Groeger, C Wandrey, in: AN Collins, GN Sheldrake, J Crosby, eds, *Chirality in Industry: The Commercial Manufacture and Applications of Optically Active Compounds*, 371–97, Wiley, Chichester, 1992.

43 J Crosby, *Tetrahedron* 1991, **47**, 4789–846.

44 W Eveleens, J Spaans, HG Wissink, *European Patent 9285 CA, 93, 95006p*, 1980.

45 R Scott, CD Armitage, *German Patent 2946652 CA 1980, 185974g*, 1980.

46 S Cavicchioli, C Giordano, F Uggeri, *J Org Chem* 1987, **52**, 3018–27.

47 C Giordano, C Graziano, *J Org Chem* 1989, **54**, 1470–73.

48 T Ohashi, J Hasegawa, J Synth, *Org Chem Japan* 1987, **45**, 331–45.

49 M Shimazaki, J Hasegawa, K Kan, K Nomura, Y Nose, H Kondo, T Ohashi, K Watanabe, *Chem Pharm Bull* 1982, **30**, 3139–46.

50 N Cohen, WF Eichel, RJ Lopresti, C Neukom, G Souci, *J Org Chem* 1976, **41**, 3505–11.

51 RA Sheldon, HJ Zeegers, HJM Houbiers, LA Hulsof, *Chimica Oggi* 1991, 35–47.

52 DM Floyd, AM Fritz, CM Cimarusti, *J Org Chem* 1982, **47**, 176–8.

53 T Ohta, T Kimura, N Sato, S Nozoe, *Tetrahedron Lett* 1988, **29**, 4303–5.

54 TN Salzmann, RW Ratcliffe, BG Christensen, FA Bouffard, *J Am Chem Soc* 1980, **102**, 6161–3.

55 M Shiozaki, T Ishihara, T Hiraoka, H Yanagisawa, *Tetrahedron Lett* 1981, **22**, 5205–8.

56 HU Blaser, *Chem Rev* 1992, **92**, 935–52.

57 S Knowles, MJ Sabacky, *J Chem Soc Chem Commun* 1968, 1445–6.

58 TP Dang, HB Kagan, *J Chem Soc Chem Commun* 1971, 481.

59 R Noyori, *Chem Soc Rev* 1989, **18**, 187–208.

60 T Ohta, H Takaya, M Kitamura, K Nagai, R Noyori, *J Org Chem* 1987, **52**, 3176–8.

61 M Kitamura, I Kasahara, K Manabe, R Noyori, H Takaya, *J Org Chem* 1988, **53**, 708–10.

62 T Nishi, M Kitamura, T Ohkuma, R Noyori, *Tetrahedron Lett* 1988, **29**, 6327–30.

63 M Kitamura, M Tokunaga, R Noyori, *J Am Chem Soc* 1995, **117**, 2931–2.

64 T Ohkuma, M Koizumi, H Doucet, T Pham, M Kozawa, K Murata, E Katayama, T Yokozawa, T Ikariya, R Noyori, *J Am Chem Soc* 1998, **120**, 13529–30.

65 R Noyori, M Suzuki, *Science* 1993, **259**, 44–5.

66 H Takahashi, S Sakuraba, H Takada, K Achiwa, *J Am Chem Soc* 1990, **112**, 5876–8.

67 H Takeda, T Tachinami, M Aburatani, H Takahashi, T Morimoto, K Achiwa, *Tetrahedron Lett* 1989, **30**, 363–6.

68 KG Watson, YM Fung, M Gredley, GJ Bird, WR Jackson, H Gountzous, BR Mathews, *J Chem Soc Chem Commun* 1990, 1018–9.

69 H Cotton, T Elebring, M Larsson, L Li, H Sörensen, S von Unge, *Tetrahedron: Asymmetry* 2002, **11**, 3819–25.

70 ER Francotte, P Richert, *J Chromatogr A* 1997, **769**, 101–7.

71 F Charton, RM Nicoud, *J Chromatogr A* 1995, **702**, 97–112.

72 RM Nicoud, A Seidel-Morgenstern, *Isolation and Purification* 1996, **2**, 165–77.

73 G Biressi, O Ludemann-Hombourger, M Mazzotti, RM Nicoud, M Morbidelli, *J Chromatogr A* 2000, **876**, 3–15.

74 JN Kinkel, M Schulte, RM Nicaud, F Charton, in: *Chiral Europe 1995, Symp Proc*, 121, London, 1995.

75 J Blehaut, *10th Int Symp on Chiral Discrimination*, Vienna, 1998.

76 DW Guest, *J Chromatogr A* 1997, **760**, 159–62.

77 E Francotte, *Analytica Conf*, 96, Munich, 1996.

78 O Reiser, Ch Bubert, R Reumer, P Kreitmeier, M Schulte, A Meudt, *German Patent 198 58 893-3*, 1998.

79 S Nagamatsu, K Murazumi, S Makino, *J Chromatogr A* 1999, **832**, 55–65.

80 M Schulte, R Ditz, RM Devant, JN Kinkel, F Charton, *J Chromatogr A* 1997, **769**, 93–100.

81 J Strube, A Jupke, A Epping, H Schmidt-Traub, M Schulte, RM Devant, *Chirality* 1999, **11**, 440–50.

82 M Schulte, R Devant, *German Patent P199 31 755-0*, 1999.

83 RM Nicoud, M Bailly, JN Kinkel, R Devant, T Hampe, E Küsters, in: RM Nicoud, ed., *Simulated Moving Bed: Basics and Applications*, 65–8, INPL, Nancy, 1993.

84 D Seebach, M Hoffmann, AR Sting, JN Kinkel, M Schulte, E Küsters, *J Chromatogr A* 1998, **796**, 299–307.

85 M Negawa, F Shoji, *J Chromatogr* 1992, **590**, 113–7.

86 CB Ching, BG Lim, EJD Lee, SC Ng, *J Chromatogr* 1993, **634**, 215–9.

87 MJ Gattuso, S Makino, Poster presentation at *Chiral 94*, Reston, VA, 1994.

88 LS Pais, JM Loureiro, AE Rodrigues, *J Chromatogr A* 1998, **827**, 215–33.

89 E Küsters, G Gerber, F Antia, *Chromatographia* 1995, **40**, 387–93.

90 RM Nicoud, G Fuchs, P Adam, M Bailly, E Küsters, FD Antia, R Reuill, E Schmid, *Chirality* 1993, **5**, 267–71.

91 G Fuchs, RM Nicoud, M Bailly, *Proc 9th Int Symp on Preparative and Industrial Chromatography*, 205–20, INPL, Nancy, 1992.

92 E Cavoy, MF Deltent, S Lehoucq, D Maggiano, *J Chromatogr A* 1997, **769**, 49–57.

93 O Ludemann-Hombourger, RM Nicoud, M Bailley, *Sep Sci Technol* 2000, **35**, 1285–305.

94 O Ludemann-Hombourger, G Pigorini, RM Nicoud, DS Ross, G Terfloth, *J Chromatogr A* 2002, **947**, 59–68.

95 ER Francotte, *J Chromatogr A* 2001, **906**, 379–97.

96 ER Francotte, *J Chromatogr* 1994, **666**, 565–601.

97 ER Francotte, in: HY Aboul-Enein, IW Wainer, eds, *The Impact of Stereochemistry on Drug Development and Use*, 633–83, Wiley, New York, 1997.

98 AP Foucault, *J Chromatogr A* 2001, **906**, 365–78.

99 B Chankvetadze, C Yamamoto, Y Okamoto, *Chem Lett* 2002, 1176–7.

100 B Chankvetadze, C Yamamoto, Y Okamoto, *J Chromatogr A* 2001, **922**, 127–37.

101 NM Maier, L Nicoletti, M Lämmerhofer, W Lindner, *Chirality* 1999, **11**, 522–8.

102 WH Pirkle, WE Brown, *J High Resolut Chromatogr* 1994, **17**, 629–33.

103 A Berthod, X Chen, JP Kulman, DW Armstrong, F Gasparrini, I D'Acquaria, C Villani, A Carotti, *Anal Chem* 2000, **72**, 1767–80.

104 B Stephan, A Mannschreck, NA Voloshin, NV Volbushko, VI Minkin, *Tetrahedron Lett* 1990, **44**, 6335–338.

105 I Fitos, M Simonyi, *Chirality* 1992, **4**, 21–3.

106 L Oliveros, C Minguillon, P Franco, AP Foucault, in: A Berthold, ed., *Countercurrent Chromatography, The Support-Free Liquid Stationary Phase*, 331–51, Elsevier, New York, 2002.

107 P Franco, J Blanc, WR Oberleitner, NM Maier, W Lindner, C Minguillon, *Anal Chem* 2002, **74**, 4175–183.

108 AM Stalcup, KH Gahm, SR Gratz, RC Sutton, *Anal Chem* 1998, **70**, 144–48.

109 M Lanz, J Caslavska, W Thormann, *Electrophoresis* 1998, **19**, 1081–90.

110 RM Sutton, SR Gratz, AM Stalcup, *Analyst* 1998, **123**, 1477–80.

111 P Glukhovskiy, G Vigh, *Anal Chem* 1997, **71**, 3814–20.

112 B Chankvetadze, N Burjanadze, D Bergenthal, G Blaschke, *Electrophoresis* 1999, **20**, 2680–5.

113 D Kaniansky, E Simunicova, E Ölvecka, A Ferancova, *Electrophoresis* 1999, **20**, 2786–93.

114 P Glukhovskiy, G Vigh, *Electrophoresis* 2000, **21**, 2010–5.

115 G Vigh, P Glukhovskiy, Lecture presented at *HPLC 2001*, Maastricht, 2001.

116 E Schneiderman, SR Gratz, AM Stalcup, *J Pharm Biomed Anal* 2002, **27**, 639–50.

117 E Gil-Av, B Feibush, R Charles-Siegler, *Tetrahedron Lett* 1966, 1009–15.

118 M Juza, H Jakubetz, H Hettescheimer, V Schurig, *J Chromatogr B* 1999, **735**, 93–102.

119 V Schurig, *J Chromatogr A* 2001, **906**, 275–99.

120 Anonymous, *Chirality* 1992, **4**, 338.

121 O Trapp, G Schoetz, V Schurig, *J Pharm Biomed Anal* 2002, **27**, 497–505.

122 G Schoetz, O Trapp, V Schurig, *Anal Chem* 2002, **72**, 2758–64.

123 AM Krstulovič, G Rossey, JP Porziemsky, D Long, I Chekreum, I, *J Chromatogr* 1987, **411**, 461–5.

124 SG Chang, GL Reid, S Chen, CD Chang, DW Armstrong, *Trends Anal Chem* 1993, **12**, 144–53.

125 HY Aboul-Enein, V Serignese, J Bojarski, *J Liquid Chromatogr* 1993, **16**, 2741–9.

126 J Dingenen, in: G Subramanian, ed., *A Practical Approach to Chiral Separations by Liquid Chromatography*, 115–81, VCH, Weinheim, 1994.

127 C Weinz, G Blaschke, HM Schiebel, *J Chromatogr B* 1997, **690**, 233–42.

128 B Chankvetadze, C Yamamoto, Y Okamoto, *Comb Chem High Throughput Screening* 2000, **3**, 497–508.

129 M Meyring, B Chankvetadze, G Blaschke, *J Chromatogr A* 2000, **876**, 157–67.

130 Y Okamoto, E Yashima, *Angew Chem* 1998, **110**, 1072–95.

131 J Haginaka, *J Pharm Biomed Anal* 2002, **27**, 357–72.

132 TAG Noctor, in: G Subramanian, ed., *A Practical Approach to Chiral Separations by Liquid Chromatography*, 357–96, VCH, Weinheim, 1994.

133 DW Armstrong, YB Tang, SS Chen, YW Zhou, C Bagwill, JR Chen, *Anal Chem* 1994, **66**, 1473–84.

134 DW Armstrong, YB Liu, KH Ekborgott, *Chirality* 1995, **7**, 474–97.

135 CJ Welch, T Szczerba, SR Perrin, *J Chromatogr A* 1997, **758**, 93–8.

136 K Tachibana, A Ohnishi, *J Chromatogr A* 2001, **906**, 127–54.

137 P Piras, C Roussel, J Pierot-Sanders, *J Chromatogr A* 2001, **906**, 443–58.

138 K Ikeda, T Hamasaki, H Kohno, T Ogawa, *Chem Lett* 1989, 1089–90.

139 A Ishikawa, T Shibata, *J Liquid Chromatogr* 1993, **16**, 859–78.

140 H Kanazawa, Y Kunito, Y Matsushima, S Ohkubo, F Mashige, *J Chromatogr A* 2000, **871**, 181–8.

141 DS Hage, *J Chromatogr A* 2001, **906**, 459–81.

142 A Mandl, L Nicoletti, M Lammerhoefer, W Lindner, *J Chromatogr A* 1999, **858**, 1–11.

143 P Franco, M Lammerhofer, RM Klaus, W Lindner, *J Chromatogr A* 2000, **869**, 111–27.

144 Y Okamoto, R Aburatani, S Miura, K Hatada, *J Liquid Chromatogr* 1987, **10**, 1613–28.

145 K Kimata, R Tsubai, K Hosoya, N Tanaka, *Anal Methods Instrument* 1993, **1**, 23.

146 E Yashima, H Fukaya, Y Okamoto, *J Chromatogr A* 1994, **677**, 11–9.

147 L Oliveros, P Lopez, C Minguillon, P Franco, *J Liquid Chromatogr* 1995, **18**, 1521–32.

148 E Francotte, *PCT WO 96/27615*, 1996.

149 E Francotte, D Huynh, *J Pharm Biomed Anal* 2002, **27**, 421–9.
150 T Kubota, T Kusano, C Yamamoto, E Yashima, Y Okamoto, *Chem Lett* 2001, 724–5.
151 KM Fried, AE Young, S Usdin Yasuda, IW Wainer, *J Pharm Biomed Anal* 2002, **27**, 479–88.
152 T Fukushima, T Santa, H Homma, SM Al-Kindy, K Imai, *Anal Chem* 1997, **69**, 1793–9.
153 G Lamprecht, T Kraushofer, K Stoschitzky, W Lindner, *J Chromatogr B* 2002, **740**, 219–26.
154 J Haginaka, H Sanbe, *Anal Chem* 2000, **72**, 5206–10.
155 G Blaschke, B Chankvetadze, *J Chromatogr A* 2000, **875**, 3–25.
156 S Zaugg, W Thormann, *J Chromatogr A* 2000, **875**, 27–41.
157 G Scriba, *J Pharm Biomed Anal* 2002, **27**, 373–99.
158 A Amini, *Electrophoresis* 2001, **22**, 3107–30.
159 J Rudolf, G Blaschke, *Enantiomer* 1999, **4**, 317–23.
160 M Meyring, B Chankvetadze, G Blaschke, *Electrophoresis* 1999, **20**, 2425–31.
161 B Knoche, G Blaschke, *Cirality* 1994, **6**, 221–4.
162 K Nishimura, Y Hashimoto, S Iwasaki, *Chem Pharm Bull* 1994, **42**, 1157–9.
163 B Chankvetadze, I Kartozia, C Yamamoto, Y Okamoto, *J Pharm Biomed Anal* 2002, **27**, 467–78.
164 B Chankvetadze, G Endresz, G Blaschke, *Electrophoresis* 1994, **15**, 804–8.
165 G Schulte, B Chankvetadze, G Blaschke, *J Chromatogr A* 1997, **771**, 259–66.
166 B Chankvetadze, G Endresz, G Blaschke, *J Cap Electrophoresis* 1995, **2**, 235–40.
167 A Aumatell, RJ Wells, DK Y Wang, *J Chromatogr A* 1994, **686**, 293–307.
168 L Chankvetadze, I Kartozia, C Yamamoto, B Chankvetadze, G Blaschke, Y Okamoto, *Electrophoresis* 2002, **23**, 486–93.
169 L Chankvetadze, I Kartozia, C Yamamoto, B Chankvetadze, G Blaschke, Y Okamoto, *J Sep Sci* 2002, **25**, 653–60.
170 B Chankvetadze, I Kartozioa, J Breitkreutz, Y Okamoto, G Blaschke, *Electrophoresis* 2001, **22**, 3327–34.
171 T Eriksson, S Björkemann, B Roth, H Björk, P Höglund, *J Pharm Pharmacol* 1998, **50**, 1409–16.
172 RL Barnhill, AC Mc Dougall, *J Am Dermatol* 1982, **7**, 317–23.
173 D Stirling, *Ann Rep Med Chem* 1995, **30**, 319–27.
174 JM Gardner-Medvin, *Exp Opin Invest Drug* 1996, **5**, 829–41.
175 C Marwick, *J Am Med Ass* 1997, **278**, 1135–7.
176 G Blaschke, M Meyring, C Mühlenbrock, B Chankvetadze, *Il Farmaco* 2002, **57**, 551–4.
177 EC Soo, AB Salmon, WJ Lough, *Chem Britain* 1999, **15 March**, 220–4.
178 M Dankova, D Kaniansky, S Fanali, F Ivanyi, *J Chromatogr A* 1999, **838**, 31–43.
179 JP Quirino, S Terabe, *Science* 1998, **282**, 465–8.
180 JP Quirino, S Terabe, *Anal Chem* 1999, **71**, 1638–44.
181 J Palmer, NJ Munro, J Landers, *Anal Chem* 1999, **71**, 1679–87.
182 RL Sheppard, X Tong, J Cai, JD Henion, *Anal Chem* 1995, **67**, 2054–8.
183 G Schulte, S Heitmeier, B Chankvetadze, G Blaschke, *J Chromatogr A* 1998, **800**, 77–82.
184 S Fanali, C Desiderio, G Schulte, S Heitmeier, D Strickmann, B Chankvetadze, G Blaschke, *J Chromatogr A* 1998, **800**, 69–76.
185 S Cherkaoui, J-L Veuthey, *J Biomed Pharm Anal* 2002, **27**, 615–26.
186 B Chankvetadze, G Schulte, G Blaschke, *Enantiomer* 1997, **2**, 157–79.
187 B Chankvetadze, *Electrophoresis* 2002, **23**, 4022–35.
188 J Magnusson, H Wan, LG Blomberg, *Electrophoresis* 2002, **23**, 3013–9.
189 M Meyring, B Chankvetadze, G Blaschke, *J Chromatogr A* 2000, **876**, 157–67.

190 E Carlsson, H Wikström, DW Armstrong, PK Owens, *J Chromatogr A* 2000, **897**, 349–63.
191 S Fanali, S Rudaz, J-L Veuthey, C Desiderio, *J Chromatogr A* 2001, **919**, 195–203.
192 B Chankvetadze, I Kartozia, C Yamamoto, Y Okamoto, G Blaschke, *J Pharm Biomed Anal* 2002, **30**, 1897–906.
193 B Chankvetadze, I Kartozia, Y Okamoto, G Blaschke, *J Sep Sci* 2001, **24**, 635–42.
194 I Agranat, H Caner, J Caldwell, *Nat Rev Drug Discov* 2002, **1**, 753–68.

8
Affinity Chromatography

Gerhard K. E. Scriba

8.1
Introduction

According to the IUPAC definition the term "affinity chromatography" characterizes the particular variant of chromatography in which the unique specificity of analyte and ligand interactions is utilized for the separation. The ligand is immobilized on a solid support, whereas the analyte (a protein in most cases) is specifically adsorbed while passing through the column. Elution from the affinity support is achieved in a separate step using defined conditions. While originally referring to biological or biospecific interactions such as the binding between an enzyme and a substrate or an antibody with an antigen, the term is now also used when ligands of nonbiological origin are employed. Examples are ligands such as boronates, metal ion complexes or synthetic dyes.

The technique may be dated back to 1910 when Starkenstein isolated α-amylase via adsorption onto starch [1], but the modern concept and the term affinity chromatography was introduced by Cuatrecasas et al. in 1968 [2]. At the same time the cyanogen bromide activation of polysaccharides was developed which enabled the coupling of nucleophiles to an affinity matrix [3]. Many other coupling techniques have been developed since, and many activated supports and matrices with bound ligands have been commercialized. It has been estimated that over 60% of the protein purification protocols published involve affinity chromatography [4]. Due to the specificity, target molecules can be isolated from a very complex mixture of often closely related compounds with a purification yield of hundreds to several thousands in a single step. At the same time the compound of interest can also be concentrated. These advantages have led to a widespread application of affinity chromatography in biochemistry, pharmaceutical sciences, as well as clinical and environmental chemistry. While only two papers on the technique are listed in *Chemical Abstracts* and Medline in 1968 when the technique was established, the databases show 1383 (*Chemical Abstracts*) and 666 hits (Medline) in 2001 when searching for the term "affinity chromatography". To name just a few, several books [5, 6], book chapters [7, 8] as well as an entire issue of the *Journal of Biochemical and Biophysical Methods* [9] have been recently published on the tech-

Molecular Biology in Medicinal Chemistry. Edited by Th. Dingermann, D. Steinhilber, G. Folkers

ISBN: 3-527-30431-2

nique. This chapter briefly summarizes the principles and the most important affinity chromatographic methods with regard to the use of these techniques in medicinal chemistry and/or purification of pharmaceutical proteins as well as analytical applications. For a more comprehensive treatment of specific topics the reader is referred to the literature cited above.

8.2 Principles of Affinity Chromatography

While "conventional" stationary phases separate analytes based on charge or hydrophobicity/hydrophilicity, affinity chromatography is based on the specific interaction of a defined region of a biomolecule with a ligand. Biomolecules such as proteins are uniquely characterized with regard to the distribution of charge, hydrophobicity and hydrophilicity resulting from the amino acid residues exposed to the surface or active sites. The ligand, which may be a small molecule or another macromolecule, interacts in a complimentary manner with the respective binding sites. The complex is formed by a combination of electrostatic, hydrogen bonding, hydrophobic and van der Waals' interactions. As often only the active form of a biomolecule exhibits the proper binding sites, the technique may also be used to separate the active form from inactive conformations. Examples of ligand–solute pairs exploited in affinity chromatography are listed in Tab. 8.1.

The principle of affinity chromatography is outlined in Fig. 8.1. The ligand is immobilized to a solid support that is placed in a chromatographic column. Upon application of the sample, only the molecule capable of specific interactions with the ligand is retained on the column while the other components are eluted with the waste fractions. The elution of the bound molecule is achieved by disruption of the solute–ligand interactions due to a change of the eluent, such as increasing

Tab. 8.1 Examples of ligand–solute complexes used in affinity chromatography.

Ligand	*Retained solute(s)*
Antibodies	drugs, hormones, peptides, proteins, cells, viruses
Substrates, inhibitors, cofactors, coenzymes	enzymes, receptors, carrier proteins
Lectins	polysaccharides, glycoproteins, glycolipids, cell surface receptors
Nucleic acids	nucleic acid-binding proteins, complementary base sequence
Protein A, Protein G	immunoglobulins
Triazine dyes	nucleotide-binding proteins and enzymes
Metal chelates	peptides and proteins with histidine, cysteine and/or tryptophan residues, His-fusion proteins
Boronates	saccharides, glycoproteins

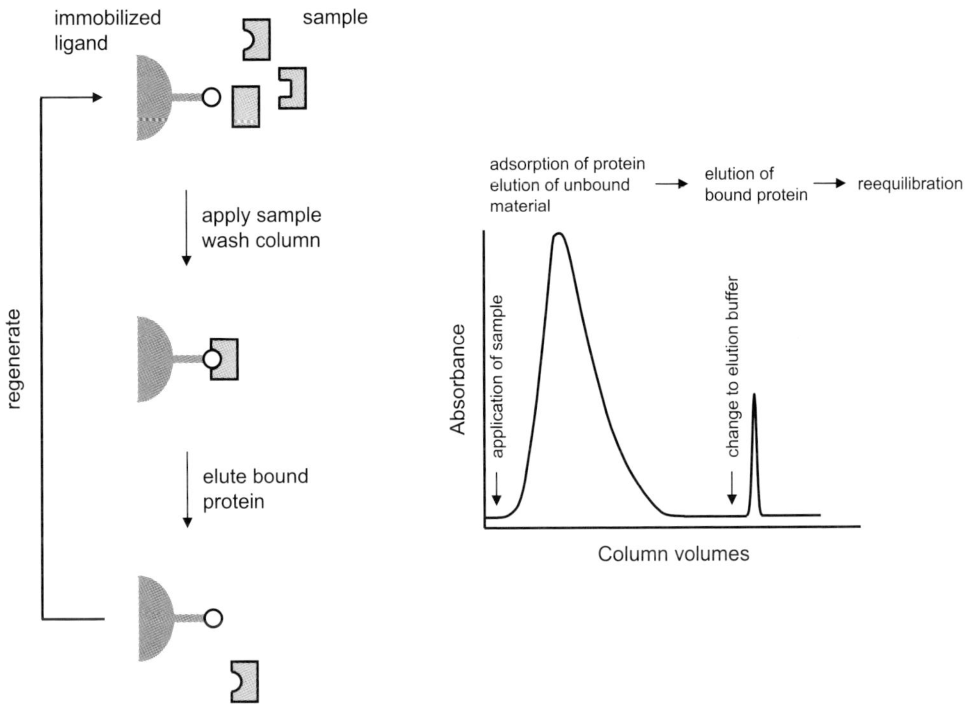

Fig. 8.1 Principles of affinity chromatography.

the ionic strength of the buffer or addition of the ligand to the mobile phase. The affinity column may be reused after suitable regeneration.

The retention and elution of a compound on an affinity column depends on the strength of the solute–ligand interactions, the amount of the immobilized ligand, and the kinetics of the association and dissociation of the complex. Assuming a single binding site resulting in a 1:1 complexation between a solute, S, and a ligand, L, the complex formation can be described by:

$$[S] + [L] \underset{k_D}{\overset{k_A}{\rightleftarrows}} [S \cdot L] \tag{1}$$

$$K_A = \frac{k_A}{k_D} = \frac{[S \cdot L]}{[S] + [L]} \tag{2}$$

$$K_D = \frac{1}{K_A} \tag{3}$$

[S] is the concentration of the solute in the mobile phase, [L] and [S · L] represent the surface concentrations of the ligand and the solute–ligand complex, respectively. k_A and k_D are the rate constants of the association and dissociation reactions, while K_A is the association constant for the complex. The dissociation constant K_D is the reciprocal value of K_A. The retention of the solute at equilibrium is described by:

$$k' = \frac{K_A \cdot m_L}{V_m} = \frac{t_r}{t_m} - 1 \tag{4}$$

where k' is the capacity factor of the solute, m_L is the moles of ligand in the column, V_m the void volume of the column, t_r is the retention time of the solute and t_m is the void time of the column. Thus, the capacity factor of a certain solute depends on the strength of the association constant and the amount of ligand in the column. Assuming an association constant (K_A) of 10^8–10^9 M^{-1} and a concentration of the ligand (m_L/V_m) of 10^{-5} M, k' becomes 1000–10,000, translating to an elution time of the bound analyte up to several days. The only way to achieve the elution within a reasonable time frame is to alter the conditions in such a manner that the association constant is considerably lower under the experimental conditions. This may be obtained either by changing the elution buffer in a single step or in a gradient. For most practical purposes, the association constant K_A should be equal to or exceed 10^5–10^6 M^{-1}, corresponding to a dissociation constant K_D of 1–10 μM, in order to ensure good adsorption of the analyte. However, a K_A in the 10^3–10^6 M^{-1} range may also cause reasonable retardation of the protein of interest if the ligand density is sufficiently high. On the other hand, having interactions exceeding 10^{10}–10^{11} M^{-1} harsh conditions for elution of the bound protein may be required which cause denaturation of the protein. Dissociation constants in the range of 10^{-3}–10^{-5} M (translating to $K_A = 10^3$–10^5 M^{-1}) are generally suitable to elute proteins under mild, nondenaturating conditions.

8.3 The Ligand

The choice of the affinity ligand is of central importance to the affinity matrix as it affects the binding specificity, the immobilization procedure and the costs of the matrix. A good ligand should satisfy the following criteria [7]:

- The ligand must form reversible complexes with the protein to be isolated or separated with an appropriate specificity for the intended application.
- The stability of the complex should be high enough to ensure sufficient chromatographic retardation of the protein.
- The complex should be easy to dissociate by a change of the buffer without affecting the stability of the protein to be isolated or the ligand.
- The ligand must have suitable properties that allow immobilization on the support.
- Ideally, the ligand should also be cheap and easily available.

Ligands may be grouped based on the nature of the interactions into biospecific, biomimetic and pseudospecific ligands [4]. Examples for biospecific ligands are antibodies, substrates or coenzymes, biomimetic ligands comprise the dyes, while the nature of the interactions with metal chelators is pseudospecific. However, ligands are more commonly classified as either monospecific or group specific with further subdivision according to the molecular weight into low-molecular-weight and macromolecular ligands.

8.3.1 Monospecific Ligands

Monospecific ligands such as substrates, hormones or antibodies bind the complementary enzymes, receptors, carrier proteins or antigens in a highly specific manner so that only a single or a very small number of proteins of a particular cell extract or physiological fluid is complexed. Thus, a separate affinity matrix is required for each given protein. For example, vitamin B_{12} binds only to the intrinsic factor from gastric juices or to transcobalamin II from plasma [10]. Antibodies are very useful monospecific macromolecular ligands in affinity chromatography, especially when a compound without an immediate binding site has to be purified. They can be used to isolate virtually anything from peptides to whole cells. As a result of modern hybridoma technology, monoclonal antibodies have largely replaced polyclonal antibodies. In principle, monoclonal antibodies also ensure a constant supply of uniform antibodies with high batch-to-batch reproducibility. Advantages include high binding affinities, in most cases at least 10-fold higher than polyclonal antibodies, resulting in purification factors of several thousand in a single step. Disadvantages are the high costs and the general susceptibility to proteolytic degradation. As a consequence, crude extracts should not be applied directly to affinity columns based on monoclonal antibodies. Generally, monospecific ligands bind more strongly and require harsher elution conditions than group-specific ligands, whose complexes can usually be dissociated under mild conditions. An example of extremely tight binding is the biotin–avidin complex with a K_A of 10^{15} [11].

8.3.2 Group-specific Ligands

Group-specific ligands are compounds that bind a family or a class of related proteins. Low-molecular-weight ligands such as enzyme cofactors and analogs comprise the largest group of ligands currently applied (Tab. 8.2). They may be of biological or nonbiological origin. In particular, adenine nucleotides and analogs such as 5′-AMP and the triazine dyes Cibacron Blue F3G-A and Procion Red HE-3B have been most effectively used for the purification of NAD^+- and $NADP^+$-dependent dehydrogenases and kinases, but other proteins such as albumin or interferons are also bound. Despite the relatively broad specificity, high purification yields were obtained using specific elution protocols or exploiting ternary complex formation using cofactors and substrates.

Tab. 8.2 Examples of group-specific ligands.

Ligand	*Examples of bound molecules*
Low-molecular-weight ligands	
5′-AMP	NAD^+-dependent dehydrogenases, ATP- and cAMP-dependent kinases
Cibacron Blue F3G-A	NAD^+-dependent dehydrogenases, nucleic acid binding proteins, albumin, interferon, α_2-macroglobulin, coagulation factors
Procion Red HE-3B	$NADP^+$-dependent enzymes, aldehyde reductase, dihydrofolate reductase, carboxy-peptidase G, interferon, plasminogen and plasminogen activator
benzamidine	trypsin, thrombin, coagulation factor Xa
arginine	serine proteases, coagulation factors, plasminogen, plasminogen activator
lysine	rRNA, plasminogen and plasminogen activator
Macromolecular ligands	
heparin	coagulation factors VII, IX, XI, XII, XIIa, thrombin, antithrombin III, complement factors, lipases, lipoproteins, RNA and DNA polymerases, steroid receptors, hepatitis B surface antigen, interferon
calmodulin	ATPases, protein kinases, phosphodiesterases, neurotransmitters, interferon
lectins	glycoproteins, membrane proteins, glycolipids, lipoproteins, polysaccharides, hormones, α_1-antitrypsin, interferon
Protein A	polyclonal and monoclonal antibodies
Protein G	polyclonal and monoclonal antibodies

Group-specific macromolecular ligands include lectins, heparin or the bacterial Proteins A and G (Tab. 8.2). Lectins interact with sugars which makes them an excellent tool to purify soluble and membrane glycoproteins such as hormones, plasma proteins, antigens, antibodies or blood group substances. Even organelles and cells may be bound. Lectins differ in their specificity towards saccharides depending on the source of the lectin. Concanavalin A and lentil lectin bind the commonly occurring α-D-mannose and β-D-glucose, and sterically related residues. The two lectins differ in their binding strength with concanavalin A forming the stronger complexes in most cases. Glycoproteins that contain a large number of *N*-acetylglucosamine such as α_2-macroglobulin, α_2-acid glycoprotein or mucins may be purified using wheat germ lectin. The sulfated polysaccharide heparin is often used for the isolation of plasma coagulation factors and other plasma proteins, but also to purify enzymes such as endonucleases or RNA and DNA polymerases as well as steroid receptors. The bacterial Proteins A and G are used most frequently in purification steps of antibodies. Both proteins differ in their specificity towards individual immunoglobulin subclasses from human and animal sources. Protein G is a cell surface protein from group G streptococci while Protein A is isolated from *Staphylococcus aureus*. Recombinant Protein A is also used.

8.3.3 Ligand Development

While displaying high selectivity, ligands of natural origin are often hampered by a low scale-up potential due to their limited availability. A new possibility for the design of highly specific, non-natural affinity ligands is offered by the development of combinatorial chemistry and biology often in combination with computational modeling techniques, X-ray and nuclear magnetic resonance spectroscopy [12].

Phage-display technologies allow the identification of peptide ligands for a given target molecule out of a huge library of different peptides expressed on the surface of bacteriophages. Presentation of the peptide library on the surface of bacteriophages ("phage display") as a fusion of peptide and a phage coat protein allows the physical link between the presented peptide and the DNA sequence coding for its amino acid sequence. Diversity of the peptides/proteins can be introduced by combinatorial mutagenesis of the fusion gene. Extremely large numbers of phages can be constructed, replicated, selected and amplified in a process called "biopanning". The libraries are incubated with a target molecule either as an immobilized target or prior to capture of the complex on a solid support. As in affinity chromatography, noninteracting peptides and proteins are washed off, and the interacting molecules are subsequently eluted. The interacting phage-displayed peptides or proteins can be amplified by bacterial infection to increase their copy number. The screening/amplification process can be repeated to further enrich those library members with relatively higher affinity to the target. The result is a final peptide population that is dominated by sequences that bind the target best. A ligand identified in this way can be subsequently immobilized to a support and applied to the isolation of the target molecule from a sample. Using this approach, ligands derived from domains of Protein A were selected from phage-displayed combinatorial libraries and applied to affinity chromatography isolation of human immunoglobulin A from plasma [13], humanized monoclonal antibodies [14] and recombinant human clotting factor VIII [15].

A similar approach to peptide phage display exploits oligonucleotide-based libraries which allow the identification of specific oligonucleotide sequences, known as aptamers, that bind to a desired target molecule with high affinity and specificity [16, 17]. The high affinity of aptamers is attributed to their ability to "incorporate" molecules into their structure and to integrate into the structure of large molecules such as peptides [18]. While largely unstructured in solution, aptamers fold upon binding of their target molecule. The nucleotide sequences are identified by a process called systematic evolution of ligands by exponential enrichment (SELEX). Binding sequences are isolated from random libraries using an affinity column with the intended target molecule and subsequently amplified by polymerase chain reaction (PCR). This is followed by several further cycles of the same treatment. The result is an oligonucleotide population that is dominated by sequences that bind the target best. A DNA-aptamer specific for human L-selectin was effectively applied in the purification of a recombinant human L-selectin–immunoglobulin fusion protein from Chinese hamster ovary cell-conditioned medium [19].

While very suitable as lead ligands or for small-scale laboratory purification, ligands of biological origin suffer from several general drawbacks for large-scale applications. They require purification of their own as they may be contaminated with host DNA or viruses and show lot-to-lot variability. In addition, sterilization procedures may cause degradation and instability of the ligand may cause contamination of the end-product with potentially toxic and/or immunogenic leachates. Synthetic ligands, on the other hand, are well-defined compounds that can be prepared on a large scale without problems. Combinatorial chemistry in combination with modeling techniques and X-ray as well as nuclear magnetic resonance (NMR) spectroscopy led to the design and development of new ligands. Screening of a combinatorial synthetic peptide library has resulted in dimeric and tetrameric tri- and tetrapeptides that proved to be useful ligands for the isolation of immunoglobulins and monoclonal antibodies [20, 21]. An example where phage display was used for identification of a lead followed by optimization via solid-phase peptide synthesis in combination with molecular modeling led to the development of a pentapeptide as ligand for the binding region of a monoclonal antibody directed against a steroid. The immobilized peptide was subsequently used for the purification of the antibody [22].

De novo design of affinity ligands has also been achieved by retrieving structural information of the target protein from X-ray crystallographic, NMR or homology data and identifying suitable binding sites either as an active site, a site involved in binding a natural ligand or as a region or motif on the protein surface. So far three approaches have been realized [23]: (1) investigation of the structure of natural protein–ligand interactions as a template on which a biomimetic ligand is modeled, (2) design of a ligand complementary to exposed amino acid residues in the target site and (3) directly mimicking natural biological recognition interactions. Examples involving combinatorial chemistry can be found in the *de novo* design of triazine scaffold-based ligands. The template approach was used for designing affinity ligands modeled on Protein A for the isolation and purification of human immunoglobulin G [24, 25]. Design by complementarity was applied to the development of a ligand for a recombinant insulin precursor [26], while mimicking of biological interactions was realized by designing an artificial lectin as an affinity absorbent for glycoproteins [27]. Some of these affinity media have been commercialized.

8.4 The Affinity Matrix

8.4.1 The Support Material

The solid support holding the ligand in the column controls numerous factors such as the immobilization chemistry and flow characteristics. An ideal material should meet several criteria [7]:

- The support should be macroporous to allow free interaction with large molecules.
- The support should be neutral and hydrophilic to prevent unspecific interaction with the matrix.
- The support should contain functional groups suitable for derivatization and attachment of the ligand using a wide variety of chemical reactions.
- The support should be chemically stable under the often harsh conditions during derivatization and regeneration (including sterilization) of the column.
- The support should be physically stable under the hydrodynamic pressure during chromatography.
- The support should be readily available at low cost.

Matrices of natural origin are based on spontaneously gel-forming polysaccharides which comply with most of the characteristics stated above. Agarose, a linear polysaccharide composed of alternating 1,3-linked D-galactose and 2,4-linked 3,6-anhydro-L-galactose, was introduced in the first publication in 1968 [2] and is still the most frequently applied support due to its considerable gel strength and relative biological inertness. The main drawback of native agarose is its chemical and physical instability, but this has been largely overcome by chemical crosslinking. Further support materials include cellulose, crosslinked dextrans, polyacrylamide, polymerized tris(hydroxymethyl)acrylamide (Trisacryl), hydroxyalkyl methacrylate gels, vinyldimethyl azlactone-*N*,*N*′-methylene-bis(acrylamide) copolymers, polystyrene derivatives as well as derivatized silica gel, controlled pore glass beads and ceramics. Mixed agarose–acrylamide and dextran–acrylamide copolymers are also employed.

Based on the support, affinity chromatography may be divided into a low-performance and a high-performance technique. Low-performance affinity utilizes nonrigid gels with large particle sizes. Most of the carbohydrate and polymer based organic materials fall within this category. Due to the relatively low backpressure, the columns may even be operated under gravity flow or using peristaltic pumps. In high-performance affinity chromatography the support consists of a rigid material such as derivatized silica gel, controlled pore glass, hydroxylated polystyrene or highly crosslinked agarose beads with a small particle size that withstand the high pressures applied under standard high-performance liquid chromatography (HPLC) conditions. This technique is mainly employed for analytical applications.

8.4.2 Immobilization of the Ligand

The individual support materials differ also with respect to the derivatization methods available. The majority of procedures for the immobilization of the ligands have been developed for hydroxyl groups present in the carbohydrate-based supports. One of the advantages of agarose is the fact that it retains its macroporous structure in organic solvents. As the majority of organic reactions are performed in organic media, these reactions can be applied to agarose as well. This may be another reason for the popularity of this material.

The immobilization reactions are determined by the functional groups present in the solid support and the ligand as well as the reaction conditions that are tolerated by the support and the ligand. A protein generally requires milder reaction conditions than a small organic molecule. In most cases, the procedure consists of three steps: (1) Activation of the support to create reactive functional groups, (2) coupling of the ligand and (3) deactivation of residual reactive groups of the support by a large excess of a low-molecular-weight compound. Examples of common immobilization methods used in affinity chromatography are summarized in Tab. 8.3. In most cases an electrophilic group is introduced into the matrix which reacts with nucleophilic groups of the ligand such as amino, thiol or hydroxyl groups.

Tab. 8.3 Immobilization methods in affinity chromatography.

Functional group	*Immobilization chemistry*
Amino	cyanogen bromide (CNBr)
	carbonyl diimidazole (CDI)
	1-ethyl-3-[3-dimethylaminopropyl) carbodiimide (EDC)
	N-hydroxy succinimide esters
	reductive amination
	epoxy or bisoxirane activation
	tosyl chloride
	tresyl chloride
	divinylsulfone
	2-fluoro-1-methylpyridinium toluene-4-sulfonate (FMP)
	azlactone activation
	cyanuric chloride
Hydroxyl	carbonyl diimidazole (CDI)
	epoxy or bisoxirane activation
	divinylsulfone
	cyanuric chloride
Thiol	iodoacetyl
	bromoacetyl
	maleimide
	pyridyl disulfide
	epoxy or bisoxirane activation
	divinylsulfone
	2-fluoro-1-methylpyridinium toluene-4-sulfonate (FMP)
	5-thio-2-nitrobenzoic acid (TNB)-thiol
Aldehyde	hydrazide method
	reductive amination
Carboxyl	carbonyl diimidazole (CDI)
	1-ethyl-3-[3-dimethylaminopropyl) carbodiimide (EDC)
Active hydrogen (CH)	diazonium formation
	Mannich condensation

Compiled from references [7, 28, 29].

Coupling between electrophilic groups of the ligand and nucleophilic groups on the matrix is less frequently employed. Oxidation of diols with potassium periodate can be performed to create an aldehyde prior to the immobilization of the ligand. The applied reactions depend on the support employed, silica gel and glass beads requiring the introduction of suitable groups before coupling can be achieved. Details on the activation and coupling procedures can be found in the literature [5–7, 28, 29]. Blocking of the remaining active groups after attachment of the ligand may be necessary in order to avoid reactions with sample components. In addition, the blocking groups should exhibit low nonspecific binding. Small molecules such as aminoethanol, mercaptoethanol, mercaptoethylamine or tris(hydroxymethyl)aminomethane are often used. Many activated supports are also available from numerous suppliers who also supply detailed information on the coupling procedures.

The choice of the immobilization chemistry plays an important role for the activity of the final affinity column. Improper or too general reaction conditions may result in steric hindrance or a variety of possible orientations of a ligand on the gel. So-called multisite attachment refers to attachment through more than one reactive group of the ligand. Although a more stable linkage compared to a single-site attachment is obtained, multisite attachment may lead to distortion or denaturation of the active site of protein ligands. Random orientation means attachment through functional groups located in different sites of the ligand. Although a certain amount of the ligands is attached in the proper orientation, others may be bound differently resulting in steric hindrance or blocking of the active site. Overall, reduced ligand activity is observed for these materials. Improper orientation also results in reduced activity of the material due to hindrance or blocking of the active site.

Even if proper orientation is ensured, steric hindrance may still occur when the ligand is attached too close to the surface of the support or to other ligands. Steric hindrance by neighboring molecules can be reduced by lowering the ligand coverage while a suitable spacer enlarges the distance between the support and the ligand. Different spacer molecules and chemistries are available [28, 29]. The spacer may be either attached first to the support followed by coupling of the ligand or the spacer can be incorporated into the ligand with subsequent immobilization on the matrix. The spacer should not contribute to unspecific interactions with any of the sample components. Frequently employed spacer molecules in affinity chromatography include 6-aminocapronic acid, diaminodipropylamine, 1,6-diaminohexane, ethylene diamine, succinic acid, glutaraldehyde and aminated epoxides.

Attachment of the ligands to the support should be achieved via a bond as stable as possible to prevent leakage of the ligand during the operation of the column. However, under certain circumstances, it may be useful if the ligand is attached via a bond that may be cleaved if required. An example is the disulfide bond that can be formed and cleaved under mild conditions by thiol–disulfide exchange reactions [30].

Designing an affinity medium includes the three key steps: (1) selection of the matrix, (2) selection of the ligand and (if necessary) a spacer, and (3) the coupling method. Many standard protocols are available from the literature or from manufacturers; however, particularly for new ligands, specific methods have to be examined in order to ensure good binding capacity of the final column. Affinity matrices containing various group-specific ligands are also available from several commercial suppliers.

A new immobilization strategy is offered by monolithic stationary phases characterized by high mass-transfer velocities. In this case a ligand is conjugated to one of the monomers of a polymerization mixture. *In situ* polymerization yields an affinity monolith with a tailored pore structure and defined ligand presentation [31].

8.5 Principles of Operation

Binding and elution conditions for a given analyte depend on the nature of the interactions between the ligand and the target molecule. Too low affinity will result in poor yields as the analyte may be washed through or leak from the column. Too strong binding can reduce yields due to insufficient elution. Sample preparation methods prior to the application of the sample onto the column should ensure that the sample is particle free and does not contain components interfering with the binding of the target molecule. The column also has to be equilibrated with the binding buffer before use. When working with weak interactions that reach the equilibrium slowly, lower flow rates have to be applied compared to strong, quickly equilibrating affinity interactions. As affinity chromatography is a binding technique, sample volume does not affect the separation as long as the experimental conditions ensure sufficiently strong binding between ligand and analyte.

After all unwanted material has been washed off the column, elution of the bound compounds can be performed. No general scheme is applicable for all affinity media. The ligand–analyte complexes are maintained by a combination of electrostatic and hydrophobic interactions as well as hydrogen bonds. Conditions that weaken these interactions lead to desorption of the bound compounds. In addition, the stability of the ligand and the target molecules have to be considered carefully, especially in the case of proteins. Often a compromise has to be made between the harshness of the conditions in order to achieve good elution and the risk of denaturing a protein.

The elution conditions in affinity chromatography may be classified into selective and nonselective conditions. Selective conditions are used in the case of group-specific interactions. In these cases, the buffer contains compounds that compete either for binding to the affinity ligand on the column or for binding to the target molecule. Selective elution is often carried out under essentially the same solvent conditions as those used for sample application.

Nonselective elution conditions are often employed for highly specific interactions. These include changes of the pH, the ionic strength or the polarity of the elution buffer. The pH determines the degree of ionization of charged groups of the ligand and/or the bound analyte so that a change in pH directly affects the binding sites, by reducing their affinity or altering the conformation. A steep decrease of the buffer pH is the most frequently used method for eluting strongly bound material. The chemical stability of the protein and the affinity matrix limit the pH that can be used. An increase of the ionic strength of the buffer will elute proteins that are bound by predominantly electrostatic interactions. This is a mild desorption technique, particularly for enzymes, as 1 M NaCl is usually sufficient. Continuous or step gradients can also resolve different proteins. Reduction of the polarity can be achieved by addition of dioxane or ethylene glycol to the eluent. In the case of strong binding dominated by hydrophobic interactions, elution can be achieved by using so-called chaotropic agents that affect the structure of the proteins such as urea, guanidine hydrochloride or detergents. However, these are drastic conditions that are likely to denature proteins. Low concentrations of detergents are often included in the entire purification process when handling membrane proteins or during the purification of antibodies in order to suppress nonspecific hydrophobic adsorption or aggregation. Combinations of selective and nonselective conditions are also applied.

The reuse of affinity adsorbents depends on the nature of the sample as well as the stability of the ligand and the matrix with regard to elution and cleaning conditions. Careful re-equilibration of the column before reuse is generally necessary. In addition, carbohydrate-based supports and protein ligands are prone to microbial degradation and column fouling may also occur from residual cell debris, lipids and proteins. These must be effectively removed. In addition, sanitation procedures to remove bacteria, viruses and pyrogens can be required, but such harsh conditions may not be applicable for protein ligands, for example. Storage of the columns at low temperatures is recommended.

8.6 Modes of Affinity Chromatography

The broad scope of affinity interactions has generated a large variety of special techniques that are recognized by their own nomenclature (Tab. 8.4). Not all of them are based on the unique specificity between a ligand and the target molecule, but rather on more general phenomena; others describe specific applications. The major modes will be briefly discussed.

8.6.1 Bioaffinity Chromatography

Bioaffinity chromatography or biosorption affinity chromatography refers to techniques that use a molecule of biological origin as affinity ligand. This was the first type of affinity chromatography [2] and represents the most diverse category of the technique. Tables 8.1 and 8.2 list several ligands, and many more have been described in the literature. They include small molecules like amino acids, pep-

Tab. 8.4 Modes of affinity chromatography and related techniques.

Bioaffinity chromatography
Immunoaffinity chromatography
Lectin-affinity chromatography
Avidin–biotin-affinity chromatography
Dye-ligand-affinity chromatography
Immobilized metal-ion-affinity chromatography
Hydrophobic interaction chromatography
Thiophilic chromatography
Hydrophobic charge induction chromatography
Covalent affinity chromatography
Weak-affinity chromatography
Receptor-affinity chromatography
Affinity repulsion chromatography
Perfusion affinity chromatography
Centrifuged affinity chromatography
Molecular imprinting chromatography
Membrane affinity chromatography
Affinity partitioning
Affinity precipitation
High-performance affinity chromatography
Affinity electrophoresis
Affinity capillary electrophoresis

tides, carbohydrates, nucleotides cofactors, substrates and inhibitors, as well as large molecules such as proteins, polysaccharides and nucleic acids. Some formats utilizing a specific interaction are also recognized as a separate method, such as lectin-affinity chromatography based on the interaction between lectins and carbohydrates and avidin–biotin-affinity chromatography that utilizes the highly specific binding of biotin by avidin.

8.6.2
Immunoaffinity Chromatography

Immunoaffinity chromatography can also be regarded as a subcategory of bioaffinity chromatography as the ligand, an antibody, is usually of biological origin. Methods that immobilize antigens for antibody purification are sometimes included under this term. Depending on the nature of the antigen, the ligand may or may not be of biological origin. The high selectivity of antigen–antibody interactions and the ability to produce antibodies against a wide range of solutes has made immunoaffinity chromatography a powerful tool for purification and analysis of compounds. Examples include drugs, hormones, peptides, enzymes, recombinant proteins, antibodies, receptors, organelles, cells and viruses. In addition to preparative purposes, immunoaffinity chromatography has received increasing attention as an analytical technique.

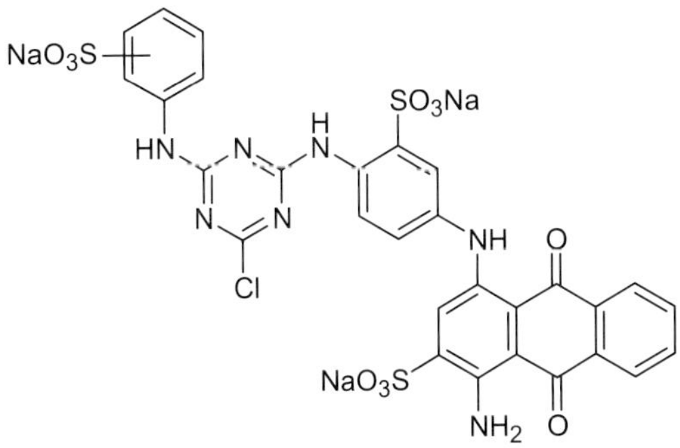

Fig. 8.2 Structure of Cibacron Blue F3G-A.

8.6.3
Dye-ligand-affinity Chromatography

Dye-ligand-affinity chromatography utilizes triazinyl-based reactive textile dyes as ligands. While many natural affinity ligands are expensive due to high costs of production and purification, only available in limited quantities and other drawbacks, synthetic dyes can be obtained in large quantities at a low price enabling easy scale-up of an affinity column. In addition, these compounds exhibit an excellent chemical and biological stability. Cibacron Blue F3G-A (Fig. 8.2) was the first to be used as an affinity ligand [32] and many related dyes have been used since. The dyes are immobilized on the support via reactive chloride(s) on the heterocyclic triazine moiety. It has been shown in many kinetic studies that triazinyl dyes interact with the binding sites of nucleoside ligands such as NAD^+, NADH, $NADP^+$, NADPH, ATP or GTP. To date, molecules isolated by dye-ligand-affinity chromatography include enzymes such as kinases, dehydrogenases, endonucleases, hydroxylases, phosphodiesterases and many others, but also clotting factors, interferons or serum albumin.

Although textile dyes can interact with proteins with a remarkable degree of specificity, in some cases their interaction with a large number of seemingly unrelated proteins inevitably compromises their binding specificity and endows them with a serious drawback. Further concerns over purity, leakage and toxicity of the commercial dyes have limited their use in the biopharmaceutical area. Rational molecular design techniques including X-ray crystallography, molecular modeling and combinatorial chemistry have led to the development of "biomimetic dye-ligands" [33] that display a high specificity for a given protein, but also retain most of the advantages of the commercial dyes.

8.6.4
Immobilized Metal Ion-affinity Chromatography (IMAC)

In IMAC or metal-chelate-affinity chromatography, a technique introduced in 1975 [34], a metal ion is complexed with an immobilized chelating agent. The metal ion forms coordination bonds with amino acids with electron donor groups such as histidine, tryptophan or cysteine. Since the interactions between the metal ions and the amino acid side-chains are readily reversible, they can be utilized for adsorption and subsequently be disrupted using mild, nondenaturating conditions. Commonly used chelators include iminodiacetic acid, nitrilotriacetic acid, carboxymethylated aspartic or *N,N,N′*-tris(carboxymethyl)ethylenediamine. Selectivity in IMAC is determined by the metal ions which can be divided based on their preferential reactivity towards nucleophiles into hard, intermediate and soft ions. Hard metal ions are Fe^{3+}, Al^{3+} and Ca^{2+} that show preference for oxygen. Soft metal ions such as Cu^{+}, Hg^{2+} or Ag^{+} prefer sulfur, while the intermediate ions Cu^{2+}, Ni^{2+}, Zn^{2+} and Co^{2+} coordinate nitrogen, oxygen and sulfur.

Despite the fact that many amino acid residues with free electron pairs can participate in binding, the actual protein retention in practical IMAC is based primarily on the availability on histidyl residues on the protein surface. The commonly used intermediate metal ions can be aligned with increasing specificity: $Cu^{2+} < Ni^{2+} < Zn^{2+} \leq Co^{2+}$ [35]. Thus, more proteins will be retained by Cu^{2+} and the number will be decreased if Ni^{2+}, Zn^{2+} or Co^{2+} are employed due to their more stringent requirements for binding sites. Examples of proteins that have been purified by IMAC due to exposed histidine residues are human serum proteins, interferon, lactoferrin and tissue plasminogen activator [36]. The preference for histidine residues has led to the development of histidine-affinity tags for recombinant peptides and proteins [35, 36]. These tags contain multiple histidines fused to either the C- or N-terminus. While several tags consisting of 1–8 peptide repeats containing histidine have been evaluated, today the by far most widely used histidine tags consist of six consecutive histidine residues and commercial expression vectors for these tags are available. Histidine tags seem to be compatible with all expression systems used today so that histidine-tagged proteins can be successfully produced in prokaryotic and eukaryotic cells, intracellularly or as secreted proteins.

As a rather general technique for purifying proteins, IMAC is one of the most popular methods for the isolation of recombinant proteins at present. It has been estimated that more than 50% of the recombinant proteins expressed in prokaryotic host cells are purified by IMAC [35]. Polyhistidine extensions usually do not compromise the biological function of a protein and do not exhibit high immunogenity. However, cleavage of the tag may be required, especially in the case of therapeutic proteins, and this may not always be easily accomplished. Enzymatic cleavage is generally applicable, but the purification of the resulting mixture of cleaved fusion protein, enzyme, uncleaved protein and other cleavage products may present a problem. In addition, the toxicity of the metal ions leaching from the IMAC column has to be considered. Moreover, metal ions often catalyze oxidation reactions that may lead to oxidative degradation of proteins. The sulfur-containing amino acids cysteine and methionine are especially prone to oxidation reactions.

8.6.5 Hydrophobic Interaction Chromatography (HIC)

HIC is based on the association of hydrophobic groups on the surface of proteins with weakly hydrophobic ligands immobilized on chromatographic supports.

HIC is carried out in aqueous solutions and this distinguishes the technique from reversed-phase chromatography where organic solvents are used for the elution of the proteins. In addition, the ligand density in reversed-phase chromatography is higher so that the stationary phase can be regarded as a continuous phase, whereas in HIC the ligands interact individually with the target molecules. Commonly employed ligands include butyl, octyl or phenyl residues often linked via a glycidyl ether. Oligoethyleneglycol and hydroxypropyl ligands are also used.

The retention of the solutes is controlled by the type and concentration of the salt used and organic additives that change the polarity of the solvent. Addition of ethylene glycol to the buffer changes the overall structure of the water slightly towards a structure resembling an organic phase, thereby decreasing the interaction between the protein and the ligands. The influence of the salts on the binding can be considered in terms of the salt's position in the Hoffmeister series, ranking anions and cations as lyotropic or chaotropic. Lyotropic salts such as ammonium sulfate or potassium phosphate will decrease the water solubility of proteins and thereby increase their retention in HIC; chaotropic salts will desorb the proteins. In practice, desorption of a protein will be achieved by lowering the salt concentration. Proper choice of the salt is an important factor determining the specificity of the separations [37]. After loading the sample on the column using a buffer with a salt concentration that is sufficiently high to effect binding of the target protein, but sufficiently low to avoid precipitation of the protein, unbound sample components are washed off the column. Subsequently, the salt concentration is decreased in steps or continuously in the form of a gradient to achieve selective elution of the protein of interest. Other factors that need to be controlled in order to obtain reproducible results include pH, temperature and additives that influence the stability of the protein. Generally, HIC is a mild method due to the stabilizing influence of the salts and recoveries are often high.

Thiophilic interaction chromatography can be considered a HIC-related method. The matrix carries linear ligands containing sulfur atoms. The original thiophilic structures were obtained from vinylsulfone activated supports and 2-mercaptoethanol. Both the functional groups, sulfone and thioether, present in the ligand structure appear to act in a cooperative manner to exert protein binding affinity. More recent developments, especially on an industrial scale, involve mostly ligands composed of heterocycles with nitrogen and/or sulfur atoms [38]. This mode of separation has been mainly applied to antibody purification from natural and recombinant sources due to a certain specificity of the sorbents towards this class of proteins.

In its behavior, thiophilic interaction chromatography resembles HIC as adsorption of proteins is promoted by highly concentrated salts and elution occurs when the salt concentration is lowered. However, several differences between the techniques indicate distinct modes of protein adsorption [38]. Hydrophobic interactions do not occur with thiophilic sorbents since thioethylsulfone structures do not possess a pronounced hydrophobicity. In contrast to HIC, the interaction between ligand and protein is relatively independent of the temperature and sodium chloride promotes desorption of the proteins. While albumin is strongly adsorbed by hydrophobic ligands, thiophilic sorbents show a distinct preference for immunoglobulins.

Hydrophobic charge induction chromatography (HCIC) is a recent technique of protein adsorption employing weak base or acid ligands that are uncharged at physiological pH [39]. Adsorption is based on hydrophobic association, while desorption results from ionic repulsion between the ligand and the adsorbed protein when the pH is changed in an appropriate direction. Enhanced selectivity for antibodies can be obtained when using ligands that combine a traditional thiophilic site with heterocyclic ionizable structures such as 4-mercaptoethyl pyridine [40]. This ligand (pK_a 4.8) is uncharged under neutral conditions and becomes charged by lowering the pH to 4–5, resulting in desorption as a consequence of the repulsion forces between the sorbent and the positively charged antibody.

8.6.6 Covalent Chromatography

Covalent chromatography, unlike other affinity techniques, involves the formation and breaking of covalent bonds between the solute and the ligand. Protein ligands are often attached via N- or C-groups with the aim of forming stable bonds that are resistant to the chromatographic conditions and difficult to release without destroying the protein. The only functional group with available chemistry forming a stable covalent bond that can also be split under mild conditions is the thiol group [30]. Thus, covalent chromatography has been almost exclusively applied to the isolation of thiol-containing peptides. This technique is also referred to as thiophilic adsorption chromatography by some authors.

While thiol groups can participate in a number of chemical reactions due to their nucleophilicity, the so-called thiol–disulfide exchange is utilized in covalent chromatography. The reaction can be described by:

$$R{-}S{-}S{-}R + 2R_1{-}SH \rightarrow 2R{-}SH + R_1{-}S{-}S{-}R_1 \qquad (5)$$

The reaction may be regarded either as a nucleophilic displacement or as a redox process as the oxidation state of the individual sulfur atoms changes in the course of the reaction. Thiol–disulfide exchange plays an important role in biochemical processes such as the biosynthesis of proteins, protein aggregation and ligand–receptor interactions. In practice, the ligand is a mixed reactive disulfide consisting of an aliphatic thiol and an aromatic thiol such as 2-thiopyridine. Upon application, a thiol-containing protein becomes immobilized on the solid phase as a result of the thiol–disulfide reaction. Desorption of the protein is achieved using a large excess of an aliphatic low-molecular-weight thiol in the mobile phase. More recently, matrices with thiolsulfonates and thiolsulfinates have also been employed for the reversible immobilization of thiols [30]. As can be deducted from the reaction mechanism, covalent chromatography serves to isolate proteins with native thiol groups or proteins where thiol groups can be generated such as by reduction of existing disulfide bonds.

8.6.7
Membrane Affinity Chromatography

The term membrane affinity chromatography refers to the format of the support. While in "normal" affinity chromatography columns are employed to hold the affinity matrix during the chromatographic process, membranes are employed in the present mode. The limitations of packed-bed column liquid chromatography such as time-consuming packing process, high pressure drop in the column and mass transfer restrictions can be overcome using membranes. In porous beads mass transfer is restricted by the slow diffusion of the analytes into the (dead end) pores of the material. Membranes consist of consecutive pore channels leaving film diffusion to the membrane surface in the interior of the channels as the only transport resistance. Film diffusion is usually orders of magnitude faster than pore diffusion, shifting the limitations of the adsorption process to the ligand–solute interactions [41]. These are identical to those in column chromatography and generally very fast processes. The macroporous structure and the fact that even stacked membranes are thin compared to packed beds in chromatography results in low pressure drops allowing increased flow rates and consequently shorter separation times. The required binding capacity is achieved by using membranes with a sufficient internal surface area. The advantage of high flow rates can turn into a disadvantage as the contact time between of solution may be too short. A single sheet of membrane can possess a large theoretical capacity, but the apparent capacity may be much lower due to the limited sample interaction time. This limitation is further enhanced when eluting the target molecule in the case of strong affinity interactions as the contact time may not be sufficient to produce rapid elution of bound material resulting in diluted recovered sample. Approaches to overcome these limitations include the use of multiple layers of membranes in the case of insufficient binding or recycling of the elution buffer in the case of slow desorption.

The chemical composition of the membranes is quite variable. Materials based on cellulose and its derivatives, polyamide, polyacrylamide, polyacrylonitrile, polysulfone, polyvinylidene, polystyrene, polypropylene, polyethylene and others, are employed. These materials can be conjugated with virtually any affinity ligand with chemistries similar to those in the column format [42, 43]. Likewise, all affinity chromatographic modes can be performed in the membrane format. Activated and ligand-coupled membranes are commercially available, and many applications have been described [43, 44].

8.7
Applications of Affinity Interactions

Affinity interactions may be exploited for both, preparative and analytical applications. For preparative purposes the format will be low-pressure chromatography or membrane technology in most cases, while HPLC, capillary electrophoresis or biosensors will be used primarily for analytical applications. In addition, analytical affinity techniques are not only used to determine a specific molecule in pharmaceutical, chemical, clinical or environmental analysis, but also for ligand identification and to study ligand–solute interactions.

8.7.1
Purification of Pharmaceutical Proteins

In addition to blood products and proteins of human or animal origin, recombinant proteins now comprise a significant and growing number of pharmaceutical drugs due to the advances in biotechnology. Regardless of the source, animal or human tissue, eggs, blood plasma, mammalian or bacterial cell cultures, or transgenic animals or plants, all these products are of biological origin and require extensive purification. Depending on the origin, impurities include bacteria, viruses, lipopolysaccharides (endotoxins), prions, DNA, RNA, host cell proteins, carbohydrates, phospholipids, etc. According to the criteria of the US Food and Drug Administration (FDA) as well as the European Agency for the Evaluation of Medicinal Products (EMEA), the final biotechnological product has to be a well-characterized product with defined purity, efficacy, stability, pharmacokinetics, pharmacodynamics, toxicity and immunogenity [23]. It has to be analyzed for contaminants mentioned above originating from the source; however, therapeutic proteins often comprise a mixture of isoforms originating from posttranslational modification as well as misfolding, aggregation and other chemical or physical degradation reactions.

Protein purification may vary from a simple one-step precipitation procedure to large-scale validated production processes. In most cases more than one step will be necessary to remove all impurities. Downstream processing of feedstocks, from whatever source, usually involves three major steps: capture, intermediate purification and polishing, sometimes also referred to as CIPP [45]. Following the preparation of the extract and clarification from cell debris by procedures such as filtration or liquid/liquid extraction, the objective of the capture step is the isolation, concentration and stabilization of the protein. The bulk of the impurities are removed during subsequent intermediate purification, while the final high purity is achieved in the final polishing step. All three steps often involve chromatographic procedures; the polishing step is sometimes composed of two (chromatographic) purification procedures. Affinity chromatography is most effectively used during capture and intermediate purification, often reducing these two steps to one if highly selective ligands are employed. Other frequently used chromatographic techniques in protein purification include gel filtration, ion-exchange chromatography or reversed-phase chromatography. Knowledge about the properties of the target protein and the impurities will help during purification development.

One of the crucial points in the production of protein pharmaceuticals is the transfer of bench-scale purification protocols. As much as 50–80% of the total production costs of a protein drug are due to purification procedures [23]. Many of the well-established biochemical or molecular biological techniques do not necessarily comply with labor- and cost-effective process technology. Thus, process economics, quality assurance and regulatory compliance have shifted conventional purification protocols involving precipitation with salt, temperature, pH or high-molecular-weight polymers to selective techniques based on affinity chromatography. However, this approach also has its drawbacks. When using an affinity ligand of natural origin, the FDA requires that the ligand used for the production of a biological product meets the same requirements as the final product itself. Thus, extensive purification of the ligand may be necessary. In addition, inevitable leaching of the ligand will occur during chromatography. In addition to toxicological aspects of the ligand, one has to keep in mind that the ligand will be associated with the product as selective complex formation is an inherent feature exploited in the purification. These complexes may be difficult to remove from the product. On the other hand, it is possible to remove pyrogens and endotoxins by affinity chromatography [46]. Biological ligands may be expensive and thus increase production costs. Some of these disadvantages may be overcome using synthetic ligands, which can also be highly selective [23]. Current limitations concerning the sample throughput in affinity columns due to high back pressure and mass-transfer kinetic limitations may be overcome using membrane technology [41].

8.7.1.1 Plasma Proteins

Therapeutic proteins such as albumin, coagulation factors and immunoglobulins have been isolated from human blood for many years. Despite the advent of recombinant products, isolation from human blood is still the most important source of such products. Although traditional plasma fractionation methods based on ethanol precipitation steps are still employed for the production of the current therapeutic albumin and immunoglobulin G preparations, affinity chromatography has been introduced in the industrial preparation of the coagulation factors VIII, IX and XI, the von Willebrand factor, fibronectin, antithrombin III, inter-α-trypsin inhibitor and plasma-derived protein C [47, 48]. While affinity chromatography is used to capture factor VIII and factor IX using immobilized heparin or monoclonal antibodies as affinity ligands, and antithrombin III through capture by a heparin ligand from relatively crude preparations, the technique is used as a downstream polishing step in the case of the other products. Sometimes affinity chromatography is used to remove impurities such as fibronectin from the van Willebrand factor using a gelatin ligand. Van Willebrand factor does not bind to gelatin and is recovered in the breakthrough fraction, while fibronectin is adsorbed to the column. No commercial process uses affinity chromatography for the isolation of albumin, although the protein can be bound to dye ligands.

8.7.1.2 Recombinant Proteins

Therapeutic recombinant proteins, also called biotech pharmaceuticals, are produced by recombinant DNA technology. They comprise a large market of an estimated US$16–17 billion in 2001 [49]. Currently, more than 50 different recombinant proteins are on the market, while over 40 are in phase III and more than 60 in phase II clinical trials. Monoclonal antibodies are the fastest growing segment with over 20 marketed products, and more than 20 and 45 products in phase III and phase II, respectively [49]. Biotech pharmaceuticals include insulin, growth hormones, erythropoietin, interferons, interleukins, granulocyte colony-stimulating factor, enzymes, clotting factors, vaccines and monoclonal antibodies. While not being disclosed due to propriety rights, the production of recombinant proteins certainly includes affinity chromatography steps. Moreover, the technique is most widely used on the laboratory scale in research and development.

Two general modes may be distinguished. In the first mode the "native" protein interacts with the ligand, while the second approach employs so-called affinity tags fused to the protein of interest in which the tag binds to the ligand while the protein itself does not exhibit binding. Fusion protein technology has become an important tool in recombinant protein production. A variety of expression vectors with different tag sequences have been designed for fusion to almost any target protein that can be cloned and expressed in host cells. Interactions that have been utilized as a basis of fusion protein affinity include interactions of enzymes with substrates or inhibitors, protein, protein binding carbohydrate–protein interactions, biotin-binding domains, antigen–antibody interactions, charged amino acids for charge-based methods and poly(His) residues for metal–chelate binding (Tab. 8.5) [36, 50]. Each tag has its own advantages and disadvantages, and selection has to be made based on the protein of interest and the affinity technique to be used.

Tab. 8.5 Common affinity tags for recombinant proteins.

Fusion tag	*Size*	*Fusion to C- or N-terminus*
Enzymes		
glutathione-S-transferase	26 kDa	N
β-galactosidase	116 kDa	C, N
Polypeptide-binding domains		
IgG-binding domain of Protein A	14–31 kDa	N
IgG-binding domain of Protein G	28 kDa	C
Carbohydrate-binding domains		
maltose-binding protein	40 kDa	N
cellulose-binding domain	111 amino acids	N
Streptavidin-biotin binding domains		
biotin-binding domain	8 kDa	N
Strep-tag	9 amino acids	C, N
Antigenic epitopes		
FLAG™-tag	8 amino acids	N
Charged amino acids		
poly(Arg)	5–15 amino acids	C
poly(Asp)	5–16 amino acids	C
Metal–chelate interaction		
poly(His)	4–9 amino acids	C, N

Compiled from references [36, 50].

Demands for a tag are that the segment should not interfere with the natural folding of the protein or inhibit its natural function or active site and it should be exposed on the surface of the protein for maximum interaction with the ligand. In addition, the tag should be suitable for mild and preferably inexpensive affinity procedures, and should be removed easily. Currently, the poly(His)-tag is predominantly used in recombinant protein technology as it appears to be compatible with all expression systems used today [36]. Using suitable affinity beads, e.g. magnetic beads, tag strategies may also be used in robotic sample processing for automated protein purification.

The question of whether to remove a tag or not depends on the final use of the protein. For characterization of proteins or diagnostic purposes it may be possible to leave the tag attached; however, for pharmaceutical applications precise removal of the fusion peptide will be necessary in most cases to ensure product authenticity. In addition, tags may be immunogenic, specifically the FLAG™-tag which has been designed to be immunogenic. Thus, suitable cleavage sites for proteases have been engineered into the fusion proteins. However, this requires secure removal of the enzyme and the cleaved tag from the protein of interest, and imposes additional purification steps and costs. In addition, proteolytic degradation of the product may occur. Thus, while being an excellent laboratory technique, the use of affinity tags may not be suitable for large-scale industrial production of proteins in all cases.

Apart from their use in the affinity purification of recombinant proteins, some tags can also be applied to specific detection in immunochemical assays under denaturing conditions of Western blots as well as in the native state in assays such as ELISA, providing a useful analytical tool for monitoring expression levels of the protein or for tracing a protein during its purification process.

8.7.2
Analytical Applications

Affinity chromatography may be used in different formats to separate and analyze specific solutes. A chromatographic separation of a substance that interacts with an affinity ligand can be the first step in quantifying its concentration in solution. Systems involving high-performance affinity chromatography (HPAC) and affinity capillary electrophoresis have been constructed to speed this process and automate the measurements. Traditional immunoassays can be designed around immobilized affinity ligands using formats such as beads, gels or plates. The electronic detection of affinity interactions led to the development of affinity biosensors that not only determine the concentrations of the analytes, but also quantitative characteristics of affinity interactions. In addition, the technique can be adapted to high-throughput screening (HTS) formats.

Affinity techniques are also employed for the specific removal of components from a sample that interfere with the performance of the analytical system. One of the major difficulties in analyzing the peptides and proteins in human serum is the wide range of concentrations of proteins in the sample. Human serum albumin ranges from approximately 55 to 70%, while γ-immunoglobulin ranges from about 8 to 25%. Removal of these proteins can be achieved with commercially available "affinity depletion" matrixes containing antibodies or Protein G which significantly reduces the protein load of the sample allowing the analysis of minor components.

"Affinity extraction" refers to the situation where an analyte is specifically isolated and concentrated by affinity chromatography from a sample prior to analysis by another method either off-line or online. The off-line mode typically employs low-performance affinity columns packed in disposable cartridges. In the online mode, a HPLC column containing the affinity matrix is connected via a switching valve to the analytical column.

The analytical techniques are increasingly used in drug discovery to identify ligands (ligand fishing) and lead structures, and in early ADME (adsorption, distribution, metabolism and excretion) of drug development in order to study ligand–analyte interactions, such as drug–protein binding, including stereochemical aspects, to determine complex stoichiometry, to estimate binding constants and binding kinetics. Of course, the methods can also be utilized in later stages of drug development for the analysis of the drug candidates.

8.7.2.1 **HPAC**

HPAC combines the specificity of ligand analyte interactions with the speed, efficiency and automation of HPLC. Virtually any ligand can be immobilized on supports able to withstand the high pressures applied in HPLC. Examples of affinity supports that are suitable for these conditions include modified silica gel or glass, azlactone beads and hydroxylated polystyrene media. The stability and efficiency of these supports allow them to be used with standard HPLC equipment. Although HPLC instrumentation makes HPAC more expensive to perform compared to low-performance affinity chromatography, HPAC can be a powerful tool for analytical determinations in addition to its potential for high-speed purification of compounds.

Probably the best-known example of an analytical application of affinity chromatography is the determination of glycated hemoglobin in clinical chemistry as a marker of the time-averaged mean blood glucose levels in diabetes control. The glycated variant can be separated from the native hemoglobin using immobilized aminophenylboronic acid as ligand, and relative ratios of glycated and nonglycated fractions are quantified by measuring the absorbance of hemoglobin at 414 nm. Further clinical applications were recently summarized [51]. HPAC using immobilized Protein A as ligand has been applied to the analysis of immunoglobulins from various matrices such as plasma or cell culture media [52], and lectin affinity chromatography was exploited for the determination of isoenzymes or the glycoform microheteregeneity of glycoproteins [53]. However, the majority of recent applications of analytical affinity chromatography were published in the area of drug protein interactions. Specifically, HPAC was used to analyze drug protein (albumin) binding including stereospecific aspects [54, 55], but the determination of enzyme substrate interactions is also possible [56].

In the zonal elution technique, the most frequently used method to study drug interactions with a protein stationary phase in HPAC, the change of the retention time or the capacity factor k' is monitored as a function of the concentration of a competing agent in the mobile phase. The frontal elution method is another technique used in HPAC. A solution with a known concentration of the solute is constantly administered to the column. As the solute binds to the ligand, the ligand becomes saturated and the amount of compound eluting from the column increases, forming a breakthrough curve [54]. Qualitative and quantitative information can be obtained from these experiments including comparison of the relative binding affinities, competitive displacement of an analyte by other compounds or the determination of equilibrium and rate constants.

High-performance immunoaffinity chromatography (HPIAC) is becoming an increasingly popular tool for the analysis of biological and synthetic compounds. In addition to its use for the direct detection of analytes in chromatographic assays and immunoextraction of analytes, immobilized antibody or antigen columns are applied to various types of immunoassays. This approach, also known as chromatographic immunoassay, is particularly suitable for determining trace analytes that do not produce a readily detectable signal by themselves. This is overcome in chromatographic immunoassays using a labeled antibody or analog of the analyte for indirect detection. Common formats include competitive assays, sandwich assays and one-site immunometric assays [57]. Fab fragments can also be used instead of antibodies. The labels consisted primarily of enzymes or fluorescent tags.

As in "stationary" assays, the principle of competitive binding immunoassays involves incubation of the sample with a fixed amount of labeled analyte in the presence of a limited amount of antibodies. The assay can be performed by simultaneous injection detecting either the label that elutes with the wash fractions or analyzing the labeled compound that elutes in the dissociation step. In sequential injection, the label is injected after the sample and analyzed either in the nonretained or retained material. In the displacement format, the column is first saturated with the labeled analog followed by injection of the sample, analyzing the amount of labeled compound that is washed off the column by local dissociation and reassociation of the complex. In sandwich or two-site immunometric assays, two different antibodies that bind to the analyte are used. The first is attached to the solid support and used for the extraction of the target molecule, while the second antibody contains a label and is added to the sample either before or after injection on the column. The second antibody places a label on the analyte to allow its detection. In one-site immunometric assays, the sample is incubated with a known excess of labeled antibodies. This mixture is subsequently applied to a column with an immobilized analog of the analyte that serves to extract antibodies that were not bound to the sample analyte. Detection is performed by measuring either the nonretained labeled or the antibody that is later dissociated from the column in the elution step. Many compounds, including enzymes, hormones and proteins, but also synthetic molecules such as theophylline, have been analyzed by HPIAC [57]. Postcolumn immunodetection for monitoring a specific solute eluting from an HPLC column by employing a postcolumn reactor and an immobilized antibody or antigen column connected to the exit of the analytical HPLC column is also possible.

8.7.2.2 Affinity Capillary Electrophoresis

Affinity capillary electrophoresis is an analytical technique in which the migration patterns of interacting molecules in an electrical field are recorded. The method is a specific variant of capillary electrophoresis used to identify specific binding and to estimate binding constants provided that the interactions are not inhibited by the separation conditions and that complexed molecules can be distinguished from uncomplexed species [58]. Thus, the interaction must change the molecular size, shape, charge or any other parameter responsible for a change in the electrophoretic separation system. Advantages of using capillary electrophoresis include the wide range of analytes that can be analyzed, speed of analysis, and the low consumption of sample and chemicals. A disadvantage is the often unsatisfactory concentration detection limit, especially when using UV detection. In principle, the method can also be transferred to chip technology.

Ligands may interact with the analyte either before and/or during electrophoresis runs. The latter case represents the classical affinity electrophoresis method that is analogous to affinity chromatography. For the analysis of weak- to intermediate-affinity interactions, ligands are used in free solution or immobilized to the capillary wall and the analytes move through constant concentrations of the ligands. Upon interaction with the ligand a shift of the migration time of the analyte depending on the concentration of the ligand is observed. This method does not require a known concentration of the sample nor do the compounds have to be pure. Compound libraries may be analyzed. Strong interactions are typically studied using capillary electrophoresis as a method to separate and quantitate bound and free analyte in a preincubated mixture in one step.

Affinity capillary electrophoresis has been used to analyze the binding constants, binding kinetics and binding stoichiometry of numerous ligand–analyte complexes including peptide–peptide, peptide–oligonucleotide, peptide–heparin, metal ion–protein and antigen–antibody interactions [58]. Migration shift affinity capillary electrophoresis can be utilized to screen compound mixtures for affinity towards a ligand, especially when coupled to mass spectrometry [59], and for structure–function studies. In addition, affinity capillary electrophoresis is a valuable tool for the determination of drug–protein binding, including stereochemical aspects.

A growing area of analytical applications is capillary electrophoresis-based immunoassays, also called affinity probe capillary electrophoresis. In these techniques an analyte is allowed to react with an antibody and the resulting complex is separated from the free analyte by capillary electrophoresis. Two distinct modes have been described. In the direct or noncompetitive assay, the affinity probe typically labeled with a fluorophore is added to the sample for binding of the target molecule. Detection of the complex is a direct measure of the target in the sample solution. In the competitive assay, a fluorescent-labeled target and a limited amount of antibody are added to the sample, allowing the labeled target and the target in the sample to compete for the antibody. The resulting complexed and free labeled molecules are separated and detected, the relative amounts of free and bound labeled compound depending on the concentration of the analyte in the sample. Capillary electrophoresis immunoassays have proven as a versatile method for a variety of analytes and sample matrices [60]. Drugs, hormones, peptides, proteins, toxins, and also viruses and bacteria have been determined. Analytes can be measured directly in serum or urine as well as cell and tissue samples. Interestingly, the first immunoassay for scrapie, the prototype of spongiform encephalopathy, was also based on capillary electrophoresis [61]. In addition, the immunoassays may be transferred to chips allowing extremely short analysis times. An example is the development of a chip-based capillary electrophoresis immunoassay for the determination of theophylline in serum where the electrophoretic run is completed within 35 s [62].

8.7.2.3 Affinity Biosensors

Advances in immobilization techniques and innovations in bioelectronics have produced new detection devices based on the specific interaction between receptors, enzymes or antibodies with their specific target molecules, creating new detection devices for such diverse fields as diagnostics, therapeutics, process control, environmental and waste control, and kinetic analysis of the interaction of biological substances. Although not based on chromatographic or electrophoretic separation principles, these sensors employ affinity interactions for their function. In addition, biosensors are increasingly used in many areas of drug discovery including target identification, ligand fishing, lead selection, assay development, early ADME studies and quality control [63].

Affinity biosensors are affinity ligand-based biosensor solid-state devices in which the ligand–target molecule interaction is coupled to a transducer that converts the biological reaction into an electronic output signal as schematically shown in Fig. 8.3. The immobilized ligand may be any (bio)specific molecule that interacts with a target molecule. It can even be an intact living cell that acts with specific substances in solution with which it comes in contact. The transducer "senses" the subtle chemical changes that take place between the immobilized ligand and the

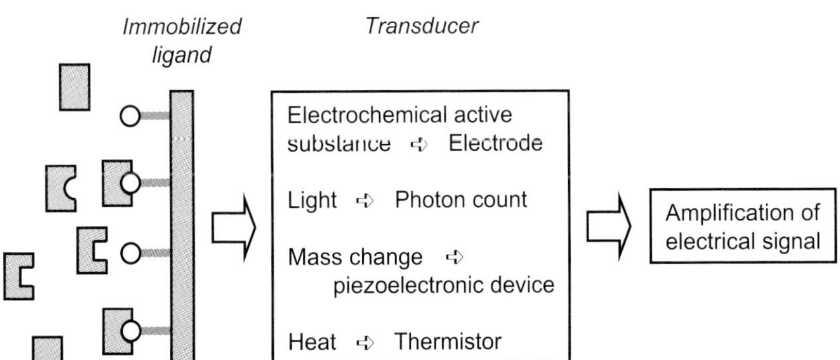

Fig. 8.3 Schematic principle of biosensors.

analyte. The detection process may involve electrical effects such as potentiometric changes or amperometric fluctuations, optical effects such as light absorption or scattering, changes in density or mass, as well as thermal differences. The electronic signals are subsequently amplified and recorded. Among the detection systems, the optical system of surface plasmon resonance is probably the most popular system and several commercial instruments are available [64]. Surface plasmon resonance detects changes in the refractive index in the immediate vicinity of the surface layer of the sensor chip due to the binding of molecules to surface immobilized ligands. This change is monitored in real-time to determine the concentration of a bound analyte, the affinity of the analyte towards the ligand, and the association and dissociation kinetics of the interaction (Fig. 8.4). An extremely wide range of molecules can be analyzed ranging from cells and bacteriophages to low-molecular-weight compounds with relative masses as low as 200 Da. High- and low-affinity interactions may be investigated.

The chemistries for ligand immobilization are comparable to those used for the preparation of affinity columns. A specific procedure includes the chemisorption of sulfide-containing spacers to gold surfaces. The stability of these linkages exceeds that of covalent silane bonds with glass. Subsequent attachment of the ligand can be achieved through reactive groups of the spacer. Dithiobis-succinimidyl propio-

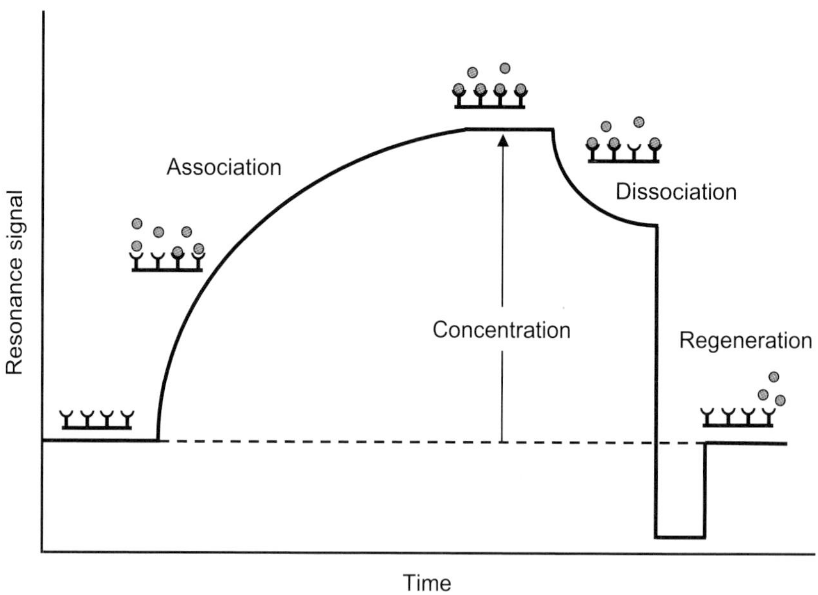

Fig. 8.4 Typical binding curve of an optical biosensor. The analyte concentration is derived from the response level at equilibrium of the curve, while binding kinetics can be calculated from the observed association rate and dissociation rate, respectively. Regeneration of the ligand by step elution is usually required as many interactions have considerable half-lives. The binding cycle is repeated using different concentrations of the analyte to generate robust data for fitting to the appropriate binding algorithm.

nate is a popular spacer for chemisorption-based immobilization protocols. In addition, chips with preactivated surfaces are commercially available.

Virtually any molecule and even whole cells can be immobilized in biosensors for the analysis of a given target compound. There are also approaches where transmembrane receptors have been reconstituted in liposomes or lipid layers attached or adsorbed to the chip surface. When analyzing for specific molecules, so-called immunosensors containing immobilized antibodies are frequently employed in clinical and environmental chemistry, but also for the detection of bacteria and viruses in monitoring of food and hygiene processes [65]. There has also been a report on the development of a biosensor consisting of a monoclonal antibody with affinity for pyrogenic lipopolysaccharide for the detection of pyrogens in water for injection having a lower detection limit than the existing regulatory limits of the United States Pharmacopeia [66]. Due to several advantages over antibodies, "aptasensors" containing immobilized aptamers appear to be an emerging class of biosensors [67].

At present, optical biosensors are primarily used in drug development to confirm hits from fluorescence-, chemoluminescence- or radiometric-based HTS and for lead optimization. In addition to identifying binding molecules, biosensors allow the investigation of the kinetics of the biospecific interactions between ligands and target molecules in real-time [68]. Biosensors can also be used to identify binding partners during ligand fishing experiments including combinatorial libraries or the screening of tissue and cell extracts. Once a binding substance has been found, amino acid sequencing can be performed by mass spectrometry using the biosensor as a micro-preparative affinity purification device [63]. Interfacing biosensors with matrix-assisted laser desorption/ionization time-of-flight mass spectrometry can also be used to study sequential binding events, competitive binding systems as well as the functional and structural characterization of proteins including posttranslational modifications in proteomics [69]. Integration of mass spectrometry and biosensor detection can provide extra information about target function and binding specificity that can accelerate the development of new drugs. In addition to drug-target identification, biosensors were utilized to monitor antibody and cytokine production, to determine drug–protein binding in early ADME assessment, and in quality control as the assays can be validated according to the International Conference on Harmonization (ICH) guidelines of Technical Requirements for Registration of Pharmaceuticals for Human Use. The technology has further been applied to serological analysis of clinical samples to determine antibody titers [63].

References

1 EV Starkenstein, *Biochem Z* 1910, **24**, 191–209.

2 P Cuatrecasas, M Wilchek, CB Anfinsen, *Proc Natl Acad Sci USA* 1968, **61**, 636–43.

3 R Axen, J Porath, S Ernbach, *Nature* 1967, **214**, 1302–4.

4 CR Lowe, *Adv Mol Cell Biol* 1996, **15B**, 513–22.

5 P Bailon, GK Ehrlich, W-J Fung, W

Berthold, *Affinity Chromatography, Methods and Protocols, Methods in Molecular Biology 147*, Humana Press, Totowa, NJ, 2000.
6 T Kline, ed., *Handbook of Affinity Chromatography*, Dekker, New York, 1993.
7 J Carlsson, J-C Janson, M Sperrman, in: J-C Janson, L Ryden, eds, *Protein Purification: Principles, High-Resolution Methods and Applications*, 2nd edn, 375–442, Wiley, New York, 1998.
8 J Pearson, in: G Subramanian, ed., *Bioseparation and Bioprocessing*, vol. 1, 113–24, Wiley-VCH, Weinheim, 1998.
9 *J Biochem Biophys Methods* 2001, **49**, 1–743.
10 RH Allen, PW Majerus, *J Biol Chem* 1972, **247**, 7709–17.
11 EA Bayer, M Wilchek, *Methods Biochem Anal* 1980, **26**, 1–45.
12 CR Lowe, *Curr Opin Chem Biol* 2001, **5**, 248–56.
13 J Rönnmark, H Grönlund, M Uhlen, P-A Nygren, *Eur J Biochem* 2002, **269**, 2647–55.
14 GK Ehrlich, P Bailon, *J Biochem Biophys Methods* 2001, **49**, 443–54.
15 K Nord, O Nord, M Uhlen, B Kelly, C Ljungquist, P-A Nygren, *Eur J Biochem* 2001, **268**, 4269–77.
16 C Tuerk, L Gold, *Science* 1990, **249**, 505–10.
17 A Ellington, J Szostak, *Nature*, 1990, **346**, 818–22.
18 T Hermann, D Patel, *Science* 2000, **287**, 820–5.
19 TS Romig, C Bell, DW Drolet, *J Chromatogr B: Biomed Sci Appl* 1999, **731**, 275–84.
20 G Palombo, M Rossi, G Cassani, G Fassina, *J Mol Recognit* 1998, **11**, 247–9.
21 G Fassina, M Ruvo, G Palombo, A Verdoliva, M Marino, *J Biochem Biophys Methods* 2001, **49**, 481–90.
22 A Murray, RG Smith, K Brady, S Williams, RA Badley, MR Price, *Anal Biochem* 2001, **296**, 9–17.
23 CR Lowe, AR Lowe, G Gupta, *J Biochem Biophys Methods* 2001, **49**, 561–74.
24 R Li, V Dowd, DJ Stewart, SJ Burton, CR Lowe, *Nat Biotechnol* 1998, **16**, 190–5.
25 SF Teng, K Sproule, A Hussain, CR Lowe, *J Mol Recognit* 1999, **12**, 67–75.
26 K Sproule, P Morrill, JC Pearson, SJ Burton, KR Hejnaes, H Valore, S Ludvigsen, CR Lowe, *J Chromatogr B* 2000, **740**, 17–33.
27 UD Palanismy, DJ Winzor, CR Lowe, *J Chromatogr B* 2000, **746**, 265–81.
28 PDG Dean, WS Johnson, FA Middle, *Affinity Chromatography: A Practical Approach*, IRL Press, Oxford, 1985.
29 GT Hermanson, KA Mallia, PK Smith, *Immobilized Affinity Ligand Techniques*, Academic Press, San Diego, 1992.
30 J Carlsson, F Batista-Viera, L Ryden, in: J-C Janson, L Ryden, eds, *Protein Purification: Principles, High-Resolution Methods and Applications*, 2nd edn, 343–73, Wiley, New York, 1998.
31 R Hahn, A Podgornik, M Merhar, E Schallaun, A Jungbauer, *Anal Chem* 2001, **73**, 5126–32.
32 P Roschlau, B Hess, *Hoppe-Seyler's Z Physiol Chem* 1972, **353**, 441–3.
33 CR Lowe, SJ Burton, NP Burton, WK Alderton, JM Pitts, JA Thomas, *Trends Biotechnol* 1992, **10**, 442–8.
34 J Porath, J Carlsson, I Olsson, G Belfrage, *Nature* 1975, **258**, 598–9.
35 GS Chaga, *J Biochem Biophys Methods* 2001, **49**, 313–34.
36 V Gaberc-Porekar, V Menart, *J Biochem Biophys Methods* 2001, **49**, 335–60.
37 K-O Eriksson, in: J-C Janson, L Ryden, eds, *Protein Purification: Principles, High-Resolution Methods and Applications*, 2nd edn, 283–309, Wiley, New York, 1998.
38 E Boschetti, *J Biochem Biophys Methods* 2001, **49**, 361–89.
39 SC Burton, DRK Harding, *J Chromatogr A* 1998, **814**, 71–81.
40 E Boschetti, *Trends Biotechnol* 2002, **20**, 333–7.
41 J Thömmes, M-R Kula, *Biotechnol Prog* 1995, **11**, 357–67.

42 DK Roper, EN Lightfoot, *J Chromatogr A* 1995, **702**, 3–26.
43 H Zou, Q Luo, D Zhou, *J Biochem Biophys Methods* 2001, **49**, 199–240.
44 E Kline, *J Membr Sci* 2000, **179**, 1–27.
45 *Protein Purification Handbook*, Amersham Biosciences, Little Chalfont [Online publication].
46 FB Anspach, *J Biochem Biophys Methods* 2001, **49**, 665–81.
47 T Burnouf, M Radosevich, *J Biochem Biophys Methods* 2001, **49**, 575–86.
48 T Burnouf, H Goubran, M Radosevich, *J Chromatogr B* 1998, **715**, 65–80.
49 A Loffet, *J Peptide Sci* 2002, **8**, 1–7.
50 A Einhauer, A Jungbauer, *J Biochem Biophys Methods* 2001, **49**, 455–65.
51 DS Hage, *Clin Chem* 1999, **45**, 593–615.
52 S Ohlson, *Trends Biotechnol* 1989, **7**, 179–86.
53 PR Satish, A Surolia, *J Biochem Biophys Methods* 2001, **49**, 625–40.
54 DS Hage, *J Chromatogr B* 2002, **768**, 3–30.
55 DS Hage, J Austin, *J Chromatogr B* 2000, **739**, 39–54.
56 R Freitag, *J Chromatogr B* 1999, **722**, 279–301.
57 DS Hage, *J Chromatogr B* 1998, **715**, 3–28.
58 NHH Heegaard, RT Kennedy, *Electrophoresis* 1999, **20**, 3122–33.
59 Y-H Chu, YM Dunayevskiy, DP Kirba, P Vouros, BL Karger, *J Am Chem Soc* 1996, **118**, 1201–4.
60 NHH Heegaard, RT Kennedy, *J Chromatogr B* 2002, **768**, 93–103.
61 MJ Schmerr, AL Jenny, MS Bulgin, JM Miller, AN Hamir, RC Cutlip, KR Goodwin, *J Chromatogr A* 1999, **853**, 207–214.
62 NH Chiem, DJ Harrison, *Clin Chem* 1998, **44**, 591–8.
63 MA Cooper, *Nat Rev Drug Discov* 2002, **1**, 515–28.
64 CL Baird, DG Myszka, *J Mol Recognit* 2001, **14**, 261–8.
65 PB Luppa, LJ Sokoll, DW Chan, *Clin Chim Acta* 2001, **314**, 1–26.
66 JA Taylor, G Bareett, W Lorch, DC Cullen, *Anal Lett* 2002, **35**, 213–25.
67 CK O'Sullivan, *Anal Bioanal Chem* 2002, **372**, 44–8.
68 M Favash, EM Towler, TJ Fisher, *Curr Opin Biotechnol* 1998, **9**, 97–101.
69 RW Nelson, D Knedelkov, KA Tubbs, *Electrophoresis*, 2000, **21**, 1155–63.

9
Nuclear Magnetic Resonance-based Drug Discovery

Ulrich L. Günther, Christina Fischer and Heinz Rüterjans

9.1
Introduction

Drug design faces new challenges with the large amounts of proteins produced in the post-genomic era. Improved and faster structure determination by X-Ray crystallography and nuclear magnetic resonance (NMR) spectroscopy provides the basis for new possibilities in rational structure-based drug design. Hence, screening of proteins against libraries of small molecules has become an important issue. While high-throughput screening (HTS) methods allow hundreds of thousands of scans in a short time, there is an increasing need for smart screening technologies that can detect compounds which slip through the broad filter of HTS. Smart screening which may identify aspects of protein function becomes increasingly important as new targets are derived from genomic databases. Although X-ray analysis can determine protein structures without any size limitation faster than NMR spectroscopy, the latter has a strong advantage for a more detailed analysis of ligand interactions. Despite the complexity of some NMR parameters, there are sensitive and easily accessible parameters that are subject to changes in corresponding spectra of proteins after ligand binding. Many of these parameters could be utilized to develop fast screening protocols. There are two principal types of techniques used in NMR-based ligand screening. In structure–activity relationship (SAR) by NMR, the resonances of the protein are observed. Changes in the protein structure induced by ligand binding which cause chemical shift changes are utilized to detect the location of the interaction with the ligand in the protein. The technique is used for initial screening and for subsequent refinement. SAR by NMR requires isotopic labeling of the protein and is usually limited to proteins with a molecular weight of below 30–40 kDa. This limit has recently been extended to proteins of molecular weights of about 100 kDa by using novel labeling strategies.

Alternatively, the NMR spectrum of the ligand can be observed. This second class of techniques provides binding information about the ligand. Many NMR parameters of the ligand are utilized to detect the interaction with the protein. Some of these techniques allow the mapping of binding epitopes, thus providing

Molecular Biology in Medicinal Chemistry. Edited by Th. Dingermann, D. Steinhilber, G. Folkers

ISBN: 3-527-30431-2

important information for drug design. Other ligand-based methods have greater potential in HTS.

The two approaches are complementary, providing information about the two different sides of the interaction. The first part of this chapter describes SAR by NMR and the second part techniques for observing ligands. Several review articles have described different aspects of NMR-based screening [1–7]. A recent review gives a comprehensive view of the field with a large amount of technical detail [8] for NMR experts. It is the intention of this chapter to provide a general overview over the most important techniques used in NMR screening.

9.2
SAR by NMR

SAR by NMR, which was first proposed in 1996 by Fesik et al. [9], is based on the observation of chemical shift changes in ^{1}H-^{15}N-heteronuclear single quantum spectroscopy (HSQC) spectra caused by ligand interactions. The signals in ^{1}H-^{15}N-HSQC spectra represent protein backbone amides (and NH_2 groups in Asn and Gln side-chains). Since amide ^{1}H and ^{15}N chemical shifts are highly sensitive to the conformation of the amino acids, and electrostatic interactions of the backbone NH and CO atoms [10], small rearrangements in proteins initiated by ligand binding can cause significant chemical shift changes. It is important to define a standard threshold for the value of chemical shift perturbations. Hajduk scored chemical shift changes $\Delta\delta$ as significant if the overall value for both spectral dimensions exceeded 0.1 ppm [11]:

$$\Delta\delta = \sqrt{(\Delta\delta(^1\text{H}))^2 + (0.2 \cdot \Delta\delta(^{15}\text{N}))^2} > 0.1 \text{ ppm} \quad (1)$$

This assumption is almost identical to our previous definition (12):

$$\Delta\delta = 5 \cdot |\Delta\delta(^1\text{H})| + |\Delta\delta(^{15}\text{N})| > 4.5 \text{ ppm} \quad (2)$$

Both relations represent approximately a chemical shift perturbation with a value similar to the typical width of a signal in a HSQC spectrum.

Figure 9.1 shows example HSQC spectra obtained for increasing concentrations of ligands. The two spectra were recorded for the N-terminal SH2 domain of phosphadidylinisitol-3-kinase using two different ligands, phosphotyrosine (ptyr, left) and a peptide ligand that binds with a micromolar dissociation constant [12]. Clearly, the two different ligands cause a different set of changes in the spectrum representing different interactions of the ligand with the protein. The larger peptide (EEEpYMPME-NH_2) causes additional effects compared to Ac-ptyr-NH_2 and larger chemical shift perturbations.

In SAR by NMR, several ligands are added simultaneously to one protein and chemical shift perturbations are deconvoluted in subsequent experiments with single ligands. The process of lead discovery and ligand optimization is depicted in

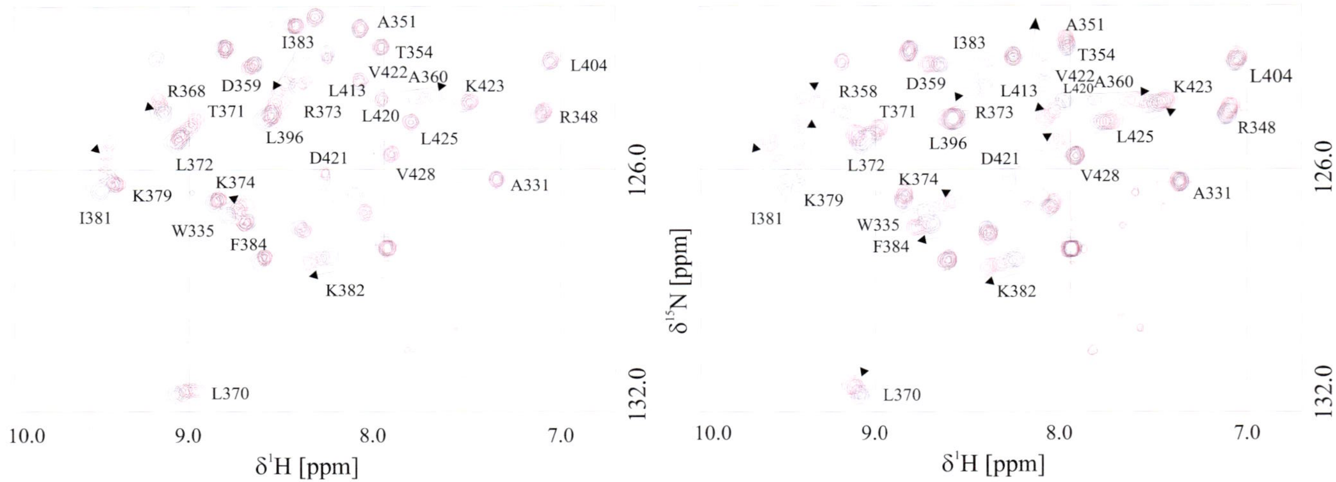

Fig. 9.1 HSQC spectra of the p85 N-SH_2 of PI3-kinase recorded for different concentrations (blue → red) of ptyr (left) and a high-affinity peptide (right). The indicated changes of chemical shifts are proof of characteristic differences between the two ligands. Fewer signals are affected for ptyr and chemical shift perturbations are smaller.

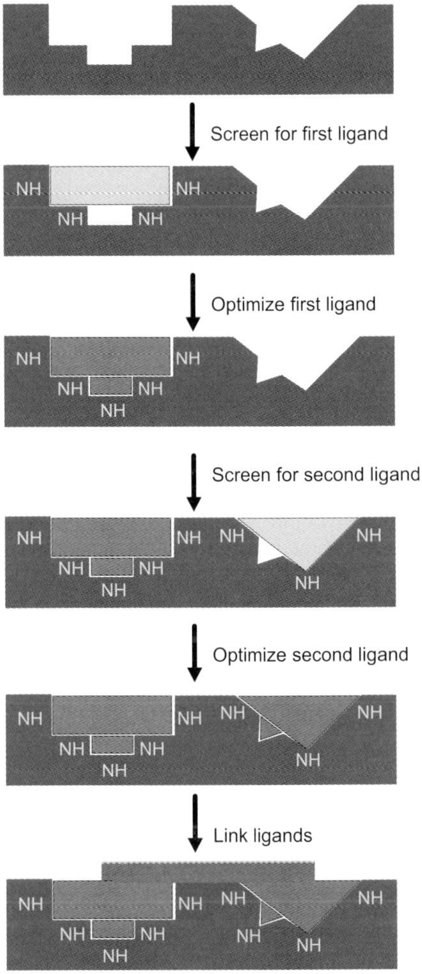

Fig. 9.2 Schematic representation of SAR by NMR. Initial screening detects low-affinity lead compounds for two binding sites. After the affinity of the lead compounds has been optimized, the two fragments are linked in a manner that preserves binding at both sites. Adapted after [9].

Fig. 9.2. Initial screening yields low-affinity ligands that subsequently have to be optimized. Leads for two different binding sites must be identified in the initial screening process. Two optimized fragments are linked to form a larger ligand with increased affinity for the protein. The combination with structural information about the protein aids the synthetic adaptation of the ligand to the protein-binding site. The original publication of SAR by NMR by Fesik et al. [9] presented binding experiments with FK506-binding protein (FKBP). A series of low-affinity compounds were identified for two different binding sites. Four binding analogs were compared to probe the structural requirements. Finally, two molecules were tethered using linkers of different length. Inhibitors with nanomolar affinity were obtained in this process.

The principle of SAR by NMR was commonly known and occasionally utilized long before it was introduced as a screening method. In some earlier experiments, SH2 domain interactions were probed using HSQC and total correlation spectroscopy (TOCSY) spectra [12–14]. However, Fesik's publication on SAR by NMR initiated the development of NMR screening as a technology to discover low-affinity lead compounds.

For initial screening, the signals in the HSQC spectrum are usually not assigned to the amino acid residues of the protein. Thus the information obtained from initial SAR by NMR runs is solely of a qualitative nature to determine whether any residue is involved in an interaction with a ligand present in the solution. For the refinement stage, it is important to have the signals of the HSQC assigned to the residues. Backbone assignments of proteins are nowadays routine operations in NMR laboratories and can be achieved within few days to weeks for proteins with a molecular size below 30 kDa [15, 16]. With assignments available, it is possible to map the binding region in the structure where a ligand interacts with the protein. Although the residues which show effects in SAR by NMR are not necessarily in direct contact with the ligand, the chemical shift perturbation pattern usually provides information that is important for drug design. Secondary effects caused by changes in the electrostatic environment of the protein backbone or by other conformational rearrangements in the protein provide equally important information for drug design.

The example shown in Fig. 9.1 will be used to demonstrate the potential of the method to assess the contribution of different sections of a ligand to the interaction with the protein. SH2 domain interactions with phosphopeptides have originally been described by a prong and socket model (17) where the peptide binds with its phosphotyrosine in one pocket of the protein and with the +1 and +3 residues after phosphotyrosine in a second binding pocket surrounded by large loops. The chemical shift perturbations observed for the ptyr interaction are displayed in blue on the ribbon diagram of the structure of the p85 N-SH2 domain in Fig. 9.3(A, see p. 248). These effects include the main Arg residues which coordinate the phosphotyrosine phenyl ring and surrounding residues. Extending the ligand to Ac-pYM-NH_2 (Fig. 9.3B) shows additional effects caused by the M in +1 position after the ptyr (red) while most chemical shift perturbations seen for ptyr are preserved (blue with identical values of $\delta\Delta$, cyan with altered $\delta\Delta$). Extending the bound ligand to EEEpYMPME-NH_2 causes additional effects in the second binding pocket (red in the left part of the structure shown in Fig. 9.3C). A few additional chemical shift perturbations are observed in the pytr binding region and the value of most chemical shift changes in the ptyr binding pocket is again altered (cyan) compared to those for ptyr. The extension of chemical changes for the high-affinity peptide reflects the binding regions where the peptide interacts with the protein. In addition, the changes in the ptyr region show that the fit of the ptyr depends on the interactions on the other side of the protein. This view has been further confirmed by studying the binding of other peptides and synthetic ligands [12]. This example demonstrates that it can be useful to start with peptide ligands derived from natural binding partners in SAR by NMR. In the case of the SH2 domain, the small peptide was derived from the middle T antigen. Small peptides are often good starting points for drug design. Figure 9.4 (see p. 249) depicts a possible scheme how SAR by NMR can be utilized when natural binding partners are available as a starting point. Splitting the peptide into segments, optimizing the individual subunits and re-linking the modified pieces is a commonly used approach.

9.2.1 Sample Preparation

Protein samples for SAR by NMR must be overexpressed and ^{15}N-labeled. This is routinely achieved by either expressing the protein using ^{15}N-labeled algal lysate or by using M9 minimal medium with $^{15}NH_4Cl$ as the sole source of nitrogen. The techniques used for isotopic labeling are reviewed in Chapter 10. $^{15}NH_4Cl$ is available at a relatively low cost which allows protein labeling in large quantities for SAR by NMR. Samples utilized for SAR by NMR should have a concentration of 0.3 mM or above to obtain good HSQC spectra in 10–30 min. The ligands added to the protein solution must have a slightly higher concentration each (1 mM). This imposes a limit on the technique because the total concentration of ligands should not exceed 5–10 mM. Hence, the number of compounds is limited to about 10 per sample. The maximum throughput for such a system is approximately 100 samples and 1000 compounds a day. Problems may arise due to the low solubility of organic molecules (often below 1 mM).

Protein solutions should be buffered at pH 5–7. At higher pH values the exchange of NH protons with water protons may cause significantly decreased signal intensities. Between 5 and 10% D_2O is commonly added to the protein solution for the deuterium lock used to stabilize the magnetic field. Stock solutions of ligands must be prepared with carefully adjusted pH to avoid pH-induced chemical shift changes. A high concentration of salts in the ligand solution must also be avoided. It is advisable to prepare stock solutions in dimethylsulfoxide (DMSO) in order to obtain concentrations of around 100 mM. This allows the addition of small volumes (3–5 μl) to the buffered protein solution. For compounds that are not soluble in water, DMSO may also be added to the protein solution. Most proteins are still active with 20–30% DMSO present in the buffer solution.

9.2.2 Quantitative Evaluation

9.2.2.1 Binding Affinities

Series of HSQC spectra obtained for different ligand concentrations bear significant information about the local binding affinity. The amount of chemical shift change observed in HSQC spectra for a protein titrated with a ligand depends on the affinity and the reaction rates for a particular ligand. The picture shown in Fig.

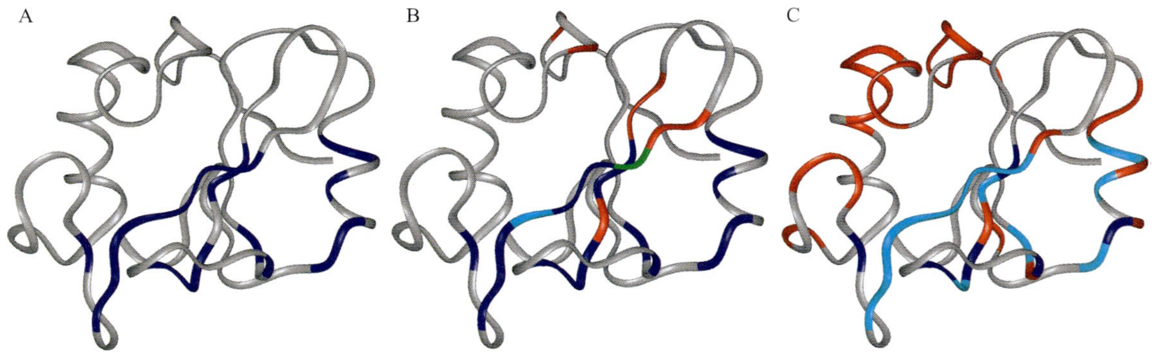

Fig. 9.3 Ribbon diagrams of the p85 N-SH2 of PI3 kinase with color-coding indicating chemical shift perturbations for different ligands. (A) Ac-pY-NH_2, (B) pYM-NH_2 and (C) EEEpYMPME-NH_2. (A) Chemical shift changes for the Ac-pY-NH_2-SH2 complex. Residues are colored in blue when they experience a chemical shift perturbation of both nuclei (eq. 2) of at least 0.45 p.p.m. In (B) and (C), chemical shifts which were not altered compared to (A) are marked in blue, red marks depict new chemical shift changes not observed in AcpY-NH_2 and the green is used for residues which showed chemical shift changes in (A), but not in (B) or (C).

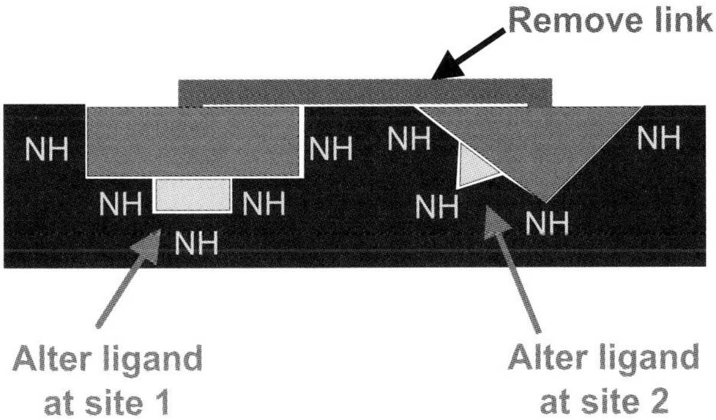

Fig. 9.4 Schematic representation of ligand design using SAR by NMR starting from natural binding partners of a protein. The shape of the bound molecule is gradually altered at both binding sites and the link is removed to study the binding independently at both sites. A high-affinity ligand may be obtained by tethering the two fragments with an alternative linker.

9.3 where peaks move upon the addition of ligand is typical for the fast exchange of the free and bound form of the protein.

The relevant exchange rate is the rate of the dissociation (k_{off}) of the protein–ligand complex. For simple first-order binding reactions of the type $P + L \underset{k_{\text{off}}}{\overset{k_{\text{on}}}{\rightleftharpoons}} PL$, the dissociation constant is $K_D = k_{\text{off}}/k_{\text{on}}$. Consequently low affinity is correlated to high rates. Peak shifts for increasing ligand concentration are only observed for low-affinity interactions ($K_D > 10^{-6}$ mM).

For ligands with high affinity where the dissociation rate of the complex is low HSQC spectra will show "slow exchange" features. In this case resonances disappear at the position of the free protein and new resonances build up successively with a chemical shift of the protein complex. This behavior is a consequence of the low dissociation rate of high-affinity complexes. In such titrations, the signals of the protein complex must often be reassigned, whereas signals in fast-exchange spectra can be assigned by following chemical shift changes for subsequent additions of small amounts of ligand.

In the fast-exchange limit, the binding affinity can be calculated from the amount of chemical shift perturbations and the corresponding ligand concentrations.[1)] This requires that the interaction is a simple one-step binding reaction ($P + L \leftrightarrow PL$). This assumption is not always valid. Recent reports show that the mechanism of protein ligand binding may be connected to the formation of intermediates [18]. For more complex mechanisms, binding parameters can be ob-

1) $K_D = ([P]_0 - [PL])([L]_0 - [PL])/[PL]$ where $[P]_0$ and $[L]_0$ are the concentrations of the free protein and the ligand and $[PL]$ is the concentration of the protein–ligand complex. $[PL]$ can be estimated from chemical shift perturbations which can be evaluated from chemical shift perturbations ($[PL] = |(\delta_{\text{free}} - \delta_{\text{obs}})/\Delta\delta_{\text{total}}|$, where δ_{free} and δ_{obs} are the chemical shifts of the free protein and that of the complex at a given ligand concentration $[PL]$, and $\Delta\delta_{\text{total}}$ is the total chemical shift perturbation).

tained by simulating the line shapes of the signals in the HSQC spectra [18, 19]. Such a detailed analysis goes far beyond a routine analysis in a screening procedure. However, a detailed analysis of protein–ligand interactions may prove beneficial in the process of the synthetic improvement of the ligand in drug design.

9.2.2.2 Structure Determination from Chemical Shift Perturbations

SAR by NMR has so far been used for a qualitative evaluation of ligand interactions. However, Bonvin et al. have recently described a protein-docking approach named HADDOCK (high ambiguity driven protein–protein docking) that can utilize chemical shift perturbations to calculate the structure of protein–ligand and protein–protein complexes. This method has been validated for two different proteins for which the structures of the complexes were available. For an EIN–HPr complex, the RMSD at the interface between the X-ray structure and the structure calculated by HADDOCK was 2.75 Å; for an E2A–HPr complex, the RMSD at the interface was 2.1 Å. Cleary, HADDOCK is an important link between SAR by NMR and structure calculations, and provides fast access to the structure of protein complexes when HSQC filtrations are available.

9.2.3 Advanced Technologies in SAR by NMR

Using cryoprobe technology that has been developed recently for NMR instrumentation, protein concentrations can be lowered to 50 μM [20]. Screening of small molecules with cryoprobes is possible at concentrations of 50 μM each in mixtures of 100 compounds while keeping the total concentration of ligands below 5 mM. The 10-fold gain in throughput increases the total number of screened compounds to 10,000 per day. The reduced solubility requirement for ligands (now 50–10 μM) also increases the variability of compounds suitable for screening. In addition, the affinity cut-off for compounds that can be detected is reduced to $K_D \sim 0.15$ mM compared to $K_D \sim 1$ mM for higher protein concentrations. This difference is a consequence of the lower relative amount of protein in the bound form for lower protein concentrations. The lower cut-off has the disadvantage that low-affinity compounds are not detected any more. For HTS, the reduced hit rate reduces the need for extensive deconvolution of mixtures. Considering deconvolution, a total throughput of approximately 200,000 compounds per month can be achieved.

Transverse relaxation optimized spectroscopy (TROSY) [21] makes significantly larger proteins of 30–50 kDa accessible to SAR by NMR. Chemical shift degeneracy becomes problematic in large proteins. Samples with selective labels for individual amino acids can help to reduce the number of overlapping signals. Alternatively, the use of ^{13}C- rather than ^{15}N-labeled proteins has been explored for NMR screening [22]. This approach is highly promising if only methyl groups are labeled because the number of signals is reduced to amino acids that contain methyl groups. Relatively narrow lines are obtained due to the favorable relaxation properties of methyl carbons. In addition, the signal intensities of methyl groups are higher because the protons contribute to one signal. The group of Fesik has developed a cost-effective expression protocol using the labeled amino acid precursors [3,3′-αC]-α-ketoisovalerate and [3-^{13}C]-α-ketobutyrate for which an inexpensive synthesis route has been described [22].

When added to the growth media, these compounds are biosynthetically transformed, specifically and efficiently, into isoleucine and valine/leucine, respectively. For proteins with a molecular weight of less than 30 kDa, ^{1}H-^{13}C-based screening was shown to be 3 times more sensitive than ^{1}H-^{15}N-TROSY-based screening (without using cryoprobe technology), providing good spectra of 50 μM samples in around 10 min. With the use of cryoprobes, sample concentrations of 15 μM become accessible for small proteins. For proteins with a molecular weight above 40 kDa, the ^{13}C-based approach was more sensitive than TROSY, but still not sufficiently sensitive for screening. Screening of even larger proteins becomes practical when samples are perdeuterated, i.e. all nonexchangeable protons are replaced by ^{2}H. For ^{2}H/^{13}C-labeled proteins with a molecular weight of around 120 kDa, ^{1}H-^{13}C-HSQC spectra of 0.3 mM proteins were obtained in 30 min using cryoprobe technology.

Fesik reports that methyl-^{13}C-labeling using [3,3′-^{13}C]-α-ketoisovalerate and [3-^{13}C]-α-ketobutyrate is almost as cost-effective as ^{15}N-labeling and significantly cheaper than using uniform ^{13}C-labeling or methyl-^{13}C-labeling using commercially available [U-^{13}C]-α-ketobutyrate and [U-^{13}C]-α-ketoisovalerate [22]. Cell-free synthesis of proteins may become an interesting alternative for complex expression protocols if labeled amino acids can be produced at a low cost.

9.2.4 Data Evaluation

The interpretation of the actual screening using thousands of spectra requires software that can visualize changes in spectra in a simple graphical manner. Superimposing spectra and manual inspection is impractical for large amounts of data. Clearly, more sophisticated computational methods are required.

The most common algorithm used to study variations in series of spectra is the using principal component analysis (PCA). PCA is a classical statistical method used to reduce the dimensionality of problems, and to transform interdependent coordinates into significant and independent ones. PCA transforms the original data (e.g. intensities and signal positions in a spectrum) into a set of "scores" for each sample, measured with respect to the principal component axes ("loadings"). The principle component scores replace the original variates and are ordered with successive principle components accounting for decreasing amounts of variance, and orthogonal, with no correlation between the scores on different axes. As a result of these properties, a small number of principle components can replace the many original variates without much loss of information.

Scatter plots of the scores on the first few principle component loadings provide an excellent means of visualizing and often show clustering of similar samples, separation of different sample types or the presence of outliers. Plots of the loadings themselves may be used to explore which compounds are most responsible for separating samples into groups: the most important compounds tend to corre-

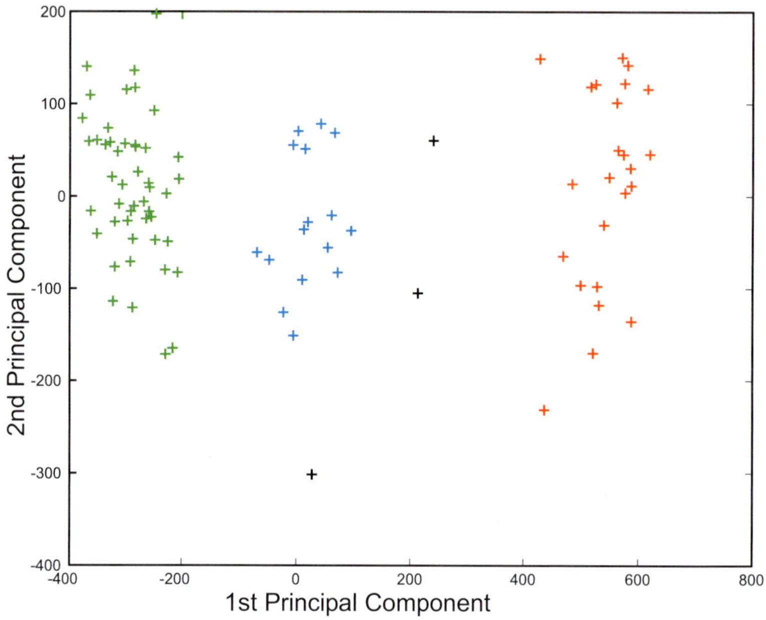

Fig. 9.5 Plot of the first against the second principal component for 92 HSQC spectra recorded for the trigger factor from *M. genitalium* (23). The four groups identified by PCA are marked in green, blue, black and red.

spond to high absolute loading values. Usually the first four principle components are relevant.

Figure 9.5 shows a typical scatter plot obtained for an SAR by NMR run of 92 spectra for the trigger factor from *Mycoplasma genitalium* [23]. Here, the spectra cluster in three independent classes.

It has been possible to distinguish groups of ligands in plots of principal components that show specific binding and pH-related effects [24]. Unfortunately, PCA is sensitive to all kinds of changes in a spectrum, including baseline artifacts and intensity changes of signals.

The evaluation of SAR by NMR data at the stage of ligand refinement is usually straightforward. When ligands are titrated into the protein solution the signals shift gradually and chemical shift perturbations can be attributed to subtle changes in the structure of the ligand. Software developers have provided tools to trace shifting resonances and store the results in a database.

9.3 Monitoring Small Molecules

In addition to the powerful methods based on protein NMR spectra, numerous techniques based on ligand spectra have been developed. Most of these techniques depend on the condition of fast exchange between the protein and the ligand. In the case of fast exchange and with an excess of ligand, the resonances of the ligand in solution will partially show spectral properties of the ligand bound to the protein.[2)]

A spectral property which is transferred in fast exchange is the line width of resonances. The signal of the ligand bound to the protein will be broadened because the signals of the protein have a much larger line width than those of small ligand molecules. The large line width of protein signals is a direct consequence of the long autocorrelation time of large proteins in solution.[3)] In the same manner, the transfer of several relaxation properties which are typical for large proteins has been used to study interactions and to screen for ligand binding. In the same way diffusion properties are transferred from the protein–ligand complex to the free ligand and can be used in screening. The most commonly used techniques for observing ligand resonances are based on magnetization transfer employing the nuclear Overhauser effect (NOE). These include transferred NOE, NOE pumping and reverse pumping, saturation transfer (STD-NMR) and NOE with water molecules on the protein surface [water-ligand observed via gradient spectroscopy (waterLOGSY)].

These techniques do not require isotopic labeling of the protein, the required amount of protein is usually very small and there is no upper limit size for the protein. Instead, relatively pure solutions must be used because any impurity contributes to the signal. Unfortunately, ligands with very high affinity have low off-rates from the complex and score therefore as non-binders.

Competition experiments can help to circumvent this problem. Another limiting factor is the solubility of the ligand because most of these methods require an excess concentration of ligand. Some of these experiments are useful for HTS, some provide more detailed information on ligand binding and some can be used to map the binding epitope of the ligand. Relaxation-based methods will be covered briefly, whereas techniques employing magnetization transfer will be described in greater detail.

9.3.1 Relaxation Methods

In NMR spectroscopy the term relaxation is used to describe the process that restores equilibrium magnetization and random phase. Two relaxation mechanisms must be distinguished, transverse and longitudinal relaxation, with respective relaxation rates R_2 and R_1. Transverse relaxation can be associated with fluctuating electrical fields which originate from the overall tumbling and internal motion of

2) For fast exchange and equal populations the exchange rate must be $k/\Delta\nu > \pi/\sqrt{2}$, where k is the rate of the interaction and $\Delta\nu$ is the chemical shift difference of the free and the bound state of the ligand. For protein–ligand interactions the rate that determines the exchange is the off-rate of the protein–ligand interaction ($\mathrm{P} + \mathrm{L} \underset{k_{\mathrm{off}}}{\overset{k_{\mathrm{on}}}{\rightleftharpoons}} \mathrm{PL}$).

3) The transverse relaxation rate R_2 is large for high molecular mass proteins with long correlation times τ_c. The line width LW is directly proportional to the relaxation rate ($\mathrm{LW} = R_2/\pi$).

the molecule. Since large molecules tumble slowly in solution the correlation time τ_c of their rotational motion is relatively long, causing large relaxation rates R_2 and thus also large line widths of NMR signals ($LW = R_2/\pi$). Compounds interacting with a receptor with a large molecular weight show broadened lines and increased R_2 values. A more detailed description of relaxation theory is beyond the scope of this chapter and several excellent reviews are available [25–28].

A ligand bound to a protein adopts the relaxation properties of the complex. In the case of fast exchange, free ligand molecules preserve the relaxation properties from the bound state for a short time. An excess of ligand must be available to observe signals of the ligand. The broadening of resonances is a typical property of spectra of small molecules in the presence of a large protein. This is reflected in higher transverse relaxation rates. For this reason the line width and the transverse relaxation rate can be used as a criteria for ligand screening. Broadening is more pronounced for large proteins or proteins bound to a matrix. Bound ligands can be identified using T_2 filter experiments (29). Usually, spectra with and without protein must be compared.

Broadening of ligand signals induced by binding to a protein has been used frequently in the past. Sykes et al. studied the interaction of the 85-kDa extracellular domain of the epidermal growth factor receptor (EGFR-ED) with transforming growth factor (TGF)-α [30]. By measuring line widths and transverse relaxation rates of 14 methyl proton signals of TGF-α for different TGF-α EGFR-ED, the concentrations binding kinetics were also determined.

The line width of methyl protons showed a regional dependence indicating which parts of TGF-α were involved in the interaction. This study indicates that this technique provides localized information in the ligand that is important for a detailed description of protein–ligand interactions.

The potential of relaxation editing for screening has been demonstrated by Fesik et al. using FKBP and a nine-compound mixture from which one ligand with a dissociation constant of 200 μM was identified [31]. This method is not practical for efficient ligand screening as differences between spectra with and without ligand must be used to extract the relevant effects. T_2 and $T_{1\rho}$ have been commonly used to filter spectral components arising from short relaxation times [32]. A broad range of relaxation methods including longitudinal relaxation, transverse relaxation in the presence of spin labels and double quantum relaxation was recently reviewed (8).

9.3.2 Diffusion Methods

Diffusion techniques rely on the fact that the translational diffusion depends on the size of the protein. According to the Stokes–Einstein equation $D = kT/6\pi\eta r$, where k is the Boltzmann constant, T is the temperature, η is the viscosity and r is the radius of the molecule, the diffusion coefficient depends inversely on the size of a molecule. This principle has been used to analyze mixtures of compounds in combinatorial chemistry [33]. Shapiro et al. combined diffusion-editing techniques with TOCSY spectra to obtain 2-D spectra with little peak overlap and valuable information on the structure of the ligands (DECODES, diffusion encoded spectroscopy) [33].

Diffusion coefficients are measured in NMR by applying a linear magnetic field gradient for a short time. This gradient applies different magnetic fields at different positions of the sample and therefore spatially dependent dephasing of the signal which can be reversed by applying the same spatially encoded gradient after a spin-echo. Rephasing does not work for molecules that have "traveled" since the application of the first gradient. Consequently, large molecules with small diffusion coefficients will be affected to a smaller degree by the diffusion filter than small molecules. The diffusion coefficient can be determined by measuring signal intensities depending on the diffusion time and gradient strength. In the case of fast exchange, the free ligand may carry diffusion properties from the complex. In principle two types of diffusion experiments have been used, pulsed field gradient spin-echo (PFG-SE) (34) and pulsed field gradient stimulated-echo (PFG-STE). Loss of magnetization due to transverse relaxation is reduced in the latter case. PFG-SE has the disadvantage that only half of the signal is observed. A sequence with bipolar gradients and a longitudinal eddy current delay (LED) is commonly used to minimize eddy current artifacts (35).

For small molecules bound to larger partners, the diffusion coefficient is reduced compared to non-binding molecules of a similar size. The observed diffusion coefficients will be time-averaged according to the rate of exchange. Shapiro et al. used the expression "affinity NMR" for this approach because it is in some way reminiscent of separation by affinity chromatography. Shapiro employed affinity NMR for quinine with a mixture of nine compounds from which one could be selected by analyzing 1-D spectra and 2-D DECODES experiments [36].

Shapiro also applied affinity NMR to study the binding of tetrapeptides to the glycopeptide vancomycin [37] and to study the binding of small molecules to DNA [38]. In these reported cases the ligand and protein were present in equimolar amounts at a concentration of 0.5–1 mM. More recently, Shapiro demonstrated that diffusion NMR can even be used to map a binding epitope. This is based on the fact that longitudinal magnetization in STE experiments [34] is subject to cross-relaxation between the protein and the ligand.

Through the NOE, magnetization is transferred between the protons in the complex. This effect induces changes in signal intensity when applying low and strong gradient strengths and for long diffusion times, because the signal intensities decay with a different rate during the diffusion period. Quantitative analysis was obtained by plotting the signal intensity against the square gradient strengths g^2 [39].

Hajduk proposed to use differences between various spectra to overcome problems arising from signal broadening of the ligand and from background signals of the protein. In the first step, STE spectra of the compound mixture in the presence of the protein at high and low gradient strength were subtracted. The result was again subtracted from a STE spectrum of the chemical mixture in the absence of the protein recorded at low gradient strength. The result is a spectrum which shows only signals of the small molecules that bind to the protein. This cumbersome approach that required three different spectra was suitable to identify a 20 μM inhibitor for stromelysine from a mixture of nine compounds. This approach is also limited because bound ligands may have different chemical shifts in fast exchange when equal amounts of protein and ligand are used.

Diffusion NMR can be used to directly determine the affinity of ligands without the need of titration with ligands. For free and bound ligand in fast exchange the observed diffusion coefficient is a time-average mean of the free and bound species. The fraction of the bound ligand can be obtained from the diffusion constants of the protein D_{bound}, of the free ligand D_{free} and that of the ligand in the mixture by $D_{obs} = x_{free} D_{free} + x_{bound} D_{bound}$ assuming that the diffusion constants of the complex and free protein are the same.

Diffusion experiments are clearly an important tool in drug screening and, to a limited degree, for epitope mapping. The main drawback is the relatively low sensitivity of the experiment which requires equimolar amounts of protein and ligand at concentrations of ~50 μM. Even larger amounts of substances are needed for low-affinity compounds as the diffusion effect becomes very small.

9.3.3
Methods Involving Magnetization Transfer

9.3.3.1 Transferred NOE (TrNOE)

The transferred NOE (TrNOE) effect was originally described by Bothner-By [40, 41]. Peters proposed to use the TrNOE for screening compound mixtures [42]. The TrNOE combines the NOE between adjacent spins in the ligand with chemical exchange between bound and free ligand.[4] Small molecules induce positive NOEs; large molecules induce negative NOEs with much larger magnitude. When a ligand is in fast exchange between its bound and free form it can transfer the negative NOE from the protein complex to the population of free molecules. In the NOESY spectrum, negative signals appear from ligands that bind to the protein, while signals from weakly bound compounds are positive or disappear. The quantitative dependence of the TrNOE effect on the mixing time, the amount of ligand bound, and the free and bound correlation times of protein and ligand is well understood [43]. The TrNOE effect benefits from large proteins that cause a strong negative NOE. TrNOEs are obtained from regular NOESY spectra. Signals arising from the protein are usually not observed for large proteins. Background signals arising from the protein can be suppressed by a T_2 or a $T_{1\rho}$ filter (spin-lock).

Peters et al. used TrNOE measurements to screen oligosaccharides for deglycosylated E-selectin/human immunoglobulin G [44]. Initial NOESY spectra of the ligand mixture were recorded for different temperatures until all NOE signals were positive. The compound mixture showed strong negative NOEs in the presence of protein. Eight compounds could be excluded based on missing TrNOE-NOE signals in characteristic regions of the spectrum. From the two remaining com-

4) The NOE effect itself is based on longitudinal dipole–dipole cross-relaxation of adjacent spins. Usually signals are observed for protons that are less than 6 Å apart.

pounds, a sialyl Lewis$_x$ mimetic was identified for which the bioactive conformation could be determined from TrNOESYs. The identification of binding epitopes was impeded by severe spin diffusion that spreads the magnetization over all protons of a compound.

Peters et al. used a 3-D TOCSY-TrNOESY experiment to overcome severe overlap in a mixture of 15 carbohydrates [45]. Although this more time-consuming approach is useful to deconvolute mixtures of compounds, it is not suitable for screening.

The TrNOE method is broadly applicable for large proteins with dissociation constants between 10^{-7} and 10^{-3} M. Disadvantages of this method are connected with the need for relatively high ligand concentrations which requires highly soluble compounds, long recording times and the large amount of spin diffusion which prohibits epitope mapping.

9.3.3.2 Saturation Transfer

STD-NMR can be considered as a natural improvement of TrNOE to overcome many of its problems. Although the STD effect has been known for a long time [46], it has only recently been used for ligand screening [47]. The principle of STD-NMR is illustrated in Fig. 9.6. Initially a proton resonance of the protein is saturated (on-resonance saturation). For large molecules, the saturation effect spreads quickly over the entire protein by flip-flop energy transitions between adjacent protons.[5] Magnetization transfer to the bound ligand causes successive saturation of the ligand residues. It is important that the saturation pulse affects only resonances of the protein, but no resonances of the ligand. This guarantees that saturation transfer stems from the protein and not from resonances within the ligand. Typically, methyl resonances with chemical shifts below 0 ppm are saturated. In a reference experiment the saturation pulse is applied at a frequency where none of the protein or ligand protons resonate (typically 30 ppm, off-resonance saturation). This reference spectrum shows no saturation effect. To minimize subtraction artifacts the difference between the two spectra is obtained by measuring alternating on- and off-resonance transients that are then added using a phase cycle. The pulse program used for these two experiments is depicted schematically in Fig. 9.7. A more detailed pulse program is described in [48]. Background protein signals can be removed by applying a spin-lock pulse that acts as a $T_{1\rho}$ filter.

Meyer et al. determined the degree of saturation depending on the ligand excess by binding *N*-acetylglucosamine to wheat germ agglutinin. The saturation effect increases with the ratio of ligand to protein concentration. An increase was observed beyond a ratio of 210:1.

Sufficiently strong STD effects were observed for a ratio of 60:1. Fast dissociation rates also favor large STD effects. The necessity of a large ligand excess allows STD measurements with very low protein concentrations. Using a 100-fold excess of

5) The rate of the flip-flop transitions responsible for the distribution of the saturation effect throughout the protein is directly proportional to the correlation time τ_c of the protein. For this reason saturation spreads faster in larger macromolecules. Typically saturation times of around 2 s are applied.

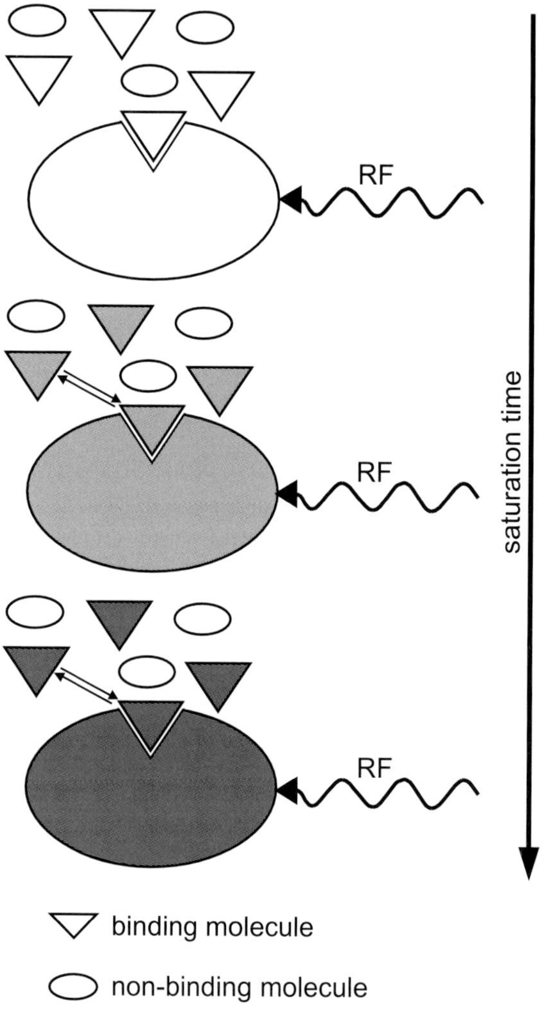

Fig. 9.6 Schematic illustration of the STD-NMR experiment. Protein resonances are saturated selectively without affecting resonances of the ligand. Increasing saturation is represented by darker grey tones. Compounds that interact with the protein (triangles) will show the saturation effect. Small molecules that are released in fast exchange rates between the complex and the free ligand can be detected. Molecules that bind to the protein show reduced signal intensities as a consequence of saturation transfer. Nonbinding molecules (ovals) show no effect.

ligand and a 1 μM ligand solution, the concentration of the protein can be as low as 10 nM.

The STD principle can be combined with 2-D homonuclear [47] and heteronuclear [49] experiments. STD-TOCSY was presented by Meyer et al. employing a

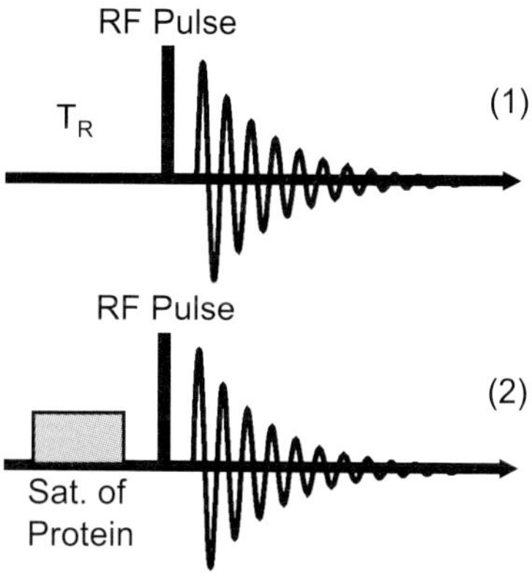

Fig. 9.7 Schematic representation of the pulse sequence used for STD-NMR. Two spectra, one with and the other without saturation of protein resonances are subtracted.

mixture of seven oligosaccharides and wheat germ agglutinin. In oligosaccharides with six sugar subunits, different groups showed significantly different STD effects. Those sugar protons in direct contact with the protein experienced a stronger degree of saturation. Vogtherr used STD-heteronuclear multi-quantum correlation (HMQC) [49] to deconvolute a library of oligosaccharides. The potential of STD-NMR for epitope mapping has recently been demonstrated for methyl β-D-galactoside binding to a 120-kDa lectin agglutinin [48]. Strong intensity differences were observed for the different protons of the methyl β-D-galactoside indicative of their different contribution to the interaction with the protein. In the same work competition studies were performed which made STD-NMR accessible to ligands with nanomolar dissociation constants and slow dissociation rates. STD-NMR has also been used in combination with HR-MAS solid-state NMR spectroscopy [50].

An example spectrum is presented in Fig. 9.8 where a phenolic compound was bound to a dehydrogenase. The spectrum in Fig. 9.8(A) depicts the STD spectrum recorded with saturation at −0.45 p.p.m. (arrow). Figure 9.8(B) shows a reference spectrum of the same compound. The spectra are scaled to equal intensity of the aromatic signals at ~7 ppm. The aliphatic signals at 3.8, 2.2 and 2 ppm all have a reduced relative intensity. This result is in agreement with the assumption that the phenyl group is the driving force for the interaction with this protein. Figure 9.8(C) is a negative control with no protein present in the mixture. This spectrum was recorded with the same parameters as the STD spectrum in Fig. 9.8(A).

The disadvantages of STD-NMR are the doubled recording time for the difference method and the fact that tight binders score as non-binders, causing false negatives in screening. The necessity to have a large excess of ligand limits STD-NMR to reasonably soluble compounds (above 10 mM). To avoid subtraction artifacts, the spectrometer and probe-head must be calibrated carefully, which is only possible with recent developments in instrumentation.

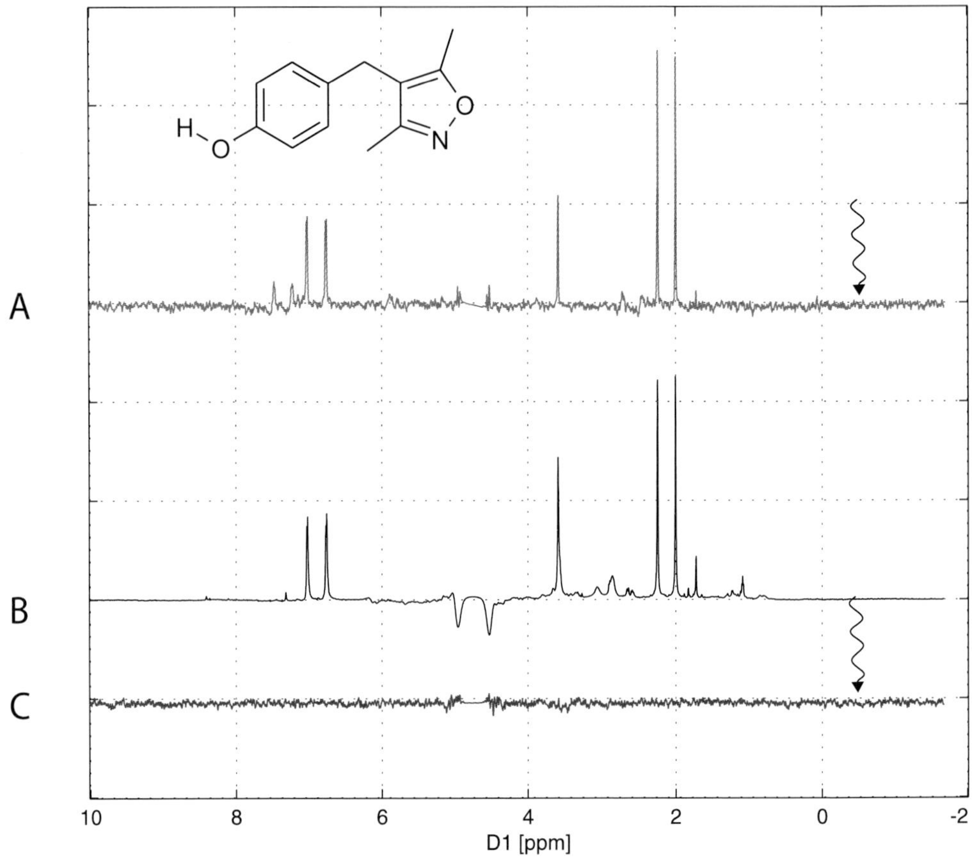

Fig. 9.8 (A) STD-NMR spectrum of a phenolic compound bound to a 72-kDa dehydrogenase. (B) Reference spectrum for the same compound. (C) Control STD-NMR spectrum with no protein present in the solution.

9.3.3.3 **NOE Pumping**

NOE pumping is closely related to STD-NMR. In STD-NMR, saturation is transferred to the ligand; in NOE pumping, magnetization is transferred from the protein to the ligand [51]. Magnetization can also be transferred from the ligand to the protein in an experiment called reverse NOE pumping (RNP). For NOE pumping, the signals of the protein are selected by a diffusion filter which cancels the signals of the ligand. Magnetization is transferred during a mixing time by intermolecular cross-relaxation. Ligand molecules in fast exchange that experienced magnetization transfer from the protein are detected. All other molecules will not show signals. This has been demonstrated using human serum albumin and a mixture of the three compounds – salicylic acid, L-ascorbic acid and glucose. Only the signals of salicylic acid were observed in the NOE pumping spectrum [51]. This ligand was not detected in a pure diffusion edited experiment without NOE transfer.

The ligand spectrum for a reverse NOE experiment [52] is selected employing a transverse relaxation filter that eliminates the signals of the protein and inverts the ligand signals. This filter exploits the much faster transverse relaxation time T_2 of the protein compared to the slower T_2 of the ligand. During the mixing time the ligand magnetization is partially transferred to the protein via intermolecular cross-relaxation. A reference experiment in which no magnetization transfer takes place is subtracted. In the difference spectrum only those signals of compounds which interact with the protein are observed. This experiment was used for human serum albumin with glucose and a mixture of unbranched fatty acids as presumable ligands. Only the fatty acids were observed in the difference spectrum. Intensity ratios between different ligands were used to rank the affinities of different compounds [52]. Reverse NOE pumping is the more sensitive experiment of the two.

NOE pumping and RNP are considered to be very sensitive experiments for primary NMR screening. Epitope mapping is limited by spin diffusion.

9.3.3.4 WaterLOGSY and ePHOGSY

WaterLOGSY uses water molecules bound to the protein–ligand interface to transfer magnetization from the protein to the ligand [53, 54]. Water at the interface may be involved in hydrogen bridges between the protein and the ligand. It is also possible that hydrophobic hydration at the surface of the protein in the neighborhood of the ligand takes the role of a transmitter. The residence time of water molecules in protein cavities ranges from a few nanoseconds to several hundred microseconds [55–61]. This is sufficiently long to build up NOEs and sufficiently short to keep the water resonances in fast exchange where only one signal is observed for the bound and free water resonance. The principle of waterLOGSY is illustrated in Fig. 9.9. The NOE transfers magnetization from the bulk water to the protein. The sign of this magnetization is unchanged from the starting magnetization. Another mechanism for the transfer of magnetization from the bulk water to the protein is chemical exchange of water protons with labile protons of the protein. This process also conserves the sign of the magnetization, i.e. these two processes add constructively to the transfer of magnetization from the bulk water to the protein. In a second step, an NOE between the bound water and the ligand protons develops. This NOE is negative for the slowly tumbling protons in the protein–ligand complex, and positive and of lower magnitude for small molecules such as the free ligand in solution. Since an excess of ligand is used in this experiment, it is always the signal of the ligand that is observed. Depending on the dissociation constant of the complex, either the contribution of negative NOEs typical for the protein or positive NOEs for small molecules will prevail. This allows a fast distinction between molecules that bind to the protein and those that do not bind because the NOEs of the two will have opposite signs in the spectra [53].

The most effective experiment used to measure the waterLOGSY effect is called ePHOGSY [62, 63]. This experiment starts with selective inversion of water proton resonances and dephasing of all other resonances. A NOE to the protein and to the ligand protons builds up from the selected water protons. Ligand titration experiments were performed to derive binding constants [53] using ePHOGSY experi-

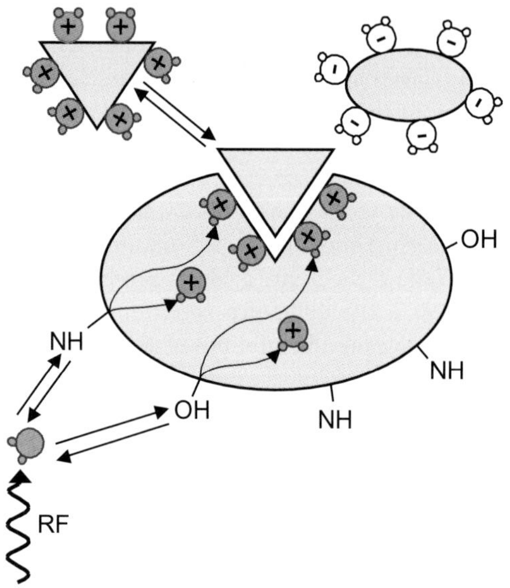

Fig. 9.9 Schematic illustration of the ePHOGSY experiment. Magnetization is transferred from bulk water to the protons of the protein by NOE or by exchange of labile protons indicated by arrows. The ligand in solution shows the transfer of magnetization from bulk water. The sign of the signal differs for molecules that bind the protein (+) and those that experience transfer of magnetization from bulk water without binding the protein (−).

ments. The potential of waterLOGSY has been demonstrated for human serum albumin in the presence of tryptophan and for cyclin-dependent kinase 2 (cdk2) binding two ligands of undisclosed nature [53]. An ePHOGSY experiment for a dehydrogenase bound to 4-*n*-butyl-phenol is demonstrated in Fig. 9.10. The mixture also contained some imidazole and DMSO; neither interact with the protein and consequently negative signals are observed.

ePHOGSY experiments have a potential in primary NMR screening. The experiment is very sensitive, devoid of artifacts, and requires only low protein and ligand concentrations. Signals of non-binding ligands have opposite signs compared to signals of ligands that bind with high affinity. The dissociation constant of complexes may be as low as 0.1 μM. Unfortunately, this method cannot be used for epitope mapping because the intensity of signals is a complex function of the residence time of water and the distance between protons in the ligand to water molecules. However, it is a very sensitive method for ligand screening.

9.3.4 Sample Preparation

Compared to protein samples used for classical SAR by NMR, there are different requirements for samples used in ligand-detecting techniques. Obviously, there is

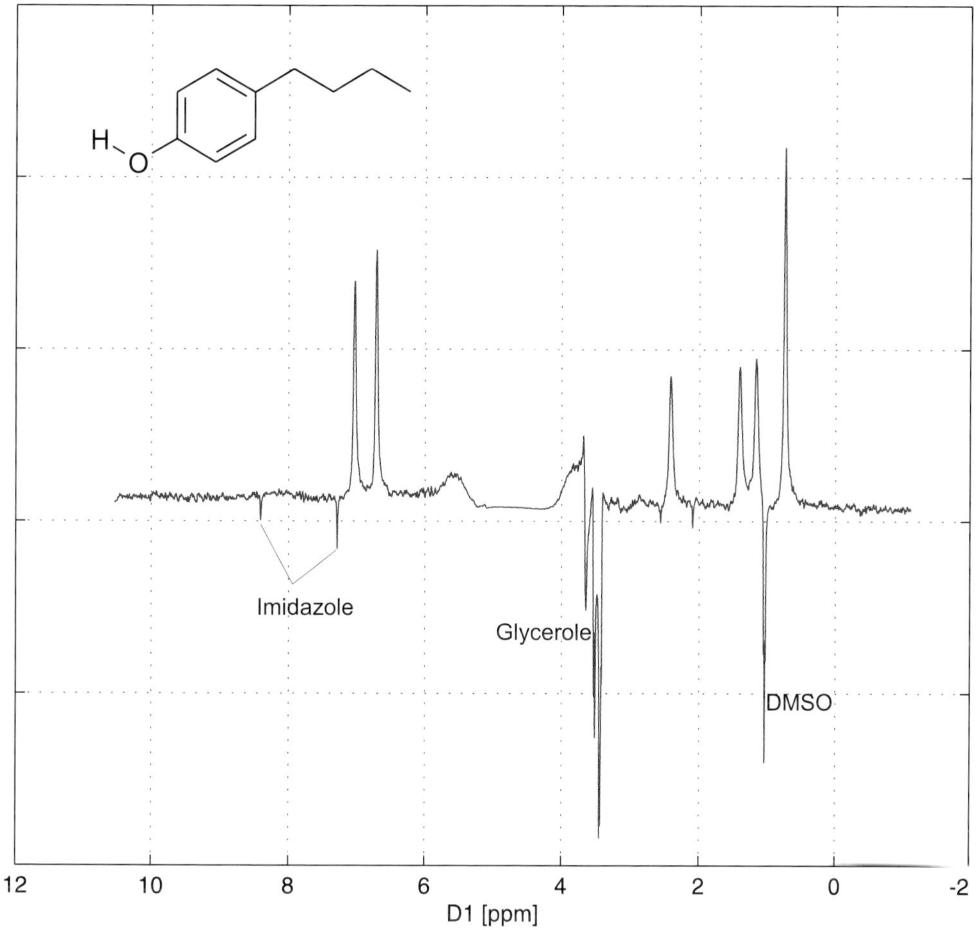

Fig. 9.10 An ePHOGSY experiment for a dehydrogenase bound to 4-*n*-butyl-phenol.

no need for isotopic labeling and there is no need to overexpress the protein. Even purification of the protein is only necessary if the substances to be analyzed bind to any other protein in the solution. There are hardly any restrictions with respect to the pH of the solution because no labile protons are examined. However, the presence of small molecules other than the ligands of interest may interfere. As a typical example, glycerol, often used in high concentrations to stabilize proteins, causes resonances of high intensity that may bury other signals underneath. The need to avoid signals from other compounds also imposes a stringent restriction on the choice of buffers to those without nonlabile protons. Buffers with labile protons in fast exchange with water such as phosphate and hydrogen carbonate can be used.

Small molecules are often dissolved in DMSO and added in small quantities from DMSO stock solutions (few microliters). In particular, for relatively insoluble compounds, improved solubility can be achieved when the solvent contains DMSO. Most proteins keep their function in aqueous solutions with DMSO concentrations of 20–30%. Deuterated DMSO has to be used for this purpose. In the same way, other deuterated solvents may be added. In the case of d^6-DMSO a small sharp residual DMSO signal is observed at around 2.2 ppm which is often a good reference for a non-binding substance. Tetramethylsilane (TMS) or 2,2-dimethyl-2-silapentane-5-sulfonic acid (DSS) may be added as a chemical shift standard.

9.4 Conclusions

In the past 7–8 years NMR methods have entered the field of HTS of compounds as ligands for proteins. This development is astonishing considering the relatively low sensitivity of the basic NMR experiment compared to highly sensitive methods such as fluorescence spectroscopy. Recent hardware development certainly played an important role in the achievements of NMR screening methods. The pace of NMR development is still fast, with continuously increasing magnetic fields and new technologies which enhance the sensitivity and stability of current NMR spectrometers. Cryoprobe technology has boosted sensitivity and pushed NMR experiments to an order of magnitude lower sample concentrations.

The typical SAR by NMR experiment is now accessible for protein concentrations as low as 50 μM. However, experiments observing protein spectra are currently limited to relatively small proteins of molecular weights up to 40 kDa. Molecular weights of 100 kDa will become feasible with more sophisticated labeling strategies. With regard to the ligand, numerous techniques are now available with no limits on the size of the protein. The requirement with respect to the dissociation constant of the protein–ligand complex is somewhat limiting for these techniques. Ligand-based methods have a considerable potential for ligand screening (ePHOGSY, NOE pumping), epitope mapping (transferred NOE, STD-NMR) and have even been used to determine the conformation of bound ligands. The real value of NMR in drug design is the wealth of information that can be obtained from a few measurements. In this sense NMR provides a valuable addition to traditional HTS and computational methods.

Acknowledgments

We thank H. Thole, J. Messinger and E. Finner from Solvay Pharmaceuticals for supplying previously unpublished results on dehydrogenases shown in this article. We thank D. Jacobs and M. Vogtherr for providing SAR by NMR spectra of the trigger factor from *Mycoplasma genitalium*. Some of the work presented was supported by the NMR Large Scale Facility Frankfurt.

References

1 G Roberts, NMR spectroscopy in structure-based drug design, *Curr Opin Biotechnol* **10**, 42–7, 1999.

2 G Roberts, Applications of NMR in drug discovery, *Drug Discov Today* **5**, 230–40, 2000.

3 J Moore, NMR screening in drug discovery, *Curr Opin Biotechnol* **10**, 54–8, 1999.

4 P Hajduk, R Meadows, S Fesik, NMR-based screening in drug discovery, *Q Rev Biophys* **32**, 211–40, 1999.

5 M Shapiro, J Wareing, NMR methods in combinatorial chemistry, *Curr Opin Chem Biol* **2**, 372–5, 1998.

6 J Peng, C Lepre, J Fejzo, N Abdul-Manan, J Moore, Nuclear magnetic resonance-based approaches for lead generation in drug discovery, *Methods Enzymol* **338**, 202–30, 2001.

7 T Diercks, M Coles, H Kessler, Applications of NMR in drug discovery, *Curr Opin Chem Biol* **5**, 285–91, 2001.

8 B Stockmann, C Dalvit, NMR screening techniques in drug discovery and drug design, *Prog Nucl Magn Spectrosc* **41**, 187–231, 2002.

9 SB Shuker, PJ Hajduk, RP Meadows, SW Fesik, Discovering high-affinity ligands for proteins: SAR by NMR, *Science* **274**, 1531–4, 1996.

10 AC de Dios, JG Pearson, E Oldfield, Secondary and tertiary structural effects on protein NMR chemical shifts: an *ab initio* approach, *Science* **260**, 1491–6, 1993.

11 P Hajduk, S Boyd, D Nettesheim, V Nienaber, J Severin, R Smith, D Davidson, T Rockway, S Fesik, Identification of novel inhibitors of urokinase via NMR-based screening, *J Med Chem* **43**, 3862–6, 2000.

12 U Günther, Y Liu, D Sanford, W Bachovchin, B Schaffhausen, NMR analysis of interactions of a PI-3 kinase SH2 domain with phosphotyrosine peptides reveals interdependence of major binding sites, *Biochemistry* **35**, 15570–81, 1996.

13 M Hensmann, GW Booker, G Panayotou, J Boyd, J Linacre, M Waterfield, ID Campbell, Phosphopeptide binding to the N-terminal SH2 domain of the p85α subunit of PI 3′-kinase: a heteronuclear NMR study, *Protein Sci* **3**, 1020–30, 1994.

14 J Rizo, Z Liu, L Gierasch, ^{1}H and ^{15}N resonance assignments and secondary structure of cellular retinoic acid-binding protein with and without bound ligand, *J Biomol NMR* **4**, 741–60, 1994.

15 G Wider, Structure determination of biological macromolecules in solution using nuclear magnetic resonance spectroscopy, *Biotechniques* **29**, 1278–94, 2000.

16 G Clore, C Schwieters, Theoretical and computational advances in biomolecular NMR spectroscopy, *Curr Opin Struct Biol* **12**, 146–53, 2002.

17 G Waksman, SE Shoelson, N Pant, D Cowburn, J Kuriyan, Binding of a high affinity phosphotyrosyl peptide the src SH2 domain: crystal structures of the complexed and peptide-free forms, *Cell* **72**, 779–90, 1993.

18 U Günther, T Mittag, B Schaffhausen, Probing src homology 2 domain ligand interactions by differential line broadening, *Biochemistry* **41**, 11658–69, 2002.

19 U Günther, B Schaffhausen, NMRKIN: simulating line shapes from two-dimensional spectra of proteins upon ligand binding, *J Biomol NMR* **22**, 201–9, 2002.

20 P Hajduk, T Gerfin, J Boehlen, M Haberli, D Marek, S Fesik, High-throughput nuclear magnetic resonance-based screening, *J Med Chem* **42**, 2315–7, 1999.

21 K Pervushin, R Riek, G Wider, K Wüthrich, Attenuated T_2 relaxation by mutual cancellation of dipole–dipole coupling and chemical shift anisotropy indicates an avenue to NMR structures of very large biological macromolecules in solution,

Proc Natl Acad Sci USA **94**, 12366–71, 1997.

22 PJ Hajduk, DJ Augeri, J Mack, R Mendoza, J Yang, SF Betz, SW Fesik, NMR-based screening of proteins containing ^{13}C-labeled methyl groups, *J Am Chem Soc* **122**, 7898–904, 2000.

23 M Vogtherr, D Jacobs, T Parac, M Maurer, A Pahl, K Saxena, H Rüterjans, C Griesinger, K Fiebig, NMR solution structure and dynamics of the peptidyl-prolyl *cis–trans* isomerase domain of the trigger factor from *Mycoplasma genitalium* compared to FK506-binding protein, *J Mol Biol* **318**, 1097–115, 2002.

24 A Ross, G Schlotterbeck, W Klaus, H Senn, Automation of NMR measurements and data evaluation for systematically screening interactions of small molecules with target proteins, *J Biomol NMR* **16**, 139–46, 2000.

25 J Engelke, H Rüterjans, Recent developments in studying the dynamics of protein structures from ^{15}N and ^{13}C relaxation time measurements, in: B Krishna, LJ Berliner, eds, *Biological Magnetic Resonance, Vol. 17: Structural Computation and Dynamics in Protein NMR*, 357–418, Kluwer Academic/Plenum, New York, 1999.

26 R Ernst, G Bodenhausen, A Wokaun, *Principles of Nuclear Magnetic Resonance in One and Two Dimensions. International Series of Monographs on Chemistry 14*, Oxford Science Publications, Oxford, 1987.

27 MW Fischer, A Majumdar, ERP Zuiderweg, Protein NMR relaxation: theory, application and outlook, *Prog Nucl Magn Reson Spectrosc* **33**, 207–72, 1998.

28 P Luginbühl, K Wüthrich, Semi-classical nuclear spin relaxation theory revisited for use with biological macromolecules, *Prog Nucl Magn Reson Spectrosc* **40**, 199–247, 2002.

29 D Rabenstein, T Nakashima, G Bigam, A pulse sequence for the measurement of spin-lattice relaxation times of small molecules in protein solutions, *J Magn Reson* **34**, 669–74, 1979.

30 D Hoyt, R Harkins, M Debanne, M O'Connor-McCourt, B Sykes, Interaction of transforming growth factor-α with the epidermal growth factor receptor: binding kinetics and differential mobility within the bound TGF-α, *Biochemistry* **33**, 15283–92, 1994.

31 P Hajduk, E Olejniczak, S Fesik, One-dimensional relaxation- and diffusion-edited NMR methods for screening compounds that bind to macromolecules, *J Am Chem Soc* **119**, 12257–61, 1997.

32 T Scherf, J Anglister, A $T_{1\rho}$-filtered two-dimensional transferred NOE spectrum for studying antibody interactions with peptide antigens, *Biophys J* **64**, 754–61, 1993.

33 M Lin, M Shapiro, Mixture analysis in combinatorial chemistry, application of diffusion-resolved NMR spectroscopy, *J Org Chem* **61**, 7617–9, 1996.

34 E Stejskal, J Tanner, Spin diffusion measurement: spin echoes in the presence of a time-dependent field gradient, *J Chem Phys* **42**, 288–92, 1965.

35 S Gibbs, C Johnson, A PFG NMR experiment for accurate diffusion and flow studies in the presence of eddy currents, *J Magn Reson* **93**, 395–402, 1991.

36 M Lin, MJ Shapiro, JR Wareing, Diffusion-edited NMR-affinity NMR for direct observation of molecular interactions, *J Am Chem Soc* **119**, 5249–50, 1997.

37 K Bleicher, M Lin, M Shapiro, J Wareing, Diffusion edited NMR: screening compound mixtures by affinity NMR to detect binding ligands to vancomycin, *J Org Chem* **63**, 8486–90, 1998.

38 R Anderson, M Lin, M Shapiro, Affinity NMR: decoding DNA binding, *J Comb Chem* **1**, 69–72, 1999.

39 J Yan, A Kline, H Mo, E Zartler, M Shapiro, Epitope mapping of ligand–receptor interactions by diffusion NMR, *J Am Chem Soc* **124**, 9984–5, 2002.

40 P Balaram, A Bothner-By, J Dadok, Negative nuclear Overhauser effects as probes of macromolecular structure, *J Am Chem Soc* **94**, 4015–7, 1972.

41 P Balaram, A Bothner-By, E Breslow, Localization of tyrosine at the binding site of neurophysin II by negative nuclear Overhauser effects, *J Am Chem Soc* **94**, 4017–8, 1972.

42 B Meyer, T Weimar, T Peters, Screening mixtures for biological activity by NMR, *Eur J Biochem* **246**, 705–9, 1997.

43 AP Campbell, BD Sykes, Theoretical evaluation of two-dimensional transferred Nuclear Overhauser Effects, *J Magn Reson* **93**, 9377–92, 1991.

44 D Henrichsen, B Ernst, J Magnani, W-T Wang, B Meyer, T Peters, Bioaffinity NMR spectroscopy – identification of an E-selectin antagonist in a substance mixture by transfer NOE, *Angew Chem Int Ed* **38**, 98–102, 1999.

45 L Herfurth, T Weimar, T Peters, Application of 3D-TOCSY-trNOESY for the assignment of bioactive ligands from mixtures, *Angew Chem Int Ed* **39**, 2097–9, 2000.

46 S Forsen, R Hoffman, Study of moderately rapid chemical exchange reactions by means of nuclear magnetic resonance double resonance, *J Chem Phys* **39**, 2892–901, 1963.

47 M Mayer, B Meyer, Characterization of ligand binding by saturation transfer difference NMR spectra, *Angew Chem Int Ed* **35**, 1784–8, 1999.

48 M Mayer, B Meyer, Group epitope mapping by saturation transfer difference NMR to identify segments of a ligand in direct contact with a protein receptor, *J Am Chem Soc* **123**, 6108–17, 2001.

49 M Vogtherr, P Peters, Application of NMR based binding assays to identify key hydroxy groups for intermolecular recognition, *J Am Chem Soc* **122**, 6093–9, 2000.

50 J Klein, R Meinecke, M Mayer, B Meyer, Detecting binding affinity to immobilized receptor proteins in compound libraries by HR-MAS STD NMR, *J Am Chem Soc* **121**, 5336–7, 1999.

51 A Chen, MJ Shapiro, NOE pumping: a novel NMR technique for identification of compounds with binding affinity to macromolecules, *J Am Chem Soc* **120**, 10258–9, 1998.

52 A Chen, MJ Shapiro, NOE Pumping. 2. A high-throughput method to determine compounds with binding affinity to macromolecules by NMR, *J Am Chem Soc* **122**, 414–5, 2000.

53 C Dalvit, G Fogliatto, A Stewart, M Veronesi, B Stockman, Water-LOGSY as a method for primary NMR screening: practical aspects and range of applicability, *J Biomol NMR* **21**, 349–59, 2001.

54 C Dalvit, P Pevarello, M Tato, M Veronesi, A Vulpetti, M Sundstrom, Identification of compounds with binding affinity to proteins via magnetization transfer from bulk water, *J Biomol NMR* **18**, 65–8, 2000.

55 G Otting, K Wüthrich, Studies of protein hydration in aqueous solution by direct NMR observation of individual protein-bound water molecules, *J Am Chem Soc* **111**, 1871–5, 1989.

56 G Otting, E Liepinsh, BT Farmer, II, K Wüthrich, Protein hydration studied with homonuclear 3D proton NMR experiments, *J Biomol NMR* **1**, 209–15, 1991.

57 G Otting, E Liepinsh, K Wüthrich, Protein hydration in aqueous solution, *Science* **254**, 974–80, 1991.

58 V Denisov, B Halle, Hydrogen exchange rates in proteins from water ^{1}H transverse magnetic relaxation, *J Am Chem Soc* **124**, 10264–5, 2002.

59 V Denisov, B Halle, Protein hydration dynamics in aqueous solution: a comparison of bovine pancreatic trypsin inhibitor and ubiquitin by oxygen-17 spin relaxation dispersion, *J Mol Biol* **245**, 682–97, 1995.

60 V Denisov, B Halle, Protein hydration dynamics in aqueous solution, *Faraday Discuss* **1996**, 227–44, 1996.

61 V Denisov, B Halle, J Peters, H Horlein, Residence times of the buried water molecules in bovine pancreatic trypsin inhibitor and its G36S mutant, *Biochemistry* **34**, 9046–51, 1995.

62 C Dalvit, Homonuclear ^{1}D and ^{2}D NMR Experiments for the observation of solvent–solute interactions, *J Magn Reson B* **112**, 282–8, 1996.

63 C Dalvit, Effective multi-solvent suppression for the study of organic solvents with macromolecules, *J Biomol NMR* **11**, 437–44, 1998.

10
^{13}C- and ^{15}N-Isotopic Labeling of Proteins

Christian Klammt, Frank Bernhard and Heinz Rüterjans

10.1
Introduction

Nuclear magnetic resonance (NMR) spectroscopy has experienced a tremendous growth over the past decade, and has been developed as a widely used technique with enormous potential for the study of the structure and dynamics of biological macromolecules. In particular, structural analysis of proteins has greatly benefited from recent advancements of high field solution NMR. Improvements in the hardware and the development of NMR spectrometers with field strengths up to 900 MHz considerably increased the sensitivity and reduced the requirements for high amounts of samples. Furthermore, major milestones in the study of proteins have been the use of isotope labeling and the development of multidimensional techniques. The structure of small proteins below a limit of approximately 10 kDa could be analyzed without the need for special isotope labeling after the development of homonuclear ^{1}H two-dimensional (2-D) experiments in the late 1970s and early 1980s. However, beyond that limit, the increased complexity due to line broadening and ^{1}H chemical shift overlap prevents detailed structural analysis of larger proteins by these techniques. The structural analysis of large proteins therefore requires multinuclear labeling and special strategies in order to replace the prevalent NMR inactive isotopes ^{14}N and ^{12}C in proteins by the NMR active isotopes ^{13}C and ^{15}N [1]. Protocols have been developed to generate either uniformly $^{13}C/^{15}N$-labeled proteins or to introduce the label only at specific sites. In addition, heteronuclear multidimensional NMR experiments have been designed that utilize the relatively large one- and two-bond scalar couplings introduced by the label [2, 3]. Using the ^{13}C and/or ^{15}N chemical shifts, the overlapping ^{1}H resonances can now be routinely separated into three or four dimensions. The combination of these techniques addresses the limitations imposed on the study of large proteins and permits structure determination of proteins up to a molecular weight of approximately 22 kDa [4, 5]. This limit could be extended up to 40 kDa by the random incorporation of ^{2}H into nonexchangeable sites of proteins [6–9].

In this chapter, recently developed strategies for the incorporation of ^{15}N and ^{13}C labels into proteins resulting in highly sensitive spectra and facilitating struc-

Molecular Biology in Medicinal Chemistry. Edited by Th. Dingermann, D. Steinhilber, G. Folkers

ISBN: 3-527-30431-2

tural analysis of high-molecular-weight systems are discussed. Conventional labeling strategies require the overproduction of proteins in bacterial or eukaryotic expression systems and we give an overview of the most advanced systems as they are indispensable prerequisites for the efficient incorporation of isotopes into proteins. In addition to structural studies, NMR plays an increasingly important role in the description of protein dynamics, and relationships between dynamics and function. We therefore also address techniques of selective isotope labeling for the study of protein side-chain dynamics and protein–ligand interactions. The second part of the chapter focuses on newly developed *in vitro* techniques, utilizing cell-free extracts for the generation of protein samples. While cell-free production of proteins in an analytical scale has been possible for several years, preparative protein production using extracts from *Escherichia coli* or other sources has become one of the most powerful tools for the production of labeled samples suitable for NMR analysis. We therefore will especially address the recent advancements in cell-free expression and labeling strategies.

10.2
Expression Systems for the *In Vivo* Incorporation of ^{13}C and ^{15}N Labels into Proteins

Apart from technical barriers, the quality and stability of protein samples often limit their analysis by NMR technologies. The relatively low sensitivity of most NMR spectrometers requires sample concentrations starting from 0.2 mM and many proteins aggregate long before they reach that limit. In addition, samples must remain stable for at least several days until a complete set of measurements is finished. In order to obtain optimal resolution, NMR analysis usually requires relatively high temperatures of 15–25°C, which could also affect protein stability. Due to the high costs involved in the preparation of multinuclear-labeled proteins, one sample should be used for several experiments and aggregation, denaturation or even degradation of the proteins due to spurious amounts of proteases or due to autoproteolysis needs to be reduced to a minimum. Reasonably high yields of sample preparations are therefore required, which in conventional *in vivo* expression systems should reach a minimum of several milligrams of protein per liter of culture medium. In order to obtain the best protein production rates, the choice of the optimal expression system is therefore of primary importance. The most commonly used organism for labeled recombinant protein production is still the enterobacterium *E. coli*. An immense variety of expression systems and vectors have been designed for protein production in *E. coli*, and this large selection is certainly one of the major advantages for choosing an *E. coli* expression system. The predominant eukaryotic systems for protein production are based on the highly productive species *Pichia pastoris*. Labeling of proteins in mammalian cells is only rarely considered because of the high costs due to low yields concomitant with the need for expensive labeling precursors.

Labeled protein samples were initially generated by growing the cells in defined minimal media supplemented with specifically labeled compounds. Common label precursors include $^{15}NH_4Cl$, $(^{15}NH_4)_2SO_4$, $K^{15}NO_3$ and $Na^{15}NO_3$ for nitrogen labeling, and $NaH^{13}CO_3$, [^{13}C]glycerol, [^{13}C]glucose and [^{13}C]acetate for carbon labeling. A variety of supplements such as trace element mixtures and vitamin cocktails have been tested to enhance the cellular growth in minimal media and to improve the productivity. Furthermore, single- or double-labeled mixtures of amino acids or complex algal or microbial hydrolyzates have been added to culture media [10–13] in order to produce uniformly labeled proteins. In order to reduce the isotope costs, cells may be grown in unlabeled rich media to high cell densities, harvested and then suspended in low volumes of labeled defined medium for protein production. High levels of isotope incorporation concomitant with a 4- to 8-fold increase in yield were obtained in *E. coli* with this technique [14, 15].

10.2.1
Protein Production and ^{13}C and ^{15}N-labeling in *E. coli* Expression Systems

Techniques for isotope enrichment in *E. coli* have been available for several years [16–18] and the vast majority of labeled proteins analyzed by NMR spectroscopy are still being produced by heterologous expression in various *E. coli* hosts yielding up to 50% of the total cell protein. Despite considerable improvements in eukaryotic and cell-free expression systems, bacterial expression remains the fastest and most economical system for the production of proteins. Standard methods of generating heteronuclear-labeled samples in *E. coli* use M9 minimal medium [19] or modified derivatives of it – supplemented with [^{13}C]glucose and [^{13}C]glycerol for carbon labeling, and $(^{15}NH_4)_2SO_4$ and $^{15}NH_4Cl$ for nitrogen labeling. An important advantage when using *E. coli* expression is the elaborated selection of vectors and host strains available. Upon overproduction of proteins, it is often essential that the transcription of the target gene can be repressed as tightly as possible until the cells reach the most suitable growth phase for induction. Leaky expression prior to induction could either favor the accumulation of mutations in the target gene in order to suppress the production of unwanted proteins, or even damage or kill the cells due to toxic effects.

The most popular inducible promoters in *E. coli* expression systems are listed in Tab. 10.1. Repression of the strong *E. coli lac* promoter is performed by interaction

Tab. 10.1 Commonly used promoters for protein overproduction in *E. coli*.

Promoter	*Origin*	*Repression*	*Induction*
P_{lac}	*E. coli*	LacI	isopropyl β-thiogalactoside (IPTG)
P_{ara}	*E. coli*	AraR	L-arabinose
P_{tet}	*E. coli*, Tn10	TetR	tetracycline
$\phi1$–$\phi10$	phage T7 of T7 RNA polymerase	repression/inactivation	requires T7 RNA polymerase, IPTG
λ_{PL}	phage λ	λ *cI* repressor	heat inducible, temperature shift of 42°C

of its repressor LacI with specific operator sequences in P_{lac}, thus preventing the *E. coli* RNA polymerase from binding. The LacI protein is released from its operator by the addition of isopropyl β-thiogalactoside (IPTG), a nonmetabolizable analog of the natural inducer. A disadvantage of the *lac* promoter is its relatively high background expression. This could be partly suppressed by an increasing copy number of LacI in expression hosts either by introducing additional copies of the *lacI* gene on a plasmid or by increasing the expression of the chromosomal *lacI* gene by an up mutation of its native promoter, called *lacI*^{q}. Several derivatives, like the *tac* or *trc* promoter, have been constructed by fusion of P_{lac} with other promoters, e.g. from the tryptophan operon.

In expression systems based on the highly specific T7 promoters, the protein production is induced by the supply of T7 RNA polymerase [20]. In principal, the T7 RNA polymerase can be provided in several ways. A common method is the expression in specially designed strains carrying a chromosomal copy of T7 gene 1, encoding T7 RNA polymerase. This construct, termed "DE3", represents a transcriptional fusion of gene 1 with the IPTG-inducible *lac* promoter. To ensure tight repression, gene 1 is accompanied by an extra copy of the *lacI*^{q} gene. In order to eliminate any background of T7 RNA polymerase due to leaky expression, the plasmid-borne gene encoding for T7 lysozyme can be provided. T7 lysozyme efficiently binds and inactivates T7 RNA polymerase, and this expression system even allows the production and labeling of toxic proteins in *E. coli*.

While the structural analysis of the native protein is predominantly the major task of a labeling approach, most samples were initially produced as translational fusions to other proteins or to artificial peptide-tags (Tab. 10.2). Those strategies can provide valuable advantages for the fast and efficient purification of the labeled protein or for its stabilization [21]. Furthermore, the addition of a suitable fusion partner to the N-terminus of the target protein can significantly enhance its production. Fusion proteins usually include a recognition site for a highly specific protease like enterokinase or the TEV protease in the linker between the two proteins. If necessary, the fusion partners could then be cleaved off from the target proteins after purification, leaving the nonmodified labeled protein for analysis.

Tab. 10.2 Popular tags and carrier proteins for protein overproduction in fusion systems.

Fusion	*Effect on solubility*[a]	*Purification*
Poly(His)$_6$-tag	no effect	immobilized metal chelate chromatography
Strep-tag	no effect	affinity chromatography
Maltose-binding protein	++	affinity chromatography
NusA	++	
Glutathione S-transferase	+	affinity chromatography
Thioredoxin	+	
Protein G (B1 domain)	+	
λ head protein D	+	
Protein A (Z domain)	+	affinity chromatography

[a] Effect on the solubility of the target protein.

The most frequent peptide-tag used for an easy purification is a poly(His)$_6$-tag added to the N- or C-terminus of the protein. Such small tags usually do not disturb the conformation or activity of proteins and their removal after purification might not even be necessary. Proteins containing a poly(His)$_6$-tag can easily be purified from a crude cell extract by forming a chelate complex with metal ions like Ni^{2+} or Cu^{2+} immobilized on a chromatography column. The imidazole moieties of adjacent histidine residues are crucial for firm chelate formation and the bound proteins can later be eluted from the column with a competing imidazole gradient. The production of proteins with a poly(His)$_6$-tag has become a routine technique for the heterologous expression of recombinant genes in any host.

Only approx. 25% of overproduced proteins are initially suitable or stable enough for structural analysis [22] and most experiments must be optimized in order to obtain sufficient yields of correctly folded proteins. General parameters of optimization are the selection of suitable expression hosts and vectors, variation of growth conditions and inducer concentrations, construction of fusion proteins, and coexpression of helper proteins. A frequently encountered problem upon heterologous expression in *E. coli* is the formation of inclusion bodies – large aggregates of unfolded inactive protein. Protein targeting as insoluble inclusion bodies could be a strategy to circumvent bactericidal activities of the overproduced protein [23]. Several protocols have been described to solubilize inclusion bodies after purification in strong denaturants and to refold denatured proteins into their correct 3-D conformation [24, 25]. However, for each protein the refolding strategy has to be optimized and might represent an intensive laborious trial-and-error procedure giving less than satisfactory results. Thus, it is often intended to maximize the production of proteins in completely soluble form. Fortunately, in many cases, the formation of inclusion bodies during expression can be prevented by C-terminal linkage of the target protein to a highly soluble fusion partner. Several carrier proteins are able to enhance the solubility of an otherwise insoluble proteins (Tab. 10.2); in systematic screens for proteins suitable to confer solubility on insoluble target proteins fused to their C-terminal ends, the *E. coli* maltose-binding protein MalE [26] and NusA [27] proved to be the most efficient. As large carrier proteins have to be removed through specific proteases prior to NMR analysis, the cleavage of the fusion proteins can reintroduce problems with solubility and the released target proteins sometimes tend to precipitate soon after they are no longer covalently attached to a carrier. In addition, the use of large carrier proteins would waste a considerable amount of the precious label precursors. Smaller carrier proteins like the phage λ head protein D [28] or domains of *Staphylococcus* Protein A [29] or *Streptococcus* Protein G [30] might therefore be used as noncleavable tags to enhance the stability of smaller proteins or peptides. Fusions with the B1 domain of Protein G even improved NMR spectra of labeled proteins [30]. As well as enhancing solubility, carrier proteins can be used to facilitate the production of small peptides in *E. coli* and to stabilize them against degradation [23].

Misfolding can be a particular problem with the expression of eukaryotic proteins in *E. coli*, especially if they have several disulfide bonds, consist of multiple subunits or contain prosthetic groups. A popular strategy to address such problems

Tab. 10.3 Modifications of *E. coli* expression systems.

Problems/requirements	***Modifications***
Disulfide bridge formation	expression in *trx*$^-$ and *gor*$^-$ strains, providing an oxidizing environment in the cytoplasm
	coexpression of disulfide isomerases like DsbA, DsbC or PDI
	periplasmic expression using signal peptides
Chaperone-dependent folding	coexpression of chaperones like GroELS and/or DnaKJ/GrpE
Rare codon usage	coexpression of corresponding tRNAs

is the coexpression of accessory helper proteins (Tab. 10.3). If folding of the overproduced protein depends on chaperones, the increase of chaperone concentrations by coexpression of GroEL/ES, DnaKJ/GrpE or other chaperone systems could be highly advantageous. Low production rates due to translational pausing or misincorporation of amino acids as a result of rare codon usage of the overproduced protein could be optimized by the coexpression of corresponding tRNAs. A frequent problem, which in many cases still cannot be completely solved when using *E. coli* expression systems, is the correct formation of disulfide bonds in proteins. Strategies for the optimization of disulfide bond formation *in vivo* include the export of overproduced proteins into the oxidizing periplasm by addition of appropriate export signal sequences, expression in specially designed strains lacking thioredoxin and glutathione reductase and thus providing an oxidizing cellular environment [31], and coexpression of disulfide isomerases like the DsbA and DsbC proteins of *E. coli* or the eukaryotic chaperone protein disulfide isomerase (PDI).

10.2.2
^{13}C- and ^{15}N-labeling of Proteins in *P. pastoris*

Yeast cells combine many advantages of prokaryotic and eukaryotic expression systems. They are small, robust and easy to handle. The growth rate of yeast cells is much faster compared with other eukaryotic cells and the fermentation of yeast cells is usually finished within a couple of days. Furthermore, yeasts do not require complex supplements in the growth media and can be propagated in simple inexpensive media or on agar plates like *E. coli* cells. Their excellent growth on defined minimal media make them especially attractive for the production of labeled proteins for NMR analysis. In addition, due to the secretion capabilities of *P. pastoris*, recombinant proteins might be easily purified in a single step [32]. Many proteins of eukaryotic origin need to undergo specific posttranslational modifications like glycosylation or lipidation, disulfide bridge formation, or posttranslational processing. The modifications are often essential for the stability or enzymatic activity of proteins [33], or for the adoption of their native conformation. The failure to form correct disulfide bridges can retard or even block the folding process of proteins, resulting in the formation of inclusion bodies or in their degradation. The production of modified proteins cannot be addressed in a satisfactory manner by using prokaryotic hosts like *E. coli*. Expression in yeast cells like the methylotrophic *P. pastoris* may therefore be the method of choice when searching for a fast-growing host with the potential for the production of modified and glycosylated proteins [34, 35]. However, the fact that the protein modification pattern obtained after expression in yeast might differ from that usually present in the native protein has to be considered, e.g. yeasts are only capable of attaching mannose-rich glycans, representing the core polysaccharide of glycoproteins.

Labeling approaches should take into account that heterologous protein production in *P. pastoris* is correlated with a high cell density. Up to 100-fold increased yields can be obtained by fermentation of the cells in bioreactors as compared to shake flask cultures [36–38]. The induction of heterologous protein production should be started after reaching high cell densities. Yields of more than 100 mg protein/l culture medium have been reported [34, 38]. In shake flasks, the reported yields of uniformly labeled heterologous proteins vary from 2 mg/l in case of the Vaccinia virus complement control protein [39] to up to 27 mg/l for tick anticoagulant protein [36]. The average protein yield in yeast using shake flasks therefore only reaches the lower limit of comparable protein production in *E. coli*. It is furthermore important to realize that if compared with protein production in *E. coli*, much higher amounts of labeled isotopes are required to obtain a reasonable growth of the yeast cells, and at least 5- to 10-fold higher concentrations of $(^{15}NH_4)_2SO_4$ and [^{13}C]glucose are routinely used when growing the cells in shake flasks. To produce 90 mg of a ^{13}C-labeled fragment of thrombomodulin in *P. pastoris*, 143 g [^{13}C]glycerol or 162 g [^{13}C]glucose had to be used by growing the cells in a 1.25-l fermentor [34]. Therefore, based on economic aspects, labeling approaches with yeasts are only competitive in cases when the proteins cannot be produced in *E. coli*.

In defined minimal medium, the sole nitrogen source for *P. pastoris* usually is NH_4OH, which in addition serves as a base to neutralize considerable amounts of acid produced by the yeast during growth [34]. However, due to economic reasons, NH_4OH has to be replaced by $(^{15}NH_4)_2SO_4$ for ^{15}N-labeling of proteins in *P. pastoris*. The required high cell density demands the addition of high amounts of $(^{15}NH_4)_2SO_4$ and more than 40-folds concentration as compared with fermentations of *E. coli* had to be used. To avoid any negative effects on cell growth arising from the increased ionic strength of the medium through the formation of K_2SO_4, the $(^{15}NH_4)_2SO_4$ had to be added in several portions at different times [38], and optionally also in combination with an exchange of medium before induction of protein expression [34].

For ^{13}C-labeling of proteins using the strong *AOX1* promoter, two usually carbon sources are required during fermentation – glycerol or glucose to obtain a high cell density and methanol for the induction of the *AOX1* promoter. Although *P. pastoris* is able to grow on methanol as the sole carbon source, the slow doubling time would prevent high cell densities and requires the use of alternate carbon sources like glycerol or glucose. While glucose could certainly help to increase the cell density, it represses the strong *AOX1* promoter, leading to a decrease in heterologous protein production. However, [^{13}C]glucose could be used in labeling experiments as an initial carbon source until the desired high cell density is reached. The cells may then be further fermented with [^{13}C]glycerol as the carbon source and induced with [^{13}C]methanol [34]. Up to 30% of the carbon from the labeled protein originates from methanol and therefore it is crucial to induce the protein production with [^{13}C]-methanol in order to obtain fully labeled proteins, especially when using a yeast strain with a *mut*$^{+}$ genotype [34].

10.2.3 ^{13}C- and ^{15}N-labeling of Proteins by Expression in Chinese Hamster Ovary (CHO) Cells

Compared to yeast cells, CHO cells represent the alternative of a more advanced system enabling the generation of labeled samples containing all of the possible modifications specific to eukaryotic proteins. Like with yeast cells, recombinant proteins can be efficiently secreted into the medium and protein production rates of more than 100 mg/l from CHO cells have been reported [40]. However, mammalian cells usually require amino acids, vitamins, cofactors and in many cases a complex serum as essential additives to the medium. To reduce costs, CHO cells could also be grown by replacing the highly expensive isotopically labeled amino acids with labeled bacterial hydrolyzates or algal extracts and sufficient yields of recombinant protein highly enriched with isotopes can be obtained [10]. To avoid dilution of the isotope label with unlabeled amino acids from the serum, a dialysis step should be included to remove low-molecular-weight compounds from the serum [10, 41]. A further important and probably cost-intensive factor is the requirement of relatively high concentrations of labeled glutamine, which is essential for the metabolic pathways of several amino acids and nucleic acid components. Although CHO cells represent an expensive host for the generation of labeled proteins, they might be the best choice for the overexpression of complex glycoproteins or of proteins that cannot be produced in *E. coli* [41–43].

10.2.4 ^{13}C- and ^{15}N-labeling of Proteins in Other Organisms

A further option to produce labeled glycosylated proteins can be the overproduction in the slime mould *Dictyostelium discoideum*. This host is able to feed on bacterial cells like *E. coli* that could then be pre-enriched with isotopic labels using standard protocols. Suitable expression vectors for *D. discoideum* are available and a glycosylated 16-kDa protein has already been labeled with ^{13}C/^{15}N to a high extent [44]. As the growth rate of *D. discoideum* is rather low, the complete experiment took more than 10 days. However, if compared to labeling approaches using yeast cells as a host, lower amounts of labeled precursors are needed to obtain sufficient protein for NMR measurements, and *D. discoideum* might therefore be considered as an economic alternative to produce labeled glycoproteins.

While the previously discussed expression systems require relatively expensive mixtures of labeled precursors, the photoautotrophic cyanobacterium *Anabaena* sp.

can be grown in defined media containing $Na^{15}NO_3$ and $NaH^{13}CO_3$ as the sole nitrogen and carbon sources. A 24-kDa domain of the *E. coli* gyrase B subunit was successfully overproduced in *Anabaena* sp. by using a specially designed shuttle vector system and more than 90% $^{13}C/^{15}N$ label incorporation was obtained [45]. The yield of the protein (approximately 3–6 mg/l) was found to be comparable to that obtained in *E. coli*. It is noteworthy that the expressed gene was under control of the *E. coli tac* promoter, which is also functional in *Anabaena*. While ^{15}N-labeling in *Anabaena* can be carried out with standard growth protocols, ^{13}C-labeling is problematic under aerobic conditions because of the ability of endogenous CO_2 fixation, resulting in incomplete ^{13}C-labeling of less than 30% [45]. However, ^{13}C-enrichment of more than 90% could be obtained after growing the *Anabaena* cells with nitrogen gas aeration under controlled anaerobic conditions. The fermentation of *Anabaena* requires illumination and the duration is about 5 times longer compared to *E. coli* because of the slower doubling time. However, as less expensive labeled precursors can be used, the cost could be reduced to only about 10%. Heterologous protein production in *Anabaena* might therefore be an option for proteins with a low production rate for which higher volumes of medium have to be used. In addition, it has to be considered that the non-protonated nature of the final carbon supply might also be advantageous and cost-effective for the generation of perdeuterated protein samples necessary for the structural analysis of large proteins or dynamic studies.

Although heterologous protein production is the most common approach, small peptide products have also been labeled in their natural host. A crucial prerequisite is that the organisms can be adapted to grow on defined media supplemented with suitable labeled precursors. Examples of successful approaches are the labeling of cyclosporin A in *Tolypocladium inflatum* [46], alamethicin in *Trichoderma viride* [47], bellenamine in *Streptomyces nashvillensis* [48] and cyanophycin in *Synechocystis* sp. [49].

10.2.5 Strategies for the Production of Selectively ^{13}C- and ^{15}N-labeled Proteins

NMR spectra of uniformly labeled proteins become increasingly complex with the increasing size of the proteins. Several methods have been established to selectively label only specific domains, amino acids or even single nuclei of the protein in order to simplify the spectral analysis. Using *in vivo* labeling techniques, this can be done by feeding the host cells with specifically designed label precursors, which are obtained in most cases by chemical synthesis. The resulting spectral simplification facilitates the unambiguous assignment of resonances in large proteins. Choosing the correct labeling strategy can therefore be crucial to accelerate the assignment of resonances in proteins.

10.2.5.1 Selective Labeling of Amino Acids

In principal, any amino acid residue in an overproduced protein can be specifically labeled by providing excessive amounts of the amino acid containing the desired label in the growth medium. Resonances of single amino acids can be easily identified by comparison with spectra of a protein with other labeled amino acids and such approaches could considerably accelerate the resonance assignments of large proteins. However, the high costs for purified labeled amino acids prevent the application of this technique for routine use of structure determinations. The selective labeling strategy is suited for the dynamic and functional analysis of selected amino acid residues e.g. during ligand interactions. Specific labeling is also useful in the analysis of protein denaturation. Unfolded proteins generally have a low dispersion of resonances, but the tracking of a limited number of labeled residues, ideally well distributed throughout the protein, could enable the analysis of the unfolding process. In a previous study, selective labeling with [^{15}N]isoleucine was used to detect folding intermediates of a complex protein [50].

As an alternatively to specific labeling of amino acids, in reverse isotope labeling an excess of one or several nonlabeled amino acids is added to a growth medium in combination with commonly used general labeling compounds like $^{15}NH_4Cl$ or [^{13}C]glucose. An interesting application is the reverse labeling of aromatic residues like phenylalanine or tyrosine as their side-chains are frequently involved in ligand-binding interfaces or positioned in hydrophobic cores of proteins, making distance restraints for the environment of these residues highly valuable [51, 52]. The reverse isotope-labeling approach clearly depends on the complexity of the analyzed protein and is generally useful for smaller proteins containing only few of the analyzed amino acid residues. The application of amino acid-specific or selective isotope-labeling strategies is limited by scrambling effects of the isotope label to other types of residues. Specific auxotrophic mutants of *E. coli* can be used for the overproduction and specific labeling of recombinant proteins in order to minimize this problem, [53].

10.2.5.2 **Specific Isotope Labeling with ^{13}C**

Amino acids labeled at selected carbon positions can be added to defined growth media for their incorporation into the recombinant protein. These partially labeled compounds were usually generated by chemical synthesis and introduced to optimize the spectral resolution of large proteins in specific NMR experiments. Aromatic side-chain protons may be difficult to assign if the number of aromatic residues increases. Phenylalanine residues ^{13}C-labeled only at the ε position could thus help to rapidly assign the aromatic ring protons [54]. Furthermore, fully ^{13}C-labeled proteins have undesirable $^{13}C^{\alpha}$–$^{13}C^{\beta}$ scalar couplings. Incorporation of amino acids labeled in the ^{13}C backbone overcome this problem [55]. Fractional ^{13}C-labeling of amino acids can be achieved by growing the protein-producing cells with a defined mixture of ^{13}C- and ^{12}C-labeled carbon sources [56]. The analysis of internal side-chain dynamics within proteins in uniformly ^{13}C-labeled proteins is also complicated due to interference effects between different contributing relaxation interactions as well as by contributions from ^{13}C–^{13}C scalar and dipolar couplings. To solve this problem, proteins can be synthesized with an alternating ^{12}C–^{13}C–^{12}C-labeling pattern by an elaborate isotope labeling procedure using either [2-^{13}C]glycerol or [1,3-$^{13}C_2$]glycerol as the sole carbon source [57]. Further carbon-selective labeling strategies are highly valuable in combination with selective protonation to analyze the structure of large proteins [52].

10.2.5.3 Segmental Isotope Labeling

Despite advances in heteronuclear multidimensional NMR techniques, the increased complexity of spectra due to a lack of resolution and increased overlap of signals having similar chemical shifts still limits the structural analysis of large proteins. However, a promising approach to further extend the actual size limit of the structural evaluation of proteins by NMR spectroscopy is the segmental- or block-labeling technique. In principal, partially labeled full-length proteins can be produced by the posttranslational *in vitro* ligation of nonlabeled domains together with a labeled domain obtained in different expression experiments. Resonances of single domains of larger proteins may be sequentially assigned and the complete structure of the protein could be obtained by a combinatorial approach, dividing a large target protein into parts of manageable size. In contrast to a structural elucidation of isolated protein domains, one should bear in mind that the structure of labeled domains with the segmental-labeling technique is analyzed in the context of the complete protein, thus providing information of the structural and functional interaction between domains while preserving the overall structural character of the protein. The ligation process is catalyzed by self-splicing enzymes – the inteins [58]. Inteins are insertion sequences cleaved off after translation through self-excision, leaving the flanking protein regions, the exteins, joined together through native peptide bond formation.

Depending on the nature of the intein and on the solubility of the target domains, several modified protocols of the segmental labeling strategy have been established.

- The intein is cut in the middle of the sequence within a flexible loop region and the two fragments are expressed separately as fusions to the target domains. The isolated denatured proteins are mixed and the two intein fragments efficiently associate during refolding, resulting in the ligation of the two target domains [59, 60]. This strategy is especially applicable if the intein fusions are expressed as insoluble inclusion bodies and if an efficient protocol for refolding is available. Using different inteins at each splice junction, even proteins containing central labeled domains can be generated [61].
- The chemical ligation of folded proteins under native conditions uses an ethyl α-thioester at the C-terminal end of a recombinant protein, generated by cleavage of an intein fusion, for the ligation with a second protein or peptide containing a cysteine residue at its N-terminal end [62]. Both ligation partners can be combined in a correctly folded conformation at a physiological pH and this strategy principally could be extended to link more than two domains together.
- The mini intein from the *dnaE* gene of the cyanobacterium *Synechocystis* sp. comprises less than 200 amino acid residues, and is divided into N- and C-terminal fragments that are expressed separately. If combined after purification, the two fragments efficiently interact under native conditions to catalyze the

splicing and ligation reaction of covalently attached proteins [63, 64]. Target domains to be ligated have to be expressed as fusion N- and C-terminal to the intein fragments.

Although *in vitro* ligation yields as high as 90% have been obtained, several potential problems should be considered when starting a segmental labeling approach. While most enzymatically important residues are located within the inteins, an efficient excision and ligation mechanism also requires some residues from the extein, and the reaction is therefore not completely independent of the sequence of newly formed protein junctions [64]. The C-terminal extein requires a cysteine, threonine or serine residue at its N-terminal end. This sequence requirement must not affect the structure or activity of the analyzed protein. Refolding and ligation conditions must be optimized for each individual protein, and the splicing/ligation junction should be located in a loop region ensuring high flexibility. In principal, block labeling might apply best to proteins composed of independently folded domains having distinct biochemical properties.

10.3
Cell-free Isotope Labeling

Cell-free protein synthesis is an attractive and promising alternative to the conventional technologies for protein production using bacterial or eukaryotic cell cultures. In contrast to *in vivo* gene expression methods where protein synthesis is carried out in the cellular context surrounded by cell walls and membranes, cell-free protein synthesis provides a completely open system. This allows direct control and access to the reaction at any time. Compounds to improve protein production or to stabilize recombinant proteins, e.g. chaperones, chemicals, detergents or protease inhibitors, can easily be added without considering side-effects of the cellular metabolism or transport problems through the cell membrane. Most cellular functions with the exception of transcription and translation need not be maintained during cell-free protein expression. In principal, applications can therefore be extended to proteins or conditions that would not be tolerated by living organisms. While *in vivo* protein production is often limited by the formation of insoluble inclusion bodies or protein instability caused by intracellular proteases, the cell-free system offers new possibilities for the synthesis of complex proteins (Tab. 10.4). Protein folding and stability can be promoted by direct addition of chaperones, PDIs or necessary cofactors. In contrast, overexpression of chaperones *in vivo* can lead to cell filamentation or other undesirable phenotypes that can be detrimental for the viability of *E. coli* cells and protein expression [65]. Reaction parameters such as pH, redox potential and ionic strength can be determined without concern for harmful effects on the growth and viability of cells, and with the certainty that these parameters will directly influence the relevant reactions. This new opportunity of *in vitro* gene expression allows full control and high flexibility of conditions, and offers new potentials for difficulties associated with cyto-

Tab. 10.4 Potential advantages and disadvantages of cell-free protein synthesis.

Potential advantages	*Disadvantages*
Completely open system (easy access) → customization of reaction conditions to synthesized protein	Often low productivity
Incorporation of labeled, glycoslylated, modified or unnatural amino acids	Complex system compared to *in vivo* protein synthesis
Direct translation of PCR products → high-throughput screening	Expensive reaction compared to *in vivo* expression
Production of toxic proteins	Small number of commercial systems and kits
Production of membrane bound proteins	Difficult standardization because of multitude of reaction components
Expression of proteins requiring cofactors	High costs
Easy addition of chaperones or PDI	
Miniaturization (e.g. 50 μl reactions)	
Working without living organisms → no growth restrictions	
Allowing a direct isolation of products to shorten time required for preparing purified proteins	

toxicity, proteolytic degradation or improper folding and aggregation of synthesized proteins. The production of cytotoxic proteins [66], membrane proteins [67, 68] as well as the production of functional antibodies using PDI and chaperones [69] or functional viruses [70] had been reported. Most of all, cell-free protein synthesis offers completely new possibilities for the incorporation of labeled [71–74], glycosylated [75, 76], modified or unnatural amino acids [77]. Another promising potential for cell-free synthesis can be found in its suitability for high-throughput expression of proteins by direct translation of linear polymerase chain reaction (PCR) products [78]. Current limitations of cell-free systems are connected with their high complexity, high costs and often low productivity.

10.3.1
Components of Cell-free Expression Systems

In cell-free protein production, all components involved in gene expression and protein synthesis have to be added to the reaction mixture (Fig. 10.1). Components like DNA, nucleotide triphosphates (NTPs), messenger RNA (mRNA), transfer RNA (tRNA), aminoacyl-tRNA synthetases (ARSases), polymerases, ribosomes, transcription and translation factors like initiation factors (IFs), elongation factors (EFs) or release factors (RFs) and amino acids have to be optimized with regard to their concentration, and optimal salt and pH environment. The process of transcription/translation requires large amounts of free energy supplied by the hydrolysis of the triphosphates ATP and GTP. Therefore, *in vitro* protein synthesis requires an ATP-regenerating energy system to maintain the triphosphate concentration. For this purpose, high-energy phosphate donors such as acetyl phosphate

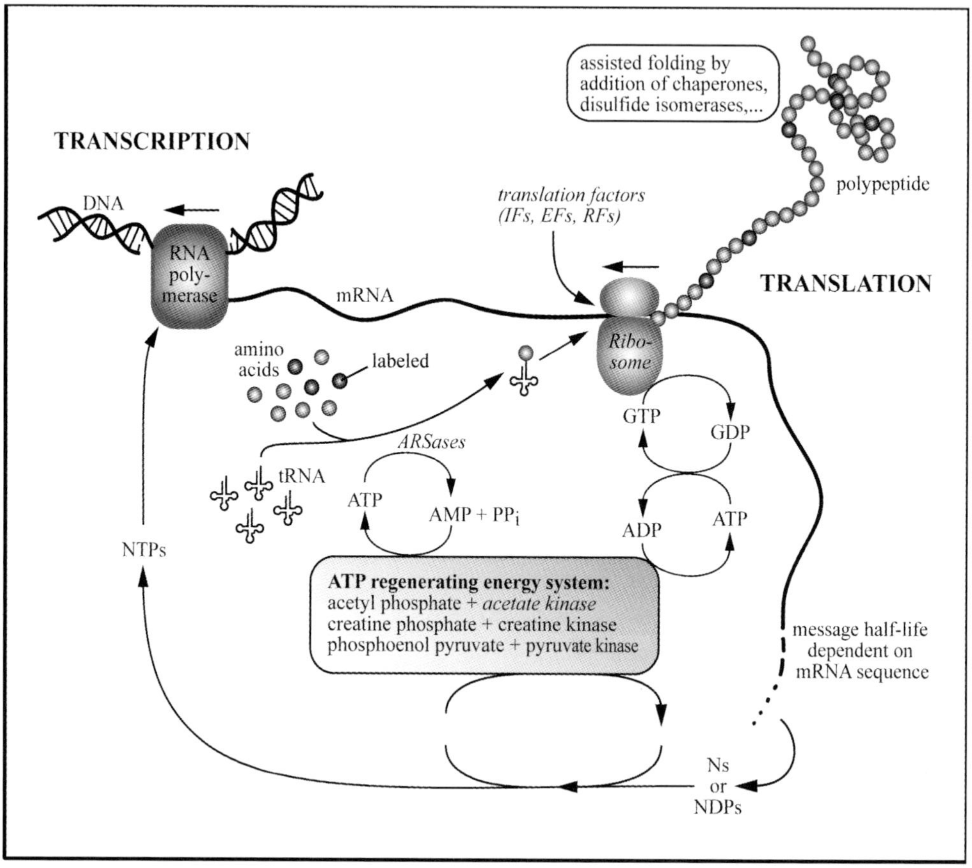

Fig. 10.1 Schematic to illustrate cell-free protein synthesis, showing the coupled process of transcription and translation in a bacterial system. Initially, cells are grown, lysed and the cell extract is prepared containing ribosomes, translation factors like initiation factors (IFs), elongation factors (EFs) and release factors (RFs), acetate kinase and aminoacyl-tRNA synthetases (ARSases). Substrates like amino acids, the energy-regenerating system components or nucleoside triphosphates (NTPs) and salts are then added to the extract, and protein synthesis is initiated by adding the template DNA. The DNA is transcribed by an added RNA polymerase. Added tRNA is loaded with amino acids by ARSases and they are used in the translation of mRNA. The incorporation of some stable isotope-labeled amino acids (dark) can easily be done in the cell-free system, leading to a selective isotope-labeled protein. Regeneration of ATP and GTP and even the NTPs in the cell-free system is achieved by an ATP-regenerating energy system based on the hydrolysis of high-energy substrates in the presence of their kinases. Chaperones can easily be added to the reaction mixture to assist the folding of the target protein.

(AcP), creatine phosphate (CrP) or phosphoenol pyruvate (PEP) in the presence of their kinases (acetate kinase, creatine kinase and pyruvate kinase) have been used.

A crucial component is a cell extract based on crude cell lysate which contains the necessary reaction components such as IFs, EFs, RFs, ARSases, tRNA and

Tab. 10.5 Bacterial and eukaryotic cell-free expression systems.

Bacterial systems		***Eukaryotic systems***	
Escherichia coli		***Wheat germ***	***Rabbit reticulocyte***
S30 extract preparation	***S100 extract preparation***	***Wheat germ extract***	***Rabbit reticulocyte lysates***
Supernatant fraction at 30,000 *g* centrifugation of *E. coli* extract preincubated to detract endogenous DNA and mRNA	Supernatant fraction at 100,000 *g* centrifugation deprived of all nucleic acids, e.g. by DEAE cellulose treatment	Directly used for expression of endogenous or exogenous templates	Treated with micrococcal Ca^{2+} dependent RNase
Contains: RNA polymerase, ribosomes, tRNAs, ARSases, translation factors	Contains: RNA polymerases, ARSases and translation factors Ribosomes, tRNAs added	Contains: Ribosomes, tRNAs, ARSases, translation factors	Contains: Ribosomes, tRNAs, ARSases, translation factors
24–38°C (optimum at 37°C)[a]	24–38°C (optimum at 37°C)[a]	20–27°C up to 32°C[a]	30–38°C[a]
[79, 82, 83, 132, 133][b]	[80, 81][b]	[85–87][b]	[82, 89, 98, 113][b]

[a] Reaction temperature.
[b] References.
To be added:
DNA (plasmid with appropriate promotor for SP6, T7 RNA-polymerase) + polymerase (bacteriophage SP6 or T7 RNA polymerase) or mRNA.
Amino acids.
Energy components: ATP and GTP.
NTP-regeneration system: PEP + PK or CP + CrP or AcP.
Formyltetrahydrofolate or its congener, e.g. methenyltetrahydrofolate or folinic acid.
SH-compound: mercaptoethanol or dithiothreitol.
Mg^{2+} and K^+ in optimal concentrations.
In some cases:
Ca^{2+} and NH_4^+ in optimal concentrations.
cAMP.
Polyamines, e.g. spermidine.

enzymes for energy regeneration like acetate kinase. Cell-free expression systems are classified according to the origin of their extract. In principle, functional *in vitro* systems can be prepared from any cell type, but many factors contribute to the protein production efficiency. The most common *in vitro* reactions are based on extracts made from *E. coli*, wheat germ or rabbit reticulocyte lysates (Tab. 10.5). *E. coli* extracts consist of the so-called S30 supernatant fraction, named after the soluble fraction when centrifuged at 30,000 g, containing endogenous ribosomes, enzymes like acetate kinase and factors necessary for transcription and translation, ARSases, tRNA and mRNA. Endogenous mRNA is removed from the ribosomes during preincubation of the crude cell extract in a "run-off" step and destroyed by endogenous ribonucleases [79]. Another way to deploy bacterial extracts is described in the Gold–Schweiger system [80, 81]. Ribosomes are added to the supernatant fraction of a S100 extract especially purified from endogenous amino acids and nucleic acids by ion-exchange chromatography. This system provides a very low background due to endogenous synthesis and better-controlled conditions at the expense of more complicated preparation.

The *E. coli* system functions well in a temperature range of 24–38°C with an optimum at 37°C [79, 82, 83]. Wheat germ extract possesses a low level of endogenously expressed messengers and therefore can be directly used for expression of endogenous [84] or exogenous templates. In the wheat germ system, the optimum temperature is in the range 20–27°C [85, 86], but can be increased to up to 32°C for higher expression of some templates [87]. Reticulocyte extract is prepared by directly lysing blood cells of anemic rabbits; this increases the number of proerythrocytes or reticulocytes that are subsequently treated with micrococcal Ca^{2+}-dependent RNase to remove endogenous mRNA [88]. This system works in a temperature range of 30–38°C [89]. With regard to the reaction temperature, it should be noted that apart from any effect on the enzymatic process of transcription/translation and mRNA degradation, the temperature affects the folding of the synthesized protein. To assist the folding of proteins, chaperones like the GroEL/ES system [90] or DnaK and DnaJ [69] can be added to the reaction mixture. Usually, lower temperatures will lead to higher yields of recombinant protein in the absence of chaperones, but not necessarily in their presence [91]. Furthermore, to assist disulfide bond formation, PDI can be added to the transcription/translation mixture [69].

Less common translation systems based on yeast extracts [92], mammalian cells like human HeLa and mouse L-cells [93] or on tobacco chloroplasts, which strongly depend on exogenously added mRNA [94], have high levels of degradation and relatively low protein yields. With regard to the cell-free labeling of proteins, only *E. coli* and wheat germ extracts have been used. Using a very elaborate approach, Shimizu et al. developed a cell-free translation system reconstructed from purified poly(His)-tagged translation factors [95]. Their system, termed the "protein synthesis using recombinant elements" (PURE) system, contains 32 individually purified components with high specific activity, allowing efficient protein production. An advantage of the PURE system, apart from the absence of inhibitory substances such as nucleases, proteases and enzymes that hydrolyze nucleoside triphosphates, is the simple purification of the synthesized protein by removing tagged protein factors by affinity chromatography.

One limitation of the cell-free system is the degradation of exogenous added mRNA. Various RNase activities present in cell extracts usually restrict the lifetime of mRNA, and subsequently the efficiency of protein synthesis is inhibited. This problem could be solved by the periodical reintroduction of messengers into the reaction mixture in a simple translation system [96]. Coupled transcription–translation systems, where mRNA is continuously synthesized from DNA templates added to the reaction mixture, can be advantageous to translation systems containing presynthesized messengers. Direct transcription in the reaction mixture may be executed from appropriate promoters by endogenous *E. coli* RNA polymerase, or by exogenous phage RNA polymerase in bacterial systems [97] or in eukaryotic systems [82, 98]. To avoid rapid messenger degradation, especially in *E. coli* cell-free systems, partially ribonuclease-depleted extracts or RNase inhibitors are used. A good choice of template DNA in the prokaryotic system is circular plasmid DNA. In the wheat germ system with lower nuclease activities, both plasmid DNAs [99] and linear PCR fragments function well [100–102]. The translational efficiency of mRNA depends on its structural features. Most cDNA sequences can be sufficiently well expressed without addition of translational enhancers. The sequence of interest only needs to be provided with a favorable Kozak sequence in eukaryotic cell-free systems and the Shine–Dalgarno sequence in prokaryotic cell-free systems, which are the respective sequences upstream of the ATG codon responsible for translation initiation.

A further problem in cell-free protein synthesis is the high consumption of biochemical energy provided by ATP and GTP. Creatine phosphate concomitant with creatine kinase is usually used for ATP and GTP regeneration in eukaryotic cell-free systems, whereas the combination of PEP and pyruvate kinase, acetyl phosphate and acetate kinase or a combination of both has been applied for bacterial *in vitro* protein synthesis. The acetyl phosphate energy system may have the advantage that the ATP level is maintained twice as long as in the presence of PEP. Since acetate kinase is present at sufficient levels in bacterial extracts, it does not need to be added exogenously in the *E. coli* system [103]. Studies of the biochemical energy levels in different cell-free systems observed a high rate of triphosphate hydrolysis to mono- and diphosphates during protein synthesis in wheat germ extracts [104] as well as in an *E. coli* S30 extract [105]. It is reported that more than 80% of ATP and GTP hydrolysis in the wheat germ system initially occurs independently of protein synthesis and it was suggested that acid phosphatases are responsible for the nonspecific hydrolysis of the nucleotide triphosphates [86, 104].

Recent extensive studies to increase the protein yield of cell-free reactions have focused on the composition of the reaction mixture, especially the amino acids composition [106, 107], and methods to prepare the cell extract. For example, endogenous phosphatases have been removed by using S30 extract prepared from the spheroplasts of *E. coli* [108] or by immunodepletion of the phosphatase [109]. Approaches to concentrate the extract components by ultrafiltration [110] or by dialysis [74] have also been reported.

10.3.2
Cell-free Expression Techniques

In cell-free reactions carried out in a "batch" mode, the reaction conditions change as a result of substrate consumption and the accumulation of products. Translation stops as soon as any essential substrate is exhausted, or any product or by-product reaches an inhibiting concentration. Actually, the bacterial cell-free systems are active only for 10–30 min at 37°C. Systems based on rabbit reticulocyte lysates or wheat germ extract are typically capable of working for up to 1 h. However, *in vitro* protein-synthesizing systems in batch mode work well for most analytical purposes, but short lifetimes and low productivities limit their application for the

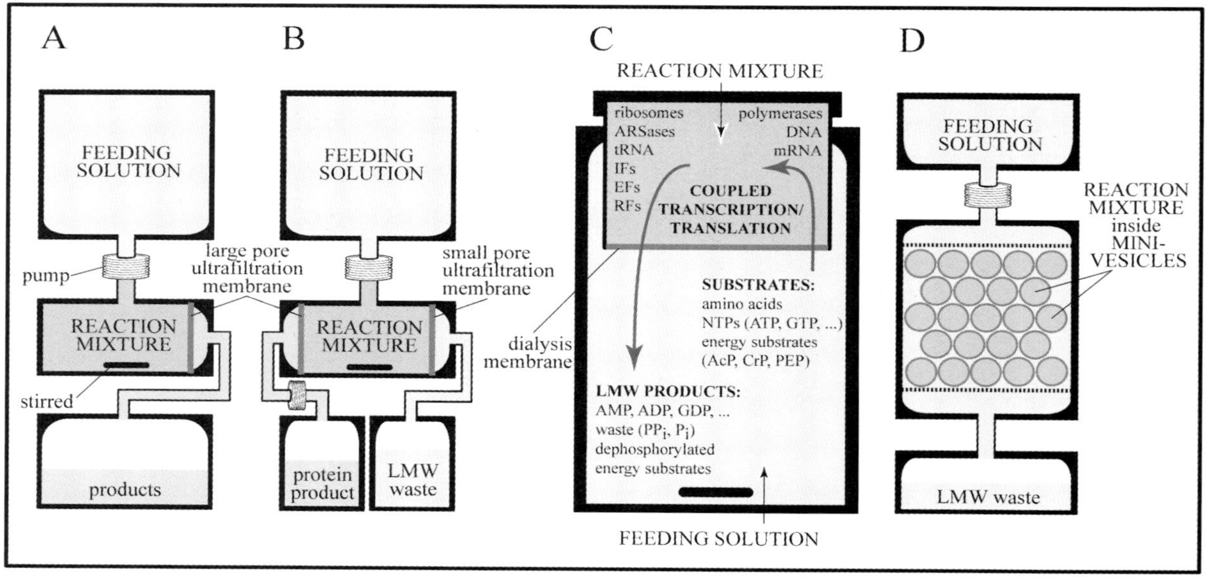

Fig. 10.2 Illustration of cell-free reactors for continuous flow (A and B) and continuous-exchange (C and D) cell-free expression systems. (A) Schematic drawing of a direct-flow CFCF reactor where substrates are supplied and products are removed by the flow of a feeding solution, forced by a pump. (B) Y-flow CFCF reactor, containing two ultrafiltration membranes of different pore size, separating protein product and low-molecular-weight waste outflows [123]. (C) CECF reactor design with explanation of feeding solution (light grey) and reaction mixture components (dark grey). Compartments are separated by a semipermeable dialysis membrane, providing the diffusion exchange of substrates and low-molecular-weight products, and the retention of reaction mixture components. (D) Flow-exchange column reactor consisting of semipermeable mini-vesicles packed in a column, containing the reaction mixture surrounded by feeding solution, changed by flow through the column. Low-molecular-weight products and substrates are changed by diffusion through the mini-vesicle membrane [113]. Modified after Shirokov *et al.* [134].

synthesis of preparative amounts of protein. The reasons for the low yield are degradation of mRNA, depletion of nucleotide triphosphates and accumulation of their hydrolyzates. Prolonged reaction times in cell-free expression were first achieved by Spirin et al. [111–113] by using a continuous-flow cell-free (CFCF) translation device (Fig. 10.2A). The basic idea is to continuously supply amino acids, energy-regenerating components (AcP, CrP or PEP) and NTPs in a feeding solution, and continuously remove small molecule byproducts (mainly products of triphosphate hydrolysis like inorganic phosphates and nucleoside monophosphates) by active (forced) flow of the feeding solution across an ultrafiltration membrane (molecular weight cut-off in the range of 10–300 kDa). In this case, all products, including the protein synthesized, are continuously removed from the reaction compartment if the pore size is large enough (Fig. 10.2A). Permanent stirring of the reaction mixture and in some set-ups upright flow of the feeding solution are applied to minimize membrane clogging [114]. The CFCF system can function for more than 20 hours and results in preparative protein expression of about 0.1–1 mg protein/ml reaction volume or higher. The template for this system can either be mRNA [113, 115], DNA transcribed by endogenous bacterial RNA polymerase or added phage RNA polymerase [82, 113, 114], or self-replicating RNA [116]. Using the CFCF system, proteins like bacteriophage MS2 coat protein [111], brome mosaic virus coat protein [111], calcitonin polypeptide [113], globin [117], functionally active dihydrofolate reductase (DHFR) [112, 113, 118, 119], chloramphenicol acetyltransferase (CAT) [105, 113, 114, 116, 120, 121] interleukin (IL)-2 [122] and IL-6 [115] have successfully been synthesized.

The important advantage of the CFCF system is the continuous removal of synthesized proteins from the reaction mixture, which can result in 80–85% purity of the protein product as previously demonstrated in the outflow of a bacterial CFCF system synthesizing DHFR [119] and in a wheat germ CFCF system synthesizing IL-6 [115]. The synthesized protein diffuses out of the CFCF reactor in a large volume of the effluent and is quite diluted. To reduce the dilution effect, the so-called Y-flow reactor with a split outflow has been proposed [123]. The Y-flow reactor (Fig. 2B) has two membranes with different pore sizes. Initially the low-molecular-weight products are removed through a small-pore membrane at a high rate and the synthesized protein is subsequently collected through a large-pore membrane at a low rate. Here, the flow of the protein product is controlled by a separate pump. Nevertheless, a number of laboratories attempting CFCF expression have had difficulties in establishing this complex system. The main problems are RNA degradation when using bacterial extracts, even in the coupled transcription/translation mode, and low efficiency of initiation complex formation, which might cause leakage and therefore loss of some translation components by ultrafiltration or the blockage of the ultrafiltration membrane [91].

However, there is an alternative way of carrying out prolonged protein synthesis, namely by using diffusion instead of pumping to supply substrates and remove low-molecular-weight products (Fig. 2C). This set-up, using a reaction mixture separated from a feeding solution by applying a dialysis membrane, is called a continuous-exchange cell-free system (CECF) [124] or a semicontinuous-flow cell-free system (SFCF) [121]. The simplest device for the continuous supply of substrates and the removal of low-molecular-weight products by passive exchange with a feeding mixture is a dialysis bag [124]. In practice, simple homemade dialysis bags or standard commercial dialyzers, such as the MicroDialyzer® and DispoDialyzer® from Spectrum, can be used successfully. The pore size of the reactor membranes is usually in the range of 10–50 kDa. However, better performance of reactors with a larger pore membrane has been reported [74]. Stirring of either the feeding solution or both, the feeding and reaction mixture, is necessary for efficient supply of substrates.

Using this approach, Kim and Choi [121] reported the synthesis of 1.2 mg/ml of CAT over 14 h in *E. coli* S30 extract. The product yield, quantified by ELISA, exceeded the yield of the analogous batch reaction by 10–12 times. More recently, Kigawa et al. [74] succeeded in synthesizing CAT and Ras proteins in amounts up to 6 mg/ml over 18 h in their version of the bacterial CECF system, and Madin *et al.* [96] reached yields of 1–4 mg/ml for several functionally active proteins like DHFR, green fluorescent protein (GFP), luciferase and RNA replicase of tobacco mosaic virus (TMV). Figure 10.3 describes the cell-free synthesis of GFP in a CECF method using a MicroDialyzer®.

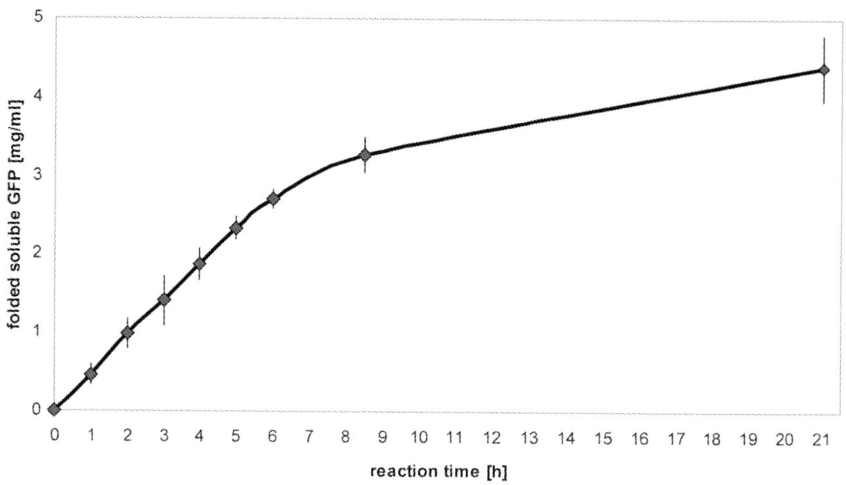

Fig. 10.3 Cell-free synthesis of GFP in a bacterial CECF system using a 100-µl MicroDialyzer®. The kinetic points display the mean of three reactions and error bars indicating the deviation of the mean. The reaction was carried out at 30°C using a 17-fold amount of feeding mixture compared to the reaction mixture volume and under the following reaction conditions: 0.05% NaN_3, 2% phosphoenolglycol, 196 µM folinic acid, 2 mM 1.4-dithiothreitol, 1.2 mM ATP, 0.8 mM each of CTP, GTP and UTP, 20 mM PEP, 20 mM acetyl phosphate, 1 mM of each amino acid, except the amino acids arginine, aspartic acid, cysteine, glutamic acid, methionine and tryptophan, for which 2 mM was used, 100 mM HEPES–KOH (pH 8.0), 2.8 mM EDTA, 1× complete protease inhibitor (Roche), 280 mM K^+, 13 mM Mg^{2+}, 40 µg/ml pyruvate kinase, 500 µg/ml tRNA (*E. coli*), 3 U/µl T7-RNA polymerase, 0.3 U/µl RNasin, 35% S30 extract (*E. coli*) and 15 µg/ml plasmid.

The advantage of the CECF system is the accumulation of the synthesized product in the reaction mixture. The synthesis of proteins and polypeptides fused with GFP provide a direct and demonstrative way to visualize product accumulation by fluorescence of the GFP moiety. The synthesis of a HIV protein, the so-called Nef antigen, fused with GFP [125] and an antibacterial polypeptide Cecropin P1 fused with GFP [126], both in the bacterial CECF T7 transcription/translation system, have been reported.

Reactors combining both exchange and flow have also been developed. In one version, the reaction mixture is encapsulated into polysaccharide minivesicles that can be packed into a column, where feeding solution is passed through the column (Fig. 10.2D). In this case, product–substrate exchange across the vesicle walls takes place during the flow [113]. More sophisticated versions of the CECF reactor are being developed in order to meet the demands of scientists and biotechnologists. The first commercial CECF reactor has recently been launched on the market by Roche Diagnostics. The Roche CECF system promises a high protein yield of more than 2 mg/ml. Other commercial systems focus on the batch mode, where recent improvements have been made. For example, a novel NTP regeneration system, avoiding accumulation of inorganic phosphate by adding pyruvate oxidase, which generates AcP from pyruvate and inorganic phosphate directly in the reaction mixture, has been proposed by Kim and Swartz [127]. Later they showed that addition of oxalate, a potent inhibitor of PEP synthetase, substantially increases the yield of CAT synthesis through the enhanced supply of ATP by about 47% [128]. Furthermore, Kim and Swartz developed a "fed-batch" mode where coordinated addition of PEP, magnesium, and the amino acids arginine, cysteine and tryptophan resulted in a final concentration of cell-free synthesized CAT that was more than 4-fold compared to a batch reaction [107]. As a result of these improvements it became possible to synthesize about 350 μg/ml CAT [107] or 450 μg/ml recombinant DNA human protein thrombopoietin [106] in a batch reaction.

10.3.3 Specific Applications of the Cell-free Labeling Technique

Conventional *in vivo* labeling techniques are often accompanied by low protein yields due to retarded growth in a minimal medium and, in the case of selective isotope labeling, by scrambling effects that drastically reduce the efficiency and selectivity of labeling. A major advantage of cell-free labeling techniques therefore is the absence of any scrambling effects or metabolic conversion of labeled amino acids [71, 73, 129]. The selective incorporation of isotope-labeled amino acids in cell-free synthesized proteins for NMR research has already been demonstrated by several groups [71, 73, 74, 129–131]. Similarly, in the reverse isotope-labeling technique, most amino acids are single or dual (^{15}N, ^{13}C) labeled, except for a few residues [74]. Figure 10.4(A) illustrates the 2-D ^{1}H–^{15}N heteronuclear single-quantum correlation spectroscopy (HSQC) NMR spectra of the cell-free reverse isotope-labeled C-terminal DNA-binding domain of the transcriptional response regulator RcsB (cRcsB) which has been uniformly ^{15}N-labeled except for aspar-

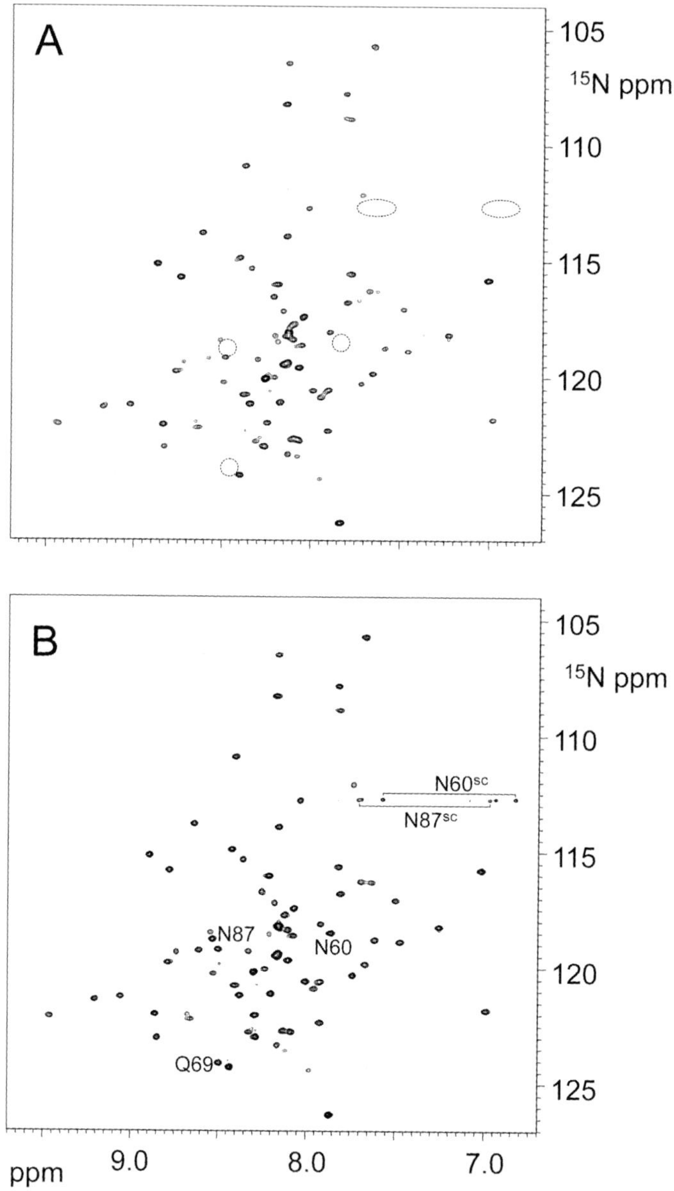

Fig. 10.4 Comparison of the 2-D ^{1}H–^{15}N HSQC spectra of the *in vitro* and *in vivo* generated purified C-terminal domain of the bacterial transcriptional regulator RcsB (cRcsB) using a Bruker DRX-800 MHz NMR spectrometer with a cryoprobe. (A) Cell-free ^{15}N-reverse isotopic-labeled cRcsB synthesized in an optimized CECF system using a DispoDialyzer®. All amino acids except asparagine (N) and glutamine (Q) were ^{15}N-labeled. As expected, the N and Q residues (circles) are absent in the 2-D ^{1}H–^{15}N HSQC spectrum. (B) Uniformly ^{15}N isotopic-labeled cRcsB expressed *in vivo*. This spectrum is in good agreement with that of the reverse cell-free isotopically labeled cRcsB, and N and Q residues can easily be identified. (NMR spectra kindly provided by Frank Löhr.)

agine (N) and glutamine (Q) residues. The uniformly ^{15}N-labeled cRcsB expressed *in vivo* is shown in Fig. 10.4(B). The 2-D $^{1}H-^{15}N$ HSQC spectrum of the labeled cRcsB protein synthesized *in vitro* is consistent with that of the uniformly labeled protein synthesized *in vivo* (only the asparagine and glutamine peaks are missing).

Cell-free expression is highly suited for the generation of protein samples labeled only at distinct residues. Site-specific labeling is extremely useful to simplify the observation and resonance assignment procedures for specified amino acid residues of particular interest, and can also be used for analyzing local structures of large proteins and protein–protein interactions. Ellman et al. demonstrated for the first time the incorporation of a particular ^{13}C-labeled residue at a suppressible termination codon by translating the modified sequence in an *in vitro* system supplemented with the charged suppressor tRNA. Subsequently it became possible to track the labeled residues upon denaturation and refolding of the protein by NMR spectroscopy [130]. Milligram quantities of site-specific isotope labeled protein can be obtained in a cell-free system involving the amber suppression strategy [73]. The *E. coli* amber suppressor tRNA can be prepared by *in vitro* transcription with T7 RNA polymerase and later aminoacylated with the appropriate purified *E. coli* amino ARSase and the cognate labeled amino acid. The codon for the selected amino acid residue in the protein was previously changed into an amber codon by standard techniques.

Segmental labeling *in vivo* is limited by specific requirements like organization of the target protein into domains, presence of specific residues and folding problems. Pavlov et al. suggested a cell-free technique based on *in vitro* translation of matrix-coupled mRNAs, which principally is devoid of any sequence and conformational requirements [72]. The size of the labeled region is controlled by codon usage and no intrinsic upper limit to the size of proteins that can be isotope-labeled in selected regions exists. This method is based on the usage of translation mixtures depleted of either one amino acid and/or its tRNA and/or its amino ARSase and consists of three steps. Using column-coupled template RNA the reaction mixture can easily be exchanged. Initially, the unlabeled N-terminal region, using an unlabeled reaction mixture, is synthesized up to the first codon without a matching amino acyl-tRNA in the extract. Here the ribosomes pause and the translation mixture can be exchanged against a mix containing isotope-labeled amino acids, now deficient for a different amino acid, tRNA or amino ARSase. Translation resumes, thereby labeling the region until the ribosomes encounter the next codon without the corresponding tRNA. In the last step, the C-terminal part is synthesized without any label and the protein is released from the ribosome. However, the technique results in low protein yields, as it can, at the very best, produce protein stoichiometric to the immobilized mRNA.

A very promising advantage of cell-free isotope labeling is the possibility of *in situ* NMR measurement as described by Guignard et al. They showed NMR analysis of *in vitro*-synthesized proteins without any chromatographic purification and with minimal sample handling using an optimized CECF reaction combined with the sensitivity of a cryoprobe [131]. As expected, they observed no cross peak for any excess ^{15}N amino acid and only the target protein was enriched with the isotope-labeled residues. They suggest a new possibility for inexpensive high-throughput protein analysis applicable in large-scale proteomics, where selectively labeled proteins can be expressed in 0.5 ml of reaction medium using small quantities of labeled amino acids and analyzed by NMR. All steps from the expression to the completed NMR spectra were done in less than 24 h.

Due to the exceptional advantages of cell-free protein synthesis, isotope labeling can be achieved for proteins that are normally difficult to express. Using cell-free labeling techniques, it might become possible to synthesize and isotope label proteins that are toxic, require cofactors or chaperones for adopting an active conformation, or even membrane proteins. Subunit labeling of oligomers, further research on disulfide bridge formation or protein oxidation should be possible in the near future.

References

1 Dötsch V, Wagner G. 1998. New approaches to structure determination by NMR spectroscopy. *Curr Opin Struct Biol* **8**, 619–23.

2 Pervushin K, Riek R, Wider G, Wüthrich K. 1997. Attenuated T2 relaxation by mutual cancellation of dipole–dipole coupling and chemical shift anisotropy indicates an avenue to NMR structures of very large biological macromolecules in solution. *Proc Natl Acad Sci USA* **94**, 12366–71.

3 Tjandra N, Bax A. 1997. Direct measurement of distances and angles in biomolecules by NMR in a dilute liquid crystalline medium. *Science* **278**, 1111–4.

4 Clore GM, Gronenborn AM. 1991. Structures of larger proteins in solution: three- and four-dimensional heteronuclear NMR spectroscopy. *Science* **252**, 1390–9.

5 Bax A. 1994. Multidimensional nuclear-magnetic-resonance methods for protein studies. *Curr Opin Struct Biol* **4**, 738–44.

6 Clore GM, Gronenborn AM. 1997. NMR structures of proteins and protein complexes beyond 20,000 M_r. *Nat Struct Biol* **4**, 849–53.

7 Kay LE, Gardner KH. 1997. Solution NMR spectroscopy beyond 25 kDa. *Curr Opin Struct Biol* **7**, 722–31.

8 Gardner KH, Kay LE. 1998. The use of ^{2}H, ^{13}C, ^{15}N multidimensional NMR to study the structure and dynamics of proteins. *Annu Rev Biophys Biomol Struct* **27**, 357–406.

9 Wider G, Wüthrich K. 1999. NMR spectroscopy of large molecules and multimolecular assemblies in solution. *Curr Opin Struct Biol* **9**, 594–601.

10 Hansen AP, Petros AM, Mazar AP, Pederson TM, Rueter A, Fesik SW. 1992. A practical method for uniform isotopic labeling of recombinant proteins in mammalian cells. *Biochemistry* **31**, 12713–8.

11 Sailer M, Helms GL, Henkel T, Niemczura WP, Stiles ME, Vederas JC. 1993. ^{15}N- and ^{13}C-labeled media from *Anabaena* sp. for universal isotopic labeling of bacteriocins: NMR resonance assignments of leucocin A from *Leuconostoc gelidum* and Nisin A from *Lactococcus lactis*. *Biochemistry* **32**, 310–8.

12 Reilly D, Fairbrother WJ. 1994. A novel isotope labeling protocol for bacterially expressed proteins. *J Biomol NMR* **4**, 459–62.

13 Brandner G, Phalip V, Kellenberger E, Zhang CC. 1999. Optimization of a cyanobacterial culture for production of isotope-labeled recombinant proteins. *Biotechnol Tech* **13**, 177–80.

14 Cai M, Huang Y, Sakaguchi K, Cloe GM, Gronenborn AM, Craigie R. 1998. An efficient and cost-effective isotope labeling protocol for proteins

expressed in *Escherichia coli*. *J Biomol NMR* **11**, 97–102.

15 Marley J, Lu M, Bracken C. 2001. A method for efficient isotopic labeling of recombinant proteins. *J Biomol NMR* **20**, 71–5.

16 Muchmore DC, McIntosh LP, Russel CB, Anderson DE, Dahlquist FW. 1989. Expression and nitrogen-15 labeling of proteins for proton and nitrogen-15 nuclear magnetic resonance. *Methods Enzymol* **177**, 44–73.

17 Ikura M, Kay LE, Bax A. 1990. A novel approach for sequential assignment of ^{1}H, ^{13}C, and ^{15}N spectra of proteins: heteronuclear triple-resonance three-dimensional NMR spectroscopy. Application to calmodulin. *Biochemistry* **29**, 4659–67.

18 Venters RA, Calderone TL, Spicer LD, Fierke CA. 1991. Uniform ^{13}C isotope labeling of proteins with sodium acetate for NMR studies: application to human carbonic anhydrase II. *Biochemistry* **30**, 4491–4.

19 Sambrook J, Fritsch EF, Maniatis T. 1989. *Molecular Cloning: A Laboratory Manual*, 2nd edn, Cold Spring Harbor Laboratory, Cold Spring Harbor, NY.

20 Studier FW, Rosenberg AH, Dunn JJ, Dubendorf JW. 1990. Use of T7 RNA polymerase to direct expression of cloned genes. *Methods Enzymol* **185**, 60–89.

21 Nilsson J, Stahl S, Lundeberg J, Uhlen M, Nygren PA. 1997. Affinity fusion strategies for detection, purification and immobilization of recombinant proteins. *Protein Express Purif* **11**, 1–16.

22 Christendat D, Yee A, Dharamsi A, Kluger Y, Savchenko A, Cort JR, Booth V, Mackereth CD, Saridakis V, Ekiel I, Kozlov G, Maxwell KL, Wu N, McIntosh LP, Gehring K, Kennedy MA, Davidson AR, Pai EF, Gerstein M, Edwards AM, Arrowsmith CH. 2000. Structural proteomics of an archaeon. *Nat Struct Biol* **7**, 903–9.

23 Majerle A, Kidric J, Jerala R. 2000. Production of stable isotope enriched antimicrobial peptides in *Escherichia coli*: an application to the production of a ^{15}N-enriched fragment of lactoferrin. *J Biomol NMR* **18**, 145–51.

24 De Bernardez CE. 1998. Refolding of recombinant proteins. *Curr Opin Biotechnol* **9**, 157–63.

25 Lilie H, Scharz E, Rudolph R. 1998. Advances in refolding of proteins produced in *E coli*. *Curr Opin Biotechnol* **9**, 497–501.

26 Kapust RB, Waugh DS. 1999. *Escherichia coli* maltose binding protein is uncommonly effective at promoting the solubility of polypeptides to which it is fused. *Protein Sci* **8**, 1668–74.

27 Davis, DD, Elisee C, Newham DM, Harrison RG. 1999. New fusion protein systems designed to give soluble expression in *Escherichia coli*. *Biotechnol Bioeng* **65**, 382–8.

28 Forrer P, Jaussi R. 1998. High-level expression of soluble heterologous proteins in the cytoplasm of *Escherichia coli* by fusion to the bacteriophage Lambda head protein D. *Gene* **224**, 45–52.

29 Jansson M, Li YC, Jendeberg L, Anderson S, Montelione BT, Nilsson B. 1996. High-level production of uniformly ^{15}N- and ^{13}C-enriched fusion proteins in *Escherichia coli*. *J Biomol NMR* **7**, 131–41.

30 Zhou P, Lugovskoy AA, Wagner G. 2001. A solubility-enhancement tag (SET) for NMR studies of poorly behaving proteins. *J Biomol NMR* **20**, 11–4.

31 Bessette PH, Aslund F, Beckwith J, Georgiou G. 1999. Efficient folding of proteins with multiple disulfide bonds in the *Escherichia coli* cytoplasm. *Proc Natl Acad Sci USA* **96**, 13703–8.

32 De Lamotte F, Boze H, Blanchard C, Klein C, Moulin G, Gautier MF, Delsuc MA. 2001. NMR onitoring of accumulation and folding of ^{15}N-labeled protein overexpressed in *Pichia pastoris*. *Protein Express Purif* **22**, 318–24.

33 Wyss DF, Wagner G. 1996. The structural role of sugars in glycoproteins. *Curr Opin Biotechnol* **7**, 409–16.

34 Wood MJ, Komives EA. 1999. Production of large quantities of isotopically labeled protein in *Pichia pastoris* by fermentation. *J Biomol NMR* **13**, 149–59.

35 Cereghino JL, Cregg JM. 2000. Heterologous protein expression in the methylotrophic yeast *Pichia pastoris. FEMS Microbiol Rev* **24**, 45–66.

36 Laroche Y, Storme V, De Meutter J, Messens J, Lauwereys M. 1994. High-level secretion and very efficient isotopic labeling of tick anticoagulant peptide (TAP) expressed in the methylotrophic yeast *Pichia pastoris. Biotechnology* **12**, 1119–24.

37 White CE, Hunter MJ, Meininger DP, White LR, Komives EA. 1995. Large-scale expression, purification and characterization of small fragments of thrombomodulin: the roles of the sixth domain and of methionine 388. *Protein Eng* **8**, 1177–87.

38 Van den Burg HA, de Wit PJGM, Vervoort J. 2001. Efficient $^{13}C/^{15}N$ double labeling of the avirulence protein AVR4 in a methanol-utilizing strain (Mut^{+}. of *Pichia pastoris. J Biomol NMR* **20**, 251–61.

39 Wiles AP, Shaw G, Bright J, Perczel A, Campell ID, Barlow PN. 1997. NMR studies of a viral protein that mimics the regulation of complement activation. *J Mol Biol* **272**, 253–65.

40 Hensley P, McDecitt PJ, Brooks I, Trill JJ, Feild JA, McNulty DE, Connor JR, Griswold DE, Kumar NV, Kopple KD. 1994. The soluble form of E-selectin is an asymmetric monomer. Expression, purification, and characterization of the recombinant protein. *J Biol Chem* **269**, 23949–58.

41 Lustbader JW, Birken S, Pollak S, Pound A, Chait BT, Mirza UA, Ramnarain S, Canfield RE, Brown JM. 1996. Expression of human chorionic gonadotropin uniformly labeled with NMR isotopes in Chinese hamster ovary cells: an advance toward rapid determination of glycoprotein structures. *J Biomol NMR* **7**, 295–304.

42 Archer SJ, Bax A, Roberts AB, Sporn MB, Ogawa Y, Piez KA, Weatherbee JA, Tsang ML, Lucas R, Zheng B, Wenker J, Torchia DA. 1993. Transforming growth factor $\beta 1$, NMR signal assignments of the recombinant protein expressed and isotopically enriched using Chinese hamster ovary cells. *Biochemistry* **32**, 1152–63.

43 Wyss DF, Dayie KT, Wagner G. 1997. The counterreceptor binding site of human CD2 exhibits an extended surface patch with multiple conformations fluctuating with millisecond to microsecond motions. *Protein Sci* **6**, 534–42.

44 Cubbedu L, Moss CX, Swarbrick JD, Gooley AA, Williams KL, Curmi PMG, Slade MB, Mabbutt BC. 2000. *Dictyostelium discoideum* as expression host: isotopic labeling of a recombinant glycoprotein for NMR studies. *Protein Express Purif* **19**, 335–42.

45 Desplancq D, Kieffer B, Schmidt K, Posten C, Forster A, Oudet P, Strub JM, Van Dorsselaer A, Weiss E. 2001. Cost-effective and uniform ^{13}C- and ^{15}N-labeling of the 24-kDa N-terminal domain of the *Escherichia coli* gyrase B by overexpression in the photoautotrophic cyanobacterium *Anabaena* sp. PCC 7120. *Protein Express Purif* **23**, 207–17.

46 Senn H, Weber C, Kobel H, Traber R. 1991. Selective ^{13}C-labelling of cyclosporin A. *Eur J Biochem* **199**, 653–8.

47 Yee AA, O'Neil JD. 1992. Uniform ^{15}N labeling of a fungal peptide: the structure and dynamics of an alamethicin by ^{15}N and ^{1}H NMR spectroscopy. *Biochemistry* **31**, 3135–43.

48 Ikeda Y, Naganawa H, Kondo S, Takeuchi T. 1992. Biosynthesis of bellenamine by *Streptomyces nashvillensis* using stable isotope labeled compounds. *J Antibiot (Tokyo)* **45**, 1919–24.

49 Suarez C, Kohler SJ, Allen MM, Kolodny NH. 1999. NMR study of the metabolic ^{15}N isotopic enrichment of cyanophycin synthesized by the cyano-

bacterium *Synechocystis* sp. strain PCC 6308. *Biochim Biophys Acta* **1426**, 429–38.

50 Biekofsky RR, Martin SR, McCormick JE, Masino L, Fefeu S, Bayley PM, Feeney J. 2002. Thermal stability of calmodulin and mutants studied by ^{1}H–^{15}N HSQC NMR measurements of selectively labeled [^{15}N]Ile proteins. *Biochemistry* **41**, 6850–9.

51 Vuister GW, Kim SJ, Wu C, Bax A. 1994. 2D and 3D NMR-study of phenylalanine residues in proteins by reverse isotope labeling. *J Am Chem Soc* **116**, 9206–10.

52 Goto NK, Kay LE. 2000. New developments in isotope labeling strategies for protein solution NMR spectroscopy. *Curr Opin Struct Biol* **10**, 585–92.

53 Waugh DS. 1996. Genetic tools for selective labeling of proteins with alpha-^{15}N-amino acids. *J Biomol NMR* **8**, 184–92.

54 Wang H, Janowick DA, Schkeryantz JM, Liu X, Fesik SW. 1999. A method for assigning phenylalanines in proteins. *J Am Chem Soc* **121**, 1611–2.

55 Coughlin PE, Anderson FE, Oliver EJ, Brown JM, Homans SW, Pollak S, Lustbader JW. 1999. Improved resolution and sensitivity of triple-resonance NMR methods for the structural analysis of proteins by use of a backbone-labeling strategy. *J Am Chem Soc* **121**, 11871–4.

56 Atreya HS, Chary KVR. 2001. Selective "unlabeling" of amino acids in fractionally ^{13}C labeled proteins: an approach for stereospecific NMR assignments of CH_3 groups in Val and Leu residues. *J Biomol NMR* **19**, 267–72.

57 LeMaster DM, Kushlan DM. 1996. Dynamical mapping of *E. coli* thioredoxin via ^{13}C NMR relaxation analysis. *J Am Chem Soc* **118**, 9255–64.

58 Perler FB. 1998. Protein splicing of inteins and hedgehog autoproteolysis: structure, function, and evolution. *Cell* **92**, 1–4.

59 Yamazaki T, Otomo T, Oda N, Kyogoku Y, Uegaki K, Ito N, Ishino Y, Nakamura H. 1998. Segmental isotope labeling for protein NMR using peptide splicing. *J Am Chem Soc* **120**, 5591–2.

60 Otomo T, Teruya K, Uegaki K, Yamazaki T, Kyogoku Y. 1999. Improved segmental isotope labeling of proteins and application to a larger protein. *J Biomol NMR* **14**, 105–14.

61 Otomo T, Ito N, Kyogoku Y, Yamazaki T. 1999. NMR observation of selected segments in a larger protein: central-segment isotope labeling through intein-mediated ligation. *Biochemistry* **38**, 16040–4.

62 Xu R, Ayers B, Cowburn D, Muir TW. 1999. Chemical ligation of folded recombinant proteins: segmental isotopic labeling of domains for NMR studies. *Proc Natl Acad Sci USA* **96**, 388–93.

63 Wu H, Hu Z, Liu XQ. 1998. Protein *trans*-splicing by a split intein encoded in a split DnaE gene of *Synechocystis* sp. PCC6803. *Proc Natl Acad Sci USA* **95**, 9226–31.

64 Evans TC, Martin D, Kolly R, Panne D, Sun L, Ghosh I, Chen L, Benner J, Liu XQ, Xu MQ. 2000. Protein *trans*-splicing and cyclization by a naturally split intein from the *dnaE* gene of *Synechocystis* species PCC6803. *J Biol Chem* **275**, 9091–4.

65 Blum P, Ory J, Bauernfeind J, Krska J. 1994. Physiological consequences of DnaK and DnaJ overproduction in *Escherichia coli*. *J Bacteriol* **174**, 7436–44.

66 Stiege W, Erdmann VA. 1995. The potentials of the *in vitro* biosynthesis system. *J Biotechnol* **41**, 81–90.

67 Falk MM, Buehler LK, Kumar NM, Gilula NB. 1997. Cell-free synthesis and assembly of connexins into functional gap junction membrane channels. *EMBO J* **16**, 2703–16.

68 Huppa JB, Ploegh HL. 1997. *In vitro* translation and assembly of a complete T cell receptor-CD3 complex. *J Exp Med* **186**, 393–403.

69 Ryabova LA, Desplancq D, Spirin AS, Pluckthun A. 1997. Functional antibody production using cell-free translation: effects of protein disulfide

isomerase and chaperones. *Nat Biotechnol* **15**, 79–84.

70 Molla A, Paul AV, Wimmer E. 1991. Cell-free, *de novo* synthesis of poliovirus. *Science* **254**, 1647–51.

71 Kigawa T, Muto Y, Yokoyama S. 1995. Cell-free synthesis and amino acid-selective stable isotope labelling of proteins for NMR analysis. *Biomol NMR* **2**, 129–34.

72 Pavlov MY, Freistroffer DV, Ehrenberg M. 1997. Synthesis of region-labelled proteins for NMR studies by *in vitro* translation of column-coupled mRNAs. *Biochimie* **79**, 415–22.

73 Yabuki T, Kigawa T, Dohmae N, Takio K, Terada T, Ito Y, Laue ED, Cooper JA, Kainosho M, Yokoyama S. 1998. Dual amino acid-selective and site-directed stable-isotope labelling of the human c-Ha-Ras protein by cell-free synthesis. *J Biomol NMR* **11**, 295–306.

74 Kigawa T, Yabuki T, Yoshida Y, Tsutsui M, Ito Y, Shibata T, Yokoyama S. 1999. Cell-free production and stable-isotope labelling of milligram quantities of proteins. *FEBS Lett* **442**, 15–9.

75 Joseph SK, Boehning D, Pierson S, Nicchitta CV. 1997. Membrane insertion, glycosylation, and oligomerization of inositol triphosphate receptors in a cell-free translation system. *J Biol Chem* **272**, 1579–588.

76 Duszenko M, Kang X, Bohne U, Homke R, Lehner M. 1999. *In vitro* translation in a cell-free system from *Trypanosoma brucei* yields glycosylated and glycosylphosphatidyl-inositol-anchored proteins. *Eur J Biochem* **266**, 789–97.

77 Park Y, Luo J, Schultz PG, Kirsch JF. 1997. Noncoded amino acid replacement probes of the aspartate aminotransferase mechanism. *Biochemistry* **36**, 10517–25.

78 Ohuchi S, Nakano H, Yamane T. 1998. *In vitro* method for the generation of protein libraries using PCR amplification of a single DNA molecule and coupled transcription/translation. *Nucleic Acids Res* **26**, 4339–346.

79 Zubay G. 1973. *In vitro* synthesis of protein in microbial systems. *Annu Rev Genet* **7**, 267–87.

80 Gold LM, Schweiger M. 1969. Synthesis of phage-specific x- and β-glucosyl transferases directed by T-even DNA *in vitro*. *Proc Natl Acad Sci USA* **62**, 892–8.

81 Gold LM, Schweiger M. 1971. Synthesis of bacteriophage-specific enzymes directed by DNA *in vitro*. *Methods Enzymol* **20**, 537–42.

82 Baranov VI, Spirin AS. 1993. Gene expression in cell-free systems on preparative scale. *Methods Enzymol* **217**, 123–42.

83 Kim DM, Kigawa T, Choi CY, Yokoyama S. 1996. A highly efficient cell-free protein synthesis system from *Escherichia coli*. *Eur J Biochem* **239**, 881–6.

84 Roberts BE, Pateron BM. 1973. Efficient translation of tobacco mosaic virus RNA and rabbit globin 9S RNA in a cell-free system from commercial wheat germ. *Proc Natl Acad Sci USA* **70**, 2330–4.

85 Anderson CW, Straus JW, Dudock BS. 1983. Preparation of a cell-free protein-synthesizing system from wheat germ. *Methods Enzymol* **101**, 635–44.

86 Kawarasaki Y, Nakano H, Yamane T. 1994. Prolonged cell-free protein synthesis in a batch system using wheat germ extract. *Biosci Biotechnol Biochem* **58**, 1911–3.

87 Tulin EE, Ken-Ichi T, Shin-Ichiro E. 1995. Continuously coupled transcription-translation system for the production of rice cytoplasmic aldolase. *Biotechnol Bioeng* **45**, 511–6.

88 Pelham HRB, Jackson RJ. 1976. An efficient mRNA-dependent translation system from reticulocyte lysates. *Eur J Biochem* **67**, 247–56.

89 Woodward WR, Ivey JL, Herbert E. 1974. Protein synthesis with rabbit reticulocyte preparations. *Methods Enzymol* **30**, 724–31.

90 Tsalkova T, Zardeneta G, Kudlicki W, Kramer G, Korwitz PM, Hardesty B. 1993. GroEL and GroES increase the specific enzymatic activity

of newly-synthesized rhodanese if present during *in vitro* transcription/translation. *Biochemistry* **32**, 3377–80.

91 Jermutus L, Ryabova LA, Pluckthun A. 1998. Recent advantages in producing and selecting functional proteins by using cell-free translation. *Curr Opin Biotechnol* **9**, 534–48.

92 Iizuka N, Sugiura M. 1997. Translation-component extract from Saccharomyces cerevisieae: effects of L-A RNA, 5′ cap, 3′ poly(A) tail on translational efficiency of mRNAs. *Methods* **11**, 353–60.

93 Tolunay HE, Yang L, Kemper WM, Safer B, Anderson WF. 1984. Homologous globin cell-free transcription system with comparison of heterologous factors. *Mol Cell Biol* **4**, 17–22.

94 Hirose T, Sugiura M. 1996. *Cis*-acting element and *trans*-acting factors for accurate translation of chloroplast *psbA* mRNAs: development of an *in vitro* translation system from tobacco chloroplasts. *EMBO J* **15**, 1687–695.

95 Shimizu Y, Inoue A, Tomari Y, Suzuki T, Yokogawa T, Nishikawa K, Ueda T. 2001. Cell-free translation reconstituted with purified components. *Nat Biotechnol* **19**, 751–5.

96 Madin K, Sawasaki T, Ogasawara T, Endo Y. 2000. A highly efficient and robust cell-free protein synthesis system prepared from wheat embryos: plants apparently contain a suicide system directed at ribosomes. *Proc Natl Acad Sci USA* **97**, 559–64.

97 Chen HZ, Zubay G. 1983. Prokaryotic coupled transcription–translation. *Methods Enzymol* **101**, 674–90.

98 Craig D, Howell MT, Gibbs CL, Hunt T, Jackson RJ. 1992. Plasmid cDNA-directed synthesis in a coupled eukaryotic *in vitro* transcription–translation system. *Nucleic Acids Res* **20**, 4987–995.

99 Nevin DE, Pratt JM. 1991. A coupled *in vitro* transcription–translation system for the exclusive synthesis of polypeptides from the T7 promotor. *FEBS Lett* **291**, 259–63.

100 Lesley SA, Brow MA, Burgess PR. 1991. Use of *in vitro* protein synthesis from polymerase chain reaction-generated templates to study interaction of *Escherichia coli* transcription factors with core RNA polymerase and for epitope mapping of monoclonal antibodies. *J Biol Chem* **266**, 2632–8.

101 Burks EA, Chen G, Georgiou G, Iverson BL. 1997. *In vitro* scanning saturation mutagenesis of an antibody binding pocket. *Proc Natl Acad Sci USA* **94**, 412–7.

102 Martemyanov KA, Spirin AS, Gudkov AT. 1997. Direct expression of PCR products in a cell-free translation/transcription system: synthesis of antibacterial peptide cecropin. *FEBS Lett* **414**, 268–70.

103 Ryabova LA, Vinokurov LM, Shekhovtsova EA, Alakhov YB, Spirin AS. 1995. Acetyl phosphate as an energy source for bacterial cell-free translation systems. *Anal Biochem* **226**, 184–6.

104 Yao SL, Shen XC, Suzuki E. 1997. Biochemical energy consumption by wheat germ extract during cell-free protein synthesis. *J Ferment Bioeng* **84**, 7–13.

105 Kitaoka Y, Nishimura N, Niwano M. 1996. Cooperativity of stabilized mRNA and enhanced translational activity in the cell-free system. *J Biotechnol* **48**, 1–8.

106 Patnaik R, Swartz JR. 1998. *E. coli*-based *in vitro* transcription/translation: *in vivo*-specific synthesis rates and high yields in a batch system. *BioTechniques* **24**, 862–8.

107 Kim DM, Swartz JR. 2000. Prolonging cell-free protein synthesis by selective reagent additions. *Biotech Prog* **16**, 385–90.

108 Kang SH, Oh TJ, Kim RG, Kang TJ, Hwang SH, Lee EY, Choi CY. 2000. An efficient cell-free protein synthesis system using periplasmic phosphatase-removed S30 extract. *J Microbiol Methods* **43**, 91–6.

109 Kawarasaki Y, Nakano H, Yamane Y. 1998. Phosphatase-immunodepleted cell-free protein synthesis system. *J Biotechnol* **61**, 199–208.

110 Nakano H, Tanaka T, Kawarasaki Y, Yamane T. 1994. An increased rate of

cell-free protein synthesis by condensing wheat germ extract with ultrafiltration membranes. *Biosci Biotechnol Biochem* **58**, 631–4.

111 Spirin AS, Baranov VI, Ryabova LA, Ovodov SY, Alakhov YB. 1988. A continuous cell-free translation system capable of producing polypeptides in high yield. *Science* **242**, 1162–4.

112 Baranov VI, Morozov IY, Ortlepp SA, Spirin AS. 1989. Gene expression in a cell-free system on the preparative scale. *Gene* **84**, 463–6.

113 Spirin AS. 1991. Cell-free protein synthesis bioreactor. In: Todd P, Sikdar SK, Beer M, eds, *Frontiers in Bioprocessing II*, 31–43. American Chemical Society, Washington, DC.

114 Kigawa T, Yokoyama S. 1991. A continuous cell-free protein synthesis system for coupled transcription–translation. *J Biochem* **110**, 166–8.

115 Volyanik EV, Dalley A, Mc Kay IA, Leigh I, Williams NS, Bustin SA. 1993. Synthesis of preparative amounts of biologically active interleukin-6 using a continuous-flow cell-free translation system. *Anal Biochem* **214**, 289–94.

116 Ryabova L, Volyanik E, Kurnasov P, Spirin AS, Wu Y, Kramer FR. 1994. Coupled replication–translation of amplifiable messenger RNA: A cell-free protein synthesis system that mimics viral infection. *J Biol Chem* **269**, 1501–5.

117 Ryabova LA, Ortleb SA, Baranov VI. 1989. Preparative synthesis of globin in a continuous cell-free translation system from rabbit reticulocytes. *Nucleic Acids Res* **17**, 4412.

118 Endo Y, Otsuzuki S, Ito K, Miura K. 1992. Production of an enzymatic active protein using a continuous flow cell-free system. *J Biotech* **25**, 221–30.

119 Kudlicki W, Kramer G, Hardesty B. 1992. High efficiency cell-free synthesis of proteins: refinement of the coupled transcription/translation system. *Anal Biochem* **206**, 389–93.

120 Ryabova LA, Morozov IY, Spirin AS. 1998. Continuous-flow cell-free translation, transcription–translation, and replication–translation system. I. *Methods Mol Biol* **77**, 179–93.

121 Kim DM, Choi CY. 1996. A semi-continuous prokaryotic coupled transcription/translation system using a dialysis membrane. *Biotechnol Prog* **12**, 645–9.

122 Kolosov MI, Kolosova IM, Alokhov VY, Ovodov SY, Alokhov YB. 1992. Preparative *in vitro* synthesis of bioactive human interleukin-2 in a continuous flow translation system. *Biotech Appl Biochem* **16**, 125–33.

123 Biryukov SV, Simonenko PN, Shirokov VA, Majorov SG, Spirin AS. 1999. Method of preparing polypeptides in cell-free system and devise for its realization. *PCT WO 00 50436.*

124 Alakov YB, Baranov VI, Ovodov SJ, Ryabova LA, Spirin AS, Morozov IJ. 1995. Method of preparing polypeptides in cell-free translation system. *US Patent 5478730.*

125 Chekulayeva MN, Kurnasov OV, Shirokov VA, Spirin AS. 2001. Continuous-exchange cell-free protein-synthesizing system: synthesis of HIV-1 antigen Nef. *Biochem Biophys Res Commun* **280**, 914–7.

126 Martemyanov KA, Shirokov VA, Kurnasov OV, Gudkov AT, Spirin AS. 2001. Cell-free production of biologically active polypeptides: application to the synthesis of antibacterial polypeptide cecropin. *Protein Expr Purif* **21**, 456–61.

127 Kim DM, Swartz JR. 1999. Prolonging cell-free protein synthesis with a novel ATP regeneration system. *Biotech Bioeng* **66**, 180–8.

128 Kim DM, Swartz JR. 2000. Oxalate improves protein synthesis by enhancing ATP supply in a cell-free system derived from *Escherichia coli. Biotechnol Lett* **22**, 1537–42.

129 Betton JM, Palibroda N, Namane A, Metzler T, Bârzu O. 2002. Selective labelling of proteins in the RTS 500 System. In Spirin AS, ed., *Cell-free Translation Systems*, 219–5, Springer, Berlin.

130 Ellman JA, Volkman BF, Mendel B, Schultz PG, Wemmer DE. 1992. Site-specific isotopic labeling of proteins for NMR studies. *J Am Chem Soc* **141**, 7959–961.

131 Guignard L, Ozawa K, Pursglove SE, Otting G, Dixon NE. 2002. NMR analysis of *in vitro*-synthesized proteins without purification: a high-throughput approach. *FEBS Lett* **524**, 159–62.

132 Lederman M, Zubay G. 1967. DNA-directed peptide synthesis *in vitro*. I. A comparison of T2 and *E. coli* DNA-directed peptide synthesis. *Biochem Biophys Acta* **149**, 253–8.

133 De Vries JK, Zubay G. 1967. DNA-directed peptide synthesis. II. The synthesis of the α-fragment of the enzyme β-galactosidase. *Proc Natl Acad Sci USA* **57**, 1010–2.

134 Shirokov V, Simonenko PN, Biryukov SV, Spirin AS. 2002. Continuous-flow and continuous-exchange cell-free translation systems and reactors. In Spirin AS, ed., *Cell-free Translation Systems*, 91–108, Springer, Berlin.

11
Application of Antibody Fragments as Crystallization Enhancers

Carola Hunte and Cornelia Münke

11.1
Introduction

After decoding the sequence of the human genome [1, 2], and as more and more genomes of pathogens, e.g. *Helicobacter pylori* [3], *Mycobacterium tuberculosis* [4] and *Plasmodium falciparum* [5], become available, hopes have been raised that drug discovery will be more empowered. Most drugs act at the level of proteins and great efforts are now put into structural genomics projects to accelerate experimental protein structure determination by means of high-throughput X-ray crystallography and nuclear magnetic resonance (NMR). High-resolution structures are used for drug discovery (1) by rational drug design, (2) by analysis of target–ligand complex structures to validate hits or rationally improve ligands that have been obtained by high-throughput screening of combinatorial libraries, or (3) by *in silico* ligand screening, which allows the reduction of experimental costs and speeds up the discovery process. Last, but not least, high-resolution structures provide crucial information for the understanding of the molecular mechanism of proteins that are present in the cell in a large variety of specific architectures – the basis for their highly diverse functions.

The most widely used method for the determination of the atomic structure of proteins larger than 30 kDa (including membrane proteins) is X-ray crystallography, which requires well-ordered three-dimensional (3-D) crystals grown from protein solution. Despite the enormous interest in structures of membrane proteins, which provide targets for the majority of drugs, structures of less than 50 integral membrane proteins have been determined since 1985 when the first membrane protein structure of the bacterial photosynthetic reaction center was presented [6]. In contrast, several thousand structures of soluble proteins have already been analyzed by X-ray crystallography. Only 13 X-ray structures of unrelated polytopic membrane proteins from the inner membrane of bacteria and mitochondria and from eukaryotic membranes are available [7]; for a regularly updated list, see www.mpibp-frankfurt.mpg.de/michel/public/memprotstruct.html. This small number is in contrast to the high proportion of integral membrane proteins predicted from an analysis of genomes of eubacterial, archaean and eukaryotic organ-

Molecular Biology in Medicinal Chemistry. Edited by Th. Dingermann, D. Steinhilber, G. Folkers

ISBN: 3-527-30431-2

isms, in which 20–30% of the open reading frames encode integral membrane proteins [8]. Obtaining well-ordered 3-D crystals of membrane proteins suitable for high-resolution structural studies is notoriously difficult due to the amphipathic surface of the macromolecules. The hydrophobic membrane-spanning region of the protein is in contact with the acyl chains of the phospholipid bilayer, whereas the polar surfaces are exposed to the polar head groups of the lipids and to the aqueous phases. For purification and crystallization purposes, membrane proteins are solubilized from the lipid bilayer in their native conformation by the use of detergents.

Detergent molecules are amphiphilic, consisting of a polar or charged head group and a hydrophobic tail. At low concentrations, they are present as monomers in aqueous solution and form a monolayer at the water–air interface. Above a certain concentration of the detergent, which is called the critical micelle concentration (CMC), self-association occurs and micelles are formed. In these spherical aggregates, the hydrophobic tails of the detergent molecules are oriented towards the micelle interior and the hydrophilic head groups point outwards. Membrane proteins are soluble in micellar detergent solutions and form protein–detergent complexes. A detergent micelle covers the hydrophobic surface of the membrane protein in a belt-like manner, where the polar detergent head groups face the aqueous phase (Fig. 11.1; for review, see [9, 10]).

Starting with detergent–protein complexes, membrane proteins can be reconstituted into a phospholipid bilayer by concomitant removal of detergent and addition of phospholipid either by dialysis or solid-phase adsorption. Under certain empirically determined conditions the respective protein forms well-ordered two-dimensional (2-D) arrays, so-called 2-D crystals, which can be used for structure determination by electron microscopy [11, 12]. In this way several structures of membrane proteins at medium resolution (around 5–10 Å) have already been obtained. Structure determination at near atomic resolution by electron microscopy is possible [13–15], but is cumbersome. The 2-D crystallization trials need less protein and it might be faster to obtain crystals. However, once good 3-D crystals are available, rapid structure determination at high resolution by X-ray crystallography is more likely.

The amphipathic surface of membrane proteins allows 3-D crystals to be formed in two ways ([16], see Fig. 11.1). Type I crystals are in principal ordered stacks of 2-D crystals that contain the ordered protein molecules within the membrane plane. Type II crystals can be formed by membrane proteins integrated in detergent micelles in a similar manner to soluble proteins. In the latter case, crystal contacts are made by the polar surfaces of those protein domains that protrude from the detergent micelle. Here, crystallization conditions are very similar to those for soluble proteins. Mixed type I and II crystals are possible, although most membrane protein crystals belong to type II.

Obviously, the detergent micelle requires space in the crystal lattice. Although attractive interactions between the micelles might stabilize the crystal packing [17, 18], they cannot contribute to rigid crystal contacts. Thus, proteins with small hydrophilic domains are especially difficult to crystallize, e.g. many transporters or

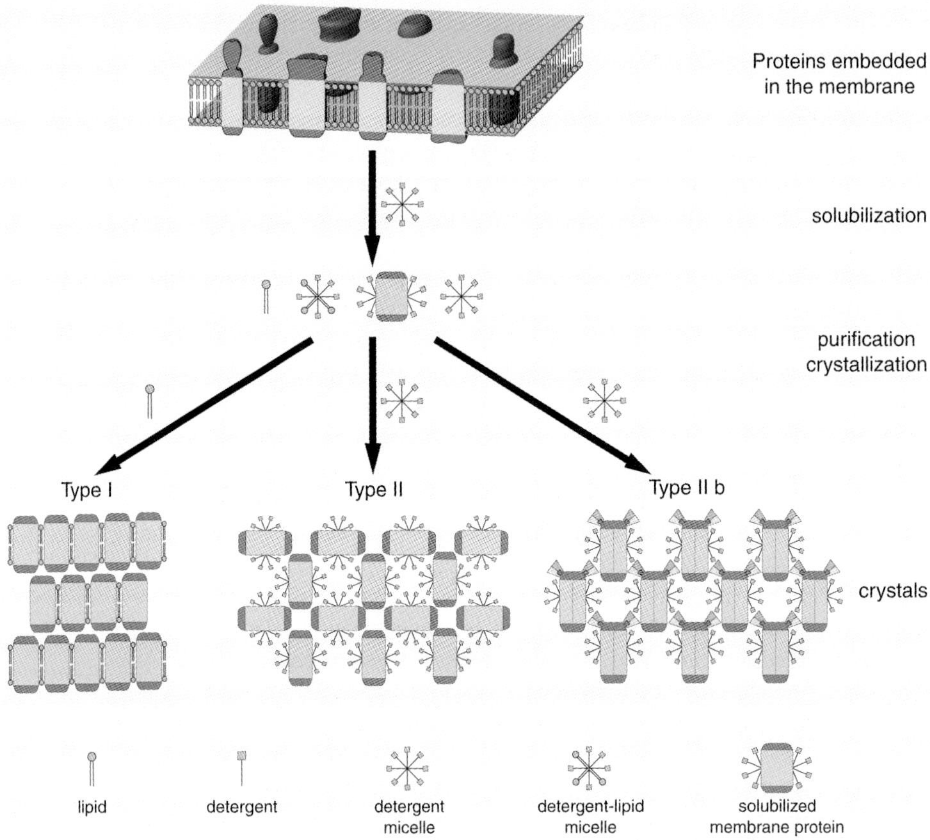

Fig. 11.1 Basic types of 3-D membrane protein crystals. Type I crystals consist of ordered stacks of 2-D crystals that are present in a bilayer. The hydrophobic surface of membrane proteins is covered by a micelle in the detergent-solubilized state. These membrane protein–detergent complexes form so-called type II 3-D crystals. Stable crystal contacts are made between the hydrophilic surfaces of the protein. In a modification of type II (type IIb), tightly and specifically bound antibody fragments add a hydrophilic domain to the complex. This surface expansion provides additional sites for crystal contacts and space for the detergent micelle (From [29].)

channels that consist mainly of transmembrane helices with very short solvent exposed loops. A strategy to increase the probability of obtaining well-ordered type II crystals is to enlarge the polar surface of the membrane protein by the attachment of polar domains with specifically binding antibody fragments. This approach, developed by Hartmut Michel et al., was first successfully applied for crystallization and structure determination of the cytochrome *c* oxidase (COX) from *Paracoccus denitrificans* with a bound antibody Fv (*fragment variable*) fragment [19, 20].

In addition to the specific detergent requirements for 3-D crystallization of membrane proteins, the second main bottleneck for structural analysis is the supply of sufficient amounts of protein. Crystallization in 3-D consumes at least 10 mg of pure protein per crystallization screen. Nearly all membrane protein structures that have been determined are of proteins that occur in high abundance in their natural environment, e.g. proteins of the respiratory chain or photosynthetic membranes. However, most membrane proteins, e.g. receptors and transporters, are present in the cell at low concentration. Using recombinant production systems, insertion of the protein into a membrane limits the production yield, unless one chooses production via inclusion body formation and refolding. The latter has been very successfully applied for the unusually stable β-barrel membrane proteins from the outer bacterial membranes [21, 22]. Good progress has been made in the development of overexpression systems for membrane proteins from bacterial sources, e.g. for the structure determination of the mechanosensitive ion channel from *M. tuberculosis* produced in *Escherichia coli* [23]. Further development of high-level overexpression systems for the production of recombinant membrane proteins from eukaryotic sources is crucial for the successful exploration of their structures [24, 25]. In addition, recombinant techniques have the advantage that proteins can be engineered with affinity tags to allow easy purification. The sequence can also be modified to provide a homogenous protein source, e.g. by removal of glycosylation sites. Furthermore, with the growing number of available genome sequences, the recombinant system allows an additional strategy: to express and purify homologous proteins from several organisms and test which protein is suitable for crystallization. For successful structure determination of the mechanosensitive channel MscL, homologs from nine prokaryotic species were subjected to crystallization trials [23]. Homologous proteins from different species might not only feature differences in essential criteria like production yield or protein stability, but small variations in surface properties might be responsible for the crystallization success, as crystal contacts are made by the protein surface.

Several reviews are available that describe the specific requirements for crystallization of membrane proteins [16, 18, 26–29]. In this chapter, we focus on describing the crystallization of membrane proteins mediated by antibody fragments, as this promising approach has already been successfully used for structure determination of five membrane protein–antibody complexes [30].

11.2
Crystallization of Membrane Proteins Mediated by Antibody Fragments

11.2.1
Overview

Following the successful antibody fragment-mediated crystallization of bacterial COX, this technique was used to determine several structures of membrane proteins that will be described below.

COX from *P. denitrificans* was crystallized as a complex with a recombinant antibody Fv fragment [19]. The structure of this complex was determined at 2.8-Å resolution, and contained all four subunits of COX and the two polypeptides of the Fv fragment [20, 31]. Fv fragments are involved in all crystal contacts. However, the lack of direct protein–protein interactions along the *c*-axis causes anisotropy in diffraction and problems in reproducing the crystals. Better-ordered crystals were obtained with a COX preparation containing only the two catalytically essential subunits I and II. The crystals diffract X-rays to below 2.5-Å resolution. The new crystal packing involves direct interaction between cytoplasmic and periplasmic surfaces of adjacent COX molecules. All further contacts are mediated exclusively by the Fv fragment. In all of these structures, the COX–Fv fragment co-complex was purified by indirect immunoaffinity chromatography (see below). The co-complex is formed by mixing solubilized membranes with Fv fragment containing periplasm and purified in a single step by streptavidin-affinity chromatography making use of the strep-tag fused to the Fv fragment [32].

The second membrane protein crystallized using this technique is the yeast cytochrome bc_1 complex (QCR), a fundamental component of the mitochondrial respiratory chain. As in the case of bacterial COX, the complex was crystallized with a recombinant Fv fragment. The latter binds to the extrinsic domain of the catalytic subunit Rip1p, the so-called Rieske protein [33]. The parental monoclonal antibody 18E11 was obtained by standard hybridoma techniques by immunizing the mouse with detergent-solubilized and purified yeast QCR (Hunte, unpublished results). The antibody does not recognize the protein in Western Immunoblot after sodium dodecylsulfate–polyacrylamide gel electrophoresis (SDS–PAGE) separation of the complex, but binds to a native discontinuous epitope.

Although this multisubunit complex (230 kDa per monomer) has a large hydrophilic domain protruding in the matrix space (about 86 kDa) and a considerable hydrophilic portion present in the intermembrane space (about 46 kDa), crystals of yeast QCR suitable for X-ray analysis were obtained only in complex with the Fv fragment. The structure of the co-complex was determined at 2.3-Å resolution [33, 34]. The bound Fv fragments result in a spacious crystal packing (space group C2) with a high solvent content of 74%, providing space for the large detergent micelle of the detergent undecylmaltopyranoside (Fig. 11.2). The main crystal contacts are mediated by the antibody fragments.

The same preparation was used to crystallize yeast QCR with its substrate cytochrome *c* bound [35]. The different crystallization conditions resulted in an altered crystal lattice (P21). Again, the Fv fragments contribute the major interactions between adjacent molecules. They neither hinder cytochrome *c* binding nor interfere with QCR activity (Hunte, unpublished results).

Crystallization and structure determination of the potassium channel KcsA was achieved with a bound *Fab* (*fragment antibody binding*) fragment [36]. The monoclonal antibody used was obtained by hybridoma technology and clones were selected that recognize the native homotetrameric, but not the monomeric, form of the channel. Crystallization was performed with proteolytic Fab fragments generated by papain proteolysis and purified by anion-exchange chromatography. The channel–Fab fragment complex crystals allowed an elegant structure solution. The

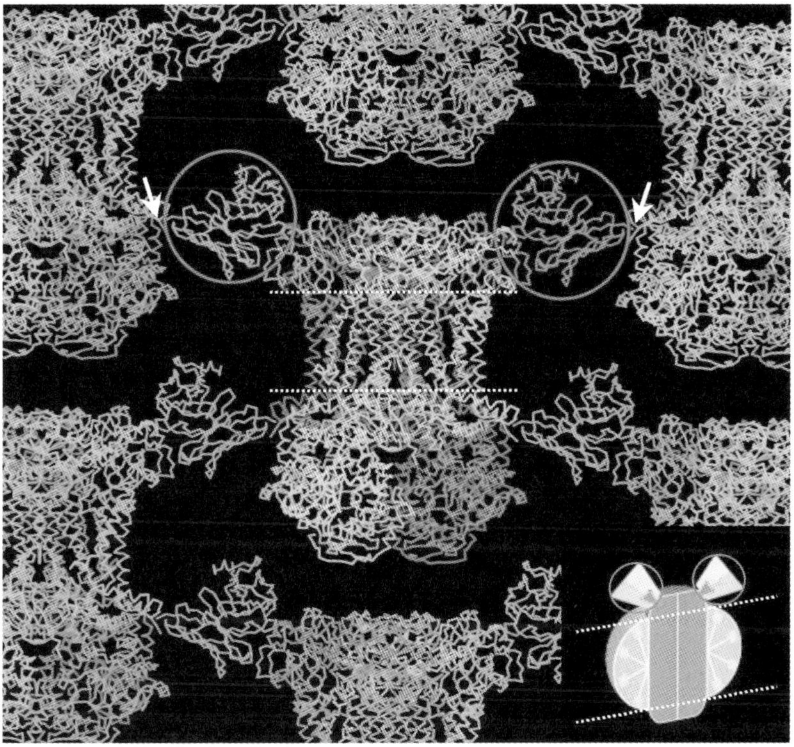

Fig. 11.2 Antibody fragments are essential for the crystallization of the yeast cytochrome bc_1 complex, as crystal contacts (white arrows) of the dimeric yeast cytochrome bc_1 complex–Fv18E11 co-complex are mediated by the Fv fragment (encircled) (from [83]). The fragment binds to the extrinsic domain of subunit Rip1p and generously provides space for the detergent micelle in which the transmembrane portion of the complex (dotted lines) is embedded, as schematically depicted in the insert.

phases were determined by molecular replacement using a published structure of another Fab fragment, as the immunoglobulin fold is highly conserved. In addition, the Fab fragment, at around 54 kDa, represents the largest portion of the complex, compared to single channel subunit of around 11 kDa. The crystals of the complex diffracted substantially better (below 2 Å) than the previous crystals of the KcsA K^+ channel alone (3.2-Å resolution [37]). The Fab fragment binds to the extracellular surface of the channel subunit. All crystal contacts are between Fab fragments, leaving space for the central molecule of the channel with its relative large detergent micelle of decylmaltoside. The size of a Fab fragment appears to be well suited for the crystallization of channels and other membrane proteins that do not contain domains of significant size protruding from the lipid bilayer.

11.2.2 Generation of Antibodies Suitable for Co-crystallization

Up to now, antibody fragments successfully used for co-crystallization of membrane proteins were derived from monoclonal antibodies obtained by classical hybridoma technology [38]. To achieve stable complex formation for crystallization, parental antibodies as well as their derivatives have to (1) bind the antigen in its native conformation, (2) have a high binding affinity to allow for stable stoichiometric complexes and (3) form a rigid complex with their native antigen. Antibodies binding to a discontinuous epitope as opposed to a linear one are preferred. The N- and C-termini of polypeptides as well as exposed loops are often recognized as epitopes in an immune response. However, these peptides are often flexible and binding of an antibody molecule would very likely introduce a mobile domain detrimental for crystallization. Specificity for a discontinuous epitope can be easily checked in Western immunoblot analysis after separation of the denatured protein by SDS–PAGE, in which the antibody should not bind to the linearized polypeptide.

11.2.2.1 Immunization

One of the first important steps for the successful generation of specific hybridoma clones is the immunization of animals with antigen to generate a sufficient antibody titer. Detergent-solubilized and purified membrane proteins have been successfully used as antigens to obtain high-affinity binders for native membrane proteins, e.g. for generating monoclonal antibody 18E11 specific for the yeast QCR [33] or various monoclonal antibodies specific for the Na^+/H^+ antiporter NhaA [39].

Generally, long-term immunization protocols are favorable to obtain many high-affinity binders. In a standard protocol applied for generating antibodies against yeast QCR and NhaA [33, 39, 40], 10-week-old female BALB/c mice were immunized by intraperitoneal injection of highly purified, detergent solubilized protein [100 μg protein in 50% (v/v) adjuvant (ABM2; Pan Biotech, Aidenbach, Germany) at a final volume of 250 μl]. The initial immunization was followed by four injections with protein suspension at 4-week intervals [50 μg protein in 50% (v/v) adjuvant (ABM1; Pan Biotech) at a final volume of 250 μl]. For the final boost the latter injection was repeated on 3 consecutive days and mice sacrificed the following day for removal of the spleen. Blood sera of the mice were taken prior to immunization and 10 days after each injection. After the third immunization, the blood serum showed clear positive signals up to a serum dilution of 10^5 in a standard enzyme-linked immunosorbent assay (ELISA) using purified NhaA as antigen [39].

It is evident that such an immunization regime in small mammals for highly homologous membrane proteins (i.e. derived from human, mouse, rat and monkey) may be less likely to produce antibodies because of self-tolerance mechanisms. However, there are several ways to stimulate the immune response, e.g. genetic immunization [41–43] or simply the use of improved lipid nanoparticle-based adjuvants [44].

11.2.2.2 Selection of Antibodies

Antibodies suitable for co-crystallization of membrane proteins have to recognize their respective antigen as a protein–detergent complex. Thus, specific requirements are imposed on the selection of conformation-specific monoclonal antibodies. The type and concentration of detergent is crucial to keep the protein in its native conformation. Some detergents may prevent proteins from binding to plastic and polystyrene surfaces used as common ELISA supports. In addition, adsorption to the solid phase can cause partial denaturation of the protein [45]. When generating monoclonal antibodies against yeast QCR, 35 clones were obtained that recognize the denatured protein in Western immunoblots and only one clone was selected that specifically recognizes the native protein (Hunte, unpublished results). There, a standard ELISA was applied in which detergent-solubilized complex was coated to the polysterene matrix. To improve the yield of native binders for membrane proteins, an easy and robust strategy, the so-called his-tag ELISA, has been devised for proper immobilization and presentation of the "native" antigen of his-tagged recombinant membrane proteins, as demonstrated for the antiporter NhaA. The protein is immobilized on Ni^{2+}-NTA-coated ELISA plates using the same buffer conditions established for the purification of the antiporter using immobilized metal-affinity chromatography [39]. Under these conditions, in the presence of the mild detergent dodecyl-β-D-maltopyranoside, the protein is properly folded and fully functional, and it can be subjected to screening for antibodies. Starting from more than 2000 hybridoma clones, implementation of his-tag ELISA (as opposed to a standard ELISA) allowed the isolation of six different monoclonal antibodies of the IgG type. All of the antibodies obtained are conformation sensitive and bind the detergent-solubilized antiporter as well as its membrane-bound form. Only one of these antibodies is positive in Western blot analysis after SDS–PAGE separation of the protein. However, this antibody (1F6) binds to a linear, but surface exposed, epitope located at the N-terminus of the enzyme [46]. Interestingly, monoclonal antibody 1F6 binds NhaA in a pH-dependent manner that mirrors the pH dependence of the activation of the protein and the associated conformational change as probed by trypsin digestion [47]. It can be used for immunoaffinity purification of the transporter from crude membrane extracts by binding NhaA to immobilized monoclonal antibody 1F6 and eluting it specifically with a pH shift.

His-tag ELISA is a powerful method that combines the sensitivity, specificity and high signal/noise ratio of a common immunosorbent assay with recombinant protein chemistry. It can easily be adapted to all recombinant proteins purified with the help of a his-tag. To our knowledge it is the first example of selection of antibodies against conformational epitopes using a detergent-solubilized membrane protein, i.e. the most common state for the growth of type II 3-D crystals.

Alternative methods for the presentation of membrane proteins in their native conformation have been reported, such as selection with lipid reconstituted samples [48], native enriched membrane fractions or intact cells [49].

Tailored screening is the key step in selecting suitable antibodies from hybridoma fusions or recombinant antibody libraries. Specifically adapted screens allow the selection of antibodies specific for a defined protein conformation. Correctly selected antibodies could help to trap a flexible protein in a fixed conformation. Co-crystallization with antibody fragments has been demonstrated to facilitate structure determination of the flexible gp120 domain [50].

11.2.2.3 Antibody Fragments

Native antibodies are unsuitable for co-crystallization attempts. They are flexible in the linker regions connecting the variable and constant domains [51], and their bivalent binding mode is in most cases undesirable. Monovalent antibody fragments can be generated by proteolytic cleavage of the whole antibody producing two Fab fragments per antibody molecule. A sufficient amount of starting material of the monoclonal antibody is either obtained by ascites production or by cell culture techniques with the producing murine hybridoma cell line. The monoclonal antibodies are then purified, e.g. using Protein A-affinity chromatography. In the following step, the Fab fragments are commonly generated with the protease papain that specifically cuts in the hinge region. Fab fragments can be purified by Protein A depletion of the split constant antibody domain, followed by size-exclusion or ion-exchange chromatography. The Fab fragments are relatively easy to obtain by proteolysis, but care has to be taken to produce homogenous fragment preparations suitable for crystallization. Residual parts of the flexible linker and in some cases glycosylation can hinder crystallization attempts. Despite these problems, a large number of those Fab fragments alone and in complex with the respective hapten or antigen, including one membrane protein (the potassium channel KcsA [36]), have been crystallized.

In the future, a promising alternative for raising monoclonal antibodies in mice is the adaptation of phage- or ribosome-display technology [48, 52, 53] to the *in vitro* selection of ligands for membrane proteins. Up to now, examples of successful selection of antibodies specific for membrane proteins via phage display of nonimmune libraries are rare. We anticipate that other strategies may be worth pursuing to establish phage display-mediated screening of antibodies against these targets, i.e. (1) generation of more diverse (more than 10^{10} different elements) antibody libraries, (2) use of a modified helper phage that allows multivalent presentation of the antibodies [54] and (3) screening for ligands after a single round of bio-panning followed by *in vitro* affinity maturation [55].

Cloning of Antibody Fragments The smallest antigen-binding domain of an antibody, i.e. the Fv fragment, consists of the variable domain of the heavy and light chain, V_H and V_L, respectively. Fv fragments cannot be obtained by proteolysis, but have to be produced as recombinant proteins. For this purpose, the genes encoding V_H and V_L have to be cloned and expressed starting from hybridoma cell lines. The cloning of Fv fragments by reverse transcriptase-polymerase chain reaction (RT-

PCR) is described here [32, 56] (Fig. 11.3). mRNA or total RNA is isolated from murine hybridoma cell lines producing the antibody of interest. In the following step, cDNA synthesis can be simply performed by using oligo(dT) as a primer. To amplify V_H and V_L genes, degenerated primers are used that match the 5′ end of framework 1 and the 3′ end of framework 4, i.e. corresponding to the N-termini of the heavy and light chain of the antibody molecule and the C-termini of the variable domains, respectively. The sequence of the framework regions is highly conserved, and a set of primer pairs was constructed to specifically amplify genes encoding for the V_λ chain, V_κ chain subclasses 1, 2, 3 and 5, V_κ chain subclass 4 and 6, and V_H [32]. These primers contain unique restriction sites, which allow insertion of the PCR product into the plasmid pASK68 that is used for production of the Fv fragment in the periplasm of *E. coli* (see below). Antisense-directed RNase H digestion can be used to suppress amplification of the nonfunctional mRNA transcripts introduced in hybridoma cell lines during fusion with the nonsecreting myeloma cell line [56]. This fast, robust approach was used to clone several antibody fragments, including the Fv fragments of monoclonal antibody 7E2 specific for COX of *P. denitrificans* [32], monoclonal antibody 18E11 specific for yeast QCR [33], and monoclonal antibodies 2C5 and 5H4 specific for the antiporter NhaA from *E. coli* [57]. However, in some cases this approach failed to amplify the desired genes. Therefore, a new set of primers was designed that accounts for the different sequences of the subgroups as listed by the Kabat databank [58] (Fig. 11.4). A specific forward primer was constructed for every subgroup of antibodies. The position of the backward primer was shifted downstream into the highly conserved constant region. Furthermore, the restriction sites were omitted for the sake of higher accuracy. PCR products were then cloned by blunt-end or TA ligation into sequencing vectors. After validation of the antibody sequence, restriction sites were introduced by PCR to match the desired expression vector. A similar strategy based on degenerated primers can be used for cloning of Fab fragments [59]. Furthermore, hybrid Fab fragments can be produced by insertion of the V_H and V_L genes into a vector that already contains constant domains of an antibody molecule [57]. Alternative PCR strategies and primer sets including cloning by rapid amplification of cDNA ends (RACE) have been published [60–62]. Often, heavy and light chain are covalently joined by introduction of a linker peptide, generating so-called sc (single-chain) Fv fragments [63] or by connecting the chains via a disulfide bridge (dsFv fragments [64]). Although these measures are intended to stabilize the recombinant fragments, they may in some cases negatively affect the binding properties of the fragment. Fragments Fv7E2 and Fv18E11, which have been used for crystallization of COX and QCR, respectively, do not contain a linker or an additional disulfide bond. See Fig. 11.4.

Production of Antibody Fragments A variety of expression systems have been used for the production of recombinant antibody fragments, e.g. *E. coli* [65], *Pichia pastoris* [66] or mammalian cells [67]. A routine procedure for the production of Fv fragments is periplasmic expression in *E. coli* [32, 65]. The encoding genes of the variable domains of heavy and light chain are inserted in the dicistronic operon of

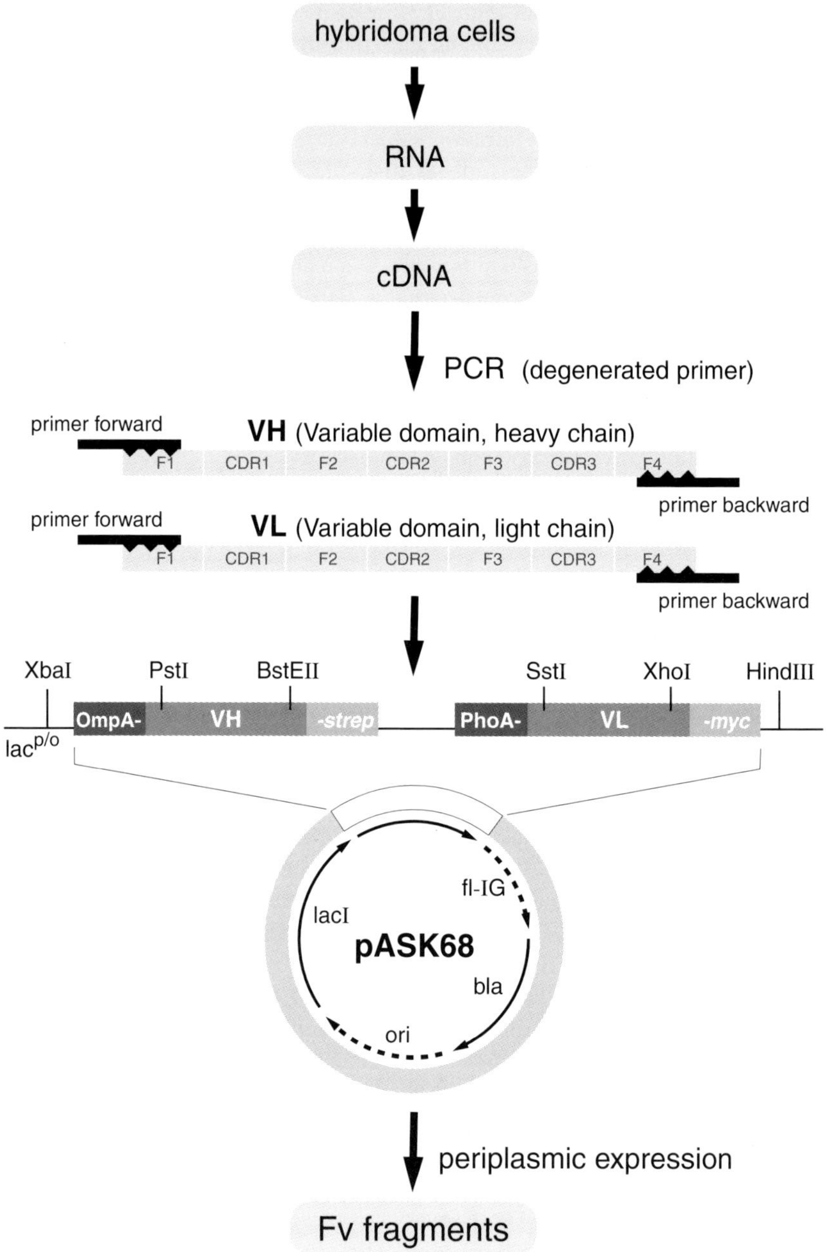

Fig. 11.3 Schematic depiction of the strategy for cloning and expression of Fv fragments from monoclonal hybridoma cell lines by RT-PCR. The genes encoding for the variable domains of the heavy and light chains (V_H and V_L, respectively) are inserted in the bicistronic plasmid pASK68 for periplasmic expression in *E. coli* [32]. The signal sequences OmpA and PhoA are used to target the polypeptides to the periplasm. A strep-tag is attached to the heavy chain for purification; the myc-tag fusion allows detection of the light chain with the monoclonal antibody 9E10.

A. Primer set I.

```
κ light chain FR1 region I,II,III,V          (SstI)
VKC1T      GAC ATT GAG CTC ACM CAG WCT CCA KYC TCC CTG
            1   2   3   4   5   6   7   8   9   10  11
κ light chain FR1 region IV,VI  (SstI)
VKC2T      GAC ATT GAG CTC ACC CAG TCT CCA GCA ATC ATG
            1   2   3   4   5   6   7   8   9   10  11
κ light chain FR4 region        (XhoI)
VKR1T      CCG TTT CAG CTC GAG CTT GGT SCC WSC WCC GAA CGT
           108 107 106 105 104 103 102 101 100  99  98  97

heavy chain FR1 region          (PstI)
VHC1T      GAG GTK MAG CTG CAG SAG TCW GGG SCT GGC YTG
            1   2   3   4   5   6   7   8   9   10  11
heavy chain FR4 region          (BstEII)
VHR1T      TGA GGA GAC GGT GAC CGT GGT CCC TTG GCC CCA
           113 112 111 110 109 108 107 106 105 104 103
```

symbol	meaning	symbol	meaning
A	Aden ine	Y	pYr imidine (C or T)
C	Cyto sine	M	A or C
G	Guan ine	W	A or T
T	Thym ine	S	C or G
R	puRine (A or G)	K	G or T

(the one-letter nucleotide symbols are used according to IUB nomenclature)

Fig. 11.4 Primer sets used for amplification of V_L- and V_H-encoding genes of murine monoclonal antibodies. (A) Degenerated primers with unique restriction sites have been designed to clone the genes compatible for insertion into the plasmid pASK68 used for periplasmic expression in *E. coli* (see Fig. 3. in [32]).

the plasmid pASK68 allowing periplasmic expression in *E. coli* under the control of the inducible *lac* promoter (see Fig. 3 in [32]). N-terminal fusion of the OmpA and PhoA signal sequence facilitate secretion of the heavy and light chain to the periplasm. A streptavidin-binding peptide (strep-tag) is appended at the C-terminus of the V_H domain for purification via streptavidin-affinity chromatography [68], while a myc-tag (recognized by the monoclonal antibody 9E10) is present at the C-terminus of the V_L domain and is used for detection of the antibody fragment in a Western blot or ELISA.

When expressing an antibody fragment that binds to a membrane protein naturally residing in the cytoplasmic membrane of the host, one should consider that the antibody may inhibit the protein activity and thereby growth of the cell. In the case of the antiporter NhaA, even though most antibodies had an inhibitory effect on the activity of NhaA [39], periplasmic expression of Fv and Fab fragments was

B. Subtype specific primer set

Light chain

κ light chain FR1 region forward primers:

	1	2	3	4	5	6	7	8	9	subgroups according to Kabat
VK-I	GAC	ATT	GTG	ATG	**W**CA	CAG	TCT	CCA	TC	I
VK-II1	GAT	GTT	**K**TG	**M**TG	ACC	CAA	ACT	CCA	C	II
VK-II2	GAT	ATT	GTG	ATG	ACT	CAG	GCT	GCA	C	II
VK-II3	GAT	ATT	GTG	AT**R**	ACC	CAG	GAT	GAA	CTC	II
VK-III	**R**A**M**	ATT	GTG	CTG	ACC	CA**R**	TCT	CC**W**	G	III
VK-IV	**S**AA	A**W**T	GT**K**	CTC	ACC	CAG	TCT	CCA	**R**C	IV
VK-V1	GA**Y**	ATC	CAG	ATG	AC**M**	CAG	**W**CT	**M**CA	TC	V
VK-V2	GA**Y**	AT**Y**	**SW**G	ATG	ACC	CAG	TCT	CC		V
VK-V3	GAC	ATT	GTG	ATG	ACC	CAG	TCT	CAC		V
VK-VI1	CAA	ATT	GTT	CTC	ACC	CAG	TCT	CCA	GC	VI
VK-VI2	GAA	AAT	GTT	CTC	ACC	CAG	TCT	CCA		VI

κ light chain constant domain backward primer:

		116	115	114	113	112	111	110	109
VK-RT	T	GGA	TAC	AGT	TGG	TGC	AGC	ATC	AGC

heavy chain

FR1 region forward primers:

	1	2	3	4	5	6	7	8		subgroups according to Kabat
VH-IA	GA**K**	GT**R**	CAG	CTT	CAG	GAG	TC**R**	GGA	C	I(A)
VH-IB	CAG	GTC	CAG	CTG	AAG	**S**AG	TCA	GG		I(B)
VH-IIA	**S**AG	GT**Y**	CAG	CTG	CA**R**	CA**R**	TCT	GG**R**	C	II(A)
VH-IIB	CAG	GTC	CAR	CTG	CAG	CAG	**Y**CT	GG		II(B)
VH-IIC	GAG	GTT	CAG	CTG	CAG	CAG	TCT	GG		II(C)
VH-IIIA	GAG	GTG	AAG	CTG	GTG	GA**R**	TCT	GG		III(A)
VH-IIIB	GAG	GTG	AAG	CTT	CTC	GAG	TCT	GG		III(B)
VH-IIIC	GAG	GTG	AAG	CTG	GAT	GAG	ACT	GG		III(C)
VH-IIID	GA**M**	GTG	**MW**G	CTG	GTG	GAG	TCT	GG		III(D)
VH-VA	**S**AG	GT**Y**	CAG	CT**K**	CAG	CAG	TCT	GGA		V(A)

CH1 region backward primers:

	137	136	135	134	133	130	129	128		
VH-G1A	CAT	GGA	GTT	AGT	TTG	GGC	AGC	AG		IgG1
VH-G1B		GGA	GTT	AGT	TTG	GGC	AGC	AG		IgG1
VH-G1C		GGA	GTT	AGT	TTG	GGC	AGC			IgG1
VH-G2a/b	GA	GGA	**R**CC	AGT	TGT	ATC	TCC	ACA	C	IgG2a/b

Abbreviations see A.

Fig. 11.4 (B) Amplification of V_H and V_L genes is considerably improved using the subgroup specific primer set. Using the same cDNA template, every primer combination is tried in parallel at different temperatures. In most cases the correct gene is amplified by a distinct primer combination.

C. Amino acid sequences of murine antibody subtypes complementary to primer sets

VL-κ	1	2	3	4	5	6	7	8	9
VK-I	D	T	V	M	T,S	Q	S	P	S
VK-II1	D	V	V,L	M,L	T	Q	T	P	
VK-II2	D	I	V	M	T	Q	A	A	
VK-II3	D	I	V	I,M	T	Q	D	E	L
VK-III	K,D	I	V	L	T	Q	S	P	
VK-IV	Q,E	I,N	V	L	T	Q	S	P	A,T
VK-V1	D	I	Q	M	T	Q	T,S	T,P	S
VK-V2	D	I	V,Q	M	T	Q	S	P	
VK-V3	D	I	V	M	T	Q	S	H	
VK-VI1	Q	I	V	L	T	Q	S	P	A
VK-VI2	E	N	V	L	T	Q	S	P	

116	115	114	113	112	111	110	109	VL-κ
S	V	T	P	A	A	D	S	VK-RT

VH	1	2	3	4	5	6	7	8	9
VH-IA	E,D	V	Q	L	Q	E	S	G	
VH-IB	Q	V	Q	L	K	E,Q	S	G	
VH-IIA	E,Q	V	Q	L	Q	Q	S	G	
VH-IIB	Q	V	Q	L	Q	Q	P,S	S	
VH-IIC	E	V	Q	L	Q	Q	S	S	
VH-IIIA	E	V	K	L	V	E	S	G	
VH-IIIB	E	V	K	L	L	E	S	G	
VH-IIIC	E	V	K	L	D	E	T	G	
VH-IIID	,D	V	Q,K,M	L	V	E	S	G	
VH-VA	E,Q	V	Q	L	Q	Q	S	G	

137	136	135	134	133	130	129	128	VH
M	S	N	T	Q	A	A		VH-G1A
M	S	N	T	Q	A	A		VH-G1B
M	S	N	T	Q	A			VH-G1C
S	S	G	T	T	D	G	C	VH-G2a/b

symbol	meaning	symbol	meaning
A	Alan ine	M	Met hionine
G	Glyc ine	S	Ser ine
L	Leuc ine	V	Val ine
P	Prol ine	D	Asp artic acid
T	Thre onine	N	Asp aragine
C	Cyst eine	E	Glu tamic acid
H	Hist idine	Q	Glu tamine
I	Isol eucine	K	Lys ine

Fig. 11.4 (C) Corresponding amino acid sequences of antibodies of different subgroups for the complementary part of primer set II. The residues of V_L and V_H are numbered according to Kabat et al. [58].

possible [57] as the transporter is vital only at certain growth conditions, e.g. at high sodium or lithium concentration, or at alkaline pH in the presence of sodium. In addition, the respective epitopes may be facing the extracellular side or not be accessible.

In bacterial expression systems, Fv and Fab fragments are normally produced in the periplasmic space, where oxidizing conditions allow the formation of the required disulphide bonds essential for proper folding and activity of the antibodies. Protein yields vary for the different antibody fragments, with maximum yields of 1.5 mg protein/l of bacterial culture for stable Fv fragments. Fab fragments are expressed at an even lower yield (0.1–0.4 mg/l of culture), most likely due to the higher complexity of their disulphide bond patterns and the larger mass. Their production creates a bottleneck for co-crystallization trials. Production of proteolytic Fab fragment is a multistep process, and is time consuming and costly. Often the yield of pure protein is below 30% of the starting protein sample. To overcome these limitations, yield optimization of recombinant Fab production would be advantageous.

Fab fragments have been produced as chimeric proteins with the C_H1 and C_L constant domains of another antibody, i.e. monoclonal antibody D1.3 specific for lysozyme, by subcloning the V_H and V_L domains of three different antibodies into the vector pASK85-D1.3 [57]. This plasmid provides a his-tag at the C-terminus of the heavy chain for purification via immobilized metal-affinity chromatography. It is designed for periplasmic expression in *E. coli* under the control of the tetracycline repressor/promoter system [68]. Monoclonal antibody 2C5 antibody fragments are produced at good levels of 1.5 and 0.5 mg/l in Fv and Fab formats, respectively. Antibody fragments of 5H4 were produced at much lower yields (0.1 and 0.03 mg/l for Fv and Fab fragments, respectively). Exchange of charged and bulky residues at the N-terminus of the variable domain of the light chain improved the folding and solubility properties of the latter antibody [69]. The yield could be considerably increased by expression of the same hybrid fragments in an oxidizing bacterial cytoplasm using the *E. coli* strain FA113 and the newly constructed expression vector pFAB1 [57]. This leads to high yield of 10–30 mg pure, functional Fab fragments per liter of *E. coli* culture and these are of suitable quality for crystallization studies. In addition, scFv fragments derived from phage display libraries have recently been successfully expressed in a similar strain co-expressing molecular chaperones [70].

11.2.3 Protein Purification and Crystallization

Starting from a good natural or recombinant protein source, the purification procedure has to ensure pure and homogenous protein preparations in sufficient yield. In general, the methods for membrane protein purification are the same as those applied for soluble proteins except that integral membrane proteins are purified as protein–detergent complexes. Membrane proteins are kept soluble in micellar detergent solutions at a usual detergent working concentration of 1.5- to 2-fold CMC. The choice of detergent for solubilization and purification is important, and should be carefully chosen. Most purification materials tolerate a wide range of detergents. Affinity purification systems have the advantage of high speed and high specificity compared to standard chromatographic procedures such as anion-exchange or size-exclusion chromatography. For specific applications, e.g. for crystallization, recombinant antibody fragments can be used in indirect immunoaffinity chromatography for one-step purification of membrane protein–antibody complexes as outlined below.

11.2.3.1 Indirect Immunoaffinity Purification

Recombinant antibody fragments are versatile tools for the structural analysis of proteins. They can be used for the preparation of membrane protein complexes, yielding highly purified material suitable for crystallization in a single immunoaffinity chromatography step [32, 71]. Compared with classical methods like ion-exchange or size-exclusion chromatography, immunoaffinity chromatography offers an excellent purification factor – usually between 100 and 10,000 – due to the high specificity of the interaction between antigen and antibody or antibody fragment. However, due to the strength of this interaction, harsh elution protocols are usually required if the antibodies or their fragments are covalently coupled to activated matrices or trapped by immobilized Protein A or G. Engineered antibody fragments, however, can be noncovalently and reversibly bound to an appropriate matrix via an affinity tag, e.g. the strep-tag [72]. The protein is consequently eluted as a co-complex with the antibody fragment (Fig. 11.5) and can be used as such for crystallization trials. This indirect immunoaffinity chromatography approach com-

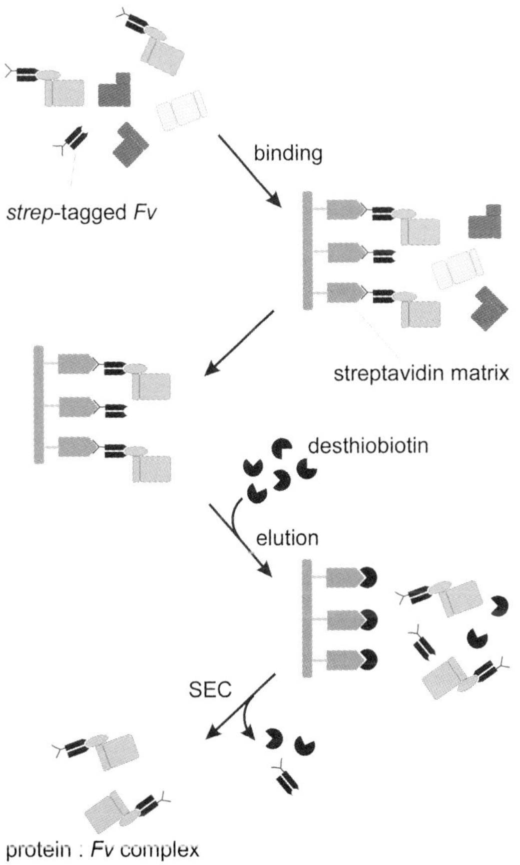

Fig. 11.5 Schematic depiction of the purification of a multisubunit membrane protein in complex with an antibody fragment (strep-tagged Fv fragment) by indirect immunoaffinity chromatography. Solubilized cellular membranes are combined with periplasmic extract of the Fv fragment-producing *E. coli* strain and the so-called co-complex is formed. The mixture is applied to a streptavidin-affinity column and the co-complex as well as the Fv fragment alone are bound via the strep-tag. They are specifically eluted by competitive displacement with an appropriate ligand after removing all other proteins in a washing step. Surplus Fv can be separated by size-exclusion chromatography resulting in pure complex of membrane protein and Fv fragment. (From [71].)

bines the advantages of antigen/antibody interaction and affinity tag techniques, i.e. high specificity and well established, fast purification protocols of broad applicability. The technique allows isolation of the protein from its natural source, as well as from expression systems in its native form without the attachment of affinity markers. Obviously, the purified protein is present in a complex with the bound antibody fragment, which should not affect the biological activity of the protein unless this is desired. COX from *P. denitrificans* [32] as well as the quinol oxidases from *P. denitrificans* and *E. coli* [73] have been efficiently purified by one-step indirect immunoaffinity chromatography using recombinant antibody fragments without affecting the functional properties of these membrane protein complexes. In contrast, to the COX co-complexes, streptavidin-affinity chromatography cannot be used for purification of the QCR–Fv fragment co-complex. A single transmembrane helix anchors subunit Rip1p at the periphery of the transmembrane portion of the complex and "pulling" at its extrinsic domain via the Fv fragment results in disintegration of QCR to a certain extent (Hunte, unpublished results). Therefore, QCR and Fv fragments were purified separately and the co-complex subsequently formed with surplus Fv fragments, which are then removed by size-exclusion chromatography [33, 71].

11.2.3.2 Crystallization

Reassuringly, membrane protein crystals are obtained with standard crystallization techniques and conditions that are routinely used for soluble proteins, and finding the optimal crystallization conditions appears not to be the limiting step (for reviews, see [27–29, 74, 75]). Thorough biochemical characterization of the protein preparation in combination with comprehensive screening for the most suitable detergent may be the most efficient strategy to cope with the difficulties of membrane protein crystallization. Protein heterogeneity has to be avoided; it can be caused by proteolysis, different oligomerization states or post-translational modifications, e.g. glycosylation. The crystallization of the water channel AQP1 was performed with fully deglycosylated protein samples [76]. In general, crystallization may be hampered by flexible termini or loops and proteolytic cleavage or recombinant products omitting these termini might be useful. For instance, crystallization of the potassium channel KcSA was achieved only after proteolytic cleavage of 35 C-terminal amino acids [36, 37]. Furthermore, restrained mutagenesis at the polar surface of the outer membrane proteins OmpA and OmpX was used to optimize crystallization [21].

Another important aspect should be mentioned here. Many membrane proteins change their conformation considerably during biological activity. For successful crystallization it may be necessary to fix certain conformations or mobile domains, e.g. by the use of specific inhibitors. The mobile extrinsic domain of subunit Rip1p of the cytochrome bc_1 complex was not resolved in the first structure of the complex from bovine mitochondria [77]. The domain's structure was only visible after its immobilization either by crystal contacts or by addition of Q_o site-specific inhibitors [33, 78]. In general, various means can be used to improve the stability of a protein. If no conditions and detergents can be found to keep the protein stable or suitable for crystallization, selection for stable mutants is advisable [79]. Antibody fragments can fix a defined protein conformation or stabilize a flexible domain to aid crystallization attempts if suitable antibodies have been selected.

Conditions applied for crystallization of membrane proteins in complex with antibody fragments comprise various pH, precipitant and salt concentrations indicating the high stability of the antigen–antibody complexes [30]. No data are available for the actual binding constants of the successful examples; however, in all cases, antibody fragments are used in molar excess during purification and stoichiometric complexes are obtained using gel filtration as a final purification step. The use of stoichiometric antibody–antigen complexes for crystallization clearly indicates the high affinity of the antibodies.

11.2.3.3 Alternative Approaches for Expansion of the Polar Surface

If a recombinant expression system is available, the addition of fusion proteins appears to be a fast alternative to enlarge the solvent-accessible surface of a membrane protein and thereby the potential for crystallization. However, protein fusions are connected via a linker region with the protein of interest. This linker will lead to the introduction of a flexible domain, a feature detrimental for crystallization. In an attempt to improve the crystal quality of membrane-bound cytochrome bo_3 ubiquinol oxidase from *E. coli*, the protein was produced as a C-terminal fusion of protein Z to subunit IV [80]. The fusion protein crystallized at similar conditions compared to the native complex, but the crystals were even less ordered. Molecular replacement trials have not been successful and protein Z could not be placed in the model. This may be due to the low resolution of the diffraction data (6 Å) or may indicate disorder of protein Z.

The fusion strategy to enlarge the hydrophilic surface could be of general help and valuable for structural genomics projects if fusion partners can be attached in a rigid manner. It has been tried to add the fusion partner not at the protein termini, but between two transmembrane helices. The soluble cytochrome b_{562} was inserted in one of the inner cytoplasmic loops of lactose permease, a secondary transport protein with 12 transmembrane helices [81]. The "red permease" has transport activity and 2-D crystals were obtained [82], but good diffracting 3-D crystals have not been reported.

Choosing physiologically stable reaction partners or stabilizing protein–protein interactions may be an alternative to increase the solvent-exposed domains of membrane proteins for crystallization.

11.3 Conclusions

In recent years, structural membrane protein research has been a rapidly evolving field, paving the way for rational drug discovery. However, the number of membrane protein structures at near atomic resolution is still small compared with those of soluble proteins. Antibody fragments are used to enlarge the hydrophilic surface of membrane proteins, thereby improving the chances for 3-D crystallization. High-resolution structures of three different membrane proteins have already been obtained with this method. All antibody fragments used for successful co-crystallization of membrane proteins so far have been derived from monoclonal antibodies obtained by classical hybridoma technology. They recognize native, nonlinear epitopes of their respective antigen and bind with high affinity. To establish this strategy as a general tool in crystallization of membrane proteins, recombinant antibody techniques have to be optimized to allow fast access to high-affinity binders. Antibody fragment-mediated crystallization may be especially helpful for membrane proteins that consist mainly of transmembrane helices with very short solvent-exposed loops, as well as for trapping flexible proteins in a fixed conformation.

Acknowledgments

This work was supported by the German Research Foundation (SFB472) and the Max Planck Society. We thank H. Pálsdóttir and E. Padan for critically reading the manuscript.

References

1 Lander ES, et al., Initial sequencing and analysis of the human genome, *Nature* 2001, **409**, 860–921.
2 Venter JC, et al., The sequence of the human genome, *Science* 2001, **291**, 1304–51.
3 Tomb JF, et al., The complete genome sequence of the gastric pathogen *Helicobacter pylori*, *Nature* 1997, **388**, 539–47.
4 Cole ST, et al., Deciphering the biology of *Mycobacterium tuberculosis* from the complete genome sequence, *Nature* 1998, **393**, 537–44.
5 Gardner MJ, et al., Genome sequence of the human malaria parasite *Plasmodium falciparum*, *Nature* 2002, **419**, 498–511.
6 Deisenhofer J, Epp O, Miki K, Huber R, Michel H, Structure of the protein subunits in the photosynthetic reaction center of *Rhodopseudomonas viridis* at 3 A resolution, *Nature* 1985, **318**, 618–24.
7 Byrne B, Iwata S, Membrane protein complexes, *Curr Opin Struct Biol* 2002, **12**, 239–43.
8 Wallin E, von Heijne G, Genome-wide analysis of integral membrane proteins from eubacterial archaean and eukaryotic organisms, *Protein Sci* 1998, **7**, 1029–38.
9 Garavito RM, Ferguson-Miller S, Detergents as tools in membrane biochemistry, *J Biol Chem* 2001, **276**, 32403–6.
10 Zulauf M, Detergent phenomena in membrane protein crystallization, in: Michel H, ed., *Crystallization of Membrane Proteins*, 54–71, CRC Press, Boca Raton, FL, 1991.
11 Walz T, Grigorieff N, Electron crystallography of two-dimensional crystals of membrane proteins, *J Struct Biol* 1998, **121**, 142–61.
12 Kuhlbrandt W, Two-dimensional crystallization of membrane proteins: a practical guide, in: Hunte C, von Jagow G, Schägger H, eds, *Membrane Protein Purification and Crystallization*, 253–84, Elsevier Science, Amsterdam, 2003.
13 Henderson R, Baldwin JM, Ceska TA, Zemlin F, Beckmann E, Down-

ING KH, Model for the structure of bacteriorhodopsin based on high resolution electron cryomicroscopy, *J Mol Biol* 1990, **213**, 899–929.

14 KUHLBRANDT W, WANG DN, FUJIYOSHI Y, Atomic model of plant light-harvesting complex by electron crystallography *Nature* 1994, **367**, 614–21.

15 MURATA K, MITSUOKA K, HIRAI T, WALZ T, AGRE P, HEYMANN JB, ENGEL A, FUJIYOSHI Y, Structural determinants of water permeation through aquaporin-1, *Nature* 2000, **407**, 599–605.

16 MICHEL H, Crystallization of membrane proteins, *Trends Biochem Sci* 1983, **8**, 56–9.

17 ROTH M, ARNOUX B, DUCRUIX A, REISS-HUSSON F, Structure of the detergent phase and protein–detergent interactions in crystals of the wild-type (strain Y) *Rhodobacter sphaeroides* photochemical reaction center, *Biochemistry* 1991, **30**, 9403–13.

18 OSTERMEIER C, MICHEL H, Crystallization of membrane proteins, *Curr Opin Struct Biol* 1997, **7**, 697–701.

19 OSTERMEIER C, IWATA S, LUDWIG B, MICHEL H, Fv fragment mediated crystallization of the membrane protein bacterial cytochrome *c* oxidase, *Nat Struct Biol* 1995, **2**, 842–6.

20 IWATA S, OSTERMEIER C, LUDWIG B, MICHEL H, Structure at 28-Angstrom resolution of cytochrome *c* oxidase from *Paracoccus denitrificans*, *Nature* 1995, **376**, 660–9.

21 PAUTSCH A, VOGT J, MODEL K, SIEBOLD C, SCHULZ GE, Strategy for membrane protein crystallization exemplified with OmpA and OmpX, *Proteins Struct Function Genet* 1999, **34**, 167–172.

22 SCHULZ GE, Beta-barrel membrane proteins, *Curr Opin Struct Biol* 2000, **10**, 443–447.

23 CHANG G, SPENCER RH, LEE AT, BARCLAY MT, REES DC, Structure of the MscL homolog from *Mycobacterium tuberculosis*: a gated mechanosensitive ion channel, *Science* 1998, **282**, 2220–6.

24 GRISSHAMMER R, TATE CG, Overexpression of integral membrane proteins for structural studies, *Q Rev Biophys* 1995, **28**, 315–422.

25 PADAN E, HUNTE C, REILANDER H, Production and purification of recombinant membrane proteins, in: HUNTE C, VON JAGOW G, SCHÄGGER H, eds, *Membrane Protein Purification and Crystallization*, 55–83, Elsevier Science, Amsterdam, 2003.

26 MICHEL H, Crystallization of membrane proteins, in: ROSSMANN MG, ARNOLD E, eds, *International Tables of Crystallography F: Macromolecular Crystallography*, 94–9, Kluwer, Dordrecht, 2001.

27 KUHLBRANDT W, 3-dimensional crystallization of membrane proteins, *Q Rev Biophys* 1988, **21**, 429–77.

28 GARAVITO RM, PICOT D, LOLL PJ, Strategies for crystallizing membrane proteins, *J Bioenerget Biomemb* 1996, **28**, 13–27.

29 HUNTE C, MICHEL H, Membrane protein crystallization, in: HUNTE C, VON JAGOW G, SCHÄGGER H, eds, *Membrane Protein Purification and Crystallization*, 143–60, Elsevier Science, Amsterdam, 2003.

30 HUNTE C, MICHEL H, Crystallisation of membrane proteins mediated by antibody fragments, *Curr Opin Struct Biol* 2002, **12**, 503–8.

31 HARRENGA A, MICHEL H, The cytochrome *c* oxidase from *Paracoccus denitrificans* does not change the metal center ligation upon reduction, *J Biol Chem* 1999, **274**, 33296–9.

32 KLEYMANN G, OSTERMEIER C, LUDWIG B, SKERRA A, MICHEL H, Engineered Fv fragments as a tool for the one-step purification of integral multisubunit membrane protein complexes, *Bio-Technology* 1995, **13**, 155–60.

33 HUNTE C, KOEPKE J, LANGE C, ROSSMANITH T, MICHEL H, Structure at 23 angstrom resolution of the cytochrome bc_1 complex from the yeast *Saccharomyces cerevisiae* co-crystallized with an antibody Fv fragment, *Struct Fold Des* 2000, **8**, 669–84.

34 LANGE C, NETT JH, TRUMPOWER BL, HUNTE C, Specific roles of tightly

bound phospholipids in the yeast cytochrome bc_1 complex structure, *EMBO J* 2001, **20**, 6591–600.

35 Lange C, Hunte C, Crystal structure of the yeast cytochrome bc_1 complex with its bound substrate cytochrome *c*, *Proc Natl Acad Sci USA* 2002, **99**, 2800–5.

36 Zhou YF, Morais-Cabral JH, Kaufman A, MacKinnon R, Chemistry of ion coordination and hydration revealed by a K^+ channel–Fab complex at 20 angstrom resolution, *Nature* 2001, **414**, 43–8.

37 Doyle DA, Cabral JM, Pfuetzner RA, Kuo AL, Gulbis JM, Cohen SL, Chait B, MacKinnon R, The structure of the potassium channel: molecular basis of K^+ conduction and selectivity, *Science* 1998, **280**, 69–77.

38 Kohler G, Milstein C, Continuous cultures of fused cells secreting antibody of predefined specificity, *Nature* 1975, **256**, 495–497.

39 Padan E, Venturi M, Michel H, Hunte C, Production and characterization of monoclonal antibodies directed against native epitopes of NhaA the Na^+/H^+-antiporter of *Escherichia coli*, *FEBS Lett* 1998, **441**, 53–8.

40 Venturi M, Hunte C, Monoclonal antibodies for the structural analysis of the Na^+/H^+ antiporter NhaA from *Escherichia coli*, *Biochim Biophys Acta Biomembr* 2003, **1610**, 46–50.

41 Tang DC, De Vit M, Johnston SA, Genetic immunization is a simple method for eliciting an immune response, *Nature* 1992, **356**, 152–4.

42 Peet NM, McKeating JA, Ramos B, Klonisch T, DeSouza JB, Delves PJ, Lund T, Comparison of nucleic acid and protein immunization for induction of antibodies specific for HIV-1 gp120, *Clin Exp Immunol* 1997, **109**, 226–32.

43 Kilpatrick KE, Cutler T, Whitehorn E, Drape RJ, Macklin MD, Witherspoon SM, Singer S and Hutchins JT, Gene gun delivered DNA-based immunizations mediate rapid production of murine monoclonal antibodies to the Flt-3 receptor, *Hybridoma* 1998, **17**, 569–76.

44 Singh M, O'Hagan DT, Recent advances in vaccine adjuvants, *Pharm Res* 2002, **6**, 715–28.

45 Goldberg ME, Djavadi-Ohaniance L, Methods for measurement of antibody/antigen affinity based on ELISA and RIA, *Curr Opin Immunol* 1993, **5**, 278–81.

46 Venturi M, Rimon A, Gerchman Y, Hunte C, Padan E, Michel H, The monoclonal antibody 1F6 Identifies a pH-dependent conformational change in the hydrophilic NH_2 terminus of NhaA Na^+/H^+-antiporter of *Escherichia coli*, *J Biol Chem* 2000, **275**, 4734–42.

47 Gerchman Y, Rimon A, Padan EJ, A pH-dependent conformational change of NhaA Na^+/H^+ antiporter of *Escherichia coli* involves loop VIII–IX, plays a role in the pH response of the protein, and is maintained by the pure protein in dodecyl maltoside, *Biol Chem* 1999, **274**, 24617–24.

48 Mirzabekov T, Kontos H, Farzan M, Marasco W, Sodroski J, Paramagnetic proteoliposomes containing a pure, native and oriented seven-transmembrane segment protein, CCR5, *Nat Biotechnol* 2000, **18**, 649–54.

49 Peipp M, Simon N, Loichinger A, Baum W, Mahr K, Zunino SJ, Fey GH, An improved procedure for the generation of recombinant single-chain Fv antibody fragments reacting with human CD13 on intact cells, *J Immunol Methods* 2001, **251**, 161–76.

50 Kwong PD, Wyatt R, Majeed S, Robinson J, Sweet RW, Sodroski J, Hendrickson WA, Structures of HIV-1 gp120 envelope glycoproteins from laboratory-adapted and primary isolates, *Struct Fold Des* 2000, **8**, 1329–39.

51 Lesk AM, Chothia C, Elbow motion in the immunoglobulins involves a molecular ball-and-socket joint, *Nature* 1988, **335**, 188–90.

52 Viti F, Nilsson F, Demartis S, Huber A, Neri D, Design and use of phage display libraries for the

selection of antibodies and enzymes, *Methods Enzymol* 2000, **326**, 480–5.

53 Hanes J, Jermutus L, Pluckthun A, Selecting and evolving functional proteins in vitro by ribosome display, *Methods Enzymol* 2000, **328**, 404–30.

54 Rondot S, Koch J, Breitling F, Dubel SA, Helper phage to improve single-chain antibody presentation in phage display, *Nat Biotechnol* 2001, **19**, 75–8.

55 de Wildt RM, Mundy CR, Gorick BD, Tomlinson IM, Antibody arrays for high-throughput screening of antibody–antigen interactions, *Nat Biotechnol* 2000, **18**, 989–94.

56 Ostermeier C, Michel H, Improved cloning of antibody variable regions from hybridomas by an antisense-directed RNase H digestion of the P3-X63-Ag8653 derived pseudogene mRNA, *Nucleic Acid Res* 1996, **24**, 1979–80.

57 Venturi M, Seifert C, Hunte C, High level production of functional antibody Fab fragments in an oxidizing bacterial cytoplasm, *J Mol Biol* 2002, **315**, 1–8.

58 Kabat EA, Wu TT, Reid-Miller M, Perry HM, Gottesman KS, Foeller C, 1991, *Sequences of Proteins of Immunological Interest*, 5th edn, US Department of Health and Human Services, NIH, Bethesda, MD.

59 Engberg J, Jensen LB, Yenidunya AF, Brandt K, Riise E, Phage-display libraries of murine antibody Fab fragments, in: Kontermann R, Dübel S, eds, *Antibody Engineering*, 65–92, Springer, Berlin, 2001.

60 Burmester J, Plückthun A, Construction of scFv fragments from hybridoma or spleen cells by PCR assembly, in: Kontermann R, Dübel S, eds, *Antibody Engineering*, 19–40, Springer, Berlin, 2001.

61 Breitling F, Moosmayer D, Brocks B, Dübel S, Construction of scFv from hybridoma by two-step cloning, in: Kontermann R, Dübel S, eds, *Antibody Engineering*, 41–55, Springer, Berlin, 2001.

62 Bradbury A, Cloning hybridoma by RACE, in: Kontermann R, Dübel S, eds, *Antibody Engineering*, 56–61, Springer, Berlin, 2001.

63 Bird RE, Hardman KD, Jacobson JW, Johnson S, Kaufman BM, Lee SM, Pope SH, Riordan GS, Whitlow M, Single-chain antigen-binding proteins, *Science* 1988, **242**, 423–6.

64 Brinkmann U, Reiter Y, Jung SH, Lee B, Pastan I, A recombinant immunotoxin containing disulfide-stabilized Fv fragment, *Proc Natl Acad Sci USA* 1993, **90**, 7538–42.

65 Skerra A, Plückthun A, Assembly of functional immunoglobulin Fv fragments in *Escherichia coli*, *Science* 1988, **249**, 1038–40.

66 Ridder R, Schmitz R, Legay F, Gram H, Generation of rabbit monoclonal antibody fragments from a combinatorial phage display library and their production in the yeast *Pichia pastoris*, *Biotechnology* 1995, **13**, 255–60.

67 Trill JJ, Shatzman AR, Ganguly S, Production of monoclonal antibodies in COS and CHO cells, *Curr Opin Biotechnol* 1995, **6**, 553–60.

68 Schmidt TGM, Skerra A, One-step affinity purification of bacterially produced proteins by means of the "Strep tag" and immobilized recombinant core streptavidin, *J Chromatogr A* 1994, **676**, 337–45.

69 Venturi M, *PhD thesis*, University of Frankfurt, 2000.

70 Jurado P, Ritz D, Beckwith J, de Lorenzo V, Fernandez LA, Production of functional single-chain Fv antibodies in the cytoplasm of *Escherichia coli*, *J Mol Biol* 2002, **320**, 1–10.

71 Hunte C, Kannt A, Antibody fragment mediated crystallization of membrane proteins, in: Hunte C, von Jagow G, Schägger H, eds, *Membrane Protein Purification and Crystallization*, 205–18, Elsevier Science, Amsterdam, 2003.

72 Schmidt TGM, Skerra A, The random peptide library-assisted engineering of a C-terminal affinity peptide useful for the detection and purification of a functional Ig Fv

fragment, *Protein Eng* 1993, **6**, 109–122.

73 Ostermann T, *PhD thesis*, University of Frankfurt, 2000.

74 McPherson SA, 1990, Current approaches to macromolecular crystallization, *Eur J Biochem* **189**, 1–23.

75 Michel H, General and practical aspects of membrane protein crystallization, in: Michel H, ed., *Crystallization of Membrane Proteins*, 73–88, CRC Press, Boca Raton, FL, 1991.

76 Sui HX, Walian PJ, Tang G, Oh A, Jap BK, Crystallization and preliminary X-ray crystallographic analysis of water channel AQP1, *Acta Crystallogr D Biol Crystallogr* 2000, **56**, 1198–200.

77 Xia D, Yu CA, Kim H, Xian JZ, Kachurin AM, Zhang L, Yu L, Deisenhofer J, Crystal structure of the cytochrome bc_1 complex from bovine heart mitochondria, *Science* 1997, **277**, 60–6.

78 Zhang ZL, Huang LS, Shulmeister VM, Chi YI, Kim KK, Hung LW, Crofts AR, Berry EA, Kim SH, Electron transfer by domain movement in stockbroker bc_1, *Nature* 1998, **392**, 677–84.

79 Bowie JU, Stabilizing membrane proteins, *Curr Opin Struct Biol* 2001, **11**, 397–402.

80 Byrne B, Abramson J, Jannson M, Holmgren E, Iwata S, Fusion protein approach to improve the crystal quality of cytochrome bo_3 ubiquinol oxidase from *Escherichia coli*, *Biochim Biophys Acta* 2000, **1459**, 449–55.

81 Prive GG, Kaback HR, Engineering the *lac* permease for purification and crystallization, *J Bioenerg Biomembr* 1996, **28**, 29–34.

82 Zhuang J, Prive GG, Werner GE, Ringler P, Kaback HR, Engel A, Two-dimensional crystallization of *Escherichia coli* lactose permease, *J Struct Biol* 1999, **125**, 63–75.

83 Hunte C, Insights from the structure of the yeast cytochrome bc_1 complex: crystallization of membrane proteins with antibody fragments, *FEBS Lett* 2001, **504**, 126–132.

Part IV
Kinetics, Metabolism and Toxicology

Molecular Biology in Medicinal Chemistry. Edited by Th. Dingermann, D. Steinhilber, G. Folkers

ISBN: 3-527-30431-2

12
Pharmacogenetics: The Effect of Inherited Genetic Variation on Drug Disposition and Drug Response

Kurt Kesseler

12.1
General

It is well known in clinical practice that there is a remarkable interindividual variation in the biological response of patients to the administration of a drug. This variability can be the result of a variety of factors such as pathogenesis and severity of the disease being treated, concomitant diseases, the patient's age, kidney and liver function, nutritional status, and drug interactions [1–3]. Despite the relevance of these factors, common variation in the genes coding drug-metabolizing enzymes (DMEs), drug transporters and drug targets/receptors – so called polymorphisms – contribute significantly to the heterogeneous response to drugs observed across populations [1].

The systematic investigation of the influence of genetic polymorphisms on interindividual variability of the biological response to drugs is termed "pharmacogenetics". In more detail, pharmacogenetics deals with the detection of hereditary differences in drug response, evaluation of the underlying molecular mechanisms at the protein and at the genetic level, establishing valid relationships between the phenotype and genotype, evaluating the clinical relevance of the polymorphic drug response, and developing test methods to identify patients which would be subject to increased therapeutic risk or low therapeutic benefit due to their genetic predisposition, before the start of drug therapy.

The first observation of inherited differences in the response to a chemical, the inability to taste phenylthiourea, was document as early in 1932 [4]. Pharmacogenetics had its beginning in the 1950s when genetically determined variations were considered to be the cause of adverse drug effects. For instance, hemolysis after antimalarial therapy was correlated with a lack of erythrocyte glucose-6-phosphate dehydrogenase activity [5]. Similarly, a deficiency in the activity of the acetylating enzyme *N*-acetyltransferase 2 (NAT2) was found to be the reason for peripheral neuropathy after isoniazid [6, 7].

The elucidation of the molecular genetic basis of inherited differences in drug response started in the early 1980s with the cloning and characterization of a poly-

Molecular Biology in Medicinal Chemistry. Edited by Th. Dingermann, D. Steinhilber, G. Folkers

ISBN: 3-527-30431-2

morphic human gene encoding the DME debrisoquine hydroxylase cytochrome P450 monooxidase II D6 (CYP2D6 [8, 9]).

Since then for an increasing number of human gene relationships between allelic variations, functional changes at the protein level and clinical outcome (drug response) have been established.

12.2
Relationships between Genotype and Phenotype

The overall pharmacological effect of a drug when administered to an individual is the result of drug metabolism and disposition, and drug–target interaction, which at the molecular level are mediated by enzymes [3, 10–14]. Genetic differences can induce changes at the enzyme level by various molecular mechanisms [12]:

- The gene encoding the enzyme could be missing completely, leading to complete absence of the enzyme.
- If the regulatory part of the gene is mutated, less mRNA and less enzyme will be formed.
- A mutation at an intron–exon boundary can lead to incorrect splicing of the pre-RNA and, thus, to the formation of an incomplete or inactive enzyme.
- A mutation within an exon can cause a nonsynonymous and nonconservative amino acid change in the protein, resulting either in complete loss or only in alteration of enzymatic activity, depending on whether the induced amino acid change is critical to enzyme activity or not.
- In contrast to the situations leading to absent or decreased enzymatic activity, the existence of multiple copies of the gene in question could resulting in an increase of the expression of the encoded enzyme.

Contrasting to these functional genetic alterations, there are also "silent" genetic changes, i.e. intron-based nucleotide changes that will not lead to any observable variation of enzyme structure/activity at all. (In fact, the majority of genomic polymorphisms belong to the silent type.)

In addition to these mechanisms, on a macroscopic level the relationship between genetic variation – the genotype – and the measurable/observable inter-individual change in drug response across a collective – the phenotype – strongly depends on the number of underlying genetic changes.

If a single genetic variation causes a trait, the phenotypes will follow a Mendelian or monogenic inheritance pattern. In this case, within the observed population, a bimodal frequency distribution will result. In particular, in family studies, the different phenotypes will clearly separate from each other.

Drug metabolism often is a monogenic trait, where plasma concentrations of a drug will only or predominantly depend on the activity of a single metabolizing enzyme [1, 3, 11]. Usually, the two phenotype cohorts in this case are divided into poor metabolizers (PMs) and extensive metabolizers (EMs) [12]. A similar scenario results if a polymorphism of a single gene determines differences in a drug receptor's sensitivity to a given drug, discriminating non-responder and responder phenotypes.

On the other hand, if a trait is the result of multiple genetic variations, then the superposition of differential molecular effects will lead to a more or less unimodal frequency distribution, in which the different phenotypes no longer clearly separate from each other. Drug response in general is polygenic in nature, because of the complex physiological cascades involved, which eventually are controlled by the interplay of a variety of enzyme-encoding genes [1, 3, 11].

12.3 Establishing Relations between Genotype and Phenotype

Until recently, the relation between phenotype and genotype has been established starting with the clinical observation of marked interindividual variation in drug response. Distribution analysis led to the discovery of phenotypes. Consequently, the potential underlying genes and their allelic variants were identified by cloning and sequencing, while the inheritance of the trait was elucidated by family and twin studies.

Being based on a clinical phenotype, this classical pharmacogenetic approach made it likely that the detected polymorphism would be causally related and clinically valid. However, although this works quite well for monogenic traits, the approach is not capable of correlating genotype to phenotype in the case of polygenic traits.

The implementation of genotype–phenotype relationships for polygenic traits requires a different and broader strategy, which forms the basic approach of the emerging discipline of pharmacogenomics. The pharmacogenomic approach is based on the fact that the human genome contains highly abundant polymorphisms, such as short tandem repeats and single nucleotide polymorphisms (SNPs), which themselves might or might not encode functional changes at the protein level, but are in linkage disequilibrium with functional alleles [15–17]. Accordingly, these common polymorphisms can act as markers for functional alleles.

Scanning a DNA region for those polymorphisms yields a map of markers, which are then tested for association with the phenotype in question. It is possible to identify associations between genetic markers and individual responses to medicines that may or may not be causal without requiring prior knowledge of the biochemical function of the gene(s) involved. Based on genotype, this approach can be applied to any trait, independent of whether it is monogenic or polygenic. It is the ultimate goal to create genome-wide maps of biomarkers, which could be associated to phenotypes by nonfamilial association studies.

For the purpose of the creation of high-density polymorphism mapping, SNPs appear to be the polymorphisms of choice [15–17]. Compared to other types of genetic variation such as variable numbers of tandem repeats (VNTRs), SNPs offer a number of advantages. SNPs are the most frequent form of polymorphisms throughout the human genome, occurring on average every 1000–2000 base pairs, thus presenting a maximum polymorphism density. They are binary, which makes them well suited to automated, high-throughput analysis. Finally, SNPs appear to have low mutation rates and, thus, are relatively stable over time.

A variety of technologies can be used for the identification and characterization of polymorphisms [18, 19]. These include

- Single-stranded conformation polymorphism (SSCP)
- Conformation sensitive gel electrophoresis
- Chemical- or enzymatic-mismatch cleavage detection
- Denaturing gradient gel electrophoresis
- Heteroduplex analysis by denaturing high-performance liquid chromatography (HPLC)
- High-throughput DNA sequencing
- Mass spectrometry (MS)
- Scanning and resequencing on oligonucleotide arrays (scanning arrays, variant detector arrays)

The creation of genome-wide SNP maps has now become feasible with the huge technical progress made in high-throughput gene analysis.

By 2001, the SNP Map Working Group had published a sequence variation map of the human genome containing 1.42 million SNPs [20]. The SNP consortium data repository currently contains nearly 1.8 million SNPs (available via the TSC Data Coordinating Center website at http://snp.cshl.org [21]).

Despite this significant progress in SNP mapping, attempts to establish associations between polygenic phenotypes and genome-wide polymorphism maps have not yet succeeded. One crucial question in this context is how many marker polymorphisms have to be included into a genome-wide association study. In principle, the number of polymorphisms to be evaluated in an association study is determined by the extent of linkage disequilibrium. However, linkage disequilibrium shows marked genomic variability, which is caused by a variety of factors such as the distance between polymorphisms, recombination rate, age of polymorphisms, population history, selection and genetic drift. Accordingly, attempts to estimate the minimum number of SNPs to be included in genome-covering association studies have produced conflicting results.

Recent simulations using a simplified general population model suggest that useful levels of linkage disequilibrium do not usually extend beyond an average distance of 3 kb (1 kb = 1000 base pairs) in the general population, implying that for whole-genome studies approximately 0.5 million SNPs would be required [22]. Experimental data have produced greatly varying linkage disequilibrium block size estimates ranging from 5 kb to several hundred kilobases, with the differences in linkage disequilibrium extent often found to be dependent on the ethnicity and/or race of the examined population [23–27]. With regard to feasibility, high-linkage disequilibrium block sizes would be advantageous. For example, a linkage disequilibrium extent of 60 kb would allow genome-wide studies with as few as 30000

SNPs [24]. However, even in this case, the number of polymorphisms to be taken into consideration exceeds current technological capabilities. In addition, even after the association between genotype and phenotype had been demonstrated, the mechanisms underlying such associations would require further studies. Accordingly, establishing genotype–phenotype relations on the basis of genome-wide association studies appears to be the preserve of the future.

In the meantime, applying the pharmacogenomic approach in a more focused manner to single or multiple "candidate genes", which are suspected to contribute to the phenotype in question, appears more feasible. Several successful examples of the candidate gene approach are known in the literature, demonstrating correlations between drug target polymorphisms to altered drug response (*vide infra*). In some cases single SNPs have been found to give sufficient association with drug response, whereas in others the correlation of traits to complete haplotypes gave better results.

12.4 Phenotyping versus Genotyping – Prediction of Individual Drug Response

Once a polymorphism is known, and the relationship between the genetic variants and the phenotype has been validated, it is possible to predict the individual drug response either by phenotyping or genotyping the subjects [18, 28–36].

In general, phenotyping can be performed by the direct measurement of the properties of any biomarker specific to a system, organ or tissue, e.g. enzyme activities or drug–receptor interactions. The general approach includes sample withdrawal and determination of biomarker activity by a relevant biochemical assay. This type of phenotyping is called functional phenotyping [18]. Although functional phenotyping allows a direct assay at the protein or enzyme level, there are limitations if enzymatic activities overlap substrate specificities, e.g. in the case of DMEs. Another disadvantage of functional phenotyping is that it often requires invasive sampling methods, e.g. liver biopsy.

In the case of DMEs, functional phenotyping can be replaced by metabolic phenotyping. Metabolic phenotyping requires the application of a probe drug that is known to be metabolized by the enzyme in question, followed by the determination of unchanged drug versus its metabolites in plasma or urine [18].

Ideally, a phenotyping procedure should be easy to perform and inexpensive. The probe drug applied should be specific for the polymorphic enzyme in question, of known metabolic pattern, easy to assay and safe. The assay method should be sensitive and simple.

Phenotyping methods were used before molecular biological tools became available and were fully validated. The advantage of phenotyping is that the results reflect the actual individual metabolic activity. However, even in the case of optimized methods, phenotyping is susceptible to confounding factors such as comedication, nutrition or disease. In addition, it requires significant effort/time for obtaining results and puts additional drug burden on the test subject.

In contrast to phenotyping, genotyping does not address actual enzyme activities, but predicts the individual drug response on the basis of identified genetic polymorphisms [35].

From a technical point of view it is relatively easy to perform genotyping with samples of genomic DNA, because the identification of the alleles of an individual at a specific gene locus can be achieved using many different assay methods, including gel electrophoresis-based genotyping methods such as polymerase chain reaction (PCR) restriction fragment length polymorphism (PCR-RFLP), allele-specific PCR, homogeneous fluorescent dye-based genotyping (e.g. the TaqMan assay or the Invader assay), mass spectrometry and DNA microarray (DNA chip) technology [18, 28–35, 37–47].

Genotyping is less time-consuming than phenotyping, is not influenced by drug–drug or food–drug interactions and there are no problems with compliance on the part of the test subject. On the other hand, only well-investigated polymorphisms can be determined, as different alleles must have been shown to encode enzymes resulting in a polymorphic metabolism, and the sequence information of wild-type and mutant alleles must be available. A correlation analysis with results from a corresponding phenotyping is always necessary to demonstrate the validity of the genotyping procedure.

The relevance of phenotyping and genotyping lies in their potential to predict before the start of a drug therapy whether or not a patient would be subject to increased therapeutic risk or low therapeutic benefit due to genetic predisposition. In practice, however, a number of challenges remain. For instance, applying a genotyping and a phenotyping procedure to the same individual can produce incongruent results. Similarly, the results of phenotyping can be influenced by the typing method and/or probe drug. Accordingly, it has been noted by numerous authors that genotyping and phenotyping procedures have to be selected and performed carefully [30, 34–36, 48].

12.5 Genetic Polymorphism in DMEs

Most of the current pharmacogenetic knowledge has been acquired in the field of DMEs. The reasons for this are that DME polymorphisms often are monogenic traits, that the resulting individual effect, e.g. the metabolic ratio, can readily be assayed and that phenotypes as characterized by interindividual differences in metabolic ratio can readily be distinguished from each other. At a minimum it is possible to distinguish individuals with normal metabolism, who by default are EMs, from PMs, who have significantly reduced DME activity. In some cases it is also possible to discriminate individuals with increased metabolism, which are classified as ultrarapid or ultraextensive metabolizers (UMs).

Potential consequences of polymorphic drug metabolism include [3, 11, 12, 49]:

- Extended pharmacological effect
- Adverse drug reactions

- Lack of prodrug activation
- Drug toxicity
- Increased effective dose
- Metabolism by alternative, deleterious pathways
- Exacerbated drug–drug interactions

DMEs are usually classified on the basis of the type of metabolic modification they catalyze. Phase I enzymes catalyze functional group transformations such as oxidations, reductions or hydrolyses. Phase II enzymes catalyze the conjugation of substances with endogenous substituents. Because Phase II DMEs are always involved in the elimination of bioactives and their metabolites, one can expect that decreased enzyme activity will lead to increased systemic drug levels which may be associated with increased therapeutic effect, but also with increased side effects and increased toxicity. In the case of phase I enzymes, the consequences of polymorphism depend on whether the DME catalyses the inactivation of an active drug, or the formation of an active metabolite from either a drug or a prodrug. The relative contribution of single DMEs to overall phase I or II metabolism is shown in Fig. 12.1.

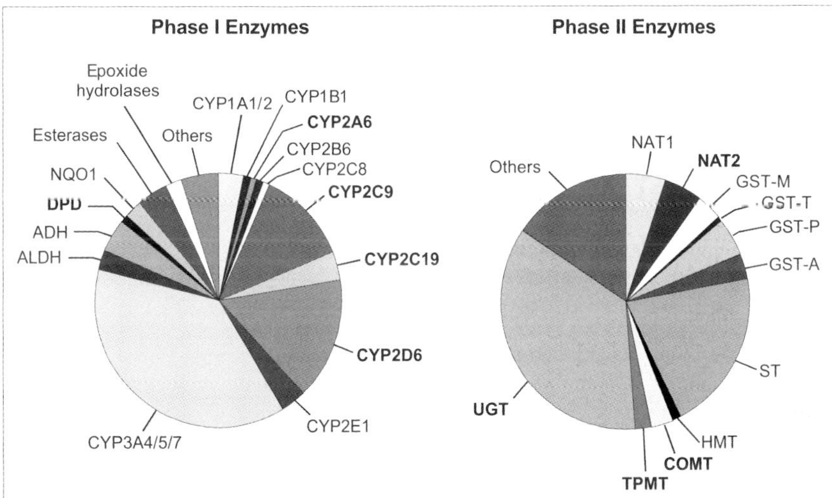

Fig. 12.1 Relative contribution of single DMEs to phase I or II metabolism of drugs (modified from [3]). The size of the section in the pie charts indicates the percentage of phase I and II metabolism of drugs that the respective enzyme contributes to. Enzymes with polymorphisms that have already been shown to be cause clinically relevant differences in drug effects are printed in bold. Abbreviations: ADH = alcohol dehydrogenase; ALDH = aldehyde dehydrogenase; CYP = cytochrome P450; DPD = dihydropyrimidine dehydrogenase; NQOH1 = NADPH:quinone oxidoreductase; COMT = catechol-*O*-methyltransferase; GST = glutathione S-transferase; HMT = histamine methyltransferase; NAT = *N*-acetyltransferase; ST = sulfotransferase; TPMT = thiopurine methyltransferase, UGT = uridine 5′-triphosphate glucuronosyltransferase.

Nearly all of the DMEs show genetic polymorphism [11]. However, not all polymorphisms have been demonstrated to be clinically relevant. In general, the clinical importance of polymorphic drug response/metabolism increases with a decrease in the therapeutic ratio of a given drug.

A number of excellent reviews on the pharmacogenetics of DMEs have been published [10–14, 49]. Therefore, within this article a detailed description about polymorphism in DMEs will focus on CYP2D6. An overview on various known polymorphisms of DMEs is given in Tab. 12.1.

12.6 Polymorphism of CYP2D6

Polymorphism of CYP2D6 was first discovered by observation of interindividual differences in the metabolism of debrisoquine and sparteine [9, 50]. Both drugs, as well as desipramine and metoprolol, have been used to phenotype subjects [36, 39, 51–58], allowing us to distinguish PMs from EMs [39, 53, 57, 59– 61]. In some cases UMs could also be discriminated [57, 59].

The human *CYP2D6* gene is encoded at the *CYP2D* locus at human chromosome 22 [62–64], which also contains two other genes, notably a related gene (*CYP2D7*), containing a mutation that disrupts the reading frame, and a pseudogene (*CYP2D8P*), which contains several mutations and does not encode a functional enzyme [62, 65]. Currently, more than 75 alleles of *CYP2D6* are known; however, many of the genetic variations do not produce measurable effects at the protein level (compilations of the various alleles of CYP2D6 and other CYPs are available at the website of the Human Cytochrome P450 (CYP) Allele Nomenclature Committee at http://www.imm.ki.se/cypalleles).

There are five alleles with silent single nucleotide changes. The allele often referred to as *CYP2D6**2 [66–68] comprises a group of alleles with various mutations resulting in the same change at the enzyme level (Arg296 → Cys and Ser486 → Thr). The enzyme encoded by *CYP2D6*2* has approximately 80% activity compared with the wild-type. Other alleles encoding a functional enzyme are *CYP2D6*33* and *CYP2D6*35* [69]. All of these alleles can be classified as functional corresponding to the activity of the enzyme encoded. Functional alleles are associated with the EM phenotype.

So far, 23 alleles have been described encoding inactive enzyme. Accordingly, these can be classified as non-functional alleles. The loss of enzyme activity is caused by various mechanisms such as frameshifts (e.g. *CYP2D6*3A* [70], *CYP2D6*6* [69, 71–73], *CYP2D6*41* [74]), splicing defects (e.g. *CYP26D6*4* [69, 70, 75–79]) or stop codon (*CYP2D6*8* [80]). In case of the *CYP2D6*5* allele the complete gene is deleted [81, 82]. Non-functional alleles are associated with the PM phenotype.

Six alleles (*CYP2D6*9* [83, 84], *CYP2D6*10A* [77], *CYP2D6*10B* [85], *CYP2D6*17* [86, 87], *CYP2D6*36* [84, 88], *CYP2D6*41* [89]) are known which encode enzyme with decreased activity compared to the wild-type. These alleles can

Tab. 12.1 Polymorphism in DMEs.

Gene/gene product	*Reference*	*Drug substrate (effect)*
Phase I DMEs		
CYP1A1	109–112	phenacetin
CYP1A2	113	acetaminophen amonafide caffeine paraxanthine ethoxyresorufin propranolol fluvoxamine
CYP1B1	114	estrogen metabolites
CYP2A6	115, 116	coumarin nicotine (cigarette addiction) halothan
CYP2B6	117	cyclophosphamide aflatoxin mephenytoin
CYP2C9	118–129	tolbutamide (hypoglycemia) warfarin (variations in anticoagulant effect of warfarin, hemorrhage) phenytoin (increased toxicity) nonsteroidal anti-inflammatories diazepam ibuprofen glipizide (hypoglycemia) losartan (decreased antihypertensive effect)
CYP2C19	37, 104, 130–137	mephenytoin (PM: toxicity increased, EM: efficacy decreased) omeprazole (PM: toxicity increased, EM: efficacy decreased, higher cure rates when given with clarythromycin) hexobarbital (PM: toxicity increased, EM: efficacy decreased) mephobarbital (PM: toxicity increased, EM: efficacy decreased) propranolol (PM: toxicity increased, EM: efficacy decreased) proguanil (PM: toxicity increased, EM: efficacy decreased) phenytoin (PM: toxicity increased, EM: efficacy decreased) citalopram (PM: toxicity increased, EM: efficacy decreased) diazepam (PM: toxicity increased, prolonged sedation, EM: efficacy decreased) Benzodiazepines
CYP2D6	9, 35, 37, 39, 51, 59, 93, 97, 99–108, 136, 138–151	β-blockers (β-blocker effect) propranolol (PM: bradycardia, hypotension) timolol (PM: decreased elimination – accumulation of parent compound resulting in toxicity, bradycardia, hypotension) bufuralol (PM: decreased elimination – accumulation of parent compound resulting in toxicity, bradycardia, hypotension) metoprolol (variation on β-blocker activity, PM: decreased elimination – accumulation of parent compound resulting in toxicity, bradycardia, hypotension)

Tab. 12.1 *(continued)*

Gene/gene product	*Reference*	*Drug substrate (effect)*
CYP2D6 cont'd		antipsychotics (tardive dyskinesia from antipsychotics, PM: toxicity increased, UM: efficacy insufficient) desipramin (PM: toxicity increased, hypotension, stupor, UM: efficacy insufficient) nortryptilin (PM: toxicity increased, postural hypotension, UM: efficacy insufficient) amitriptylin [PM: Decreased elimination of parent compound and active metabolite (active = nortryptilin), toxicity increased, postural hypotension, UM: efficacy insufficient] imipramin [PM: decreased elimination of active metabolite (=desipramine)] clomipramin (PM: decreased elimination – accumulation of parent compound resulting in toxicity) codeine [PM: Inefficacy as analgesic, decreased prodrug activation (active = morphine) narcotic side effects, dependence] debrisoquin (PM: toxicity increased, hypotension, appreciable orthostatic tension, UM: efficacy insufficient) dextromethorphan (PM: toxicity increased UM: efficacy insufficient) encainide (PM: decreased prodrug activation (active = *O*-desmethyl encainide), EM: widening of QRS complex) flecainide (PM: decreased elimination – accumulation of parent compound resulting in increased toxicity, inconsistent β-blockade, UM: efficacy insufficient) fluoxetin (PM: toxicity increased, UM: efficacy insufficient) guanoxan (PM: toxicity increased, hypotension, UM: efficacy insufficient) methoxyamphetamine (PM: toxicity increased, UM: efficacy insufficient) nortryptiline (PM: Decreased elimination – accumulation of parent compound resulting in increased toxicity, UM: efficacy insufficient) *N*-propylajmaline (PM: toxicity increased, UM: efficacy insufficient) perhexiline (PM: toxicity increased peripheral neuropathy, agranulocytosis, UM: efficacy insufficient) phenacetin (PM: toxicity increased, UM: efficacy insufficient) phenformin (PM: toxicity increased, lactic acidosis, UM: efficacy insufficient) propafenone (PM: toxicity increased, UM: efficacy insufficient) sparteine (PM: decreased elimination – accumulation of parent compound resulting in increased toxicity cardiac depression, uterine contraction, UM: efficacy insufficient) mexiletin (PM: decreased elimination – accumulation of parent compound resulting in toxicity) amiodaron indoramin perphenazin (PM: decreased elimination – accumulation of parent compound resulting in toxicity) zuclopenthixol (PM: decreased elimination – accumulation of parent compound resulting in toxicity)

Tab. 12.1 *(continued)*

Gene/gene product	***Reference***	***Drug substrate (effect)***
CYP2D6 cont'd		ithioridazin carvedilol (variations in α_1/β_2 blockade) antiarrhythmics (proarrhythmic and other toxic effects)
CYP2E1	152–156	*N*-nitrosodimethylamine acetaminophen ethanol (possible effect on alcohol consumption)
CYP3A4/A5/A7	103, 157–162	macrolides cyclosporin tacrolimus calcium channel blockers midazolam terfenadine lidocaine dapsone quinidine triazolam etoposide teniposide lovastatin alfentanil tamoxifen steroids
ALDH2	163	cyclophosphamide ethanol (slow metabolizers: facial flushing, fast metabolizers: protection from liver cirrhosis)
ALDH3	154, 156, 164	ethanol (increased alcohol consumption and dependence)
DPD	165–167	fluorouracil (possible enhanced toxicity, 5-fluorouracil neurotoxicity, myelotoxicity)
NQO1	131, 168–171	ubiquinone menadione (menadione-associated urolithiasis) mitomycin c
Plasma pseudo-cholinesterase	172	succinylcholine (slow ester hydrolysis: prolonged apnea)
Phase II DMEs		
COMT	173 175	levodopa (poor methylators: increased response) methyldopa (poor methylators: increased response) estrogens (substance abuse) ascorbic acid
GSTM1	12, 176, 177	aminochrome dopachrome adenochrome noradrenochrome

Tab. 12.1 *(continued)*

Gene/gene product	*Reference*	*Drug substrate (effect)*
GSTP1	12, 176, 178–180	13-*cis*-retinoic acid ethacrynic acid acrolein epirubicin
HMT	174, 181–184	histamine
NAT1	185, 186	*p*-aminosalicylic acid *p*-aminobenzoic acid sulfamethoxazole
NAT2	6, 7, 131, 185–189	isoniazid [EM (autosomal dominant): hepatitis, PM: (autosomal recessive): isoniazid neurotoxicity, lupus erythematosus, peripheral neuropathy, inhibition of hepatic mixed-function oxidase, increased phenytoin toxicity when coadministered] hydralazine (PM: hydralazine-induced lupus erythematosus) procainamide (PM: earlier and more frequently development of lupus erythematosus) phenelzine (PM: nausea, drowsiness) sulfasalazine (PM: severe reaction, more likely when large doses used) dapsone (PM: hematological effects) sulfonamides (hypersensitivity to sulfonamides) caffeine amonafide (rapid acetylators: efficacy decreased, toxicity increased, myelotoxicity colorectal cancer)
STs	190–194	steroids acetaminophen estrogens dopamine epinephrine naringenin
TPMT	189, 195–211	6-mercaptopurine (PM: thiopurine toxicity and efficacy, risk of second cancers, myelotoxicity, hematopoietic toxicity, bone marrow depression) 6-thioguanine (PM: thiopurine toxicity and efficacy, risk of second cancers, myelotoxicity, hematopoietic toxicity) azathioprine (PM: thiopurine toxicity and efficacy, risk of second cancers, myelotoxicity, hematopoietic toxicity)
UGT1A1	189, 212–217	irinotecan (diarrhea, myelosuppression) bilirubin flavopiridol
UGT2Bs	218, 219	opioids androgens morphine naproxen ibuprofen

Abbreviations: see Fig. 12.1.

be classified as reduced functioning and have been associated with an intermediate metabolizer phenotype.

Finally, there are also alleles in which the complete CYP2D6 gene has been duplicated or even multiplied. In those cases, where genes encoding fully active enzyme are amplified, i.e. *CYP2D6*1XN* [77, 90], *CYP2D6*2XN* [90, 91] and *CYP2D6*35X2* [92], the result is an increased synthesis of enzyme. Accordingly, these alleles are associated with an UM phenotype.

The CYP2D6 allele activity varies significantly among racial/ethnic groups [93]. The most important features are that there is a significantly higher frequency of nonfunctional alleles in Caucasian as compared with other populations (26 versus 6% in Asians and Africans). Homozygous carriers of nonfunctional alleles accordingly represent 6–7% of Caucasian populations. The most important among the nonfunctional alleles are CYP2D6*4 and CYP2D6*5.

Reduced functioning alleles are much more frequent in Asian and African than in Caucasian populations. The CYP2D6*10 allele is very abundant in Asian populations (41%), but only present at 3% in Caucasians. Inherited amplification of functional alleles exists at a rather low frequency throughout all populations.

CYP2D6 contributes significantly to the overall oxidoreductive metabolism in man (approximately 25% of all prescribed drugs, see Fig. 12.1), although it is only a relatively minor form in the human liver with an abundance of only 1.5% [10]. Since CYP2D6 is also found in the brain [94], it is likely that it acts as a defense mechanism protecting the central nervous system from environmental neurotoxins [10]. CYP2D6 catalyzes the primary metabolism of a wide variety of drugs, including tricyclic antidepressants (TCAs), antiarrhythmics, β-blockers and codeine, which accordingly are susceptible to impacts caused by DME polymorphism. The following examples will illustrate the clinical relevance of CYP2D6 polymorphism.

- The metabolism of TCAs [51, 56, 95–97, 99] such as nortryptiline, desipramine and imipramine follows a common scheme, in which hydroxylation by CYP2D6 leads to inactivation of the drug [95]. Accordingly, treatment with normal doses can put debrisoquine PMs at increased risk of toxicity, whereas in UMs standard dosage might fail to achieve the desired pharmacodynamic effect. It is also clinically important that a number of antidepressant drugs, e.g. fluoxetine, haloperidol and paroxetine, are inhibitors of CYP2D6 [98]. This could result in increased toxicity if TCAs are coadministered with such drugs.
- Similarly, β-blocking agents such as metoprolol are inactivated by CYP2D6 [100–102]. Debrisoquine PMs are subject to higher active drug concentrations and accordingly to increased β-blockade.
- In the case of the class 1c antiarrhythmic propafenone, CYP2D6 polymorphism leads to a shift in the pharmacodynamic profile [100]. Propafenone pharmacodynamically acts as both a sodium channel blocker and a β-blocker. The drug is racemic and, whereas both enantiomers exert comparable sodium channel blockade, the β-blocking activity is limited to the *S*-isomer. CYP2D6 catalyzes the hydroxylation of propafenone. The product, 5-hydroxypropafenone, is an active

metabolite, which is mainly a sodium channel blocker, but also shows limited β-blocking activity. PMs of propafenone are subject to increased β-blockade and at higher risk of severe adverse events compared to EMs.

- The drug carvedilol is a racemic mixture of two enantiomers with different pharmacodynamics and metabolism. The *S*-enantiomer shows α_1 activity as well as β_2 activity, whereas the *R*-enantiomer is primarily a β_2-blocker. The *R*-isomer is metabolized by CYP2D6 to a much higher extent than the *S*-enantiomer. As a result the α_1/β_2 relative activities are dependent on the CYP2D6 genotypes of the treated patients [103, 104].
- The analgesic effect of codeine is mediated by its active metabolite morphine. Because codeine is *O*-demethylated to morphine by CYP2D6, the desired analgesic effect is found to be dependent on the CYP2D6 genotype. Accordingly, an analgesic effect upon treatment with codein is limited to EMs [105–108] (EMs show higher efficacy than PMs as measured by baseline resting minute ventilation [106]).

12.7 Genetic Polymorphism in Drug Transporters

In contrast to DMEs, not much is known so far about polymorphisms in drug transporters [3, 11]. Membrane transporters are responsible for the absorption of drugs in the gastrointestinal tract, excretion into urine or bile, transport across the blood–brain barrier and into sites of action, such as cardiovascular tissue, tumor cells and infectious microorganisms [3]. Accordingly polymorphism of transporters will have an influence on the bioavailability of drugs. Transporters include P-glycoprotein (P-gp; MDR1), α_1 acid glycoprotein, MRP1–6 (multidrug resistance proteins) and sister P-gp (SP-gp).

12.8 Multidrug Resistance Gene – Marker versus Functional Polymorphism

The MDR1 gene encodes P-glycoprotein (P-gp), an integral membrane protein belonging to the adenosine triphosphate-binding superfamily of membrane transporters [220–229]. P-gp contributes to the excretion of xenobiotics from kidney, liver and intestine. In endothelial cells of the central nervous system it prevents the penetration of drugs across the blood–rain barrier. Overexpression of P-gp in tumor cells results in the phenomenon known as multidrug resistance to antineoplastic agents. Drugs known to be eliminated via P-gp include digoxin, anthracycline antibiotics, vinblastine, daunomycine and cyclosporine A. Polymorphism in the MDR1 gene influencing expression level or protein sequence of the P-gp could therefore have an impact on the bioavailability of these drugs.

The MDR1 gene is a 28-exon gene, the cDNA consisting of 3843 bp. So far 16 alleles have been identified; however, most of the detected polymorphisms are intronic or silent [221]. Interestingly, a silent SNP in exon 26 (C3435T) was found to be associated with the expression level of intestinal P-gp and p.o. bioavailability of

Tab. 12.2 Polymorphisms in drug transporters.

Gene/gene product	*Reference*	*Drug substrate (effect)*
MDR-1	220–229	natural product anticancer drugs vinblastine (overexpression in cancer: drug resistance) doxorubicin (overexpression in cancer: drug resistance) paclitaxel (overexpression in cancer: drug resistance) CYP3A4 substrates digoxin fexofenadine
BSEP	230	conjugates
MRPs	231–234	glutathione glucuronide conjugates sulfate conjugates nucleoside antivirals

digoxin [222]. Individuals with the TT genotype showed significantly reduced expression of P-gp levels compared to individuals genotyped CC. Thus, upon oral administration, digoxin C_{max} values were higher in case of the TT genotype compared with the CC genotype. In an investigation of the distribution of MDR1 polymorphism in a white population, the C3435T SNP appeared to be the most common polymorphism, with more than 50% heterozygous and 28% homozygous carriers of the variant. Other studies showed that C3435T polymorphism is subject to remarkable interethnic variation [223–226]. Since the C3435T SNP does not cause an amino acid change at the protein level, it is most likely a marker polymorphism for a functional polymorphism. A potential candidate is a nonsynonymous SNP in exon 16 (G2677T), which is in linkage disequilibrium with the C3435T SNP. The G2677T polymorphism has been associated with increased *in vitro* P-gp function and lower plasma concentrations of fexofenadine [226]. Since these observations are contradictory to the effects reported for digoxin, further investigations are required to elucidate the contribution of the various MDR1 polymorphisms on the bioavailability of drugs.

A few additional examples of polymorphisms of drug transporters are given in Tab. 12.2.

12.9
Genetic Polymorphisms in Drug Targets

Polymorphism of a drug target directly influences the pharmacodynamic effect of a drug. Identification and investigation of polymorphism among drug targets is a formidable task, because, at least under *in vivo* conditions, the measurable drug effect is more or less under polygenic control, making it difficult to clearly distinguish between the phenotypes involved [1, 3]. Despite this, there have been numerous attempts to establish relationships between isolated polymorphisms in a single drug-target-encoding gene and the anticipated drug response. Some of these attempts were successful, whereas others showed cross-study inconsistencies, implying that more complex genetic regulation is involved in these cases. This will be illustrated by the following examples.

12.9.1 Cholesterol Ester Transfer Protein (CETP) Polymorphism as a Biomarker for Response to Lipid-lowering Therapy

CETP, which is involved in controlling plasma levels of high-density lipoproteins (HDL), has been shown to be polymorphic. The CETP polymorphism includes a restriction polymorphism in intron 1 of the CETP gene, termed TaqIB. The presence of the restriction site is referred to as B1 and its absence is referred to as B2. In Caucasians, 30% of the population are carriers of the B1B1 genotype and 19% of the B2B2 genotype, respectively, which corresponds to frequencies of 59% for the B1 allele and 41% for the B2 allele [235].

The B1 variant of the CETP gene was found to be associated with higher plasma CETP concentrations and lower HDL cholesterol concentrations than the B2 variant. In addition, an association was found between the B1 allele and a higher degree of coronary atherosclerosis, with the most pronounced progression of atherosclerosis in the B1B1 carriers, an intermediate degree of progression in the B1B2 carriers and the least progression in B2B2 carriers. Pravastatin therapy slowed the progression of coronary atherosclerosis in B1B1 carriers, but not in B2B2 carriers (representing 16% of the patients taking pravastatin) [235].

This suggested that the CETP polymorphism appears to predict whether men with coronary artery disease would benefit from treatment with pravastatin to delay the progression of coronary atherosclerosis. However, since the TaqIB polymorphism is located at an intron of the CETP gene, a causal relationship between genotype and drug response has to be excluded. It is more likely that this polymorphism constitutes a nonfunctional marker, which is in almost complete linkage disequilibrium with the CETP/−629 polymorphism, a functional polymorphism in the promoter region of the CETP gene [236, 237].

12.9.2 β_2-Adrenergic Receptor: Predictability of Haplotype versus Single SNPs

It has been demonstrated that the bronchodilatory effect of β_2-adrenoreceptor agonists is subject to receptor polymorphism. The β_2-adrenergic receptor, a prototypic example of G-protein-coupled receptors, interacts with endogenous catecholamines and various medications. The receptor exists in a variety of tissues, and is involved in the regulation of cardiac, vascular, pulmonary and metabolic functions. Activation of the β_2-adrenoreceptor in the lung results in relaxation of airway smooth muscle and, thus, β_2-adrenoreceptor agonists are commonly used as bronchodilators in asthmatics.

Multiple mutations in the β_2-adrenoreceptor gene are known and some of them are in linkage disequilibrium. A recent study identified a total of 13 SNPs organized into 12 haplotypes [238].

There have been several attempts to correlate the response to bronchodilator therapy with different polymorphisms of the β_2-adrenoreceptor gene. For instance, the response to albuterol was shown to be highly associated with a SNP in the β_2-adrenoreceptor gene, resulting in amino acid changes in the receptor at codon 16 (Arg → Gly). The increase in FEV_1 was found to be 6.5 times greater in patients that were homozygous for 16Arg/16Arg compared to patients with the Gly/Gly genotype. This relationship, however, could not be confirmed consistently by other studies [238–242]. By applying the candidate gene approach, another group of researchers determined the responses to bronchodilator therapy with albuterol of individuals with the five most common β_2-adrenergic receptor haplotypes. They found the responses to vary more than 2-fold and to be significantly associated to haplotype pair, but not related to individual SNPs [238].

12.9.3
Polygenic Drug Target Polymorphism: Response to Clozapine Treatment

The atypical antipsychotic clopazine has been found to interact with a variety of drug receptors, e.g. adrenergic receptors, the dopamine D3 receptor and derotonin receptors. Although it could be shown that allelic variation in the serotonin neurotransmitter receptor 2A (5-HT_{2A}) is a factor in determining the response to clozapine, this cannot explain the full extent of interindividual variation [243].

A study involving multiple candidate genes was conducted in order to find associations with improved predictive values. Nineteen polymorphisms in a set of nine clozapine targeted receptor subtypes plus one neurotransmitter transporter were studied. The strongest association was found for a combination of four 5-HT_{2A} polymorphisms, one 5-HT_{2C} polymorphism, one polymorphism of the serotonin transporter promoter (5-HTTLPR) and the histamine H_2 receptor, resulting in 76% predicatability of clozapine response and a sensitivity of 95% for satisfactory response [244].

Other examples of polymorphism in drug receptors are compiled in Tab. 12.3.

12.10
Concluding Remarks

Pharmacogenetic research has demonstrated that genetic polymorphism is an important determinant of drug disposition and response in humans. Especially in the field of drug metabolism, current knowledge allows us to explain interindividual pharmacokinetic differences by genetic polymorphism. However, elucidation of the relation between individual drug response and genotype is much more demanding because of its polygenic nature, and thus cannot be achieved by application of the mechanistic pharmacogenetic approach.

Strategies that are principally capable of addressing polygenic polymorphisms are genome-wide searches for polymorphisms associated with drug effects using high-density anonymous SNP maps and the candidate gene strategy, which takes

Tab. 12.3 Polymorphisms in drug receptors.

Gene/gene product	*Reference*	*Drug substrate (effect)*
Cholesteryl ester transport protein (CETP)	235–237	pravastatin (low HDL, progression of atherosclerosis)
β_2-adrenergic receptor	238–242, 245–251	albuterol (response in asthmatics, poor control of asthma) ventolin (response in asthmatics, poor control of asthma)
5-HT_{2A}	243	clozapine and other neuroleptics (variable drug efficacy, typical antipsychotic response and long-term outcomes)
5-HT_{26}	252	clozapine (response in schizophrenics)
5-HT_6	253–255	clozapine and other neuroleptics (drug response)
Serotonin transporter (5-HTT)	256–258	antidepressants, e.g. clomipramine, fluoxetine, paroxeine, fluvoxamine (5–HT neurotransmission, antidepressant response)
Dopamine D2 D3 D4 receptors	244, 259, 260–265	antipsychotics, e.g. halopridol, clozapine, thioridazine nemorapride [antipsychotic response (D2, D3, D4), antipsychotic-induced tardive dyskinesia (D3), antipsychotic-induced acute akathisia (D3), hyperprolactinemia in females (D2)]
5-Lipoxygenase promoter (ALOX 5)	266, 251	leukotriene synthesis inhibitors (response in asthmatics)
Angiotensin-converting enzyme (ACE)	267–279	ACE inhibitors, e.g. enalapril, lisinopril, captopril (renoprotective effects, bp reduction, left ventricular mass reduction, endothelial function improvement, ACE inhibitor induced cough, cardiac indices IgA nephropathy)
Bradykinin B2 receptor	280	ACE inhibitors (ACE inhibitor-induced cough)
Stromelysin	236, 281	pravastatin (efficacy in coronary atherosclerosis and restenosis)
β-Fibrinogen	236	pravastatin (decreased progression of coronary artery disease)
Hepatic lipase	236, 282	lovastatin/colestipol or niacin/colestipol (improvement of coronary stenosis)
Sulfonylurea receptor	283	tolbutamide (sulfonylurea-induced insulin release, serum C-peptide and insulin response)
HERG channel	284	quinidine (drug-induced long QT syndrome)
KvLQT1 channel	285, 286	terfenadine (drug-induced long QT syndrome) meflaquine (drug-induced long QT syndrome) disopyramide (drug-induced long QT syndrome) cisapride (Y315C polymorphism: torsade de pointe)

Tab. 12.3 *(continued)*

Gene/gene product	*Reference*	*Drug substrate (effect)*
hKCNE2 channel	285, 287, 288	quinidine bactrim (drug induced arrhythmia) bactrim (drug induced arrhythmia) procainamide (drug induced arrhythmia) oxatomide (drug induced arrhythmia) clarithromycin (torsade de pointe)
Inositol-p1p	289	lithium (response of manic depressive illness)
Apolipoprotein E4	236, 281, 290–295	simvastatin (improved mortality) tacrine (response of Alzheimer's disease)
Ryanodine receptor	296	halothane or succinylcholine (drug-induced malignant hyperthermia)
Platelet FC receptor (FCR II)	297	heparin (heparin-induced thrombocytopenia)
Gs protein α	298	β-blockers, e.g. metoprolol (antihypertensive effect)
Glycoprotein IIIa subunit of glycoprotein IIb/IIIa receptor	299	aspirin (antiplatelet effect) glycoprotein IIb/IIIa inhibitors, e.g. abciximab (antiplatelet effect)
Estrogen receptor	300, 301	conjugated estrogens (increase of bone mineral density); hormone replacement (increase in HDL)
G protein β_3	302, 303, 197	antidepressants (response to antidepressant therapy) diuretics (diuretic effect)
α-Adducin	304	diuretics (hypotensive effect)
Angiotensinogen	305	ACE inhibitors (hypotensive effect??)
Prothrombin gene	306	oral contraceptives (G20210A polymorphism: cerebral vein thrombosis)
Factor V	307	oral contraceptives (G1691A polymorphism: cerebral vein thrombosis)
Methylguanine methyltransferase	308	carmustine (response of glioma)

advantage of existing knowledge about pharmacodynamics and disposition of a drug.

Currently, the candidate approach appears to be the more promising one, because it focuses on a manageable number of genes and polymorphisms that are likely to be important, and has yielded the first promising results (e.g. in case of response to clozapine treatment as outlined above).

Although beyond current technological capacities and reasonable resource requirements, the genome-wide pharmacogenomic approach bears the potential to discover the genetic determinants of drug response in a more comprehensive and precise fashion.

Thus, based on an improved understanding of the relation between genetic variability and therapeutic success, pharmacogenomics will allow us to predict individual response to drug therapy. Utilizing pharmacogenomic knowledge in clinical practice should make it possible to select the best-suited medication or individually adjust the dosage regimen as required by a patient's genotype. Regarding the development of new drugs, genetic preselection of patients to be recruited for clinical trials into responders and nonresponders could help to optimize clinical trials, because the elimination of nonresponders could result in a reduction of the patient numbers required to reach statistically significance.

Ultimately, the goal of this effort is it to replace the current one-medicine-fits-all strategy of drug treatment by individualized therapies with efficient and safe drugs that are aligned to the specific genetic constitution of an individual patient.

References

1 Renegar G, Rieser P, Manasco P. 2001. Pharmacogenetics: the Rx perspective. *Expert Rev Mol Diagn* **1**, 255–63.

2 Meyer UA. 1994. The molecular basis of genetic polymorphism of drug metabolism. *J Pharm Pharmacol* **46** (Suppl. 1), 409–15.

3 Evans WE, Johnson JA. 2001. Pharmacogenomics: the inherited basis for interindividual differences in drug response. *Annu Rev Genomics Hum Genet* **2**, 9–39.

4 Snyder LH. 1932. The inheritance of taste deficiency in man. *Ohio J Sci* **32**, 436–68.

5 Carson PE, Flannagan CL, Ickes CE, Alving AS. 1956. Enzymatic deficiency in primaquine-sensitive erythrocytes. *Science* **124**, 484–5.

6 Hughes HB, Biehl JP, Jones AP, Schmidt LH. 1954. Metabolism of isoniazid in man as related to occurrence of peripheral neuritis. *Am Rev Tuberc* **70**, 266–73.

7 Evans DAP, Manley KA, McKusick VA. 1960. Genetic control of isoniazid metabolism in man. *Br Med J* **2**, 485–91.

8 Gonzalez FJ, Skoda RC, Kimura S, Umeno M, Zanger UM, Nebert DW, Gelboin HV, Hardwick JP, Meyer UA. 1988. Characterization of the common genetic defect in humans deficient in debrisoquine metabolism. *Nature* **331**, 442–6.

9 Mahgoub A, Idle JR, Dring LG, Lancaster R, Smith RL. 1977. Polymorphic hydroxylation of debrisoquine in man. *Lancet* **ii**, 584–6.

10 Wolf RC, Smith G. 1999. Pharmacogenetics. *Br Med Bull* **55**, 366–86.

11 Evans WE, Relling MV. 1999. Pharmacogenomics: translating functional genomics into rational therapeutics. *Science* **286**, 487–91.

12 Wormhoudt LW, Commandeur JN, Vermeulen NP. 1999. Genetic polymorphisms of human *N*-acetyltransferase, cytochrome P450, glutathione-S-transferase, and epoxide hydrolase enzymes: relevance to xenobiotic metabolism and toxicity. *Crit Rev Toxicol* **29**, 59–124.

13 Weinshilboum R. 2003. Inheritance and drug response. *N Engl J Med* **348**, 529–37.

14 Evans WE, McLeod HL. 2003. Pharmacogenomics – drug disposition, drug targets and side effects. *N Engl J Med* **348**, 538–49.

15 Nowotny P, Kwon JM, Goate AM. 2001. SNP analysis to dissect human traits. *Curr Opin Neurobiol* **11**, 637–41.

16 Brookes AJ. 1999. The essence of SNPs. *Gene* **234**, 177–86.

17 Schafer AJ, Hawkins JR. 1998. DNA variations and the future of human genetics. *Nat Biotechnol* **16**, 33–9.

18 Shi MM, Bleavins MR, De La Iglesia FA. 2001. Technologies for

detecting genetic polymorphisms in pharmacogenomics. *Drug Metab Dispos* **29**, 591–5.

19 Mir KU, Southern ME. 2000. Sequence variation in genes and genomic DNA: methods for large scale analysis. *Annu Rev Genomics Hum Genet* **1**, 329–60.

20 International SNP Map Working Group. 2001. A map of the human genome sequence variation containing 1.42 million single nucleotide polymorphisms. *Nature* **409**, 928–33.

21 Thorisson GA, Stein LD. 2003. The SNP Consortium website: past, present and future. *Nucleic Acids Res* **31**, 124–7.

22 Kruglyak L. 1999. Prospects for whole-genome linkage disequilibrium mapping of common disease genes. *Nat Genet* **22**, 139–44.

23 Reich DE, Cargill M, Bolk S, Ireland J, Sabeti PC, Richter DJ, Lavery T, Kouyoumjian R, Farhadian SF, Ward R, Lander ES. 2001. Linkage disequilibrium in the human genome. *Nature* **411**, 199–204.

24 Collins A, Lonjou C, Morton NE. 1999. Genetic epidemiology of single-nucleotide polymorphisms. *Proc Natl Sci Acad USA* **96**, 15173–7.

25 Taillon-Miller P, Bauer-Sardiña I, Saccone NL, Putzel J, Laitinen T, Cao A, Kere J, Pilia G, Rice JP, Kwok PY. 2000. Juxtaposed regions of extensive and minimal linkage disequilibrium in human Xq25 and Xq28. *Nat Genet* **25**, 324–8.

26 Dunning AM, Durocher F, Healey CS, Teare MD, McBride SE, Carlomagno F, Xu CF, Dawson E, Rhodes S, Ueda S, Lai E, Luben RN, Van Rensburg EJ, Mannermaa A, Kataja V, Rennart G, Dunham I, Purvis I, Easton D, Ponder BAJ. 2000. The extent of linkage disequilibrium in four populations with distinct demographic histories. *Am J Hum Genet* **67**, 1544–54.

27 Abecasis GR, Noguchi E, Heinzmann A, Traherne JA, Bhattacharyya S, Leaves NI, Anderson GG, Zhang Y, Lench NJ, Carey A, Cardon LR, Moffatt MF, Cookson WOC. 2001. Extent and distribution of linkage disequilibrium in three genomic regions. *Am J Hum Genet* **68**, 191–7.

28 Cronin MT, Pho M, Debjani D, Frueh F, Schwarcz L, Brennan T. 2001. Utilization of new technologies in drug trials and discovery. *Drug Metab Dispos* **29**, 586–90.

29 Ingelman-Sundberg M. 2001. Implications of polymorphic cytochrome P450-dependent drug metabolism for drug development. *Drug Metab Dispos* **29**, 570–3.

30 Gonzalez FJ, Idle JR. 1994. Pharmacogenetic phenotyping and genotyping. Present status and future potential. *Clin Pharmacokinet* **26**, 59–70.

31 Sapolsky RJ, Hsie L, Berno A, Ghandour G, Mittmann M, Fan JB. 1999. High-throughput polymorphism screening and genotyping with high-density oligonucleotide arrays. *Genet Anal Biomol Eng* **14**, 187–92.

32 Aitman TJ. 2001. DNA microarrays in medical practice. *Br Med J* **323**, 611–5.

33 Lockhart DJ, Winzeler EA. 2000. Genomics, gene expression and DNA arrays. *Nature* **405**, 827–36.

34 Heim MH. 1992. Polymerase chain reaction and its potential as a pharmacokinetic tool. *Clin Pharmacokinet* **23**, 321–7.

35 Zühlsdorf MT. 1998. Relevance of pheno- and genotyping in clinical drug development. *J Clin Pharmacol Ther* **36**, 607–12.

36 Streetman DS, Bertino JS, Nafziger AN. 2000. Phenotyping of drug-metabolizing enzymes in adults: a review of *in-vivo* cytochrome P450 phenotyping probes. *Pharmacogenetics* **10**, 187–216.

37 Bertilsson L. 1995. Geographical/interracial differences in polymorphic drug oxidation. Current state of knowledge of cytochromes P 450 (CYP) 2D6 and 2C19. *Clin Pharmacokinet* **29**, 192–209.

38 Linder MW, Vales R. 1999. *Ann Clin Lab Sci* **29**, 140–9.

39 Broly F, Gaedigk A, Heim M, Eichelbaum M, Morike K, Meyer

UA. 1991. Debrisoquine/sparteine hydroxylation genotype and phenotype: analysis of common mutations and alleles of CYP2D6 in a European population. *DNA Cell Biol* **10**, 545–58.

40 Gaedigk A. 2000. Interethnic differences of drug-metabolizing enzymes. *Int J Clin Pharmacol Ther* **38**, 61–8.

41 Ieiri I, Tainaka H, Morita T, Hadama A, Mamiya K, Hayashibara M, Ninomiya H, Ohmori S, Kitada M, Tashiro N, Higuchi S, Otsubo K. 2000. Catalytic activity of three variants (Ile, Leu, and Thr) at amino acid residue 359 in human CYP2C9 gene and simultaneous detection using single-strand conformation polymorphism analysis. *Ther Drug Monit* **22**, 237–44.

42 Tanigawara Y, Kita T, Hirono M, Sakaeda T, Komada F, Okumura K. 2001. Identification of *N*-acetyltransferase 2 and CYP2C19 genotypes for hair, buccal cell swabs, or fingernails compared with blood. *Ther Drug Monit* **23**, 341–6.

43 Kwiatkowski RW, Lyamichev V, de Arruda M, Neri B. 1999. Clinical, genetic, and pharmacogenetic applications of the Invader assay. *Mol Diagn* **4**, 353–64.

44 Pusch W, Wurmbach HJ, Thiele H, Kostrzewa M. 2002. MALDI-TOF mass spectrometry-based SNP genotyping. *Pharmacogenomics* **3**, 537–48.

45 Kwok PY. 2002. SNP genotyping with fluorescence polarization detection. *Hum Mutat* **19**, 315–23.

46 Tsuchihashi Z, Dracopoli NC. 2002. Progress in high throughput SNP genotyping methods. *Pharmacogenomics J* **2**, 103–10.

47 Brennan MD. 2002. High throughput genotyping technologies for pharmacogenomics. *Am J Pharmacogenomics* **1**, 295–302.

48 Zaigler M, Tantcheva-Poor I, Fuhr U. 2000. Problems and perspectives of phenotyping for drug-metabolizing enzymes in man. *Int J Clin Pharmacol Ther* **37**, 1–9.

49 Meyer UA. 2000. Pharmacogenetics and adverse drug reactions. *Lancet* **356**, 1667–71.

50 Eichelbaum M, Spannbrucker N, Steincke B, Dengler JJ. 1979. Defective *N*-oxidation of sparteine in man: a new pharmacogenetic defect. *Eur J Clin Pharmacol* **16**, 183–7.

51 Spina E, Gitto C, Avenoso A, Campo GM, Caputi AP, Perucca E. 1997. Relationship between plasma desipramine levels, CYP2D6 phenotype and clinical response to desipramine: a prospective study. *Eur J Pharmacol* **51**, 395–8.

52 Frye RF, Matzke GR, Adedoyin A, Porter JA, Branch RA. 1997. Validation of the five-drug "Pittsburgh cocktail" approach for assessment of selective regulation of drug-metabolizing enzymes. *Clin Pharmacol Ther* **62**, 365–76.

53 Wanwimolruk S, Bhawan S, Coville PF, Chalcroft SCW. 1998. Genetic polymorphism of debrisoquine (CYP2D6) and proguanil (CYP2C19) in South Pacific Polynesian populations. *Eur J Clin Pharmacol* **54**, 431–5.

54 Streetman DS, Bleakley JF, Kim JS, Nafziger AN, Leeder JS, Gaedigk A, Gotschall R, Kearns GL, Bertino JS. 2000. Combined phenotypic assessment of CYP1A2, CYP2C19, CYP2D6, CYP3A, *N*-acetyltransferase-2, and xanthine oxidase with the "Cooperstown cocktail". *Clin Pharmacol Ther* **68**, 375–83.

55 Tamminga WJ, Werner J, Oosterhuis B, Brakenhoff JPG, Gerritis MGF, de Zeeuw RA, de Leij LF, Jonkman JH. 2001. An optimized methodology for combined phenotyping and genotyping on CYP2D6 and CYP2C19. *Eur J Clin Pharmacol* **57**, 143–6.

56 Lledo P. 1993. Variations in drug metabolism due to genetic polymorphism. *Drug Invest* **5**, 19–34.

57 Brøsen K, Hansen JG, Nielsen KK, Sindrup SH, Gram LF. 1993. Inhibition by paroxetine of desipramine metabolism in extensive but not in poor metabolizers of sparteine. *Eur J Clin Pharmacol* **44**, 349–55.

58 Hou ZY, Pickle LW, Meyer PS, Woosley RL. 1991. Salivary analysis for determination of dextromethor-

phan metabolic phenotype. *Clin Pharmacol Ther* **49**, 410–9.

59 Lovlie R, Daly AK, Matre GE, Molven A, Stehen VM. 2001. Polymorphisms in CYP2D6 duplication-negative individuals with the ultrarapid metabolizer phenotype: a role for the CYP2D6*35 allele in ultrarapid metabolism? *Pharmacogenetics* **11**, 45–55.

60 Schadel M, Wu D, Otton SV, Kalow W, Sellers EM. 1995. Pharmacokinetics of dextromethorphan and metabolites in humans: influence of the CYP2D6 phenotype and quinidine inhibition. *J Clin Psychopharmacol* **15**, 263–9.

61 Brøsen K, Gram LF, Haghfelt T, Bertilsson L. 1987. Extensive metabolizers of debrisoquine become poor metabolizers during quinidine treatment. *Pharmacol Toxicol* **60**, 312–4.

62 Gonzalez FJ, Vilbois F, Hardwick JP, McBride OW, Nebert DW, Gelboin HV, MEYER UA. 1988. Human debrisoquine 4-hydroxylase (P450IID1), cDNA and deduced amino acid sequence and assignment of the CYP2D locus to chromosome 22. *Genomics* **2**, 174–9.

63 Eichelbaum M, Baur MP, Dengler HJ, Osikowska-Evers BO, Tieves G, Zekorn C, Rittner C. 1987. Chromosomal assignment of human cytochrome P-450 (debrisoquine/sparteine type) to chromosome 22. *Br J Clin Pharmacol* **23**, 455–8.

64 Gough AC, Smith CA, Howell SM, Wolf CR, Bryant SP, Spurr NK. 1993. Localization of the CYP2D gene locus to human chromosome 22q13.1 by polymerase chain reaction, *in situ* hybridization, and linkage analysis. *Genomics* **15**, 430–2.

65 Kimura S, Umeno M, Skoda RC, Meyer UA, Gonzalez FJ. 1989. The human debrisoquine 4-hydroxylase (CYP2D) locus: sequence and identification of the polymorphic CYP2D6 gene, a related gene and a pseudogene. *Am J Hum Genet* **45**, 889–904.

66 Panserat S, Mura C, Gerard N, Vincent-Viry M, Galteau MM, Jacqz-Aigrain E, Krishnamoorthy R. 1994. DNA haplotype-dependent differences in the amino acid sequence of debrisoquine 4-hydroxylase (CYP2D6), evidence for two major allozymes in extensive metabolisers. *Hum Genet* **94**, 401–6.

67 Tsuneoka Y, Matsuo Y, Iwahashi K, Takeuchi H, Ichikawa Y. 1993. A novel cytochrome P-450ID6 mutant gene associated with Parkinson's disease. *J Biochem* **114**, 263–6.

68 Armstrong M, Idle JR, Daly AK. 1993. A polymorphic *Cfo*I site in exon 6 of the human cytochrome P450CYP2D6 gene detected by the polymerase chain reaction. *Hum Genet* **91**, 616–7.

69 Marez D, Legrand M, Sabbagh N, Guidice JM, Spire C, Lafitte JJ, Meyer UA, Broly F. 1997. Polymorphism of the cytochrome P450 CYP2D6 gene in a European population: characterization of 48 mutations and 53 alleles, their frequencies and evolution. *Pharmacogenetics* **7**, 193–202.

70 Kagimoto M, Heim M, Kagimoto K, Zeugin T, Meyer UA. 1990. Multiple mutations of the human cytochrome P450IID6 gene (CYP2D6) in poor metabolizers of debrisoquine. Study of the functional significance of individual mutations by expression of chimeric genes. *J Biol Chem* **265**, 17209–14.

71 Saxena R, Shaw GL, Relling MV, Frame JN, Moir DT, Evans WE, Caporaso N, Weiffenbach B. 1994. Identification of a new variant CYP2D6 allele with a single base deletion in exon 3 and its association with the poor metabolizer phenotype. *Hum Mol Genet* **3**, 923–6.

72 Evert B, Griese EU, Eichelbaum M. 1994. Cloning and sequencing of a new non-functional CYP2D6 allele: deletion of T1795 in exon 3 generates a premature stop codon. *Pharmacogenetics* **4**, 271–4.

73 Daly AK, Leathart JB, London SJ, Idle JR. 1995. An inactive cytochrome P450 CYP2D6 allele containing a

deletion and a base substitution. *Hum Genet* **95**, 337–41.

74 GAEDIGK A, NDJOUNTCHE L, BRADFORD LD, LEEDER JS. 2002. Identification of a novel non-functional CYP2D6 allele, 2D6*42, in African Americans. *Drug Metab Rev* **34** (Suppl. 1), 151.

75 GOUGH AC, MILES JS, SPURR NK, MOSS JE, GAEDIGK A, EICHELBAUM M, WOLF CR. 1990. Identification of the primary gene defect at the cytochrome P450 CYP2D locus. *Nature* **347**, 773–6.

76 HANIOKA N, KIMURA S, MEYER UA, GONZALEZ FJ. 1990. The human CYP2D locus associated with a common genetic defect in drug oxidation: a G1934A base change in intron 3 of a mutant CYP2D6 allele results in an aberrant 3′ splice recognition site. *Am J Hum Genet* **47**, 994–1001.

77 YOKOTA H, TAMURA S, FURUYA H, KIMURA S, WATANABE M, KANAZAWA I, KONDO I, GONZALEZ FJ. 1993. Evidence for a new variant CYP2D6 allele CYP2D6J in a Japanese population associated with lower *in vivo* rates of sparteine metabolism. *Pharmacogenetics* **3**, 256–63.

78 SACHSE C, BROCKMÖLLER J, BAUER S, ROOTS I. 1997. Cytochrome P4502D6 variants in a Caucasian population: allele frequencies and phenotypic consequences. *Am J Hum Genet* **60**, 284–95.

79 SHIMADA T, TSUMURA F, YAMAZAKI H, GUENGERICH FP, INOUE K. 2001. Characterization of (+/−)-bufuralol hydroxylation activities in liver microsomes of Japanese and Caucasian subjects genotyped for CYP2D6. *Pharmacogenetics* **11**, 143–56.

80 BROLY F, MAREZ D, LO GUIDICE JM, SABBAGH N, LEGRAND M, BOONE P, MEYER UA. 1995. A nonsense mutation in the cytochrome P450 CYP2D6 gene identified in a Caucasian with an enzyme deficiency. *Hum Genet* **96**, 601–3.

81 GAEDIGK A, BLUM M, GAEDIGK R, EICHELBAUM M, MEYER UA. 1991. Deletion of the entire cytochrome P450 CYP2D6 gene as a cause of impaired drug metabolism in poor metabolizers of the debrisoquine/sparteine polymorphism. *Am J Hum Genet* **48**, 943–50.

82 STEEN VM, MOLVEN A, AARSKOG NK, GULBRANDSEN AK. 1995. Homologous unequal cross-over involving a 2.8 kb direct repeat as a mechanism for the generation of allelic variants of human cytochrome P450 CYP2D6 gene. *Hum Mol Genet* **4**, 2251–7.

83 TYNDALE R, AOYAMA T, BROLY F, MATSUNAGA T, INABA T, KALOW W, GELBOIN HV, MEYER UA, GONZALEZ FJ. 1991. Identification of a new variant CYP2D6 allele lacking the codon encoding Lys-281: possible association with the poor metabolizer phenotype. *Pharmacogenetics* **1**, 26–32.

84 BROLY F, MEYER UA. 1993. Debrisoquine oxidation polymorphism: phenotypic consequences of a 3-base-pair deletion in exon 5 of the CYP2D6 gene. *Pharmacogenetics* **3**, 123–30.

85 JOHANSSON I, OSCARSON M, YUE QY, BERTILSSON L, SJOQVIST F, INGELMAN-SUNDBERG M. 1994. Genetic analysis of the Chinese cytochrome P4502D locus: characterization of variant CYP2D6 genes present in subjects with diminished capacity for debrisoquine hydroxylation. *Mol Pharmacol* **46**, 452–9.

86 MASIMIREMBWA C, PERSSON I, BERTILSSON L, HASLER J, INGELMAN-SUNDBERG M. 1996. A novel mutant variant of the CYP2D6 gene (CYP2D6*17) common in a black African population: association with diminished debrisoquine hydroxylase activity. *Br J Clin Pharmacol* **42**, 713–9.

87 OSCARSON M, HIDESTRAND M, JOHANSSON I, INGELMAN-SUNDBERG M. 1997. A combination of mutations in the CYP2D6*17 (CYP2D6Z) allele causes alterations in enzyme function. *Mol Pharmacol* **52**, 1034–40.

88 LEATHART JB, LONDON SJ, STEWARD A, ADAMS JD, IDLE JR, DALY AK. 1998. CYP2D6 phenotype–genotype relationships in African-Americans and Caucasians in Los Angeles. *Pharmacogenetics* **8**, 529–41.

89 RAIMUNDO S, FISCHER J, EICHELBAUM M, GRIESE EU, SCHWAB M, ZANGER

UM. 2000. Elucidation of the genetic basis of the common 'intermediate metabolizer' phenotype for drug oxidation by CYP2D6. *Pharmacogenetics* **10**, 577–81.

90 Dahl ML, Johansson I, Bertilsson L, Ingelman-Sundberg M, Sjoqvist F. 1995. Ultrarapid hydroxylation of debrisoquine in a Swedish population. Analysis of the molecular genetic basis. *J Pharmacol Exp Ther* **274**, 516–20.

91 Aklillu E, Persson I, Bertilsson L, Johansson I, Rodrigues F, Ingelman-Sundberg M. 1996. Frequent distribution of ultrarapid metabolizers of debrisoquine in an Ethiopian population carrying duplicated and multiduplicated functional CYP2D6 alleles. *J Pharmacol Exp Ther* **278**, 441–6.

92 Griese EU, Zanger UM, Brudermanns U, Gaedigk A, Mikus G, Morike K, Stuven T, Eichelbaum M. 1998. Assessment of the predictive power of genotypes for the *in-vivo* catalytic function of CYP2D6 in a German population. *Pharmacogenetics* **8**, 15–26.

93 Bradford LD. 2002. CYP2D6 allele frequency in European Caucasians, Asians, Africans and their descendants. *Pharmacogenomics* **3**, 229–43.

94 Gilham DE, Cairns W, Paine MJ, Modi S, Poulsom R, Roberts GC, Wolf CR. 1997. Metabolism of MPTP by cytochrome P4502D6 and the demonstration of 2D6 mRNA in human foetal and adult brain by *in situ* hybridization. *Xenobiotica* **27**, 111–25.

95 Jann MW, Cohen LJ. 2000. The influence of ethnicity and antidepressant pharmacogenetics in the treatment of depression. *Drug Metab Drug Interact* **16**, 39–67.

96 DeVane CL. 1994. Pharmacogenetics and drug metabolism of newer antidepressant agents. *J Clin Psychiatry* **55** (Suppl. 12), 38–45.

97 Daly AK, Stehen VM, Fairbrother KS, Idle JR. 1996. CYP2D6 multiallelism. *Methods Enzymol* **272**, 199–210.

98 Brøsen K, Gram LF. 1993. The pharmacogenetics of the selective serotonin reuptake inhibitors. *Clin Invest* **71**, 1002–9.

99 Bluhm RE, Wilkinson GR, Shelton R, Branch RA. 1993. Genetically determined drug-metabolizing activity and desipramine-associated cardiotoxicity: a case report. *Clin Pharmacol Ther* **53**, 89–95.

100 Lee JT, Kroemer HK, Silberstein DJ, Funck-Brentano C, Lineberry MD, Wood AJ, Roden DM, Woosley RL. 1990. The role of genetically determined polymorphic drug metabolism in the beta blockade produced by propafenone. *N Engl J Med* **322**, 1764–8.

101 Lennard MS, Tucker GT, Silas JH, Freestone S, Ramsay LE, Woods HF. 1983. Differential stereoselective metabolism of metoprolol in extensive and poor debrisoquin metabolizers. *Clin Pharmacol Ther* **34**, 732–7.

102 Zhou HH, Koshakji RP, Silberstein DJ, Wilkinson GR, Wood AJ. 1989. Altered sensitivity to and clearance of propranolol in men of Chinese descent as compared with American whites. *N Engl J Med* **320**, 565–70.

103 Zhou HH, Wood AJ. 1995. Stereoselective disposition of carvedilol is determined by CYP2D6. *Clin Pharmacol Ther* **57**, 518–24.

104 Swynghedauw B. 2001. Cardiovascular pharmacogenetics and pharmacogenomics. *J Clin Basic Cardiol* **4**, 205–10.

105 Poulsen L, Brøsen K, Arendt-Nielsen L, Gram K, Elbæk LF, Sindrup SH. 1996. Codeine and morphine in extensive and poor metabolizers of sparteine: pharmacokinetics, analgesic effect and side effects. *Eur J Clin Pharmacol* **51**, 289–95.

106 Caraco Y, Sheller J, Wood AI. 1996. Pharmacogenetic determination of the effects of codeine and prediction of drug interactions. *J Pharmacol Exp Ther* **278**, 1165–74.

107 Desmeules J, Gascon MP, Dayer P, Magistris M. 1991. Impact of environmental and genetic factors on co-

deine analgesia. *Eur J Clin Pharmacol* **41**, 23–6.

108 Sindrup SH, Brøsen K. 1996. The pharmacogenetics of codeine hypoalgesia. *Pharmacogenetics* **10**, 335–46.

109 Cascorbi I, Brockmoller J, Roots I. 1996. A C4887A polymorphism in exon 7 of human CYP1A1 population frequency, mutation linkages, and impact on lung cancer susceptibility. *Cancer Res* **56**, 4965–9.

110 Garte S. 1998. The role of ethnicity in cancer susceptibility gene polymorphisms: the example of CYP1A1. *Carcinogenesis* **19**, 1329–32.

111 Kawajiri K, Watanabe J, Hayashi S. 1996. Identification of allelic variants of the human CYP1A1 gene. *Methods Enzymol* **272**, 226–32.

112 MacLeod S, Sinha R, Kadlubar FF, Lang NP. 1997. Polymorphisms of CYP1A1 and GSTM1 influence the *in vivo* function of CYP1A2. *Mutat Res* **376**, 135–42.

113 Liang HC, Li H, McKinnon RA, Duffy JJ, Potter SS, Puga A, Nebert DW. 1996. CYP1A2$^{-/-}$ null mutant mice develop normally but show deficient drug metabolism. *Proc Natl Acad Sci USA* **93**, 1671–6.

114 Bailey LR, Roodi N, Dupont WD, Parl FF. 1998. Association of cytochrome P450 1B1 (CYP1B1) polymorphism with steroid receptor status in breast cancer. *Cancer Res* **58**, 5038–41.

115 London SJ, Idle JR, Daly AK, Coetzee GA. 1999. Genetic variation of CYP2A6, smoking, and risk of cancer. *Lancet* **353**, 898–9.

116 Pianezza ML, Sellers EM, Tyndale RF. 1998. Nicotine metabolism defect reduces smoking. *Nature* **393**, 750.

117 Code EL, Crespi CL, Penman BW, Gonzalez FJ, Chang TK, Waxman DJ. 1997. Human cytochrome P4502B6, interindividual hepatic expression, substrate specificity, and role in procarcinogen activation. *Drug Metab Dispos* **25**, 985–93.

118 Aithal GP, Day CP, Kesteven PI, Daly AK. 1999. Association of polymorphisms in the cytochrome P450 CYP2C9 with warfarin dose requirement and risk of bleeding complications. *Lancet* **353**, 717–9.

119 Furuya H, Fernandez-Salguero P, Gregory W, Taber H, Steward A, Gonzalez FJ, Idle JR 1995. Genetic polymorphism of CYP2C9 and its effect on warfarin maintenance dose requirement in patients undergoing anticoagulation therapy. *Pharmacogenetics* **5**, 389–92.

120 Kaminsky LS, Zhang Z. 1996. Human P450 metabolism of warfarin. *Pharmacol Ther* **73**, 67–74.

121 Rettie AE, Wienkers LC, Gonzales FJ, Trager WF, Korzekwa KR. 1994. Impaired *S*-warfarin metabolism catalyzed by R144C variant of CYP2C9. *Pharmacogenetics* **4**, 39–42.

122 Steward DJ, Haining RL, Henne KR, Davis G, Rushmore TH, Trager WF, Rettie AE. 1997. Genetic association between sensitivity to warfarin and expression of CYP2C9*3. *Pharmacogenetics* **7**, 361–7.

123 Taube J, Halsall D, Baglin T. 2000. Influence of cytochrome P-450 CYP2C9 polymorphisms on warfarin sensitivity and risk of oval anticoagulation in patients on long-term treatment. *Blood* **96**, 1816–9.

124 Loebstein R, Yonath H, Peleg D, Almog S, Rotenberg M, Lubetsky A, Roitelman J, Harats D, Halkin H, Ezra D. 2001. Interindividual variability in sensitivity to warfarin – nature or nurture? *Clin Pharmacol Ther* **70**, 159–64.

125 van der Weide J, Stejins LSW, van Weelden MJM, de Haan K. 2001. The effect of genetic polymorphism of cytochrome P450 CYP2C9 on phenytoin dose requirement. *Pharmacogenetics* **11**, 287–91.

126 Kidd RS, Straughn AB, Meyer MC, Blaisdell J, Goldstein JA, Dalton JT. 1999. Pharmacokinetics of chlorphenipramine, phenytoin, glipizide and nifedipine in an individual homozygous for the CYP2C9*3 allele. *Pharmacogenetics* **9**, 71–80.

127 Stubbins MJ, Harries LW, Smith G, Tarbit MH, Wolf RC. 1996. Genetic analysis of the human cytochrome

P450 CYP2C9 locus. *Pharmacogenetics* **6**, 429–39.

128 CRESPI CL, MILLER VP. 1997. The R144C change in the CYP2C9*2 allele alters interaction of the cytochrome P450 with NADPH:cytochrome 450 oxidoreductase. *Pharmacogenetics* **7**, 203–10.

129 MINERS JP, BIRKETT DJ. 1998. Cytochrome P450 CYP2C9; an enzyme of major importance in human drug metabolism. *Br J Clin Pharmacol* **45**, 525–38.

130 FURUTA T, OHASHI K, KOBAYASHI K, IIDA I, YOSHIDA H, SHIRAI N, TAKASHIMA M, KOSUGE K, HANAI H, CHIBA K, ISHIZAKI T, KANEKO E. 1999. Effects of clarythromycin on the metabolism of omeprazole in relation to CYP2C19 genotype status in humans. *Clin Pharmacol Ther* **66**, 265–74.

131 BALIAN JD, SUKHOVA N, HARRIS JW, HEWETT J, PICKLE L, GOLDSTEIN JA, WOOSLEY RL, FLOCKHART DA. 1995. The hydroxylation of omeprazole correlates with S-mephenytoin metabolism: a population study. *Clin Pharmacol Ther* **57**, 662–9.

132 TYBRING G, BÖTTIGER Y, WIDÉN J, BERTILSSON L. 1997. Enantioselective hydroxylation of omeprazole catalyzed by CYP2C19 in Swedish white subjects. *Clin Pharmacol Ther* **62**, 129–37.

133 FURUTA T, OHASHI K, KAMATA T, TAKASHIMA M, KOSUGE K, KAWASAKI T, HANAI H, KUBOTA T, ISHIZAKI T, KANEKO E. 1998. Effect of genetic differences in omeprazole metabolism on cure rates for *Helicobacter pylori* infection and peptic ulcer. *Ann Intern Med* **129**, 1027–30.

134 TANIGAWARA Y, AOYAMA N, KITA T, SHIRAKAWA K, KOMADA F, KASUGA M, OKUMURA K. 1999. CYP2C19 genotype-related efficacy of omeprazole for the treatment of infection caused by *Helicobacter pylori*. *Clin Pharmacol Ther* **66**, 528–34.

135 FERGUSON RJ, DE MORAIS SMF, BENHAMOU S, BOUCHARDY C, BLAISDELL J, IBEANU G, WILKINSON GR, SARICH TC, WRIGHT JM, DAYER P, GOLDSTEIN JA. 1998. A new genetic defect in human CYP2C19: mutation of the initiation codon is responsible for poor metabolism of S-mephenytoin. *J Pharmacol Exp Ther* **284**, 356–61.

136 MEYER UA. Drugs in special patient groups: clinical importance of genomics in drug effects. In: CARRUTHERS GS, HOFFMANN BB, MELMON KL, NIERENBERG DW, eds, *Clinical Pharmacology*, McGraw-Hill, New York, 2000, 1179–205.

137 GOLDSTEIN JA, FALETTO MB, ROMKES-SPARKS M, SULLIVAN T, KITAREEWAN S, RAUCY JL, LASKER JM, GHANAYEM BI. 1994. Evidence that CYP2C19 is the major (*S*) mephentoin 4′-hydroxylase in humans. *Biochemistry* **33**, 1743–52.

138 JOHANSSON I, LUNDQVIST E, BERTILSSON L, DAHL ML, SJOQVIST F, INGELMAN-SUNDBERG M. 1993. Inherited amplification of an active gene in cytochrome P450 CYP2D locus, as a cause of ultrarapid metabolism of debrisoquine. *Proc Natl Acad Sci USA* **90**, 11825–9.

139 BRØSEN K, GRAM LF. 1989. Clinical significance of the sparteine/debrisoquine oxidation polymorphism. *Eur J Pharmacol* **36**, 537–47.

140 CHOU WH, YAN FX, DELEON J, BARNHILL J, ROGERS T, CRONIN M, PHO M, XIAO V, RYDER TB, LIU WW, TEILING C, WEDLUND PJ. 2000. Extension of a pilot study: impact from the cytochrome P450 2D6 polymorphism on outcome and costs associated with severe mental illness. *J Clin Psychopharmacol* **20**, 246–51.

141 CHEN S, CHOU HW, BLOUIN RA, MAO Z, HUMPHRIES LL, MEEK QC, NEILL JR, MARTIN WL, HAYS LR, WEDLUND PJ. 1996. The cytochrome P450 2D6 (CYP2D6) enzyme polymorphism: screening costs and influence on clinical outcomes in psychiatry. *Clin Pharm Ther* **60**, 522–34.

142 KAPITANY T, MESZAROS K, LENZINGER E, SCHINDLER SD, BARNAS C, FUCHS K, SIEGHART W, ASCHAUER HN, KASPER S. 1998. Genetic polymorphisms for drug metabolism

(CYP2D6) and tardive dyskinesia in schizophrenia. *Schizophr Res* **32**, 101–6.

143 Arthur H, Dahl ML, Siwers B, Sjöqvist F. 1995. Polymorphic drug metabolism in schizophrenic patients with tardive dyskinesia. *J Clin Psychopharmacol* **15**, 211–6.

144 Arcavi L, Benovitz NL. 1993. Clinical significance of genetic influences on cardiovascular drug metabolism. *Cardiovasc Drugs Ther* **7**, 311–24.

145 Sallee FR, DeVane CL, Ferrell RE. 2000. Fluoxetine-related death in a child with cytochrome P4502D6 genetic deficiency. *Child Adolesc Psychopharmacol* **10**, 27–34.

146 Sindrup SH, Poulsen L, Brøsen K, Arendt-Nielsen L, Gram LF. 1993. Are poor metabolisers of sparteine/debrisoquine less pain tolerant than extensive metabolisers? *Pain* **53**, 335–9.

147 Tyndale RF, Droll KP, Sellers EM. 1997. Genetically deficient CYP2D6 metabolism provides protection against oral opiate dependence. *Pharmacogenetics* **7**, 375–9.

148 Kroemer HK, Eichelbaum M. 1995. Molecular basis and clinical consequences of genetic cytochrome P450 2D6 polymorphism. *Life Sci* **56**, 2285–98.

149 Hersberger M, Marti-Jaun J, Rentsch K, Hanseler E. 2000. Rapid detection of the CYP2D6*3, CYP2D6*4 and CYP2D6*6 alleles by tetra-primer PCR and of the CYP2D6*5 allele by multiplex long PCR. *Clin Chem* **46**, 1072–7.

150 Roberts R, Sullivan P, Joyce P, Kennedy MA. 2000. Rapid and comprehensive determination of cytochrome P450CYP2D6 poor metabolizer genotypes by multiplex polymerase chain reaction. *Hum Mutat* **16**, 77–85.

151 West WL, Knight EM, Pradhan S, Hinds TS. 1997. Interpatient variability: genetic predisposition and other genetic factors. *J Clin Pharmacol* **37**, 635–48.

152 Carriere V, Berthou F, Baird S, Belloc C, Beaune P, de Waziers I. 1996. Human cytochrome P450 2E1 (CYP2E1): from genotype to phenotype. *Pharmacogenetics* **6**, 203–11.

153 Fairbrother KS, Grove J, de Waziers I, Steimel DT, Day CP, Crespi CL, Daly AK. 1998. Detection and characterization of novel polymorphisms in the CYP2E1 gene. *Pharmacogenetics* **8**, 543–52.

154 Grove J, Brown AS, Daly AK, Bassendine MF, James OF, Day CP. 1998. The *Rsa*I polymorphism of CYP2E1 and susceptibility to alcoholic liver disease in Caucasians: effect on age of presentation and dependence on alcohol dehydrogenase genotype. *Pharmacogenetics* **8**, 335–42.

155 Lieber CS. 1997. Cytochrome P-4502E1, its physiological and pathological role. *Physiol Rev* **77**, 517–44.

156 Tanaka F, Shiratori Y, Yokosuka O, Imazeki F, Tsukada Y, Omata M. 1997. Polymorphism of alcohol-metabolizing genes affects drinking behavior and alcoholic liver disease in Japanese men. *Alcohol Clin Exp Res* **21**, 596–601.

157 Felix CA, Walker AH, Lange BJ, Williams TM, Winick NJ, Cheung NK, Lovett BD, Nowell PC, Blair IA, Rebbeck TR. 1998. Association of CYP3A4 genotype with treatment-related leukemia. *Proc Natl Acad Sci USA* **95**, 13176–81.

158 Hamzeiy H, Vahdati-Mashadian N, Edwards HJ, Goldfarb PS. 2002. Mutation analysis of the human CYP3A4 gene 5′ regulatory region: population screening using non-radioactive SSCP. *Mutat Res* **500**, 103–10.

159 Kuehl P, Zhang J, Lin Y, Lamba J, Assem M, Schuetz J, Watkins PB, Daly A, Wrighton SA, Hall SD, Maurel P, Relling M, Brimer C, Yasuda K, Venkataramanan R, Strom S, Thummel K, Boguski MS, Schuetz E. 2001. Sequence diversity in CYP3A promoters and characterization of the genetic basis of polymorphic CYP3A5 expression. *Nat Genet* **27**, 383–91.

160 Rebbeck TR, Jaffe JM, Walker AH, Wein AJ, Malkowicz SB. 1998.

Modification of clinical presentation of prostate tumors by a novel genetic variant in CYP3A4. *J Natl Cancer Inst* **90**, 1225–9.

161 Sata F, Sapone A, Elizondo G, Stocker P, Miller VP, Zheng W, Raunio H, Crespi CL, Gonzalez FJ. 2000. CYP3A4 allelic variants with amino acid substitutions in exon 7 and 12: evidence for an allelic variant with altered catalytic activity. *Clin Pharmacol Ther* **67**, 48–56.

162 Hsieh KP, Lin YY, Cheng CL, Lai ML, Lin MS, Siest JP, Huang JD. 2000. Novel mutations of CYP3A4 in Chinese. *Drug Metab Dispos* **29**, 268–73.

163 Wong RH, Wang JD, Hsieh LL, Du CL, Cheng TJ. 1998. Effects on sister chromatid exchange frequency of aldehyde dehydrogenase 2 genotype and smoking in vinyl chloride workers. *Mutat Res* **420**, 99–107.

164 Whitfield JB, Nightingale BN, Bucholz KK, Madden PA, Heath AC, Martin NG. 1998. ADH genotypes and alcohol use and dependence in Europeans. *Alcohol Clin Exp Res* **22**, 1463–9.

165 Tuchman M, Stoeckeler JS, Kiang DT, O'Dea RF, Ramnaraine ML, Mirkin BL. 1985. Familial pyrimidinemia and pyrimidinuria associated with severe fluorouracil toxicity. *N Engl J Med* **313**, 245–9.

166 Diasio RB, Beavers TL, Carpenter JT. 1988. Familial deficiency of dihydropyrimidine dehydrogenase. Biochemical basis for familial pyrimidinemia and severe 5-fluorouracil-induced toxicity. *J Clin Invest* **81**, 47–51.

167 Gonzalez FJ, Fernandez-Salguero P. 1995. Diagnostic analysis, clinical importance and molecular basis of dihydropyrimidine dehydrogenase deficiency. *Trends Pharmacol Sci* **16**, 325–7.

168 Ross D. 1996. Metabolic basis of benzene toxicity. *Eur J Haematol* **60**, 111–8.

169 Rothman N, Smith MT, Hayes RB, Traver RD, Hoener B, Campleman S, Li GL, Dosemeci M, Linet M, Zhang L, Xi L, Wacholder S, Lu W, Meyer KB, Titenko-Holland N, Stewart JT, Yin S, Ross D. 1997. Benzene poisoning, a risk factor for hematological malignancy, is associated with the NQO1 609C → T mutation and rapid fractional excretion of chlorzoxazone. *Cancer Res* **57**, 2839–42.

170 Schulz WA, Krummeck A, Rosinger I, Schmitz-Drager BJ, Sies H. 1998. Predisposition towards urolithiasis associated with the NQO1 null-allele. *Pharmacogenetics* **8**, 453–4.

171 Siegel D, McGuinness SM, Winski SL, Ross D. 1999. Genotype–phenotype relationships in studies of a polymorphism in NAD(P)H:quinone oxidoreductase 1. *Pharmacogenetics* **9**, 113–21.

172 Kalow W, Gunn DR. 1957. The relation between dose of succinylcholine and duration of apnea in man. *J Pharmacol Exp Ther* **120**, 203–14.

173 Reilly DK, Rivera-Calimlim L, VanDyke D. 1980. Catechol-*O*-methyltransferase activity: a determinant of levodopa response. *Clin Pharmacol Ther* **28**, 278–86.

174 Weinshilboum RM, Otterness DM, Szumlanski CL. 1999. Methylation pharmacogenetics: catechol-*O*-methyltransferase, thiopurine methyltransferase and histamine-*N*-methyltransferase. *Annu Rev Pharmacol Toxicol* **39**, 19–52.

175 Vandenbergh DJ, Rodriguez LA, Miller IT, Uhl GR, Lachmnn HM. 1997. High activity catechol-*O*-methyltransferase allele is more prevalent in polysubstance abusers. *Am J Med Genet* **74**, 439–42.

176 Baez S, Segara-Aguilar J, Widersten M, Johansson AS, Mannervik B. 1997. Glutathione transferases catalyse the detoxication of oxidized metabolites (o-quinones) of catecholamines and may serve as an antioxidant system preventing degenerative cellular processes. *Biochem J* **324**, 25–8.

177 Seidegard J, Ekstrom G. 1997. The role of human glutathione trans-

ferases and epoxide hydrolases in the metabolism of xenobiotics. *Environ Health Perspect* **105** (Suppl. 4), 791–9.

178 Chen H, Juchau MR. 1998. Recombinant human glutathione-S-transferases catalyse enzymic isomerization of 13-*cis*-retinoic acid to all *trans*-retinoic acid *in vitro. Biochem J* **336**, 223–6.

179 Kinnula K, Linnainmaa K, Raivio KO, Kinnula VL. 1998. Endogenous antioxidant enzymes and glutathione S-transferase in protection of mesothelioma cells against hydrogen peroxide and epirubicin toxicity. *Br J Cancer* **77**, 1097–102.

180 Satoh K, Sato R, Takahata T, Suzuki S, Hayakari M, Tsuchida S, Hatayama I. 1999. Quantitative differences in the active-site hydrophobicity of five human glutathione S-transferase isoenzymes: water-soluble carcinogen-selective properties of the neoplastic GSTP1-1 species. *Arch Biochem Biophys* **361**, 271–6.

181 Aksoy S, Raftogianis R, Weinshilboum R. 1996. Human histamine *N*-methyltransferase gene: structural characterization and chromosomal location. *Biochem Biophys Res Commun* **219**, 548–54.

182 Wang L, Thomae B, Eckloff B, Wieben E, Weinshilboum RM. 2002. Human histamine *N*-methyltransferase pharmacogenetics: gene resequencing, promoter characterization, and functional studies of a common 5′ flanking region single nucleotide polymorphism (SNP). *Biochem Pharmacol* **64**, 699–710.

183 Hirata N, Takeuchi K, Ukai K, Sakakura Y. 1999. Expression of histidine decarboxylase messenger RNA and histamine *N*-methyltransferase messenger RNA in nasal allergy. *Clin Exp Allergy* **29**, 76–83.

184 Preuss CV, Wood TC, Szumlanski CL, Raftogianis RB, Otterness DM, Girard B, Scott MC, Weinshilboum R. 1998. Human histamine *N*-methyltransferase pharmacogenetics: common genetic polymorphisms that alter activity. *Mol Pharmacol* **53**, 708–17.

185 Grant DM, Hughes NC, Janezic SA, Goodfellow GH, Chen HJ, Gaedigk A, Yu VL, Grewal R. 1997. Human acetyltransferase polymorphisms. *Mutat Res* **376**, 61–70.

186 Spielberg SP. 1996. *N*-acetyltransferases: pharmacogenetics and clinical consequences of polymorphic drug metabolism. *J Pharmacokinet Biopharm* **24**, 509–19.

187 Blum M, Demierre A, Grant DM, Heim M, Meyer UA. 1991. Molecular mechanism of slow acetylation of drugs and carcinogens in humans. *Proc Natl Acad Sci USA* **88**, 5237–41.

188 Nakamura H, Uetrecht J, Cribb AE, Miller MA, Zahid N, Hill J, Josephy PD, Grant DM, Spielberg SP. 1995. *In vitro* formation, disposition and toxicity of *N*-acetoxy-sulfamethoxazole, a potential mediator of sulfamethoxazole toxicity. *J Pharmacol Exp Ther* **274**, 1099–104.

189 Iyer L, Ratain MJ. 1998. Pharmacogenetics and cancer chemotherapy. *Eur J Cancer* **34**, 442–6.

190 Raftogianis RB, Wood TC, Otterness DM, Van Loon JA, Weinshilboum RM. 1997. Phenol sulfotransferase pharmacogenetics in humans: association of common SULT1A1 alleles with TS PST phenotype. *Biochem Biophys Res Commun* **239**, 298–304.

191 Freimuth RR, Eckloff B, Wieben ED, Weinshilboum RM. 2001. Human sulfotransferase SULT1C1 pharmacogenetics: gene resequencing and functional genomic studies. *Pharmacogenetics* **11**, 747–56.

192 Thomae BA, Eckloff BW, Freimuth RR, Wieben ED, Weinshilboum RM. 2002. Human sulfotransferase SULT2A1 pharmacogenetics: genotype-to-phenotype studies. *Pharmacogenomics* **2**, 48–56.

193 Adjei AA, Weinshilboum RM. 2002. Catecholestrogen sulfation: possible role in carcinogenesis. *Biochem Biophys Res Commun* **292**, 402–8.

194 Weinshilboum RM, Otterness DM, Aksoy IA, Wood TC, Her C, Raftogianis RB. 1997. Sulfation and sulfotransferases l: sulfotransferase

molecular biology: cDNAs and genes. *FASEB J* **11**, 3–14.

195 AARBAKKE J, JANKA-SCHAUB G, ELION GB. 1997. Thiopurine biology and pharmacology. *Trends Pharmacol Sci* **18**, 3–7.

196 SZUMLANSKI C, OTTERNESS D, HER C, LEE D, BRANDRIFF B, KELSELL D, SPURR N, LENNARD L, WIEBEN E, WEINSHILBOUM R. 1996. Thiopurine methyltransferase pharmacogenetics: human gene cloning and characterization of a common polymorphism. *DNA Cell Biol* **15**, 17–30.

197 TAI HL, KRYNETSKI EY, YATES CR, LOENNECHEN T, FESSING MY, KRYNETSKAIA NF, EVANS WE. 1996. Thiopurine S-methyltransferase deficiency: two nucleotide transitions define the most prevalent mutant allele associated with loss of catalytic activity in Caucasians. *Am J Hum Genet* **58**, 694–702.

198 BLACK AJ, MCLEOD HL, CAPELL HA, POWRIE RH, MATOWE LK, PRITCHARD SC, COLLIE-DUGUID ES, REID DM. 1998. Thiopurine methyltransferase genotype predicts therapy-limiting severe toxicity from azathioprine. *Ann Intern Med* **129**, 716–8.

199 MCLEOD HL, KRYNETSKI EY, RELLING MV, EVANS WE. 2000. Genetic polymorphism of thiopurine methyltransferase and its clinical relevance for childhood acute lymphoblastic leukemia. *Leukemia* **14**, 567–72.

200 COULTHARD SA, HALL AG. 2002. Recent advances in the pharmacogenomics of thiopurine methyltransferase. *Pharmacogenomics J* **1**, 254–61.

201 EVANS WE, HORNER M, CHU YQ, KALWINSKY D, ROBERTS WM. 1991. Altered mercaptopurine metabolism, toxic effects, and dosage requirement in a thiopurine methyltransferase-deficient child with acute lymphocytic leukemia. *J Pediatr* **119**, 985–9.

202 KRYNETSKI EY, EVANS WE. 1998. Pharmacogenetics of cancer therapy: getting personal. *Am J Hum Genet* **63**, 11–6.

203 KRYNETSKI EY, SCHUETZ JD, GALPIN AJ, PUI CH, RELLING MV, EVANS WE. 1995. A single point mutation leading to loss of catalytic activity in human thiopurine S-methyltransferase. *Proc Natl Acad Sci USA* **92**, 949–53.

204 LENNARD L. 1998. Clinical implications of thiopurine methyltransferase-optimization of drug dosage and potential drug interactions. *Ther Drug Monit* **20**, 527–31.

205 LENNARD L, VAN LOON IA, LILLEYMAN JS, WEINSHILBOUM RM. 1987. Thiopurine pharmacogenetics in leukemia: correlation of erythrocyte thiopurine methyltransferase activity and 6-thioguanine nucleotide concentrations. *Clin Pharmacol Ther* **41**, 18–25.

206 RELLING MV, HANCOCK ML, BOYETT JM, PUI CH, EVANS WE. 1999. Prognostic importance of 6-mercaptopurine dose intensity in acute lymphoblastic leukemia. *Blood* **93**, 2817–23.

207 EVANS WE, HON YY, BOMGAARS L, COUTRE S, HOLDSWORTH M, JANCO R, KALWINSKY D, KELLER F, KHATIB Z, MARGOLIN J, MURRAY J, QUINN J, RAVINDRANATH Y, RITCHEY K, ROBERTS W, ROGERS ZR, SCHIFF D, STEUBER C, TUCCI F, KORNEGAY N, KRYNETSKI EY, RELLING MV. 2001. Preponderance of Thiopurine S-methyltransferase deficiency and heterozygosity among patients intolerant to mercaptopurine or azathioprine. *J Clin Oncol* **19**, 2293–301.

208 RELLING MV, HANCOCK ML, RIVERA GK, SANDLUND JT, RIBEIRO RC, KRYNETSKI EY, PUI CH, EVANS WE. 1999. Mercaptopurine therapy intolerance and heterozygosity at the thiopurine S-methyltransferase gene locus. *J Natl Cancer Inst* **91**, 2001–8.

209 RELLING MV, RUBNITZ JE, RIVERA GK, BOYETT JM, HANCOCK ML, FELIX CA, KUN LE, WALTER AW, EVANS WE, PUI CH. 1999. High incidence of secondary brain tumors related to irradiation and antimetabolite therapy. *Lancet* **354**, 34–9.

210 SCHUTZ E, GUMMERT J, MOHR F, OELLERICH M. 1993. Azathioprine-induced myelosuppression in

thiopurine methyltransferase deficient heart transplant recipient. *Lancet* **341**, 436.

211 Yates CR, Krynetski EY, Loennechen T, Fessing MY, Tai HL, Pui CH, Relling MV, Evans WE. 1997. Molecular diagnosis of thiopurine S-methyltransferase deficiency: genetic basis for azathioprine and mercaptopurine intolerance. *Ann Intern Med* **126**, 608–14.

212 Ando Y, Saka H, Asai G, Sugiura S, Shimokata K, Kamataki T. 1998. UGT1A1 genotypes and glucuronidation of SN-38, the active metabolite of irinotecan. *Ann Oncol* **9**, 845–17.

213 Iyer L, Hall D, Das S, Mortell MA, Ramirez J, Kim S, Di Rienzo A, Ratain MJ. 1999. Phenotype–genotype correlation of *in vitro* SN-38 (active metabolite of irinotecan) and bilirubin glucuronidation in human liver tissue with UGT1A1 promoter polymorphism. *Clin Pharmacol Ther* **65**, 576–82.

214 Iyer L, Das S, Janisch L, Wen M, Ramirez J, Karrison T, Fleming GF, Vokes EE, Schilsky RL, Ratain MJ. 2002. UGT1A1*28 polymorphism as a determinant of irinotecan disposition and toxicity. *Pharmacogenomics J* **2**, 43–7.

215 Beutler E, Gelbart T, Demina A. 1999. Racial variability in the UDP-glucuronosyltransferase 7 (UGT1A1) promoter: a balanced polymorphism for regulation of bilirubin metabolism? *Proc Natl Acad Sci USA* **95**, 8170–4.

216 Iyer L, King CD, Whitington PF, Green MD, Roy SK, Tephly TR, Coffman BL, Ratain MJ. 1998. Genetic predisposition to the metabolism of irinotecan (CPT-11). Role of uridine diphosphate glucuronosyltransferase isoform lAl in the glucuronidation of its active metabolite (SN-38) in human liver microsomes. *J Clin Invest* **101**, 847–54.

217 Ramirez J, Iyer L, Journault K, Belanger P, Innocenti F, Ratain MJ, Guillemette C. 2002. *In vitro* characterization of hepatic flavopiridol metabolism using human liver microsomes and recombinant UGT enzyme. *Pharm Res* **19**, 588–94.

218 Mackenzie PI, Owens IS, Burchell B, Bock KW, Bairoch A, Bélanger A, Fournel-Gigleux S, Green M, Hum DW, Iyanagi T, Lancet D, Louisot P, Magdalou J, Chowdhury JR, Ritter JK, Schachter H, Tephly TR, Tipton KF, Nebert DW. 1997. The UDP glycosyltransferase gene superfamily: recommended nomenclature update based on evolutionary divergence. *Pharmacogenetics* **7**, 255–69.

219 Innocentil F, Iyer L, Ramirez J, Green MD, Ratain MJ. 2001. Epirubicin Glucuronidation is catalyzed by human UDP glucuronosyl transferase 2B7. *Drug Met Dispos* **29**, 686–92.

220 Choi KH, Chen CJ, Kriegler M, Roninson IB. 1988. An altered pattern of cross-resistance in multidrug-resistant human cells results from spontaneous mutations in the *mdr1* (P-glycoprotein) gene. *Cell* **53**, 519–29.

221 Cascorbi I, Gerloff T, Johne A, Meisel C, Hoffmeyer S, Schwab M, Schaeffeler E, Eichelbaum M, Brinkmann U, Roots I. 2001. Frequency of single nucleotide polymorphisms in the P-glycoprotein drug transporter MDR1 gene in white subjects. *Clin Pharmacol Ther* **69**, 169–74.

222 Hoffmeyer S, Burk O, von Richter O, Arnold HP, Brockmöller J, Johne A, Cascorbi I, Gerloff T, Roots I, Eichelbaum M, Brinkmann U. 2000. Functional polymorphisms of the human multidrug resistance gene: multiple sequence variations and correlation of one allele with P-glycoprotein expression and activity *in vivo*. *Proc Natl Acad Sci USA* **97**, 3473–8.

223 Ameyaw MM, Regateiro F, Li T, Liu X, Tariq M, Mobarek A, Thornton N, Folayan GO, Githanga J, Indalo A, Ofori-Adjei D, Price-Evans DA, McLeod HL. 2001. MDR1 pharmacogenetics: frequency of the C3435T mutation in exon 26 is significantly

influenced by ethnicity. *Pharmacogenetics* **11**, 217–21.

224 McLeod H. 2002. Pharmacokinetic differences between ethnic groups. *Lancet* **359**, 78.

225 Schaeffeler E, Eichelbaum M, Brinkmann U, Penger A, Asante-Poku S, Zanger UM, Schwab M. 2001. Frequency of C3435T polymorphism of MDR1 gene in African people. *Lancet* **358**, 383–4.

226 Kim RB, Leake BF, Choo EF, Dresser GK, Kubba SV, Schwarz UI, Taylor A, Xie HG, McKinsey J, Zhou S, Lan LB, Schuetz JD, Schuetz EG, Wilkinson GR. 2001. Identification of functionally variant MDR1 alleles among European Americans and African Americans. *Clin Pharmacol Ther* **70**, 189–99.

227 Kioka N, Tsubota J, Kakehi Y, Komano T, Gottesman MM, Pastan I, Ueda K. 1989. P-glycoprotein gene (MDR1) cDNA from human adrenal: normal P-glycoprotein carries Gly185 with an altered pattern of multidrug resistance. *Biochem Biophys Res Commun* **162**, 224–31.

228 Mickley LA, Lee JS, Weng Z, Zhan Z, Alvarez M, Wilson W, Bates SE, Fojo T. 1998. Genetic polymorphism in MDR-1, a tool for examining allelic expression in normal cells, unselected and drug-selected cell lines, and human tumors. *Blood* **91**, 1829–56.

229 Schinkel AH. 1997. The physiological function of drug-transporting P-glycoproteins. *Semin Cancer Biol* **8**, 161–70.

230 Strautnieks SS, Bull LN, Knisely AS, Kocoshis SA, Dahl N, Arnell H, Sokal E, Dahan K, Childs S, Ling V, Tanner MS, Kagalwalla AF, Németh A, Pawlowska J, Baker A, Mieli-Vergani G, Freimer NB, Gardiner RM, Thompson RJ. 1998. A gene encoding a liver-specific ABC transporter is mutated in progressive familial intrahepatic cholestasis. *Nat Genet* **20**, 233–8.

231 Borst P, Evers R, Kool M, Wijnholds J. 2000. A family of drug transporters: the multidrug resistance-associated proteins. *J Natl Cancer Inst* **92**, 1295–302.

232 Cannella G, Paoletti E, Barocci S, Massarino F, Delfino R, Ravera G, Di Maio G, Nocera A, Patrone P, Rolla D. 1998. Angiotensin-converting enzyme gene polymorphism and reversibility of uremic left ventricular hypertrophy following long-term hypotensive therapy. *Kidney Int* **54**, 618–26.

233 Keppler D, Leier I, Jedlitschky G. 1997. Transport of glutathione conjugates and glucuronides by the multidrug resistance proteins MRP1 and MRP2. *Biol Chem* **378**, 787–91.

234 Schuetz JD, Connelly MC, Sun D, Paibir SG, Flynn PM, Srinivas RV, Kumar A, Fridland A. 1999. MRP4, a previously unidentified factor in resistance to nucleoside-based antiviral drugs. *Nat Med* **5**, 1048–51.

235 Kuivenhoven JA, Jukema JW, Zwinderman AH, de Knijff P, McPherson R, Brushcke AV, Lie KI, Kastelein JJ. 1998. The role of a common variant of the cholesteryl ester transfer protein gene in the progression of coronary atherosclerosis. The Regression Growth Evaluation Statin Study Group. *N Engl J Med* **338**, 86–93.

236 Maitland-van der Zee A, Klungel OH, Stricker BHC, Verschuren WMM, Kastelein JJP, Leufkens HGM, de Boer A. 2002. Genetic polymorphisms: importance for response to HMG-CoA reductase inhibitors. *Atherosclerosis* **163**, 213–22.

237 Dachet C, Poirier O, Cambien F, Chapman J, Rouis M. 2000. New functional promoter polymorphism, CETP/−629, in cholesteryl ester transfer protein (CETP) gene related to CETP mass and high density lipoprotein cholesterol levels: role of Sp1/Sp3 in transcriptional regulation. *Arterioscler Thromb Vasc Biol* **20**, 507–15.

238 Drysdale CM, McGraw DW, Stack CB, Stephens JC, Judson RS, Nandabalan K, Arnold K, Ruano G, Liggett SB. 2000. Complex promoter and coding region $beta_2$-adrenergic

receptor haplotypes alter receptor expression and predict *in vivo* responsiveness. *Proc Natl Acad Sci USA* **97**, 10483–8.

239 Israel E, Drazen JM, Liggett SB, Boushey HA, Cherniack RM, Chinchilli VM, Cooper DM, Fahy JV, Fish JE, Ford JG, Kraft M, Kunselman S, Lazarus SC, Lemanske RF, Martin RJ, McLean DE, Peters SP, Silverman EK, Sorkness CA, Szefler SJ, Weiss ST, Yandava CN. 2000. The effect of polymoryhisms of the beta$_2$-adrenergic receptor on the response to regular use of albuterol in asthma. *Am J Respir Crit Care Med* **162**, 75–80.

240 Lima JJ, Thomason DB, Mohamed MHN, Eberle LV, Self TH, Johnson JA. 1999. Impact of genetic polymorphisms of the beta$_2$-adrenergic receptor on albuterol bronchodilator pharmacodynamics. *Clin Pharmacol Ther* **65**, 519–25.

241 Martinez FD, Graves PE, Baldini M, Solomon S, Erickson P. 1997. Association between genetic polymorphisms of the beta$_2$-adrenoceptor and response to albuterol in children with and without a history of wheezing. *J Clin Invest* **100**, 3184–8.

242 Tan S, Hall IP, Dewar J, Dow E, Lipworth B. 1997. Association between beta 2 adrenoceptor polymorphism and susceptibility to bronchodilator desensitisation in moderately severe stable asthmatics. *Lancet* **350**, 995–9.

243 Arranz MJ, Munro J, Sham P, Kirov G, Murray RM, Collier DA, Kerwin RW. 1998. Meta-analysis of studies on genetic variation in 5HT-2A receptors and clozapine response. *Schiz Res* **32**, 93–9.

244 Arranz MJ, Munro J, Birkett J, Bolonna A, Mancama D, Sodhi M, Lesch KP, Meyer JF, Sham P, Collier DA, Murray RM, Kerwin RW. 2000. Pharmacogenetic prediction of clozapin response. *Lancet* **355**, 1615–6.

245 Cockcroft JR, Gazis AG, Cross DJ, Wheatley A, Dewar J, Hall IP, Noon JP. 2000. Beta$_2$-adrenoceptor polymorphism determines vascular activity in humans. *Hypertension* **36**, 371–5.

246 Gratze G, Fortin J, Labugger R, Binder A, Kotanko P, Timmermann B, Luft FC, Hoehe MR, Skrabal F. 1999. Beta$_2$-adrenergic receptor variants affect resting blood pressure and agonist-induced vasodilation in young adult Caucasian. *Hypertension* **33**, 1425–30.

247 Hoit BD, Suresh DP, Craft L, Walsh RA, Liggett SB. 2000. Beta$_2$-adrenergic receptor polymorphisms at amino acid 16 differentially influence agonist-stimulated blood pressure and peripheral blood flow in normal individuals. *Am Heart J* **139**, 537–42.

248 Liggett SB. 2000. Beta$_2$-adrenergic receptor pharmacogenetics. *Am J Resp* **161**, 197–201.

249 Hancox RJ, Aldridge RE, Cowan JO, Flannery EM, Herbison GP, McLachlan CR, Town GI, Taylor DR. 1999. Tolerance to beta-agonists during acute bronchoconstriction. *Eur Resp J* **14**, 283–7.

250 Lipworth BJ, Hall IP, Tan S, Aziz I, Coutie W. 1999. Effects of genetic polymorphism on *ex vivo* and *in vivo* function of beta$_2$-adraenoceptors in asthmatic patients. *Chest* **115**, 324–8.

251 Palmer LJ, Silverman ES, Weiss ST, Drazen JM. 2002. Pharmacogenetics of asthma. *Am J Respir Crit Care Med* **65**, 861–6.

252 Sodhi MS, Arranz MJ, Curtis D, Ball DM, Sham P, Roberts GW, Price J, Collier DA, Kerwin RW. 1995. Association between clozapine response and allelic variation in the 5-HT_{2C} receptor gene. *NeuroReport* **7**, 169–72.

253 Joober R, Benkelfat C, Brisebois K, Toulouse A, Turecki G, Lal S, Bloom D, Labelle A, Lalonde P, Fortin D, Alda M, Palmour R, Rouleau GA. 1999. T102C polymorphism in the $5HT_{2A}$ gene and schizophrenia relation to phenotype and drug response variability. *J Psychiat Neurosci* **24**, 141–6.

254 Masellis M, Basile V, Meltzer HY, Lieberman JA, Sevy S, Macciardi

FM, Cola P, Howard A, Badri F, Nöthen MM, Kalow W, Kennedy JL. 1998. Serotonin subtype 2 receptor genes and clinical response to clozapine in schizophrenia patients. *Neuropsychopharmacology* **19**, 123–32.

255 Yu YW, Tsai SJ, Lin CH, Hsu CP, Yang KH, Hong CJ. 1999. Serotonin-6 receptor variant (C267T) and clinical response to clozapine. *NeuroReport* **10**, 1231–3.

256 Kim DK, Lim SW, Lee S, Sohn SE, Kim S, Hahn CG, Carroll BJ. 2000. Seronin transporter gene polymorphism and antidepressant response. *NeuroReport* **11**, 215–9.

257 Smeraldi E, Zanardi R, Benedetti F, DiBella D, Perez J, Catalano M. 1998. Polymorphism within the promoter of the serotonin transponer gene and antidepressant efficacy of fluvoxamine. *Mol Psychiatry* **3**, 508–11.

258 Whale R, Quested DJ, Laver D, Harrison PJ, Cowen PJ. 2000. Serotonin transporter (5-HTT) promoter genotype may influence the prolactin response to clomipramine. *Psychopharmacology* **150**, 120–2.

259 Basile VS, Masellis M, Badri F, Paterson AD, Meltzer HY, Lieberman JA, Potkin SG, Macciardi F, Kennedy JL. 1999. Association of the *MscI* polymorphism of the dopamine D3 receptor gene, with tardive dyskinesia in schizophrenia. *Neuropsychopharmacology* **27**, 17–27.

260 Cohen BM, Ennulat DJ, Centorrino F, Matthysse S, Konieczna H, Chu HM, Cherkerzian S. 1999. Polymorphisms of the dopamine D4 receptor and response to antipsychotic drugs. *Psychopharmacology* **141**, 6–10.

261 Eichhammer P, Albus M, Borrmann-Hassenbach M, Schoeler A, Putzhammer A, Frick U, Klein HE, Rohrmeier T. 2000. Association of dopamine D3-receptor gene variants with neuroleptic induced akathiasia in schizophrenic patients: a generalization of Steen's study on DRD3 and tardive dyskinesia. *Am J Med Genet* **96**, 187–91.

262 Hwu HG, Hong CJ, Lee YL, Lee PC, Lee SF. 1998. Dopamine D4 receptor gene polymorphisms and neuroleptic response in schizophrenia. *Biol Psychiatry* **44**, 483–7.

263 Kaiser R, Könneker M, Henneken M, Dettling M, Müller-Oerlinghausen B, Roots I, Brockmöller J. 2000. Dopamine D4 receptor 48bp repeat polymorphism: no association with response to antipsychotic treatment, but association with catatonic schizophrenia. *Mol Psychiatry* **5**, 418–24.

264 Scharfetter J, Chaudhry HR, Hornik K, Fuchs K, Sieghart W, Kasper S, Aschauer HN. 1999. Dopamine D3 receptor gene polymorphism and response to clozapine in schizophrenic Pakistani patients. *Eur Neuropsychopharmacol* **10**, 17–20.

265 Suzuki A, Mihara K, Kondo T, Tanaka O, Nagashima U, Otani K, Kaneko S. 2000. The relationship between dopamine D2-receptor polymorphism at the TaqlA locus and therapeutic response to nemonapride, a selective dopamine antagonist, in schizophrenic patients. *Pharmacogenetics* **10**, 335–41.

266 Drazen JM, Yandava CN, Dubé L, Szczerback N, Hippensteel R, Pillari A, Israel E, Schork N, Silverman ES, Katz DA, Drajesk J. 1999. Pharmacogenetic association between ALOX5 promoter genotype and the response to anti-asthma treatment. *Nat Genet* **22**, 168–70.

267 Jacobsen P, Rossing K, Rossing P, Tarnow L, Mallet C, Poirier O, Cambien F, Parving HH. 1998. Angiotensin converting enzyme gene polymorphism and ACE inhibition in diabetic nephropathy. *Kidney Int* **53**, 1002–6.

268 Kohno M, Yokokawa K, Minami M, Kano H, Yasunari K, Hanehira T, Yoshikawa J. 1999. Association between angiotensin-converting enzyme gene polymorphisms and regression of left ventricular hypertrophy in patients treated with angiotensin-converting enzyme inhibitors. *Am J Med* **106**, 544–9.

269 Ohmichi N, Iwai N, Uchida Y, Shichiri G, Nakamura Y, Kinoshita M. 1997. Relationship between the response to the angiotensin converting enzyme inhibitor imidapril and the angiotensin converting enzyme genotype. *Am J Hypertens* **10**, 951–5.

270 Okamura A, Ohishi M, Rakugi H, Katsuya T, Yanagitani Y, Takiuchi S, Taniyama Y, Moriguchi K, Ito H, Higashino Y, Fujii K, Higaki J, Ogihara T. 1999. Pharmacogenetic analysis of the effect of angiotensin-converting enzyme inhibitor on restenosis after percutaneous transluminal coronary angioplasty. *Angiology* **50**, 811–22.

271 Penno G. 1998. Effect of angiotensin converting enzyme (ACE) gene polymorphism on progression of renal disease and the influence of ACE inhibition in IDDM patients: findings from the EUCLID Randomized Controlled Trial. EURODIAB Controlled Trial of Lisinopril in IDDM1. *Diabetes* **47**, 1507–11.

272 Perna A, Ruggenenti P, Testa A, Spoto B, Benini R, Misefari V, Remuzzi G, Zoccali C. 2000. ACE genotype and ACE inhibitors induced renoprotection in chronic proteinuric nephropathies l. *Kidney Int* **57**, 274–81.

273 Prasad A, Narayanan S, Husain S, Padder F, Waclawiw M, Epstein N, Quyyumi AA. 2000. Insertion–deletion polymorphism of the ACE gene modulates reversibility of endothelial dysfunction with ACE inhibition. *Circulation* **102**, 35–41.

274 Sasaki M, Oki T, Iuchi A, Tabata T, Yamada H, Manabe K, Fukuda K, Abe M, Ito S. 1996. Relationship between the angiotensin converting enzyme gene polymorphism and the effects of enalapril on left ventricular hypertrophy and impaired diastolic filling in essential hypertension: M-mode and pulsed doppler echocardiographic studies. *J Hypertens* **14**, 1403–8.

275 Stavroulakis GA, Makris TK, Krespi PG, Hatzizacharias AN, Gialeraki AE, Anastasiadis G, Triposkiadis P, Kyriakidis M. 2000. Predicting response to chronic antihypertensive treatment with fosinopril: the role of angiotensin converting enzyme gene polymorphism. *Cardiovasc Drugs Ther* **14**, 427–32.

276 van der Kleij FG, Navis GJ, Gansevoort RT, Heeg JE. 1997. ACE polymorphism does not determine short-term renal response to ACE-inhibition in proteinuric patients. *Nephrol Dial Transplant* **12** (Suppl. 2), 42–6.

277 Haas M, Yilmaz N, Schmidt A, Neyer U, Arneitz K, Stummvoll HK, Wallner M, Auinger M, Arias I, Schneider B, Mayer G. 1998. Angiotensin converting enzyme gene polymorphism determines the anti-proteinuric and systemic hemodynamic effect of enalapril in patients with proteinuric renal disease. *Kidney Blood Press Res* **21**, 66–9.

278 Nakano Y, Oshima T, Watanabe M, Matsuura H, Kajiyama G, Kambe M. 1997. Angiotensin I-converting enzyme gene polymorphism and acute response to captopril in essential hypertension. *Am J Hypertens* **10**, 1064–8.

279 O'Toole L, Stewart M, Padfield P, Channer K. 1998. Effect of the insertion/deletion polymorphism of the angiotensin-converting enzyme gene on response to angiotensin-converting enzyme inhibitors in patients with heart failure. *J Cardiol Pharmacol* **32**, 988–94.

280 Mukae S, Aoki S, Itoh S, Iwata T, Ueda H, Katagiri T. 2000. Bradykinin B(2) receptor gene polymorphism is associated with angiotensin-converting enzyme inhibitor-related cough. *Hypertension* **36**, 127–31.

281 de Maat MP, Jukema JW, Ye S, Zwinderman AH, Moghaddam PH, Beekman M, Kastelein JJ, van Boven AJ, Bruschke AV, Humphries SE, Kluft C, Henney AM. 1999. Effect of the stromelysin-1 promoter on efficacy of pravastatin in coronary atherosclerosis and restenosis. *Am J Cardiol* **83**, 852–6.

282 Zambon A, Deeb SS, Brown BG, Hokanson JE, Brunzell JD. 2001. Common hepatic lipase gene promoter variant determines clinical response to intensive lipid-lowering treatment. *Circulation* **103**, 792–8.

283 Hansen T, Echwald SM, Hansen L, Møller AM, Almind K, Clausen JO, Urhammer SA, Inoue H, Ferrer J, Bryan J, Aguilar-Bryan L, Permutt MA, Pedersen O. 1998. Decreased tolbutamide-stimulated insulin secretion in healthy subjects with sequence variants in the high-affinity sulfonylurea receptor gene. *Diabetes* **47**, 598–605.

284 Priori SG, Barhanin J, Hauer RN, Haverkamp W, Jongsma HJ, Kleber AG, McKenna WJ, Roden DM, Rudy Y, Schwartz K, Schwartz PJ, Towbin JA, Wilde AM. 1999. Genetic and molecular basis of cardiac arrhythmias: impact on clinical management part III. *Circulation* **99**, 674–81.

285 Napolitano C, Schwartz PJ, Brown AM, Ronchetti E, Bianchi L, Pinnavaia A, Acquaro G, Priori SG. 2000. Evidence for a cardiac ion channel mutation underlying drug-induced QT prolongation and life-threatening arrhythmias. *J Cardiovasc Electrophysiol* **11**, 691–6.

286 Donger C, Denjoy I, Berthet M, Neyroud N, Cruaud C, Bennaceur M, Chivoret G, Schwartz K, Coumel P, Guicheney P. 1997. KVLQT1C-terminal missense mutation causes a forme fruste long-QT syndrome. *Circulation* **96**, 2778–81.

287 Abbott GW, Sesti F, Splaswki I, Buck M, Lehmann MH, Timothy KW, Keating MT, Goldstein SAN. 1999. MiRP1 forms IKr potassium channels with HERG and is associated with cardiac arrhythmia. *Cell* **97**, 175–87.

288 Sesti F, Abbott GW, Wei J, Murray KT, Saksena S, Schwartz PJ, Priori SG, Roden DM, George Jr AL, Godstein SAN. 2000. A common polymorphism associated with antibiotic-induced cardiac arrhythmia. *Proc Natl Acad Sci USA* **97**, 10613–8.

289 Steen VM, Lovlie R, Osher Y, Belmaker RH, Berle JO, Gulbrandsen AK. 1998. The polymorphic inositol polyphosphate 1-phosphatase gene as a candidate for pharmacogenetic prediction of lithium responsive manic-depressive illness. *Pharmacogenetics* **8**, 259–68.

290 Poirier J, Delisle MC, Quirion R, Aubert I, Farlow M. 1995. Apolipoprotein E4 allele as a predictor of cholinergic deficits and treatment outcome in Alzheimer disease. *Proc Natl Acad Sci USA* **92**, 12260–4.

291 Richard F, Helbecque N, Neuman E, Guez D, Levy R, Amouyel P. 1997. APOE genotyping and response to drug treatment in Alzheimer's disease. *Lancet* **349**, 539.

292 Farlow MR, Lahiri DK, Poirier J, Davignon J, Hui S. Apolipoprotein E genotype and gender influence response to tacrine therapy. *Ann NY Acad Sci* **802**, 101–10.

293 Farlow MR, Lahiri DK, Poirier J, Davignon J, Schneider L, Hui SL. 1998. Treatment outcome of tacrine therapy depends on apolipoprotein genotype and gender of the subjects with Alzheimer's disease. *Neurology* **50**, 669–77.

294 Gerdes LU, Gerdes C, Kervinen K, Savolainen M, Klausen IC, Hansen PS, Kesäniemi YA, Faergeman O. 2000. The apolipoprotein epsilon4 allele determines prognosis and the effect on prognosis of simvastatin in survivals of myocardial infarction: a substudy of the Scandinavian Simvastatin survival study. *Circulation* **101**, 1366–71.

295 Ordovasa JM, Mooser V. 2002. The APOE locus and the pharmacogenetics of lipid response. *Curr Opin Lipidol* **13**, 113–17.

296 McCarthy TV, Quane KA, Lynch PJ. 2000. Ryanodine receptor mutations in malignant hyperthermia and central core disease. *Hum Mutat* **15**, 410–7.

297 Brandt JT, Isenhart CE, Osborne JM, Ahmed A, Anderson CL. 1995. On the role of platelet Fc gamma RIIa phenotype in heparin-induced

thrombocytopenia. *Thromb Haemost* **74**, 1564–72.

298 JIA H, HINGORANI AD, SHARMA P, HOPPER R, DICKERSON C, TRUTWEIN D, LLOYD DD, BROWN MJ. 1999. Association of the G(s)alpha gene with essential hypertension and response to beta-blockade. *Hypertension* **34**, 8–14.

299 MICHELSON AD, FURMAN MI, GOLDSCHMIDT-CLERMONT P, MASCELLI MA, HENDRIX C, COLEMAN L, HAMLINGTON J, BARNARD MR, KICKLER T, CHRISTIE DJ, KUNDU S, BRAY PF. 2000. Platelet GP IIIa Pl(A) polymorphisms display different sensitivities to agonists. *Circulation* **101**, 1013–8.

300 ONGPHIPHADHANAKUL B, CHAN-PRASERTYOTHIN S, PAYATIKUL P, TUNG SS, PIASEU N, CHAILURKIT L, CHANSIRIKARN S, PUAVILAI G, RAJATANAVIN R. 2000. Oestrogen-receptor-alpha gene polymorphism affects response in bone mineral density to oestrogen in postmenopausal women. *Clin Endocrinol* **52**, 581–5.

301 HERRINGTON DM, HOWARD TD, HAWKINS GA, REBOUSSIN DM, XU J, ZHENG SL, BROSNIHAN KB, MEYERS DA, BLEECKER ER. 2002. Estrogen-receptor polymorphisms and effects of estrogen replacement on high-density lipoprotein cholesterol in women with coronary disease. *N Engl J Med* **346**, 967–74.

302 ZILL P, BAGHAI TC, ZWANZGER P, SCHÜLE C, MINOV C, RIEDEL M, NEUMEIER K, RUPPRECHT R, BONDY B. 2000. Evidence for an association between a G-protein beta 3 gene variant with depression and response to antidepressant treatment. *NeuroReport* **11**, 1893–7.

303 TURNER ST, SCHWARTZ GL, CHAPMAN AB, BOERWINKLE E. 2001. C825T polymorphism of the G protein β3-subunit and antihypertensive response to a thiazide diuretic. *Hypertension* **37**, 739–43.

304 PSATY BM, SMITH NL, HECKBERT SR, VOS HL, LEMAITRE RN, REINER AP, SISCOVICK DS, BIS J, LUMLEY T, LONGSTRETH WT, ROSENDAAL FR. 2002. Diuretic therapy, the alpha-adducin gene variant, and the risk of myocardial infarction or stroke in persons with treated hypertension. *J Am Med Ass* **287**, 1680–9.

305 HINGORANI A, JIA H, STEVENS P, HOPPER R, DICKERSON JE, BROWN M. 1995. Renin–angiotensin gene polymorphisms influence blood pressure and the response to angiotensin converting enzyme inhibition. *J Hypertens* **13**, 1602–9.

306 MARTINELLI I, SACCHI E, LANDI G, TAIOLI E, DUCA F, MANNUCCI PM. 1998. High risk of cerebral-vein thrombosis in carriers of a prothrombin-gene mutation and users of oral contraceptives. *N Engl J Med* **338**, 1793–7.

307 MARTINELLI I, TAIOLI E, BUCCIARELLI P, AKHAVAN S, MANNUCCI PM. 1999. Interaction between the G20210A mutation of the prothrombin gene and oral contraceptive use in deep vein thrombosis. *Arterioscler Thromb Vasc Biol* **19**, 700–3.

308 ESTELLER M, GARCIA-FONCILLAS J, ANDION E, GOODMAN SN, HIDALGO OF, VANACLOCHA V, BAYLIN SB, HERMAN JG. 2000. Inactivation of the DNA-repair gene MGMT and the clinical response of gliomas to alkylating agents [Erratum. 2000. *N Engl J Med* **343**, 1740]. *N Engl J Med* **343**, 1350–4.

13
Pharmacogenomics of Bioavailability and Elimination

Ingolf Cascorbi and Heyo K. Kroemer

13.1
Introduction

Interindividual drug effects are subject to substantial variability. There are multiple reasons for this based on pathophysiological factors and environmental interactions, but also genetic characteristics. As early as 1953 it had been observed that patients who received the tuberculostatic drug isoniazid excreted the acetylated product in urine at different rates. The concept of pharmocogenetics was coined in 1958 by the German geneticist Vogel, who observed that individually varying pharmacokinetics are subject to familial heredity. Since then, groundbreaking successes have been achieved in the field of pharmacogenomics. In particular, the identification of hereditary polymorphisms in genes of the cytochrome P450 (CYP) system have contributed considerably to the explanation of the individually varying pharmacokinetics of a number of drugs. Furthermore, hereditary variations in genes of membrane drug transporters have recently been discovered. Along with these factors which can influence pharmacokinetics, efforts have been undertaken to clarify the role of genetic polymorphisms in receptors or signal transduction proteins modulating drug efficacy. Pharmacogenomics tries to elucidate the complex interaction of many polymorphic genes in order to explain and to develop improved therapies, particularly for common diseases such as cardiovascular disorders and cancer. This represents a technical as well as a logistic challenge in the face of the millions of single nucleotide polymorphisms (SNPs) currently known.

13.2
Contribution of Metabolism to Drug Clearance

13.2.1
CYPs

Phase I enzymes catalyze oxidative and reductive reactions of foreign compound metabolism, as well as the transformation of certain lipids and steroids. CYPs

Molecular Biology in Medicinal Chemistry. Edited by Th. Dingermann, D. Steinhilber, G. Folkers

ISBN: 3-527-30431-2

Tab. 13.1 Substrates of polymorphic cytochrome P450 enzymes, functional important alleles and frequencies in caucasians [25, 111, 112].

CYP	***Substrates (selected)***	***Allele (relevant cDNA and/ or protein variant)***	***Functional consequence***	***Allele frequency (%)***
CYP2A6	various: cumarines, nicotine	*2 (T479A, L160H)	inactive enzyme	1–3
		*4 (gene deletion)	no enzyme	1
		*9 (T-48G)	decreased expression	5
CYP2B6	cytostatics: cyclophosphamide, ifosfamide benzodiazepines: diazepam, tenazepam, midazolam various: clopidogrel, nicotine, tamoxifen	*5 (C1459T, R487C)	decreased activity	14
		*7 (G516T, Q172H; A785G, K262R; C1459T, R487C)	decreased activity	1
CYP2C8	various: paclitaxel, rosiglitazone	*2 (A805T, I269F)	decreased activity	0
		*3 (G416A; R139K; A1196G, K399R)	decreased activity	13
CYP2C9	nonsteroidal anti-inflammatory drugs: diclofenac, ibuprofen, (*S*)-naproxen, meloxicam, piroxicam, tenoxicam oral antidiabetics: glibenclamide, glipizide, tolbutamide AT1-antagonists: irbesartane, losartane, valsartane various: amitriptyline, fluoxetine, phenytoine, sulfamethoxazole, tamoxifen, torasemide, (*S*)-warfarin	*2 (C430T, R144C)	decreased activity	8–13
		*3 (A1075C, I359L)	decreased activity	7–9
CYP2C19	proton pump inhibitors: lansoprazole, omeprazole, pantoprazole anticonvulsants: diazepam, phenytoine, (*S*)-mephenytoine antipsychotics: citalopram, clomipramine, imipramine various: moclobemide, proguanil, propranolol	*2 (G681A, splice-site mutation)	inactive enzyme	13
		*3 (G636A, stop)	inactive enzyme	0
CYP2D6	*β*-blockers: alprenolol, carvedilol, metoprolol, propranolol antiarrhythmics: propafenone, encainide, flecainide, mexiletinr, sparteine neuroleptics: haloperidol, perhexiline, perphenazine, risperidon, thioridazine antidepressants: amitriptyline, clomipramine, desipramine, fluoxetine, fluvoxamine, imipramine, maprotiline,	*2 × *N* (gene duplication)	increased activity	1–5
		*2 (R296C; S486T)	slightly decreased activity	12–21
			inactive enzyme	
		*4 (G1846A splice-site mutation)	no enzyme	4–6
		*5 (gene deletion)	no enzyme	1
		*6 (T1707del frameshift)	decreased activity	1–2
		*10 (P34S)	decreased activity	<1
		*17 (T107I; R296C)	decreased	10–20
		*41 (G-1584C)	expression	

Tab. 13.1 *(continued)*

CYP	Substrates (selected)	Allele (relevant cDNA and/ or protein variant)	Functional consequence	Allele frequency (%)
	nortriptyline, paroxetine, odeine, dextromethorphan, tramadol, amphetamine, methoxyamphetamine, debrisoquine, ondansetron, phenacetine			
CYP2E1	anesthetics: enflurane, halothane, isoflurane, methoxyflurane, sevoflurane various: paracetamol, chlorzoxazone, ethanol	*2 (R76H) *3 (V381I) *4 (V179I)	decreased activity inactive enzyme inactive enzyme	0 <1 <1
CYP3A5	widely overlapping with CYP3A4 substrates	*3 (intron 3 A6986G splice-site mutation)	no enzyme	94

are the most important and the CYP3A family contributes approximately 50% of the total CYP activity of the adult human liver. It metabolizes about 60% of all the usually prescribed drugs. The main subjects of hereditary polymorphisms are CYP2C9, 2C19 and 2D6, contributing to 30% of all metabolized drugs (Tab. 13.1). Moreover, there are known polymorphisms in the genes of CYP1A1, 1A2, 1B1, 2A6, 2B6, 2C8, 2E1, 3A4, 3A5, 3A7, 5A1 and 8A1, but their functional significance is still controversial [1].

13.2.1.1 CYP2C9

At the beginning of 1999 a study came into focus showing that patients who received a low dose of the coumarin derivative warfarin for thrombosis prophylaxis exhibited significantly more frequent episodes of hemorrhagia than a control group that received a higher dosage [2]. This apparently paradoxical finding was caused by the fact that a larger number of CYP2C9 poor metabolizers (PMs) were present in the group with a low dosage regimen. The clearance of the effective *S*-enantiomer of warfarin was significantly decreased in these individuals, resulting in augmented intermediate plasma concentrations, but not in lower concentrations as expected from the low dosage. Consequently, the synthesis of the vitamin K-dependent coagulation factor was strongly inhibited, corresponding to elevated International Normalized Ratio (INR) values and an increased tendency to bleeding episodes.

CYP2C9 belongs to a close gene cluster on chromosome 10, comprising CYP2C8, 2C9, 2C18 and 2C19. CYP2C9 deficiency is determined to a major extent by two missense point mutations that lead to the exchange of the amino acids Arg144 → Cys (CYP2C9*2) and Ile359 → Leu (CYP2C9*3) [3]. The allele frequencies in the German population are approximately 11 and 7%. Homozygous carriers who account for the phenotype of PMs have a prevalence of 3.2%. Interestingly, in Asians, an alternative Ile359 → Thr amino acid replacement was observed, termed CYP2C9*4 [4]. A further polymorphism in exon 7 (*2C9*5*; Asp360 → Glu) [5] and a premature stop codon due to a deletion polymorphism 818delA exhibiting no activity (*2C9*6*) [6] was identified in African-Americans. To date, 12 different alleles have been recognized by the Human Cytochrome P450 (CYP) Allele Nomenclature Committee.

In addition to warfarin (but to a much lower extent, phenprocumon), many nonsteroidal antiphlogistics such as diclofenac, ibuprofen and meloxicam, oral antidiabetics like tolbutamide and glimenclamide, the angiotensin receptor antagonists irbesartan and losartan as well as some other medications such as phenytoin are metabolized by CYP2C9 [7–9]. There is increasing evidence that the clearance of oral antidiabetics is directly dependent on the CYP2C9 genotype. The consequences are not always clear. After glubyride or glibenclamide intake, the area under the curve (AUC) increased nearly 3-fold, but glucose levels of heterozygous carriers of CYP2C9 variants did not differ [8]. In homozygous CYP2C9*3 carriers, these effects are more pronounced and are mirrored in an accelerated insulin response to glucose stimulation [10, 11].

13.2.1.2 **CYP2C19**

CYP2C19, the so-called mephenytoine hydroxylase, catalyzes the hydroxylation particularly of proton pump inhibitors like omeprazole and lanozoprazole. In addition to many other mutations, a splice-site mutation leads to total lack of any activity in 3–5% of Caucasians [12]. Interestingly, there is some evidence that PMs profit more from *Helicobachter pylori* eradication therapy. Only 80% were successfully treated in CYP2C19 wild-types, whereas all cases showed benefit among *CYP2C19*2* carriers [13, 14]. These intriguing results could not be confirmed in other studies [15], but similar CYP2C19-dependent effects can be seen for the treatment of gastroesophageal reflux [16]. CYP2C19 is highly polymorphic – many novel SNPs have been detected recently [17] and currently there are 15 different alleles annotated, but most of the genetic variants identified have a very low prevalence. Apart from the G681A splice-site mutation, G636A (*CYP2C19*3*) should be considered when genotyping Blacks and Orientals. However, this premature stop codon in codon 212 is rare in Caucasians.

13.2.1.3 **CYP2D6**

In addition to CYP3A, the polymorphic CYP2D6 is one of the most studied CYP enzymes. Nearly 40 years ago it was observed that the large interindividual variability of nortriptyline plasma concentration has a strong genetic background and is based on individually different metabolic rates (for review, see [18]). Later, it was shown that metabolism of the antihypertensive drug debrisoquine [19] and the antiarrhythmic drug sparteine [20] is polymorphic and metabolized by the same enzyme CYP2D6. In particular, antidepressants and neuroleptics, possessing a number of adverse effects, are also metabolized by the polymorphic sparteine/debrisoqione hydroxylase [21]. Therefore, it became rapidly clear that this hereditary trait may have severe clinical consequences.

CYP2D6 is now known to be one of the most important polymorphic genes involved in drug metabolism. Approximately 7–10% of the European population are CYP2D6 PMs who show reduced metabolism of numerous drugs like anti-arrhythmics, antidepressants, neuroleptics and some β-blockers or opiates. Among this subgroup, clinically relevant drug side-effects are more likely compared to extensive metabolizers (EMs). The occasionally observed absence of the desired effect accounts (in some cases) for the phenomenon of gene duplications, occurring in 1–3% of Middle-Europeans.

CYP2D6 belongs to a gene cluster of the highly homologous inactive pseudogenes CYP2D7 containing a single reading frame-disrupting insertion in its first exon and the real pseudogene CYP2D8 [22]. The CYP2D6 gene is mapped to chromosome 22q13.1 [23, 24] and encompasses 9 exons with an open reading frame of 1383 bp coding for 461 amino acids. The polymorphisms are well characterized and extensively described by Sachse et al. [25]. Currently, more than 73 different CYP2D6 haplotypes are recorded by the Human Cytochrome P450 (CYP) Allele Nomenclature Committee (http://www.iim.ki.se/cypalleles). The alleles may be classified on the basis of the level of activity for which they encode CYP2D6 enzymes into functional, nonfunctional and reduced function groups. One of the major primary gene defects at the CYP2D locus is *CYP2D6*4* [26], with a frequency of 20.7% in Caucasians [25]. A G1846A transition generates a shift of the splice site at the boundary of intron 3 to exon 4, consequently leading to the generation of a nonfunctional protein. The entire coding region is deleted in 4–6% of Caucasian and other ethnic populations, [27]. Hence, allele **5* is believed to have an ancient origin. Further relevant variants are a 2549A deletion in **3* (2.0%) and a 1707T deletion in **6* (0.9%), generating frame shifts. In contrast to these fatal polymorphisms, a triple-base-pair deletion in allele *9 (1.8%) does not significantly alter enzyme activity [28], and a proline to serine exchange in codon 34 is associated with lower enzyme activity and particularly decreased stability of CYP2D6.10 [29]. This variant occurs in 1–2% of Caucasians, but is the major cause of low CYP2D6 activity in Orientals [30]. In African-Blacks, *17 is one of the major reasons for low CYP2D6 activity.

Extremely high CYP2D6 activities in 1–2% of Caucasians were identified to be due to gene duplications of allele **1* (*$*1 \times 2$*) and **2* (*$*2 \times 2$*). In Northern Europe, the prevalence is below 2% [31], but in some regions of Spain, frequencies of more than 7% were observed [32]. In Arabian countries [33] as well is in the north-east Ethiopia, a prevalence of ultrarapid metabolizers (UMs) of up to 29% is reported [34]. In a few cases there were families with up to 13 gene copies identified. In contrast, in China, CYP2D6 gene duplications are very rare, but the mean metabolic ratio of debrisoquine/4-hydroxydebrisoquine is increased compared to Caucasians [35]. This is due to the high prevalence of the low active (intermediate) CYP2D6 variant *10, e.g. [36]. In Blacks, however, large heterogeneity seems to exist [37, 38].

The phenotype of intermediate metabolizers (IMs) may also be due to diminished expression rates of CYP2D6, e.g. there is convincing evidence that homozygous carriers of a C/G polymorphism 1584 bp upstream of the start codon exhibited only 50% of protein compared to carriers of the −1584G variant [39].

This variant, linked to CYP2D6*2, is termed *CYP2D6*41* and may improve the prediction of so-called IMs.

Many other alleles have been reported so far – some extremely rare and existing only in specified populations [40]. Although the metabolic ratio of, for example, dextrometorphan may vary between individuals with the same genotype by more than an order of magnitude, genotyping enables a reliable prediction of CYP2D6 PM, IM, EM and UM status.

The first case of a gene duplication was identified in a patient exhibiting extremely low plasma levels after oral treatment with nortriptyline [41, 42]. The UM phenotype of CYP2D6 has now been well established as a relevant cause of non-response to antidepressant drug therapy. Clearance of such drugs like nortripyline, desipramine, and, to some extent, imipramine and amitriptyline [43–45] evidently depends on the CYP2D6 polymorphism. Specific serotonine reuptake inhibitors like fluoxetine, citalopram or paroxetine were shown to be inhibitors of CYP2D6 [46, 47]. Interestingly, a study on side-effects like seizures and myoclonus after treatment with antidepressant drugs could not identify any patient with a CYP2D6 PM status, but half were concomitantly treated with drugs with potential inhibitory effects on CYP2D6 [48].

The effects of the CYP2D6 polymorphism on antipsychotic therapy appear to be more pronounced in neuroleptics. Compounds like perphenazine, zuclopenthixol, thioridazine, haloperidol and risperidone are metabolized to a significant extent by CYP2D6. PMs appeared to posses an elevated risk to suffering from side-effects like extrapyramidal symptoms [18, 49, 50]. Moreover, the antipsychotic efficacy seems to be influenced by the number of active copies of CYP2D6 genes [50]. The findings indicate the need for genotyping before treatment with polymorphically metabolized antipsychotics [51].

The impact of polymorphic enzymes on the treatment of cardiovascular disorders is controversial and more data are needed for verification, e.g. propafenone shows a stronger β-receptor blockage in PMs [52] and was believed to be causative for strong side-effects in a patient being a PM for CYP2D6 [53]. Flecainide, however, is partly metabolized by CYP2D6 [54], but apparently the electrophysiology of the heart was only marginally influenced by the genotype [55]. β-Blockers like metoprolol, and to some extent carvedilol, are also CYP2D6 substrates [56, 57]. Even after long-term treatment, in PMs the median adjusted metoprolol plasma concentrations were 6.2-fold higher compared with EMs [58]. From a survey on adverse events after metoprolol treatment, the same group concluded that side-effects may be due to the CYP2D6 genotype, since the frequency of PMs was elevated in an albeit small group of patients, reportedly suffering from pronounced adverse symptoms [59]. However, there is no data currently available on whether the clinical outcome of β-blocker therapy is influenced by genetic polymorphisms of CYP2D6.

The beneficial use of CYP2D6 genotyping for drug therapy was demonstrated recently on the treatment of nausea and vomiting in cancer chemotherapy with the antiemetic 5-HT_3 receptor antagonist tropisetron [60]. In a study including 42 cancer patients receiving chemotherapy, approximately 30% of all patients experienced nausea and vomiting. CYP2D6 UMs, however, had a significantly higher frequency of vomiting after treatment than all other patients. The authors concluded that antiemetic treatment with tropisetron or, to lesser extent, ondansetron could be improved by adjustment for the CYP2D6 genotype and that approximately 50 subjects would have to be genotyped to protect one patient from severe emesis.

13.2.1.4 CYP3A

The CYP3A family consists of a number of well-known drugs metabolizing CYP3A4, CYP3A5, fetal CYP3A7, as well as the recently identified CYP3A43. The interindividual variability of the total CYP3A activity accounts in part for the presence or absence of active CYP3A5. Less than a third of Caucasians and 66% of African-Blacks exhibit CYP3A5 expression, caused by a G/A DNA base exchange within intron 3. This SNP leads to alternative splicing and generation of a premature stop codon in exon 3B, resulting in the inactive CYP3A5*3 [61, 62]. This polymorphism partly explains the bimodal distribution of midazolam kinetics. Additionally, among American-Blacks, a rare G/A SNP was identified in exon 7, associated with reduced CYP3A5 activity (CYP3A5*6) [62].

Apart from CYP3A5, CYP3A4 contributes to the clearance of midazolam. Recently, a functionally significant polymorphism of CYP3A4 (L373F) was discovered. It reduces the affinity of midazolam to CYP3A4 from a K_M of 8.7 to 36.4 μM. Further novel identified SNPs lead in part to lowered or even lack of expression. However, due to the rarity of these polymorphisms, they do not contribute to the explanation of individual variability of CYP3A4 activity [63].

CYP3A can be strongly induced by drugs like rifampicin, carbamazepine, phenobarbital and others. They interact with the nuclear pregnane X receptor (PXR) followed by dimerization with the 9-*cis*-retinoic acid receptor α (RXRα) and binding to the respective promoter responsive elements. Therefore, genetic variants of the PXR gene could contribute to the variability of CYP3A activity. Indeed, PXR exhibits numerous variants – at least six of them generate amino acid exchanges. The variant 163G leads to a 15-fold induction by rifampicin compared with a 5-fold induction by 163D. Other variants such as V140M and A370T exhibit effects of minor significance [64]. These discoveries show the great importance of gene–environment interactions, since the PXR polymorphisms do not affect the basal CYP3A activity, but CYP3A induction. Likewise, in CYP3A4 polymorphisms, the prevalence of PXR SNPs is rather low, hence the observed CYP3A activity can be explained on in small part by this variable.

13.2.2 Phase II Enzymes

13.2.2.1 Arylamine *N*-acetyltransferases (NAT2)

Only a short time after implication of isoniazid in clinical treatment of tuberculosis, it was observed that considerable interindividual differences exist with regard to urinary excretion of the acetylated metabolite [65]. The responsible enzyme –

NAT2 – catalyses the acetylation of substrates like sulfamethazine and the apparent interindividual differences of isoniazid acetylation were caused by genetic polymorphisms in the *NAT2* gene [66]. Slow acetylation in humans was shown to be based on reduced activity rather than lower expression rates. *NAT2* is expressed preferably in the human liver, whereas the sister gene *NAT1* can be determined in a wide variety of different tissues.

Phenotyping studies over the last four decades have disclosed distinct ethnic differences of slow acetylator frequencies [67]. Extremes can be found between North Africa with a frequency of alleles coding for slow acetylation of 95% and the Far-East Pacific region with a frequency of only 11%. The expected partition of slow acetylators in these populations spans 90 to 1.2%. Later determinations by genotyping additionally revealed that the allelic pattern differs significantly between Caucasians, Asians, Aborigines and, particularly, Blacks [68]. In Caucasians, the predominant alleles are the slow *NAT2*5B* (40.9%) and **6A* (28.4%), and the rapid **4* (23.4%) [69]. All other alleles have a frequency lower than 3%. In Asians, within the slow acetylators, *NAT2*6A* is more common than **5B*, e.g. 23 versus 19% in Japan. Interestingly, mutation 857A, occurring in **7B*, has a frequency of 11%. The highest frequency (25%) of the low stable variant *NAT2*7B* was observed in Australian Aborigines, [70]. In Africa, apart from the frequently observed G191A mutations, a large diversity of haplotypes can be observed. Apart from the rapid **4* allele, **12A* and **13* occur in many samples obtained from different tribes.

NAT2 is responsible for the conjugation of drugs like isoniazid, dapsone, procainamide and many others. The slow NAT2 acetylator is supposed to be at higher risk for drug side-effects such as peripheral neuropathia after isoniazid treatment [71] or certain disorders such as drug-induced lupus erythematosus [72] and Stevens–Johnson's syndrome [73]. These severe diseases may be caused, in susceptible individuals, by a large number of drugs involved in acetylation metabolism – there is no association to the idiopathic form of lupus erythematosus [74]. Peripheral neuropathy provoked by isoniazid overdosage may be a major problem in ethnicities with a high frequency of slow acetylators as in Northern Africa. Low drug efficacy may be expected in rapid acetylators.

13.2.2.2 **Thiopurine-*S*-methyltransferase (TPMT)**

A rare but serious side-effect of treatment with the antimetabolites 6-mercatopurine and azathioprine is severe bone marrow depression, which may result in lethal side-effects [75]. Detoxification takes place by TPMT, a phase II enzyme which, however, is homozygously deficient in 1:300 individuals [76]. Among these patients, metabolism takes place by an alternative pathway to the 6-thioguanine nucleotide (6-TGN). The plasma concentration of 6-TGN correlates with the severity of the medication's side-effects. However, a problem arises as TPMT exhibits a high number of mutations which allow a limited degree of genotyping. So far, eight mutations that determine an amino acid exchange and a splice-site mutation are known. Apparently even more, though very rare, SNPs exist which determine a TPMT PM. Many clinics still routinely prefer *ex vivo* phenotyping procedures, but increasing knowledge about rare variations and application of techniques like denaturing high-performance chromatography will allow a reliable prediction of TMPT status by means of genotyping [77].

13.3 Drug Transporters: P-glycoprotein (P-gp)

Drug transporters belong to the ATP-binding cassette (ABC) superfamily of membrane proteins which may influence the intracellular concentration of numerous compounds in a variety of cells and tissues. Initially, it was observed that ABC transporters like P-gp were overexpressed in tumor cells, conferring the commonly known phenomenon of multidrug resistance against antineoplastic agents like paclitaxel, anthracycline, vinca alkaloids and etoposide. The gene encoding P-gp was therefore termed multidrug-resistance gene *MDR1* (*ABCB1*).

However, it is well established that P-gp is responsible for the apical transport of various other lipophilic drugs like digoxin and the β-blockers celiprolol, pafenolol and talinolol, cholesterol synthesis inhibitors like lovastatine, atorvastatine and simvastatine, the HIV protease inhibitors indinavir and saquinavir, opioids, as well as the immunosuppressant cyclosporine A. Since P-gp mediates the transport of these important compounds via membranes of the intestine or the endothelial cells of brain capillaries [78, 79], it may serve as a functional barrier against drug entry [80] or contribute to excretion (expression at the canalicular site of hepatocytes or tubular cells of the kidneys). Moreover, expression in endothelial cells of the blood–brain barrier protects against drug penetration into the central nervous system [81]. Hence, P-gp knockout mice exhibit significantly elevated disposition of P-gp substrates [82].

In humans, expression of P-gp discloses a wide interindividual variability and is subject to markedly drug–drug interactions, e.g. the antituberculous agent rifampicin is an effective inducer of P-gp as well as of MRP2 [83, 84] due to a PXR element in the MDR1 promoter region [85]. P-gp induction was also observed after co-administration of St John's Wort [86]. On the other hand, competition with verapamil may also lead to inhibition of the transporter activity [87].

Although P-gp appears to be co-regulated in many aspects with CYP3A4, there is no relationship to CYP3A4 expression, explaining the large interindividual variability [88]. Thus, the identification of genetic variants in the MDR1 gene stimulated a large number of studies attempting to investigate the correlation of P-gp expression and activity in dependence of MDR1 SNPs. MDR1 spans 28 exons and the cDNA consists of 3843 bp. Mickley et al. discovered two sense SNPs, a G2677T transversion in exon 21 and a G2995A transition in exon 24 of MDR1 [89]. These SNPs led to an Ala/Ser exchange in codon 893 and Ala/Thr exchange in codon 999, respectively. Notably, the different alleles showed similar expression in normal tissue, unselected cell lines and untreated malignant lymphomas. Later *in vitro* studies confirmed these observations [90]. Systematic sequencing of all MDR1 exons and the 5′ region in Caucasian individuals revealed several SNPs. Twelve out of 15 mutations had no effect on protein sequence [91]. Strikingly, there was a significant correlation of a silent polymorphism in exon 26 (C3435T) with intestinal P-gp expression levels and oral bioavailability of digoxin. The C3435T SNP had an allelic frequency of 53.9% in a sample of 461 German Caucasians [92], but varies significantly between different ethnic groups. The prevalence is much lower (0.17–0.27) in African Blacks, whereas values of 0.41–0.47 are given in Oriental populations [93–95]. This may be important for the probability of interindiviually different drug responses in certain populations.

In codon 893, six different genotypes may exist as a result of combinations of the three alleles 2677G, T and A with allelic frequencies of 56.7, 41.6 and 1.7%, leading to amino acids Ala, Ser or Thr, respectively. Further studies identified a number of further mutations with varying frequencies and questionable functional significance.

As mentioned above, in the initial investigation P-gp expression differed significantly depending on C3435T being in a wobble position of codon 1145 [91]. The SNP was associated with lower expression of P-gp in human duodenal mucosa and functionally with elevated digoxin plasma levels. It was assumed that 3435C leads to lower transport activity. The molecular mechanism by which C3435T influences P-gp expression is not well understood since there is no evidence for a genetic linkage disequilibrium to variants in promoter regions of the MDR1 gene or sequences that control mRNA processing. The finding, however, was supported by determination of rhodamine 123 efflux from $CD56^+$ natural killer cells [96]. This striking effect of the exon 26 SNP was also found, although it was less pronounced, in subjects in steady state with 0.25 mg digoxin/day. TT carriers had a 20% reduced AUC within the first 4 h, and the differences were more pronounced when considering C3435T and G2677T haplotypes [97]. Similar results were obtained in a French study on digoxin [98].

Possibly, such polymorphisms may play a role in patients not responding to treatment [99, 100]. Indeed, Schwab et al. demonstrated an association to inflammatory bowel disease [101]. However, the functional impact of C3435T is not consistent. There was lack of evidence in a study performed in 124 German renal transplant recipients who received a microemulsion of the P-gp substrate cyclosporine over at least 6 months [102]. AUCs did not differ depending on the C3435T polymorphism. On the other hand, the formulation contains the nonionic surfactant Cremophor EL, which is known to inhibit P-gp towards transport of paclitaxel [103, 104]. No association for C3435T was reported from a Japanese group, who evaluated whether MDR1 correlated with placenta trophoblast P-gp expression in 100 placentas obtained from Japanese women [105]. In contrast, in two further Japanese studies on digoxin kinetics, the AUC in the first 4 h was significantly higher in the CC group than in subjects homozygous for TT [106, 107]. The latter study also investigated the effects of MDR1 polymorphisms on duodenal mRNA expression, demonstrating elevated expression in the case of TT carriers compared to CC carriers, which could explain the lower digoxin levels in 3435CC carriers. Investigations in a German sample with 55 volunteers using the β-blocker talinolol as probe drug revealed that the exon 26 SNP is of minor importance. However, the amino acid exchange Ala893Thr in exon 21 correlated with significantly altered talinolol plasma levels. In carriers of the TT/TA variants of G2677T/A, the AUC values of oral talinolol were slightly, but significantly, elevated compared to carriers of at least one wild-type allele ($p = 0.014$) [108], but there were no changes in mRNA or protein expression levels with regard to any polymorphism investigated.

In summary, the functional importance of MDR1 polymorphisms remains to be elucidated. Possibly, the linkage disequilibrium between exon 26 C3435T and exon 21 Ala893 → Ser may contribute to the varying P-gp activity; however, there is still no explanation for the association of C3435T with P-gp expression as described in some studies. Therefore, it may be hypothesized that C3435T represents a genetic marker for variants not yet identified in regulatory regions of the MDR1 gene.

13.4 Conclusion

The early adaptation of a therapy regimen to genetic traits could help to avoid side-effects and improve the clinical outcome of pharmacotherapy. Genotyping instead of phenotyping with probe drugs should be preferred during an on-going pharmacotherapy, since some drugs may inhibit, for example, CYP2D6 activity. At least the major deficient alleles occurring in the particular ethnic group should be characterized for prediction of the phenotype by genotyping. The benefit of genotype-related dose adaptation is best described by studies of psychotropic drugs [18]; however, there is a need of proof-performing prospective studies [109]. Initial dosage recommendations depending on pharmacogenetic traits have been published recently for a set of antidepressants [110]. This may be an important step in an attempt to improve individual drug therapy, but more standardized clinical studies are required, testing the efficacy and side effects of genotype adapted and non adapted dosage regimens.

References

1 Ingelman-Sundberg M. Genetic susceptibility to adverse effects of drugs and environmental toxicants. The role of the CYP family of enzymes. *Mutat Res* 2001, **482**, 11–9.

2 Aithal GP, Day CP, Kesteven PJ, Daly AK. Association of polymorphisms in the cytochrome P450 CYP2C9 with warfarin dose requirement and risk of bleeding complications. *Lancet* 1999, **353**, 717–9.

3 Goldstein JA, de Morais SMF. Biochemistry and molecular biology of the human CYP2C subfamily. *Pharmacogenetics* 1994, **4**, 285–99.

4 Imai J, Ieiri I, Mamiya K, Miyahara S, Furuumi H, Nanba E, et al. Polymorphism of the cytochrome P450 (CYP) 2C9 gene in Japanese epileptic patients: genetic analysis of the CYP2C9 locus. *Pharmacogenetics* 2000, **10**, 85–9.

5 Dickmann LJ, Rettie AE, Kneller MB, Kim RB, Wood AJ, Stein CM, et al. Identification and functional characterization of a new CYP2C9 variant (CYP2C9*5) expressed among African Americans. *Mol Pharmacol* 2001, **60**, 382–7.

6 Kidd RS, Curry TB, Gallagher S, Edeki T, Blaisdell J, Goldstein JA. Identification of a null allele of

CYP2C9 in an African-American exhibiting toxicity to phenytoin. *Pharmacogenetics* 2001, **11**, 803–8.

7 Sullivan-Klose TH, Ghanayem BI, Bell DA, Zhang Z-Y, Kaminsky LS, Shenfield GM, et al. The role of the CYP2C9-Leu359 allelic variant in the tolbutamide polymorphism. *Pharmacogenetics* 1996, **6**, 341–9.

8 Niemi M, Cascorbi I, Timm R, Kroemer HK, Neuvonen PJ, Kivisto KT. Glyburide and glimepiride pharmacokinetics in subjects with different CYP2C9 genotypes. *Clin Pharmacol Ther* 2002, **72**, 326–32.

9 Kirchheiner J, Meineke I, Steinbach N, Meisel C, Roots I, Brockmoller J. Pharmacokinetics of diclofenac and inhibition of cyclo-oxygenases 1 and 2: no relationship to the CYP2C9 genetic polymorphism in humans. *Br J Clin Pharmacol* 2003, **55**, 51–61.

10 Kidd RS, Straughn AB, Meyer MC, Blaisdell J, Goldstein JA, Dalton JT. Pharmacokinetics of chlorpheniramine, phenytoin, glipizide and nifedipine in an individual homozygous for the CYP2C9*3 allele. *Pharmacogenetics* 1999, **9**, 71–80.

11 Kirchheiner J, Brockmoller J, Meineke I, Bauer S, Rohde W, Meisel C, et al. Impact of CYP2C9 amino acid polymorphisms on glyburide kinetics and on the insulin and glucose response in healthy volunteers. *Clin Pharmacol Ther* 2002, **71**, 286–96.

12 De Morais S, Wilkinson G, Blaisdell J, Nakamura K, Meyer U, Goldstein J. The major genetic defect responsible for the polymorphism of *S*-mephenytoin metabolism in humans. *J Biol Chem* 1994, **269**, 15419–22.

13 Kawamura M, Ohara S, Koike T, Iijima K, Suzuki J, Kayaba S, et al. The effects of lansoprazole on erosive reflux oesophagitis are influenced by CYP2C19 polymorphism. *Aliment Pharmacol Ther* 2003, **17**, 965–73.

14 Tanigawara Y, Aoyama N, Kita T, Shirakawa K, Komada F, Kasuga M, et al. CYP2C19 genotype-related efficacy of omeprazole for the treatment of infection caused by *Helicobacter pylori. Clin Pharmacol Ther* 1999, **66**, 528–34.

15 Miwa H, Misawa H, Yamada T, Nagahara A, Ohtaka K, Sato N. Clarithromycin resistance, but not CYP2C-19 polymorphism, has a major impact on treatment success in 7-day treatment regimen for cure of H. pylori infection: a multiple logistic regression analysis. *Dig Dis Sci* 2001, **46**, 2445–50.

16 Furuta T, Shirai N, Watanabe F, Honda S, Takeuchi K, Iida T, et al. Effect of cytochrome P4502C19 genotypic differences on cure rates for gastroesophageal reflux disease by lansoprazole. *Clin Pharmacol Ther* 2002, **72**, 453–60.

17 Blaisdell J, Mohrenweiser H, Jackson J, Ferguson S, Coulter S, Chanas B, et al. Identification and functional characterization of new potentially defective alleles of human CYP2C19. *Pharmacogenetics* 2002, **12**, 703–11.

18 Bertilsson L, Dahl ML, Dalen P, Al Shurbaji A. Molecular genetics of CYP2D6: clinical relevance with focus on psychotropic drugs. *Br J Clin Pharmacol* 2002, **53**, 111–22.

19 Mahgoub A, Idle JR, Dring LG, Lancaster R, Smith RL. Polymorphic hydroxylation of Debrisoquine in man. *Lancet* 1977, **2**, 584–6.

20 Eichelbaum M, Spannbrucker N, Steincke B, Dengler HJ. Defective *N*-oxidation of sparteine in man: a new pharmacogenetic defect. *Eur J Clin Pharmacol* 1979, **16**, 183–7.

21 Brosen K, Otton SV, Gram LF. Imipramine demethylation and hydroxylation: impact of the sparteine oxidation phenotype. *Clin Pharmacol Ther* 1986, **40**, 543–9.

22 Kimura S, Umeno M, Skoda RC, Meyer UA, Gonzalez FJ. The human debrisoquine 4-hydroxylase (CYP2D) locus: sequence and identification of the polymorphic CYP2D6 gene, a related gene, and a pseudogene. *Am J Hum Genet* 1989, **45**, 889–904.

23 Eichelbaum M, Baur MP, Dengler HJ, Osikowska-Evers BO, Tieves G, Zekorn C, et al. Chromosomal assignment of human cytochrome P-450 (debrisoquine/sparteine type) to chromosome 22. *Br J Clin Pharmacol* 1987, **23**, 455–8.

24 Gough AC, Smith CA, Howell SM, Wolf CR, Bryant SP, Spurr NK. Localization of the CYP2D gene locus to human chromosome 22q13.1 by polymerase chain reaction, *in situ* hybridization, and linkage analysis. *Genomics* 1993, **15**, 430–2.

25 Sachse C, Brockmoller J, Bauer S, Roots I. Cytochrome P450 2D6 variants in a Caucasian population: allele frequencies and phenotypic consequences. *Am J Hum Genet* 1997, **60**, 284–95.

26 Gough AC, Miles JS, Spurr NK, Moss JE, Gaedigk A, Eichelbaum M, et al. Identification of the primary gene defect at the cytochrome P450 CYP2D locus. *Nature* 1990, **347**, 773–6.

27 Gaedigk A, Blum M, Gaedigk R, Eichelbaum M, Meyer UA. Deletion of the entire cytochrome P450 CYP2D6 gene as a cause of impaired drug metabolism in poor metabolizers of the debrisoquine/sparteine polymorphism. *Am J Hum Genet* 1991, **48**, 943–50.

28 Broly F, Meyer UA. Debrisoquine oxidation polymorphism: phenotypic consequences of a 3-base-pair deletion in exon 5 of the CYP2D6 gene. *Pharmacogenetics* 1993, **3**, 123–30.

29 Nakamura K, Ariyoshi N, Yokoi T, Ohgiya S, Chida M, Nagashima K, et al. CYP2D6.10 present in human liver microsomes shows low catalytic activity and thermal stability. *Biochem Biophys Res Commun* 2002, **293**, 969–73.

30 Bertilsson L. Geographical/interracial differences in polymorphic drug oxidation. Current state of knowledge of cytochromes P450 (CYP) 2D6 and 2C19. *Clin Pharmacokinet* 1995, **29**, 192–209.

31 Dahl ML, Johansson I, Bertilsson L, Ingelman-Sundberg M, Sjoqvist F. Ultrarapid hydroxylation of debrisoquine in a Swedish population. Analysis of the molecular genetic basis. *J Pharmacol Exp Ther* 1995, **274**, 516–20.

32 Agundez JA, Ledesma MC, Ladero JM, Benitez J. Prevalence of CYP2D6 gene duplication and its repercussion on the oxidative phenotype in a white population. *Clin Pharmacol Ther* 1995, **57**, 265–9.

33 McLellan RA, Oscarson M, Seidegard J, Evans DA, Ingelman-Sundberg M. Frequent occurrence of CYP2D6 gene duplication in Saudi Arabians. *Pharmacogenetics* 1997, **7**, 187–91.

34 Aklillu E, Persson I, Bertilsson L, Johansson I, Rodrigues F, Ingelman-Sundberg M. Frequent distribution of ultrarapid metabolizers of debrisoquine in an Ethiopian population carrying duplicated and multiduplicated functional CYP2D6 alleles. *J Pharmacol Exp Ther* 1996, **278**, 441–6.

35 Johansson I, Oscarson M, Yue QY, Bertilsson L, Sjoqvist F, Ingelman-Sundberg M. Genetic analysis of the Chinese cytochrome P4502D locus: characterization of variant CYP2D6 genes present in subjects with diminished capacity for debrisoquine hydroxylation. *Mol Pharmacol* 1994, **46**, 452–9.

36 Garcia-Barcelo M, Chow LY, Chiu HF, Wing YK, Lee DT, Lam KL, et al. Genetic analysis of the CYP2D6 locus in a Hong Kong Chinese population. *Clin Chem* 2000, **46**, 18–23.

37 Griese EU, Asante-Poku S, Ofori-Adjei D, Mikus G, Eichelbaum M. Analysis of the CYP2D6 gene mutations and their consequences for enzyme function in a West African population. *Pharmacogenetics* 1999, **9**, 715–23.

38 Masimirembwa C, Hasler J, Bertilssons L, Johansson I, Ekberg O, Ingelman-Sundberg M. Phenotype and genotype analysis of debrisoquine hydroxylase (CYP2D6) in a black Zimbabwean population. Reduced

enzyme activity and evaluation of metabolic correlation of CYP2D6 probe drugs. *Eur J Clin Pharmacol* 1996, **51**, 117–22.

39 Zanger UM, Fischer J, Raimundo S, Stuven T, Evert BO, Schwab M, et al. Comprehensive analysis of the genetic factors determining expression and function of hepatic CYP2D6. *Pharmacogenetics* 2001, **11**, 573–85.

40 Ingelman-Sundberg M, Oscarson M, Daly AK, Garte S, Nebert DW. Human cytochrome P-450 (CYP) genes: a web page for the nomenclature of alleles. *Cancer Epidemiol Biomarkers Prev* 2001, **10**, 1307–8.

41 Bertilsson L, Aberg-Wistedt A, Gustafsson LL, Nordin C. Extremely rapid hydroxylation of debrisoquine: a case report with implication for treatment with nortriptyline and other tricyclic antidepressants. *Ther Drug Monit* 1985, **7**, 478–80.

42 Johansson I, Lundqvist E, Bertilsson L, Dahl ML, Sjoqvist F, Ingelman-Sundberg M. Inherited amplification of an active gene in the cytochrome P450 CYP2D locus as a cause of ultrarapid metabolism of debrisoquine. *Proc Natl Acad Sci USA* 1993, **90**, 11825–9.

43 Brosen K, Zeugin T, Meyer UA. Role of P450IID6, the target of the sparteine-debrisoquin oxidation polymorphism, in the metabolism of imipramine. *Clin Pharmacol Ther* 1991, **49**, 609–17.

44 Ghahramani P, Ellis SW, Lennard MS, Ramsay LE, Tucker GT. Cytochromes P450 mediating the *N*-demethylation of amitriptyline. *Br J Clin Pharmacol* 1997, **43**, 137–44.

45 Venkatakrishnan K, von Moltke LL, Greenblatt DJ. Nortriptyline *E*-10-hydroxylation *in vitro* is mediated by human CYP2D6 (high affinity) and CYP3A4 (low affinity): implications for interactions with enzyme-inducing drugs. *J Clin Pharmacol* 1999, **39**, 567–77.

46 Crewe HK, Lennard MS, Tucker GT, Woods FR, Haddock RE. The effect of selective serotonin re-uptake inhibitors on cytochrome P4502D6 (CYP2D6) activity in human liver microsomes. *Br J Clin Pharmacol* 1992, **34**, 262–5.

47 Alfaro CL, Lam YW, Simpson J, Ereshefsky L. CYP2D6 status of extensive metabolizers after multiple-dose fluoxetine, fluvoxamine, paroxetine, or sertraline. *J Clin Psychopharmacol* 1999, **19**, 155–63.

48 Spigset O, Hedenmalm K, Dahl ML, Wiholm BE, Dahlqvist R. Seizures and myoclonus associated with antidepressant treatment: assessment of potential risk factors, including CYP2D6 and CYP2C19 polymorphisms, and treatment with CYP2D6 inhibitors. *Acta Psychiatr Scand* 1997, **96**, 379–84.

49 Scordo MG, Spina E, Romeo P, Dahl ML, Bertilsson L, Johansson I, et al. CYP2D6 genotype and antipsychotic-induced extrapyramidal side-effects in schizophrenic patients. *Eur J Clin Pharmacol* 2000, **56**, 679–83.

50 Brockmoller J, Kirchheiner J, Schmider J, Walter S, Sachse C, Muller-Oerlinghausen B, et al. The impact of the CYP2D6 polymorphism on haloperidol pharmacokinetics and on the outcome of haloperidol treatment. *Clin Pharmacol Ther* 2002, **72**, 438–52.

51 Dahl ML. Cytochrome p450 phenotyping/genotyping in patients receiving antipsychotics: useful aid to prescribing? *Clin Pharmacokinet* 2002, **41**, 453–70.

52 Lee JT, Kroemer HK, Silberstein DJ, Funck-Brentano C, Lineberry MD, Wood AJ, et al. The role of genetically determined polymorphic drug metabolism in the beta-blockade produced by propafenone. *N Engl J Med* 1990, **322**, 1764–8.

53 Morike K, Magadum S, Mettang T, Griese EU, Machleidt C, Kuhlmann U. Propafenone in a usual dose produces severe side-effects: the impact of genetically determined metabolic status on drug therapy. *J Intern Med* 1995, **238**, 469–72.

54 Mikus G, Gross AS, Beckmann J, Hertrampf R, Gundert-Remy U,

Eichelbaum M. The influence of the sparteine/debrisoquin phenotype on the disposition of flecainide. *Clin Pharmacol Ther* 1989, **45**, 562–7.

55 Funck-Brentano C, Becquemont L, Kroemer HK, Buhl K, Knebel NG, Eichelbaum M, et al. Variable disposition kinetics and electrocardiographic effects of flecainide during repeated dosing in humans: contribution of genetic factors, dose-dependent clearance, and interaction with amiodarone. *Clin Pharmacol Ther* 1994, **55**, 256–69.

56 Oldham HG, Clarke SE. In vitro identification of the human cytochrome P450 enzymes involved in the metabolism of *R*(+)- and *S*(−)-carvedilol. *Drug Metab Dispos* 1997, **25**, 970–7.

57 Pepper JM, Lennard MS, Tucker GT, Woods HF. Effect of steroids on the cytochrome P4502D6-catalysed metabolism of metoprolol. *Pharmacogenetics* 1991, **1**, 119–22.

58 Rau T, Heide R, Bergmann K, Wuttke H, Werner U, Feifel N, et al. Effect of the CYP2D6 genotype on metoprolol metabolism persists during long-term treatment. *Pharmacogenetics* 2002, **12**, 465–72.

59 Wuttke H, Rau T, Heide R, Bergmann K, Bohm M, Weil J, et al. Increased frequency of cytochrome P450 2D6 poor metabolizers among patients with metoprolol-associated adverse effects. *Clin Pharmacol Ther* 2002, **72**, 429–37.

60 Kaiser R, Sezer O, Papies A, Bauer S, Schelenz C, Tremblay PB, et al. Patient-tailored antiemetic treatment with 5-hydroxytryptamine type 3 receptor antagonists according to cytochrome P-450 2D6 genotypes. *J Clin Oncol* 2002, **20**, 2805–11.

61 Hustert E, Haberl M, Burk O, Wolbold R, He YQ, Klein K, et al. The genetic determinants of the CYP3A5 polymorphism. *Pharmacogenetics* 2001, **11**, 773–9.

62 Kuehl P, Zhang J, Lin Y, Lamba J, Assem M, Schuetz J, et al. Sequence diversity in CYP3A promoters and characterization of the genetic basis of polymorphic CYP3A5 expression. *Nat Genet* 2001, **27**, 383–91.

63 Eiselt R, Domanski TL, Zibat A, Mueller R, Presecan-Siedel E, Hustert E, et al. Identification and functional characterization of eight CYP3A4 protein variants. *Pharmacogenetics* 2001, **11**, 447–58.

64 Hustert E, Zibat A, Presecan-Siedel E, Eiselt R, Mueller R, Fuss C, et al. Natural protein variants of pregnane X receptor with altered transactivation activity toward CYP3A4. *Drug Metab Dispos* 2001, **29**, 1454–9.

65 Mitchell R, Bell J. Clinical implications of isoniazid PAS and streptomycin blood levels in pulmonary tuberculosis. *Trans Am Clin Ass* 1957, **69**, 98–105.

66 Grant DM, Möricke K, Eichelbaum M, Meyer UA. Acetylation pharmacogenetics. The slow acetylator phenotype is caused by decreased or absent arylamine *N*-acetyltransferase in human liver. *J Clin Invest* 1990, **85**, 968–72.

67 Evans DAP. *N*-Acetyltransferases. In: Kalow W, ed., *Pharmacogenetics of Drug Metabolism*. Pergamon Press, New York, 1992, 95–197.

68 Lin HJ, Han CY, Lin BK, Hardy S. Slow acetylator mutations in the human polymorphic *N*-acetyltransferase gene in 786 Asians, blacks, Hispanics, and whites application to metabolic epidemiology. *Am J Hum Genet* 1993, **52**, 827–34.

69 Cascorbi I, Drakoulis N, Brockmöller J, Maurer A, Sperling K, Roots I. Arylamine *N*-acetyltransferase (*NAT2*) mutations and their allelic linkage in unrelated Caucasian individuals: correlation with phenotypic activity. *Am J Hum Genet* 1995, **57**, 581–92.

70 Ilett KF, Chiswell GM, Spargo RM, Platt E, Minchin RF. Acetylation phenotype and genotype in Aboriginal leprosy patients from the north-west region of Western Australia. *Pharmacogenetics* 1993, **3**, 264–9.

71 Yamamoto M, Sobue G, Mukoyama

M, Matsuoka Y, Mitsuma T. Demonstration of slow acetylator genotype of *N*-acetyltransferase in isoniazid neuropathy using an archival hematoxylin and eosin section of a sural nerve biopsy specimen. *J Neurol Sci* 1996, **135**, 51–4.

72 Reidenberg MM, Levy M, Drayer DE, Zylber-Katz E, Robbins WC. Acetylator phenotype in idiopathic systemic lupus erythematosus. *Arthritis Rheum* 1980, **23**, 569–73.

73 Wolkenstein P, Carrière V, Charue D, Bastuji-Garin S, Revuz J, Roujeau J-Ceal. A slow acetylator genotype is a risk factor for sulphonamide-induced toxic epidermal necrolysis and Stevens–Johnson syndrome. *Pharmacogenetics* 1995, **5**, 255–8.

74 Zschieschang P, Hiepe F, Gromnica-Ihle E, Roots I, Cascorbi I. Lack of association between arylamine *N*-acetyltransferase 2 (NAT2) polymorphism and systemic lupus erythematosus. *Pharmacogenetics* 2002, **12**, 559–63.

75 Krynetski EY, Evans WE. Genetic polymorphism of thiopurine S-methyltransferase: molecular mechanisms and clinical importance. *Pharmacology* 2000, **61**, 136–46.

76 Weinshilboum RM. Methylation pharmacogenetics: thiopurine methyltransferase as a model system. *Xenobiotica* 1992, **22**, 1055–71.

77 Schaeffeler E, Lang T, Zanger UM, Eichelbaum M, Schwab M. High-throughput genotyping of thiopurine S-methyltransferase by denaturing HPLC. *Clin Chem* 2001, **47**, 548–55.

78 Schinkel AH, Wagenaar E, Mol CA, van Deemter L. P-glycoprotein in the blood–brain barrier of mice influences the brain penetration and pharmacological activity of many drugs. *J Clin Invest* 1996, **97**, 2517–24.

79 Terao T, Hisanaga E, Sai Y, Tamai I, Tsuji A. Active secretion of drugs from the small intestinal epithelium in rats by P-glycoprotein functioning as an absorption barrier. *J Pharm Pharmacol* 1996, **48**, 1083–9.

80 Zhang Y, Benet LZ. The gut as a barrier to drug absorption: combined role of cytochrome P450 3A and P-glycoprotein. *Clin Pharmacokinet* 2001, **40**, 159–68.

81 Fromm MF. P-glycoprotein: a defense mechanism limiting oral bioavailability and CNS accumulation of drugs. *Int J Clin Pharmacol Ther* 2000, **38**, 69–74.

82 Gottesman MM, Hrycyna CA, Schoenlein PV, Germann UA, Pastan I. Genetic analysis of the multidrug transporter. *Annu Rev Genet* 1995, **29**, 607–49.

83 Fromm MF, Kauffmann HM, Fritz P, Burk O, Kroemer HK, Warzok RW, et al. The effect of rifampin treatment on intestinal expression of human MRP transporters. *Am J Pathol* 2000, **157**, 1575–80.

84 Greiner B, Eichelbaum M, Fritz P, Kreichgauer HP, von Richter O, Zundler J, et al. The role of intestinal P-glycoprotein in the interaction of digoxin and rifampin. *J Clin Invest* 1999, **104**, 147–53.

85 Geick A, Eichelbaum M, Burk O. Nuclear receptor response elements mediate induction of intestinal MDR1 by rifampin. *J Biol Chem* 2001, **276**, 14581–7.

86 Johne A, Brockmoller J, Bauer S, Maurer A, Langheinrich M, Roots I. Pharmacokinetic interaction of digoxin with an herbal extract from St John's wort (*Hypericum perforatum*). *Clin Pharmacol Ther* 1999, **66**, 338–45.

87 Pauli-Magnus C, von Richter O, Burk O, Ziegler A, Mettang T, Eichelbaum M, et al. Characterization of the major metabolites of verapamil as substrates and inhibitors of P-glycoprotein. *J Pharmacol Exp Ther* 2000, **293**, 376–82.

88 Schuetz EG, Furuya KN, Schuetz JD. Interindividual variation in expression of P-glycoprotein in normal human liver and secondary hepatic neoplasms. *J Pharmacol Exp Ther* 1995, **275**, 1011–8.

89 Mickley LA, Lee JS, Weng Z, Zhan Z, Alvarez M, Wilson W, et al. Genetic polymorphism in MDR-1: a tool for examining allelic expression in

normal cells, unselected and drug-selected cell lines, and human tumors. *Blood* 1998, **91**, 1749–56.

90 Kimchi-Sarfaty C, Gribar JJ, Gottesman MM. Functional characterization of coding polymorphisms in the human MDR1 gene using a vaccinia virus expression system. *Mol Pharmacol* 2002, **62**, 1–6.

91 Hoffmeyer S, Burk O, von Richter O, Arnold HP, Brockmoller J, Johne A, et al. Functional polymorphisms of the human multidrug-resistance gene: multiple sequence variations and correlation of one allele with P-glycoprotein expression and activity *in vivo*. *Proc Natl Acad Sci USA* 2000, **97**, 3473–8.

92 Cascorbi I, Gerloff T, Johne A, Meisel C, Hoffmeyer S, Schwab M, et al. Frequency of single nucleotide polymorphisms in the P-glycoprotein drug transporter MDR1 gene in white subjects. *Clin Pharmacol Ther* 2001, **69**, 169–74.

93 Ameyaw MM, Regateiro F, Li T, Liu X, Tariq M, Mobarek A, et al. MDR1 pharmacogenetics: frequency of the C3435T mutation in exon 26 is significantly influenced by ethnicity. *Pharmacogenetics* 2001, **11**, 217–21.

94 Kim RB, Leake BF, Choo EF, Dresser GK, Kubba SV, Schwarz UI, et al. Identification of functionally variant MDR1 alleles among European Americans and African Americans. *Clin Pharmacol Ther* 2001, **70**, 189–99.

95 Schaeffeler E, Eichelbaum M, Brinkmann U, Penger A, Asante-Poku S, Zanger UM, et al. Frequency of C3435T polymorphism of MDR1 gene in African people. *Lancet* 2001, **358**, 383–4.

96 Hitzl M, Drescher S, van der KH, Schaffeler E, Fischer J, Schwab M, et al. The C3435T mutation in the human MDR1 gene is associated with altered efflux of the P-glycoprotein substrate rhodamine 123 from CD56^{+} natural killer cells. *Pharmacogenetics* 2001, **11**, 293–8.

97 Johne A, Kopke K, Gerloff T, Mai I, Rietbrock S, Meisel C, et al. Modulation of steady-state kinetics of digoxin by haplotypes of the P-glycoprotein MDR1 gene. *Clin Pharmacol Ther* 2002, **72**, 584–94.

98 Verstuyft C, Schwab M, Schaeffeler E, Kerb R, Brinkmann U, Jaillon P, et al. Digoxin pharmacokinetics and MDR1 genetic polymorphisms. *Eur J Clin Pharmacol* 2003, **58**, 809–12.

99 Farrell RJ, Murphy A, Long A, Donnelly S, Cherikuri A, O'Toole D, et al. High multidrug resistance (P-glycoprotein 170) expression in inflammatory bowel disease patients who fail medical therapy. *Gastroenterology* 2000, **118**, 279–88.

100 Llorente L, Richaud-Patin Y, Diaz-Borjon A, Alvarado de la Barrera C, Jakez-Ocampo J, De La FH, et al. Multidrug resistance-1 (MDR-1) in rheumatic autoimmune disorders. Part I: increased P-glycoprotein activity in lymphocytes from rheumatoid arthritis patients might influence disease outcome. *Joint Bone Spine* 2000, **67**, 30–9.

101 Schwab M, Schaeffeler E, Marx C, Fromm MF, Kaskas B, Metzler J, et al. Association between the C3435T MDR1 gene polymorphism and susceptibility for ulcerative colitis. *Gastroenterology* 2003, **124**, 26–33.

102 von Ahsen N, Richter M, Grupp C, Ringe B, Oellerich M, Armstrong VW. No influence of the MDR-1 C3435T polymorphism or a CYP3A4 promoter polymorphism (CYP3A4-V allele) on dose-adjusted cyclosporin A trough concentrations or rejection incidence in stable renal transplant recipients. *Clin Chem* 2001, **47**, 1048–52.

103 Fjallskog ML, Frii L, Bergh J. Paclitaxel-induced cytotoxicity – the effects of cremophor EL (castor oil) on two human breast cancer cell lines with acquired multidrug resistant phenotype and induced expression of the permeability glycoprotein. *Eur J Cancer* 1994, **30A**, 687–90.

104 Meerum Terwogt JM, Malingre MM, Beijnen JH, ten Bokkel Huinink WW, Rosing H, Koopman FJ, et al. Coadministration of oral

cyclosporin A enables oral therapy with paclitaxel. *Clin Cancer Res* 1999, **5**, 3379–84.

105 Tanabe M, Ieiri I, Nagata N, Inoue K, Ito S, Kanamori Y, et al. Expression of P-glycoprotein in human placenta: relation to genetic polymorphism of the multidrug resistance (MDR)-1 gene. *J Pharmacol Exp Ther* 2001, **297**, 1137–43.

106 Moriya Y, Nakamura T, Horinouchi M, Sakaeda T, Tamura T, Aoyama N, et al. Effects of polymorphisms of MDR1, MRP1, and MRP2 genes on their mRNA expression levels in duodenal enterocytes of healthy Japanese subjects. *Biol Pharm Bull* 2002, **25**, 1356–9.

107 Sakaeda T, Nakamura T, Horinouchi M, Kakumoto M, Ohmoto N, Sakai T, et al. MDR1 genotype-related pharmacokinetics of digoxin after single oral administration in healthy Japanese subjects. *Pharm Res* 2001, **18**, 1400–4.

108 Siegmund W, Ludwig K, Giessmann T, Dazert P, Schroeder E, Sperker B, et al. The effects of the human MDR1 genotype on the expression of duodenal P-glycoprotein and disposition of the probe drug talinolol. *Clin Pharmacol Ther* 2002, **72**, 572–83.

109 Brockmoller J, Kirchheiner J, Meisel C, Roots I. Pharmacogenetic diagnostics of cytochrome P450 polymorphisms in clinical drug development and in drug treatment. *Pharmacogenomics* 2000, **1**, 125–51.

110 Kirchheiner J, Brosen K, Dahl ML, Gram LF, Kasper S, Roots I, et al. CYP2D6 and CYP2C19 genotype-based dose recommendations for antidepressants: a first step towards subpopulation-specific dosages. *Acta Psychiatr Scand* 2001, **104**, 173–92.

111 Ingelman-Sundberg M. Pharmacogenetics: an opportunity for a safer and more efficient pharmacotherapy. *J Intern Med* 2001, **250**, 186–200.

112 Lang T, Klein K, Fischer J, Nussler AK, Neuhaus P, Hofmann U, et al. Extensive genetic polymorphism in the human CYP2B6 gene with impact on expression and function in human liver. *Pharmacogenetics* 2001, **11**, 399–415.

14
Toxicogenomics: Integration of New Molecular Biological Tools in Toxicology

Wilbert H. M. Heijne, Ben van Ommen and Rob H. Stierum

14.1
Developments in Toxicology

Toxic effects on an organism induced by substances in the environment have been studied for many years. Initially, toxicologists considered the morphology and physiology of the organism, including lethal dose determination, body and organ weight, and gross pathology. The development of histopathological techniques enabled the microscopic examination of tissues and the determination of toxic effects at the cellular level. In the 1980s, molecular techniques became available for sensitive and early identification of specific molecular endpoints, and for discrimination between different types of toxicity. This resulted in a shortening of the time from administration to detection of effects and, thus, animal exposure. Changes in levels of particular proteins or metabolites in tissue, blood or urine that correlate well with certain types of toxicity are now routinely assessed. Most of the conventional clinical chemistry parameters were established empirically. They are the result of aberrant cellular processes, rather than components of the mechanism of toxicity.

For instance, leakage of enzymes through disrupted membranes is clearly a secondary process in a late stage of cell death. The single-biomolecule-based endpoints can only be applied if a preconceived notion exists on the possible mode of action of the agent under study. Hence, these tools primarily contribute to the confirmation of hypotheses. More specific markers of toxicity are needed, and might be found in the changes in the levels of gene expression, proteins and metabolites that determine the cellular response to an insult by the cell's environment. As a result of major genome sequence elucidation efforts, functional genomics technologies have emerged that provide toxicologists with new possibilities to study toxicity in an organism at the molecular level in a cell-wide approach. In theory, all cellular biomolecules can now be measured at the same time. Toxicogenomics, therefore, is defined as the application of functional genomics technologies in toxicology. In this definition, toxicogenomics integrates functional genomics with conventional toxicological methods. Figure 14.1 introduces the development of toxicology, up to the era of toxicogenomics.

Molecular Biology in Medicinal Chemistry. Edited by Th. Dingermann, D. Steinhilber, G. Folkers

ISBN: 3-527-30431-2

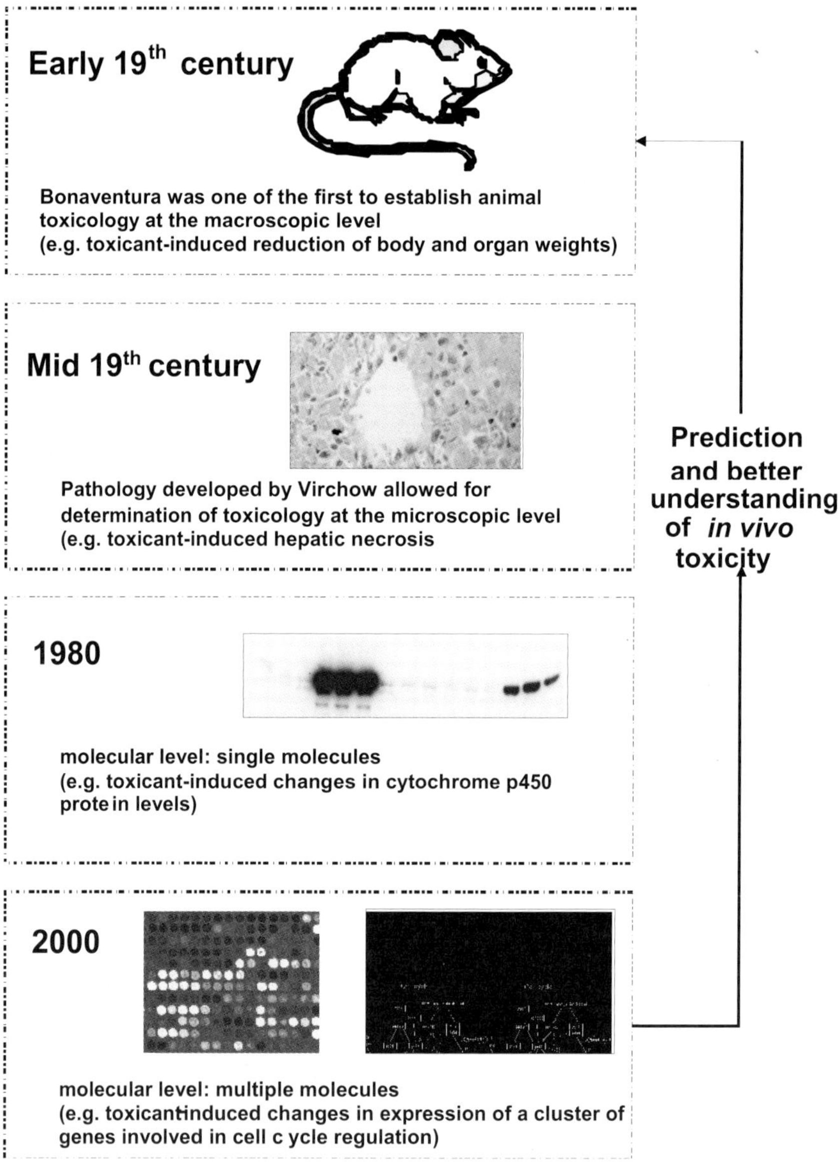

Fig. 14.1 Evolutions in toxicological sciences.

In this chapter, first, we will review the different functional genomics technologies and, subsequently, discuss applications in toxicological research. Toxicogenomics has the potential to improve the prediction of the effects of genome composition on toxicological outcome, provide a better understanding of mechanisms of toxicity, facilitate the prediction of toxicity of unknown compounds, yield mechanism-based markers of toxicity, improve interspecies and *in vitro–in vivo* extrapolations, and facilitate the toxicological subdiscipline dealing with chemical mixtures.

14.2
Functional Genomics: New Molecular Biological Tools

Functional genomics technologies measure molecular responses in an organism in a cell-wide manner. The available information of the genome sequence, and of cellular protein and metabolite contents, is used to deduce functional roles in mechanisms and cellular pathways of the different biomolecules. Genomics is the area of research that studies the genome through analysis of the nucleotide sequence, the genome structure and its composition. Transcriptomics, proteomics and metabolomics determine the expression levels of gene transcripts, proteins or metabolites, respectively.

14.2.1
Genomics

A wealth of possibilities was offered to biology and related life sciences by human and mouse genome sequencing projects, as well as efforts to sequence the genomes of other organisms like rat and monkey. The nucleic acids store the genetic information that is the basic information needed for the production of the cellular proteins. Understanding this basic genetic information is a major step towards the elucidation of molecular mechanisms involving those proteins. Moreover, information on the chromosomal organization of genes and transcriptional regulatory elements contributes to the understanding of the function of genes and their roles in specific processes. The comparison of complete genomes from different organisms enables the identification of the molecular origin that determines commonly shared as well as unique cellular mechanisms and physiological properties of different species.

Another level within genomics research is the identification of small modifications in individual genomes, single nucleotide polymorphisms (SNPs), which are responsible for many interindividual differences and for a range of genetic diseases. Interindividual differences that occur through SNPs in specific genes are of great importance in toxicological research. If the gene encoding a drug-metabolizing enzyme is slightly different, the catabolizing activity of the corresponding enzyme could be altered. The rate of activation and metabolism of a xenobiotic and the mechanisms of protection determine to what extent toxicity is found in an organism.

14.2.2
Transcriptomics

Many, but not all, cellular processes are controlled at the level of gene expression. Determination of gene expression by measurement of the mRNA levels has been found to be valuable in the prediction of protein synthesis and protein activity (the total enzyme activity towards a substrate), at least for some classes of enzymes. Traditionally, mRNA levels were determined using Northern blots with radio-labeled nucleotides. The extremely sensitive, although only qualitative, detection of a specific mRNA was possible with the development of the reverse transcription polymerase chain reaction (RT-PCR). An adapted method of this principle (quantitative RT-PCR) is also used to quantitate the change in the amount of mRNA in comparison to a constitutively expressed ("housekeeping") gene. As these techniques are very labor-intensive and not suitable for scale-up, the determination of expression levels of many genes simultaneously is not feasible. Serial analysis of gene expression (SAGE) is a scaled-up method that works by sequence determination of hundreds of mRNA fragments followed by counting the number of molecules of a specific mRNA present in a random population of mRNAs isolated from a cell. Recently, cDNA- and oligonucleotide-based microarrays were developed as a large-scale method for gene expression measurement which takes advantage of the availability of collections (libraries) of gene fragments with a known sequence and often with annotation on their (putative) function in the cell. In addition, microarrays can be generated from expressed sequence tag (EST) collections – cDNAs derived from mRNA molecules for which frequently a function has yet to be determined. As in Northern blots, the capacity of single-strand DNA and RNA to specifically hybridize to the complementary strand is used to determine mRNA levels of the gene of interest. However, instead of one gene at the time, single-strand cDNA molecules for thousands of different genes are deposited in a fixed spot on a surface (e.g. glass slide, plastic, nylon membrane), called a cDNA microarray or DNA chip. By hybridization of the cDNA micorarray with a pool of isolated mRNAs, each specific mRNA will only hybridize with the cDNA in the spot containing the complementary cDNA for this specific gene. The amount of cDNA hybridized in each spot can be detected by measurement of fluorescence that was incorporated in the sample. An overview of the process of cDNA microarray-based mRNA level measurements is shown in Fig. 14.2. In practice, two samples are always hybridized together on the cDNA microarray, where one test sample is labeled with one type of fluorophore (e.g. green fluorescence) and one reference or control sample is labeled with another type (e.g. red fluorescence). Quantification of both types of fluorescence enables the determination of a ratio of expression for each gene in the test sample with respect to the control sample. Whereas the large part of the thousands of genes measured will not show changes in expression levels when diseased or treated samples are compared to controls, the genes that are found to be induced or repressed provide a wealth of information on cellular mechanisms that are affected at the gene expression level by the disease or treatment. For more in-depth reading on cDNA microarrays technology, we refer to [1].

14.2.3 Proteomics

Not all changes in cellular mechanisms can be measured at the mRNA level. Rapid modification or subcellular redistribution of proteins already present in the cell

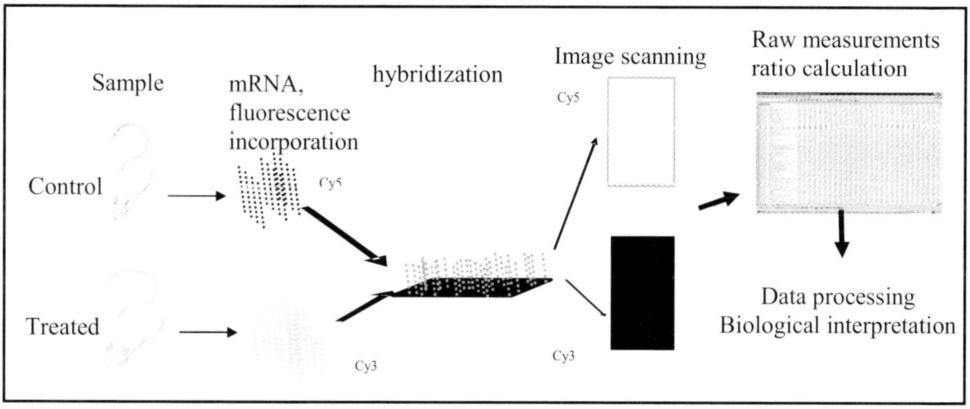

Fig. 14.2 cDNA microarray-based transcriptomics measurements. mRNA extracted from tissue samples is fluorescently labeled and hybridized to a cDNA microarray containing thousands of different cDNA molecules in fixed positions. Scanning the microarray results in two images that are quantified and used to calculate relative gene expression in one tissue with respect to a control tissue.

might be required, especially in cell-protective mechanisms of response. Technologies for protein analysis may better visualize processes that do not involve active biosynthesis or at least they will be complementary to gene expression analysis. The term "proteome" is used to denote the complete population of expressed proteins in a cell. In analogy to transcriptomics, proteomics technologies are applied for the simultaneous measurement of the thousands of proteins in a cell. As proteins predominantly act in the cellular reactions rather than the gene transcripts, proteomics will prove to be more important than measurement of the transcriptome. The measurement of the proteome is more complex than transcriptomics, where the total mRNA population can be isolated at once and, relatively easily, the thousands of transcripts that only differ in nucleotide sequence are sorted in the process of cDNA microarray hybridization. In contrast, the separation of all different proteins from a cell is a very complicated task, as all the proteins have different properties (mass, isoelectric point (p*I*), solubility, stability, etc.). Posttranslational modifications and the capability to form protein complexes cannot be ignored in proteomics. Various methods have been developed in proteomics research. The first generation of proteomics methodologies combines the relatively old technique of two-dimensional agarose gel electrophoresis with the developments of powerful automated image analysis software, advances in mass spectrometry and Internet-based global information exchange. The total protein content of a sample is first separated on the basis of the p*I* of the proteins, using a strip with an immobilized pH gradient. Different pH gradients can be chosen in order to obtain maximal separation in the area of interest. After isoelectric focusing, proteins are transferred to a polyacrylamide gel and, based on protein mass, separation is effected by standard gel electrophoresis. Separated proteins are then visualized using fluorescent or silver staining. Gels are scanned, images are analyzed with dedicated software and spot volumes are quantified. As an example, characteristic gel images are shown of protein patterns obtained from liver of controls and rats after exposure to the hepatotoxicant bromobenzene (Fig. 14.3). Subsequently, spots of interest can be isolated from the gel, purified and analyzed by means of mass spectrometry. Proteins can be identified after a specific fragmentation (e.g. digestion with trypsin), which generates peptide fragments with a very specific mass determined by matrix assisted laser desorption ionization time-of flight mass spectrometry (MALDI-TOF-MS). The fragmentation pattern is then used to identify what protein matches this pattern in a database with predicted fragment patterns of all known protein sequences. Also, peptide sequencing can be conducted by MS/MS techniques to further confirm protein identities. Without actual isolation from the gel, protein spots can be putatively identified based on a match with a previously identified protein in the exact same position on a reference gel [3].

Another proteomics method under development is applying protein microarrays in a manner comparable to the cDNA microarrays in transcriptomics. Proteins or antibodies against proteins are spotted on a solid surface and used to specifically bind labeled proteins from a sample. A major drawback in this approach in comparison to cDNA microarrays is the variation in binding affinity among proteins. Optimized conditions for binding are only applicable for part of the proteins and suboptimal binding will result in biased results. Recently, methods have been developed that apply these differences in binding affinity to sort proteins accordingly. This enables the separation of proteins based on a few characteristic properties only, but does not separate all proteins from the mixture isolated from a cell. Different surface chemistries are used to selectively absorb proteins on arrays which then are directly analyzed by mass spectrometry. Examples are ICAT®, which gives a differential display of two protein samples using liquid chromatography-MS/MS with isotope-encoded affinity tags, and SELDI (surface-enhanced laser desorption/ionization) [4].

14.2.4 Metabolomics

The occurrence of cellular processes is often reflected in the metabolite levels, which can be measured in the cell and in extracellular fluids. Biochemical reactions in the cell can be studied while, and after, taking place by monitoring the (dis)appearance of reaction products. The dynamics of cellular processes can be monitored using metabolite profiling. Processes like metabolism and biotransformation of xenobiotics can be followed, but also clearance of metabolites can be determined in the urine. Another major advantage is that noninvasive methods can be used to collect samples such as blood (plasma), urine or other body fluids. This largely increases the applicability in both human and animal experiments. More recently, nuclear magnetic resonance (NMR) analysis was also performed on tissue and cell material. Mass spectrometric techniques (gas chromatography/MS) allow the measurement of low concentrations of individual components. For global

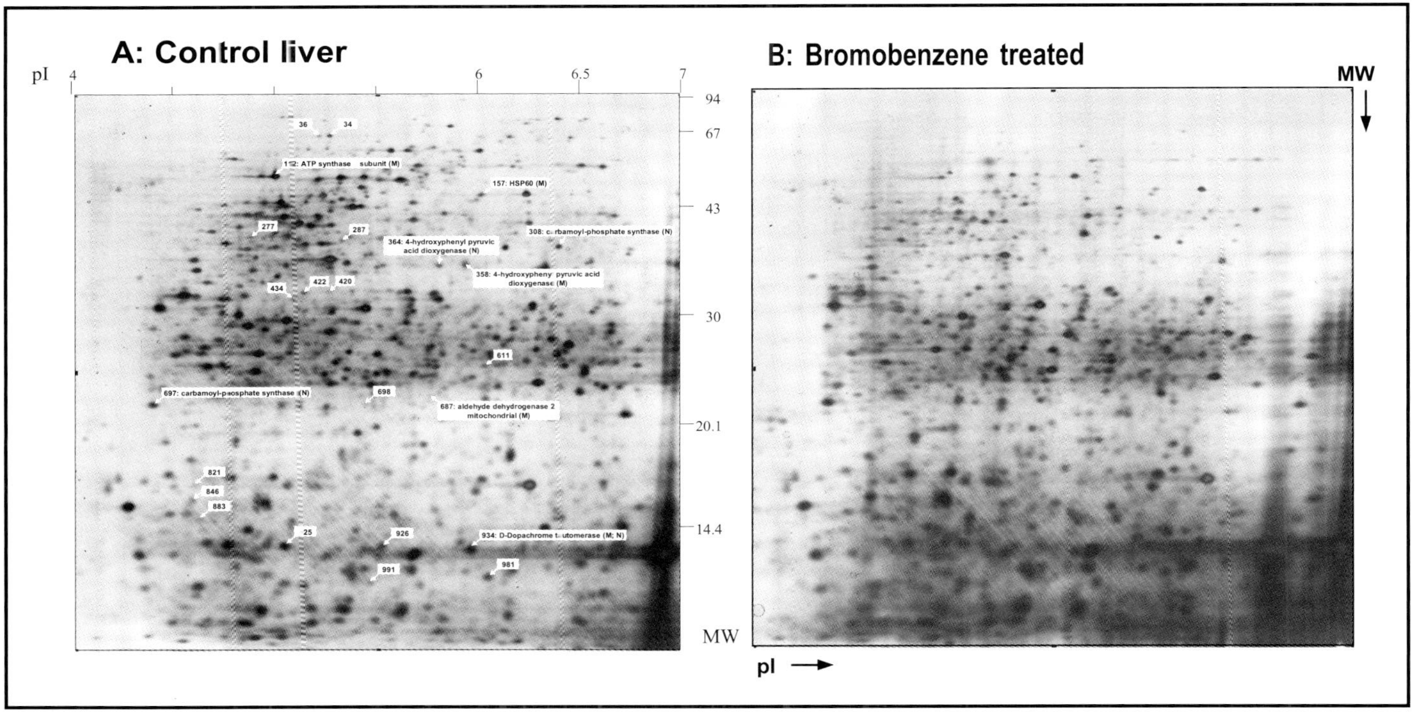

Fig. 14.3 Two-dimensional gel electrophoresis images of proteomics measurements. The p-otein patterns of rat liver without (A) and with (B) bromobenzene treatment are shown.

screening, ^{1}H-NMR spectroscopy is an attractive approach, as a wide range of metabolites can be quantified at the same time without extensive sample preparation. A spectrum is obtained with resonance peaks characteristic for all small biomolecules. In this way a metabolic fingerprint is obtained characterizing the biological fluid under study. Individual signals can be quantified and identified in the spectrum based on available reference spectra. NMR spectra of biological fluids are very complex due to the mixture of numerous metabolites present in these fluids. Variations between samples are often too small to be recognized by eye. In order to increase the comparability of NMR spectra and thereby maximize the power of the subsequent data analysis, a partial linear fit algorithm adjusts minor shifts in the spectra while maintaining the resolution. To find significant differences, multivariate data analysis is needed to explore recurrent patterns in a number of NMR spectra. A factor spectrum is used to identify the metabolite NMR peaks that differ in urine upon a treatment compared to the control (Fig. 14.4). Correlations between variables in the complex and large data sets (thousands of signals in spectra) are related to a target variable such as toxicity status. The combination of MS profiling with multivariate data analysis provides a powerful fingerprinting methodology. A combination of analytical techniques is desirable for an exhaustive analysis of a complex mixture of metabolites [5–8].

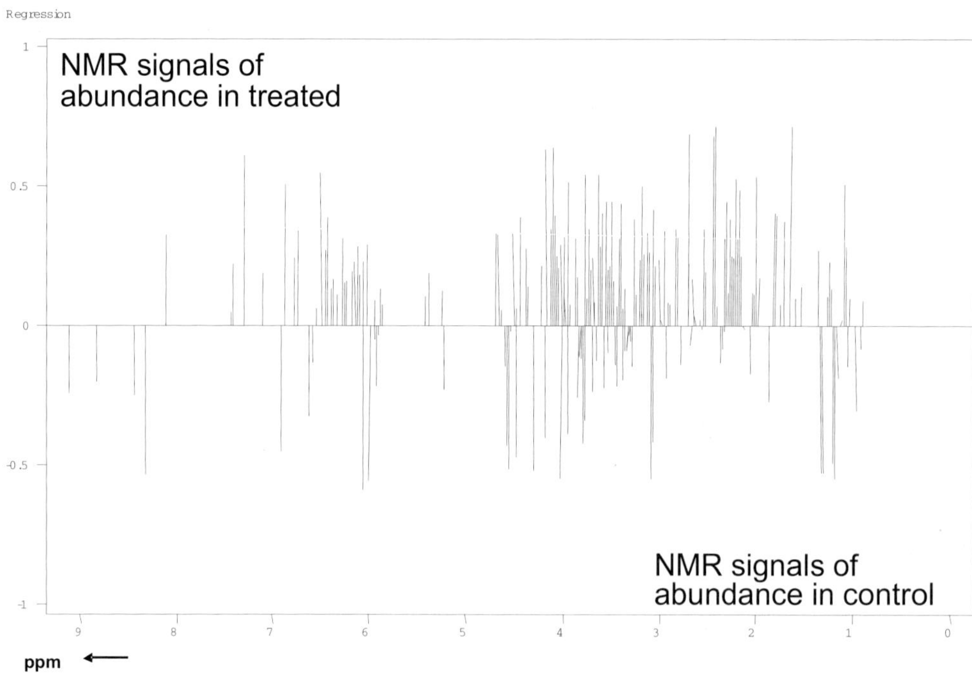

Fig. 14.4 Metabolic fingerprint. This factor spectrum visualizes the differences in the NMR patterns of rat urine with and without toxicant treatment. The peaks represent metabolites that differ in contents between the two samples.

14.2.5 Data Processing and Bioinformatics

The handling of data from large-scale experiments requires powerful data processing equipment and algorithms. Laboratory information that describes the experimental conditions should be automatically stored (e.g. the Laboratory Information Management System (LIMS)) and organized in a searchable manner. Experimental results should be linked to the experimental information and stored in an organized way using a computer application called the data warehouse. Data processing includes corrections for (mostly technical) factors that influence the data, but are not the conditions under study. Filtering out of the unreliable measurements is essential to improve the quality of the data set. Quantification of the signal-to-noise ratios and application of a lower boundary should exclude weak signals from further interpretation. Specific biases introduced in the data set, e.g. due to efficiency of incorporated fluorescence, are corrected using normalization procedures. The data processing heavily depends on assumptions of both technical and biological origin. In transcriptomics, it is frequently assumed that most of the thousands of genes do not change in expression levels upon (modest) changes in experimental conditions. Standardization, normalization or scaling are methods that are applied to transform data sets in order to be able to compare the measurements amongst each other and with measurements from external sources. Systematic bias originating from various technological sources should be corrected. After raw data processing, secondary results files are obtained that should be stored along with the description of the method of data analysis.

A new challenge for biologists is to turn large data sets with high amounts of noise and without obvious biological meaning into relevant findings, which includes selection of subsets out of the thousands of genes, proteins or metabolites that have biologically relevant characteristics. Those subsets can be subjected to elaborate further research. Mathematical techniques are applied to cluster the data into interpretable groups or structured patterns. Methods applied for this purpose, clustering algorithms like hierarchical clustering (Fig. 14.5), *K*-means clustering and self-organizing maps, calculate a measure of similarity between the expression profiles of the genes. Clusters are sets of genes that behave more similar to each other than to genes outside of the cluster. The number of clusters formed can be imposed upon the data set or can be determined by the clustering algorithms automatically. Once smaller subsets have been found that account for most of the changes introduced by the environmental conditions, the molecules in these subsets are further analyzed with respect to biological relevance. Unsupervised methods such as principal component analysis (PCA) determine intrinsic structure within data sets, without prior knowledge. With such methods, a direct comparison of data sets is made and subsets of data are formed solely on the basis of similarities of NMR spectra. Supervised methods such as partial least squares and principal component discriminants analysis are more powerful tools, which use additional information on the data set such as biochemical, histopathological or clinical data to identify differences between predefined groups. For further

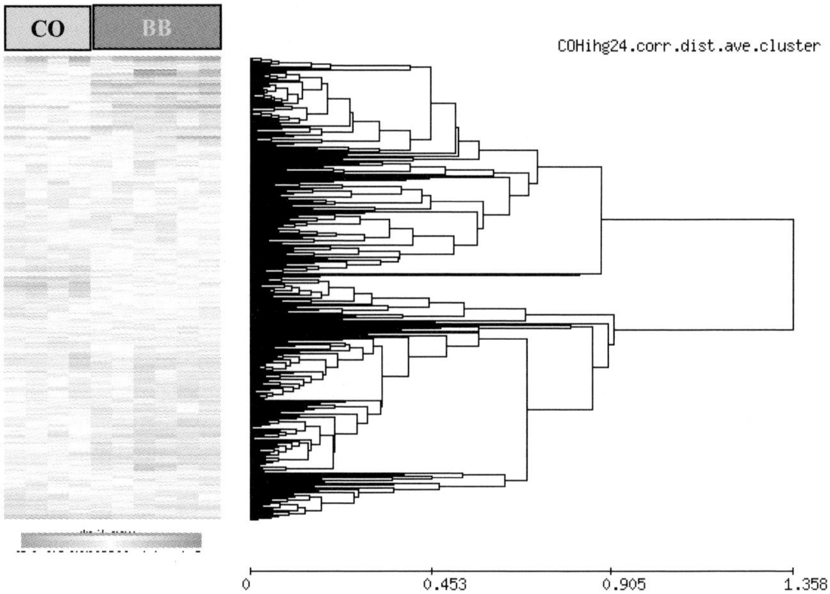

Fig. 14.5 Dendrogram of hierarchically clustered transcriptomics data of livers from untreated, corn oil control (CO)- and high bromobenzene (BB)-treated rats. Genes are ranked horizontally, while columns each represent a different sample. Grey intensities correspond with increased or decreased gene expression levels.

reading on different methods of functional genomics data analysis, we refer to [9, 10].

14.2.6
Biological Interpretation

It is not feasible to analyze the expression data one-by-one, on a single gene or protein basis. Moreover, doing so would result in a great loss of the information that resides in the coherence of the data collected in one study. The relationships between genes or proteins expressed in a certain situation regarding time, localization and experimental conditions are the most valuable information obtainable in functional genomics studies. Measurement of levels of gene expression, proteins or metabolites *per se* are not sufficient to gain full biological insights in the cell. Studying the interaction and integration of the different biological entities is of crucial importance for the understanding of biology. By mapping the interactions within a cell, a complex, dynamic model of pathways can eventually be constructed. Observed changes at the cellular/molecular level are biologically relevant only if reflected in the physiology of the organism. Functionality is assumed upon interpretation of changes in expression; however, in the end, this has to be

Tab. 14.1 Internet references

Swissprot	http://us.expasy.org/
European Bioinformatics Institute (EBI)	http://www.ebi.ac.uk/
Expression Profiler at the EBI	http://ep.ebi.ac.uk/EP/EPCLUST
NCBI's UniGene, GenBank, OMIM, etc.	http://www.ncbi.nlm.nih.gov/
GeneCards	http://bioinfo.weizmann.ac.il/cards/
KEGG pathway maps	http://www.genome.ad.jp/kegg/
Biocarta pathway maps	http://www.biocarta.com/genes/index.asp

determined at the physiological level. By looking at the (physiological) effects of specific genetic diseases, researchers can gain insight in the functionality of the particular genes that are affected. For instance, mutation of one gene may have effects on components downstream in a cellular mechanism regulated by this mutated gene. An approach is the examination of gene functions one-by-one through deletion of a gene, creating so-called knockout organism strains. A vast range of aspects of physiology and behavior are monitored without *a priori* hypotheses. Whether knowing the function of individual genes will be sufficient to understand complete mechanisms involving those genes is questionable.

The vast expansion of the world-wide web proved to be crucial for the rapid developments in functional genomics and bioinformatics. Sharing of biological data and information on a world-wide scale and in accessible formats enables the linking of previously scattered and isolated knowledge. Relying and building on the public release of the complete DNA sequence of the human genome, many sources of biological information on genes and proteins and interaction with the environment have become available, such as through Swissprot, NCBI's UniGene, GenBank, OMIM, Homologene, GeneCards, KEGG pathway maps, etc. (Tab. 14.1). Furthermore, the development of methods for literature data mining holds great promise, since innumerable amounts of data have been gathered in the past and reside passively in literature, awaiting "excavation".

Functional genomics technologies are designed as large-scale, high-throughput technologies. As a consequence, a vast amount of data is obtained, although with a typically somewhat lower level of confidence from a quantitative point of view, compared to single gene or protein measurements. Nevertheless, the technologies have already proven to be very useful for the holistic monitoring of effects, the discovery of unexpected results and the generation of hypotheses on mechanisms of action. However, major findings on changes in specific biomolecules or processes should be confirmed and further investigated with dedicated methods. This also requires the involvement of field-specific expert researchers, expertise and equipment. Techniques such as quantitative RT-PCR and Northern blotting are useful for technical validation, but, like cDNA microarrays, determine mRNA levels which have only predictive value. Western blots and immunological assays (ELISA) can be applied to confirm the presence and levels of proteins in the sample with more accuracy compared to two-dimensional gel-based proteomics technologies, at least when total protein levels are considered. A disadvantage of single gene or protein measurements is the loss of the information from circumstantial parameters, such as genes that have a relationship with the one studied. At best, levels of one or two additional reference proteins are measured under the assumption that these do not change under the conditions of study (so-called housekeeping proteins). However, genes and proteins once assumed to be present at a constant level in every condition in fact show variation in expression under certain conditions. The measurement of protein levels by itself does not provide evidence for actual involvement of the proteins in cellular mechanisms. Specific assays for enzyme activity provide an additional level of confirmation – the functional capabilities of the proteins (enzymes) in the sample. Moreover, metabolite levels provide information of biochemical reactions that have already taken place. The most biologically relevant confirmation of observed "omics" changes is to relate these changes to physiological effects. Only at this level do changes have relevant consequences for the organism. Changes in gene and protein expression or metabolite contents can only be designated as adverse effects after correlation and validation with physiological effects known to occur in toxicity.

14.3 Integration of Functional Genomics Technologies in Toxicology

Toxicity can be seen as the distortion of biological processes in an organism, organ or cell. Thousands of gene or protein expression and metabolite levels in a sample determine the specific status of that sample. Thus, by investigating the cellular/molecular mechanisms in a cell, functional genomics technologies can be of extreme value in toxicology as in other life sciences. Gene expression profiles, for instance, can be used to discriminate samples exposed to different classes of toxicants, predict toxicity of unknown compounds and study cellular mechanisms that lead to or result from toxicity [11, 12]. Indeed, transcriptomics was shown to be powerful in the mechanistic assessment of toxic responses [13]. In the following subsections, major applications of functional genomics technologies that have the potential to greatly improve toxicological sciences are discussed.

14.3.1 Mechanism Elucidation

The interactions of genes and proteins form the basis of the majority of the biological processes, and the coordinate expression of genes or proteins under specific circumstances provides an indication of the relationship between molecules in a biological mechanism. A mechanism of toxicity can thus be reconstructed at the molecular level through the observed changes in expression of genes that interact with, or influence, each other. As thousands of molecules are investigated simultaneously, the chance of identifying the target molecule that triggers a toxic effect upon interaction with the xenobiotic is greatly improved. This target molecule can be a cellular receptor that initiates an adverse effect in the cell upon binding of the (xenobiotic) ligand. Transcriptomics experiments may be designed to identify the organs or organelles in an organism that most likely form the target sites of adverse effects, without prior knowledge. Initial organism-wide assessment can help identify characteristic changes in gene expression that indicate which organs should be chosen for further toxicological examinations, and thus limit the amount of those elaborate studies.

14.3.2
Transcriptomics Fingerprinting and Prediction of Toxicity

A very advantageous application of toxicogenomics may be the recognition of toxic properties in the early stages of screening drug candidates. Drug discovery programs are developing new candidate active compounds at an extremely high rate, using technologies like combinatorial chemistry. Most of the thousands of newly synthesized compounds will never reach the market, while only a few will be selected as potential drugs, and those are evaluated in a time- and cost-consuming assessment of efficacy and toxicity. The screening of drug candidates for potential signs of toxicity may provide a useful criterion for early selection. The measurement of gene or protein expression profiles upon compound exposure can, like a fingerprint, be used to classify the compound according to similarity in profile to exposure upon known (model) toxicants. Large databases of these expression profiles enable the classification of the expression pattern from the sample of toxicological interest according to the toxic potency, type of mechanism, target organ, dose and time of exposure. A new compound can thus be classified as putatively toxic based on a transcriptional response common to known toxicants. This specific application was also termed toxicogenomics in a more narrow definition in some publications.

14.3.3
Identification of Early Markers of Toxicity

Since changes in expression of thousands of genes and proteins are measured, new specific and indicative marker genes or proteins that can detect or even predict certain types of toxicity are expected to be identified. However, to discern slightly different molecular mechanisms that lead to a similar toxicological outcome, single gene or protein toxicity markers will not be sufficient. More likely, subtly altered expression levels of many genes determine the mechanisms of toxicity, requiring precise and large-scale measurements of the pattern of expression of thousands of genes or proteins. A cell-wide pattern of gene or protein expression, in analogy to a fingerprint, can be used to discern a healthy cell from the different stages of distortion from this normal status. Using metabolite profiling has an additional advantage in toxicology and pharmacology since the (metabolites of) toxicants or drugs can be traced back in the plasma or urine to assess levels of exposure or confirm successful dosing.

14.3.4 Interspecies Extrapolation

Laboratory animals and humans, although very different in many aspects, are highly similar at the molecular level. The majority of the genes found in humans are also present in other organisms like rat or mouse. Some genes are identical in different species, while genomes of man and rodents exhibit more than 90% similarity. Using the molecular biological techniques of toxicogenomics, the similarity at the molecular level will be of great benefit for the extrapolation of results. While the physiological responses might be different, the underlying molecular mechanisms may be shared to a much higher extent between species.

14.3.5 Extrapolation from *In Vitro* Experiments to the *In Vivo* Situation

While physiological effects of toxicity might not be identified *in vitro*, the underlying molecular mechanisms eventually leading to these effects can be studied using these experiments. Depending on which *in vitro* model system is chosen, the molecular processes are more or less preserved. Additionally, the circumstances for performing experiments *in vitro* can be controlled and monitored in a more precise manner compared to *in vivo* testing.

14.3.6 Reduction of Use of Laboratory Animals

With the possibility to measure multiple parameters in the cell in great detail, the chance of observing effects of toxicity is greatly increased. Many replicate measurements might be required to significantly establish changes in a conventional, empirical marker of toxicity. It is expected that with the new molecular biology tools, effects do not have to be replicated as much if they are supported with an understanding of the cellular mechanisms. Additionally, toxicogenomics might help to identify effects in alternative models rather than in live animals. The recognition of toxicity at earlier time points relieves the necessity of labor- and cost-intensive long-term exposures. In particular, when assessing carcinogenic properties of compounds, a considerable amelioration is to be expected.

14.3.7 Combinatorial Toxicology

Harmful effects can be expected to be caused by more than one substance in complex mixtures, such as environmental pollutants. A proper analysis of the toxicity of these mixtures is currently hardly feasible. In order to assess combinatorial effects of compounds, large studies have to be designed including both exposures to individual chemicals as well as combinations. As most often empirical, descriptive markers of toxicity are monitored, discrimination between additive or synergistic effects in the different treatment groups is limited. Functional genomics technologies allow the monitoring of thousands of effects and increase the likelihood of finding different effects between exposures. Mechanisms of the combinatorial effects can be studied and/or predicted at the molecular level. When similar molecules or pathways prove to be the targets of different substances in the mixture, synergistic or additive effects might be expected. When changes are provoked in different biological pathways, direct interference would not be expected from the compounds in the mixture. However, because of the complexity of an organism, secondary effects could influence other mechanisms of toxicity. For instance, a xenobiotic-induced depletion of the intracellular antioxidant defense system could lower the threshold for various insults by other xenobiotics that have different primary target sites. At present, our laboratory is studying the physiological effects of combined exposure to food additives, in combination with a transcriptomics approach. Food additives are authorized for their use in the European Union under the condition that they do not impose any hazard upon consumers. Intake levels of individual food additives are deemed to be safe at no observed adverse effect levels (NOAEL). However, the current NOAELs used for these additives to perform risk assessment are most often based on empirical sciences, including clinical chemistry and pathology. In addition, little is known about potential health effects under circumstances when additives are taken simultaneously at or below the NOAEL. Recently, an evaluation was performed for the potential interactions occurring between more than 300 approved food additives [18]. Possible joint actions could not be excluded for four food additives, with respect to the liver as target organ. To assess the effects of a low-level exposure to mixtures of toxicants, we exposed rats for 28 days to low levels of selected food additives which were found to cause hepatic effects (induction of drug-metabolizing enzymes and liver enlargement) at higher concentrations. The effects on the liver induced by the food additives are expected to be subtle and the ability to observe changes at the level of gene expression was explored. The transcriptomics experiments for four individual compounds provide a wealth of biological information including changes in drug-metabolizing enzymes, peroxisome proliferation and DNA damage-dependent pathways. Importantly, transcriptome findings could be confirmed at the biochemical level. The findings are used for inference of interactions upon combinatorial exposure to these food additives. Thus, the application of toxicogenomics yields mechanistic insights at the molecular level that could not be identified using conventional techniques. Mixture studies with these food additives are ongoing and will be used to evaluate the hypotheses formulated from the single compound experiments.

14.4 Illustration: The Application of Toxicogenomics to Study the Mechanism of Bromobenzene-induced Hepatotoxicity

Toxicity in the liver is currently monitored on a routine basis using a wide variety of empirically determined parameters. Relative liver weight changes and gross pathological observations like color or texture of the tissue are nonspecific indicators of toxicity. Specific liver toxicity endpoints like necrosis, hypertrophy, cholestasis, steatosis, hepatitis and hyperplasia involve more specific changes at the cellular molecular level. Upon necrosis, liver cells are ruptured and membrane damage leads to leakage of intracellular enzymes to the extracellular fluids. Enzymes like alkaline phosphatase, lactate dehydrogenase and aminotransferases are commonly used as indicators of liver necrosis. The added value of functional genomics technologies in toxicology is illustrated by a study of bromobenzene-induced hepatotoxicity in rats [19]. The aim was to further elucidate the mechanism of toxicity and to identify specific maker genes and proteins of hepatotoxicity. Acute (24-h) liver toxicity was induced by bromobenzene, and protein and gene expression levels were determined in the liver. At high bromobenzene doses, hepatic cellular glutathione is depleted, evoking secondary processes like lipid peroxidation that ultimately lead to cell death. Additionally, bromobenzene toxicity has been related to the covalent binding of reactive bromobenzene metabolites to endogenous proteins, especially containing sulfhydryl groups [20]. The bromobenzene treatment altered the gene expression pattern of rat liver, as shown by PCA and cluster analysis (Fig. 14.5). The corn oil-injected controls could not be distinguished from the untreated controls. The importance of the depletion of intracellular GSH levels in bromobenzene hepatotoxicity was corroborated with the finding of the differential expression of genes involved in GSH metabolism. The GST subunits, catalyzing the conjugation of reactive metabolites to GSH, were upregulated as well as γ-glutamylcysteine synthetase, which is the rate-limiting enzyme in GSH synthesis. Upon verification, a significant increase in enzyme activity of GST was observed in the cytosolic liver fractions of rats treated with bromobenzene compared to controls. The induction of microsomal epoxide hydrolase is coherent with its role in the hydrolysis of epoxide bromobenzene intermediates. Other enzymes involved in drug metabolism were significantly altered. Many of the genes that were induced have been shown to be under transcriptional control of the Electrophile Response Element (EpRE, formerly ARE) [21, 22]. Characteristic gene inductions such as heme oxygenase-1, peroxiredoxin1, metallothioneins, ferritin and TIMP1 indicate the presence of oxidative stress. The high dose of bromobenzene clearly elicits a response well known as the acute phase response. Proteomics technology using two-dimensional gel electrophoresis was used to separate the cellular protein samples of the livers. In Fig. 14.5(A), the protein pattern obtained from an untreated animal is shown. Figure 5(B) shows a typical two-dimensional gel of liver proteins of a bromobenzene-treated rat. Spots significantly changing were identified by MS. Amongst others, bromobenzene induced an increase in mitochondrial aldehyde dehydrogenase 2, and a decrease in heat shock protein 60 and ATPase β subunit protein. Bromobenzene specifically induced a shift in the protein pattern from high-molecular-mass proteins towards more lower molecular mass and gene transcription of many factors involved in protein synthesis such as ribosomal proteins, translation initiation and elongation factors suggests a higher protein turnover rate. Urine and blood plasma were collected from bromobenzene-exposed rats. NMR spectra of urine and plasma from bromobenzene-treated rats were distinct from control samples.

14.5 Conclusion

Toxicogenomics, defined as the application and integration of transcriptomics, proteomics and metabolomics in toxicology, will serve many goals in the recognition and resolution of toxicity at the molecular level. First, it is important that the findings from functional genomics experiments are linked to common knowledge and conventional toxicological examinations. Results from our and other laboratories indicate that great progress in the field of toxicology is to be expected from the development of functional genomics technologies. Mechanisms of toxicity can be determined in more detail, at earlier time points and at lower doses compared to conventional toxicity parameters. Great advantages in toxicology are expected from toxicogenomics applications like the finding of new markers, enhanced extrapolations and assessment of combinatorial toxicity. In particular, in safety testing, with the importance of a complete monitoring of possible adverse effects induced by a compound on the complete organism, we see that functional genomics approaches will prove to be indispensable.

References

1 JL DeRisi et al. Exploring the metabolic and genetic control of gene expression on a genomic scale, *Science* 1997, **278**, 680–6.
2 M Schena et al. Quantitative monitoring of gene expression patterns with a complementary DNA microarray, *Science* 1995, **270**, 467–70.
3 A Shevchenko et al. Linking genome and proteome by mass spectrometry: large-scale identification of yeast proteins from two dimensional gels, *Proc Natl Acad Sci USA* 1996, **93**, 14440–5.
4 SP Gygi et al. Quantitative analysis of complex protein mixtures using isotope-coded affinity tags, *Nat Biotechnol* 1999, **17**, 994–9.
5 KP Gartland et al. Application of pattern recognition methods to the analysis and classification of toxicological data derived from proton nuclear magnetic resonance spectroscopy of urine, *Mol Pharmacol* 1991, **39**, 629–42.
6 JK Nicholson et al. "Metabonomics": understanding the metabolic responses of living systems to pathophysiological stimuli via multivariate statistical analysis of biological NMR spectroscopic data, *Xenobiotica* 1999, **29**, 1181–1189.
7 JD Robertson and S Orrenius, Molecular mechanisms of apoptosis induced by cytotoxic chemicals, *Crit Rev Toxicol* 2000, **30**, 609–27.
8 E Holmes et al. Metabonomic characterization of genetic variations in toxicological and metabolic responses using probabilistic neural networks, *Chem Res Toxicol* 2001, **14**, 182–91.
9 A Brazma and J Vilo, Gene expression data analysis, *FEBS Lett* 2000, **480**, 17–24.
10 MR Fielden et al. *In silico* approaches to mechanistic and predictive toxicology: an introduction to bioinformatics for toxicologists, *Crit Rev Toxicol* 2002, **32**, 67–112.
11 EF Nuwaysir et al. Microarrays and toxicology: the advent of toxicogenomics, *Mol Carcinog* 1999, **24**, 153–9.
12 S Farr and RT Dunn, Concise review: gene expression applied to toxicology, *Toxicol Sci* 1999, **50**, 1–9.
13 SJ Bulera et al. RNA expression in the early characterization of hepatotoxicants in Wistar rats by high-

density DNA microarrays, *Hepatology* 2001, **33**, 1239–1258.

14 JF Waring et al. Clustering of hepatotoxins based on mechanism of toxicity using gene expression profiles, *Toxicol Appl Pharmacol* 2001, **175**, 28–42.

15 JF Waring et al. Microarray analysis of hepatotoxins *in vitro* reveals a correlation between gene expression profiles and mechanisms of toxicity, *Toxicol Lett* 2001, **120**, 359–68.

16 HK Hamadeh et al. Gene expression analysis reveals chemical-specific profiles, *Toxicol Sci* 2002, **67**, 219–31.

17 ME Burczynski et al. Toxicogenomics-based discrimination of toxic mechanism in HepG2 human hepatoma cells, *Toxicol Sci* 2000, **58**, 399–415.

18 JP Groten et al. An analysis of the possibility for health implications of joint actions and interactions between food additives, *Regul Toxicol Pharmacol* 2000, **31**, 77–91.

19 WH Heijne et al. Toxicogenomics of bromobenzene hepatotoxicity: a combined transcriptomics and proteomics approach, *Biochem Pharmacol* 2003, **651**, 857–75.

20 YM Koen et al. Identification of three protein targets for reactive metabolites of bromobenzene in rat liver cytosol, *Chem Res Toxicol* 2000, **13**, 1326–35.

21 RS Friling et al. Xenobiotic-inducible expression of murine glutathione S-transferase Ya subunit gene is controlled by an electrophile-responsive element, *Proc Natl Acad Sci USA* 1990, **87**, 6258–62.

22 Y Tsuji et al. Coordinate transcriptional and translational regulation of ferritin in response to oxidative stress, *Mol Cell Biol* 2000, **20**, 5818–27.

Index

Molecular Biology in Medicinal Chemistry. Edited by Th. Dingermann, D. Steinhilber, G. Folkers

ISBN: 3-527-30431-2

I